W0246250

Ergebnisse der Mathematik und ihrer Grenzgebiete

3. Folge · Band 24

A Series of Modern Surveys in Mathematics

Editorial Board

E. Bombieri, Princeton S. Feferman, Stanford
M. Gromov, Bures-sur-Yvette
H.W. Lenstra, Jr., Berkeley P.-L. Lions, Paris
R. Remmert (Managing Editor), Münster
W. Schmid, Cambridge, Mass. J-P. Serre, Paris
J. Tits, Paris

Vladimir F. Lazutkin

KAM Theory and Semiclassical Approximations to Eigenfunctions

With an Addendum by A.I. Shnirelman

With 66 Figures

Springer-Verlag
Berlin Heidelberg New York
London Paris Tokyo
Hong Kong Barcelona
Budapest

Professor Vladimir F. Lazutkin

Department of Mathematical Physics
Institute of Physics
St. Petersburg State University
Ulyanov str. 1, kor. 1, 198904 Petrodvorets
St. Petersburg, Russia

Mathematics Subject Classification (1991):
35J10, 35P20, 35Q40, 58FXX, 58G25, 70HXX, 81Q05, 81Q20

ISBN-13: 978-3-642-76249-9 e-ISBN-13: 978-3-642-76247-5
DOI: 10.1007/978-3-642-76247-5

Library of Congress Cataloging-in-Publication Data. Lazutkin, V.F. (Vladimir Fedorovich) KAM theory and semiclassical approximations to eigenfunctions / Vladimir F. Lazutkin; with addendum by A.I. Shnirelman. p. cm.—(Ergebnisse der Mathematik und ihrer Grenzgebiete : 3. Folge, Bd. 24) Includes bibliographical references and index.
ISBN-13: 978-3-642-76249-9 1. Hamiltonian systems. 2. Eigenfunctions. 3. Asymptotic distribution. 4. Schrödinger operator. I. Title. II. Series.
QA614.83.L39 1993, 514'.74—dc20 93-17491

This work is subject to copyright. All rights are reserved, whether the whole or part of the material is concerned, specifically the rights of translation, reprinting, reuse of illustrations, recitation, broadcasting, reproduction on microfilm or in any other way, and storage in data banks. Duplication of this publication or parts thereof is permitted only under the provisions of the German Copyright Law of September 9, 1965, in its current versions, and permission for use must always be obtained from Springer-Verlag. Violations are liable for prosecution under the German Copyright Law.

© Springer-Verlag Berlin Heidelberg 1993
Softcover reprint of the hardcover 1st edition 1993

The use of general descriptive names, registered names, trademarks, etc. in this publication does not imply, even in the absence of a specific statement, that such names are exempt from the relevant protective laws and regulations and therefore free for general use.

Typesetting: Thomson Press (India) Ltd., New Delhi
41/3140/SPS—5 4 3 2 1 0—Printed on acid-free paper

Preface

More than ten years ago I wrote a short book "Convex billiards and Laplacian eigenfunctions" which was issued in 1981 by Leningrad State University Press. It contained a detailed account of results on caustics in convex domains in the plane and their applications to the high-frequency asymptotics of eigenfunctions of the Laplacian in those domains, both topics being of interest for physicists and mathematicians. Issued in a very limited edition in Russian, that book was almost unknown to the Western reader. Professor L.D. Faddeev proposed to publish the English-translation in Springer-Verlag. In the process of preparing a new version, I started to generalize the results to higher dimensions and applications to the Schrödinger equation. The book was growing and the affair resulted in this volume. I tried to give a comprehensive exposition of the topics involved, including all the details. The main theme is KAM tori (in the C^∞ setting) and eigenfunctions of the Schrödinger operator, which "tend" to KAM tori of the corresponding classical Hamiltonian system if $\hbar$, the Planck constant, tends to zero. So I restrict myself somewhat to what can be called the "quasiintegrable" case: the classical system in question possesses KAM tori where it is quasiintegrable, the trajectories being quasiperiodic. The behaviour of trajectories in the complement seems to be chaotic.

Professor A.I. Shnirelman kindly agreed to write an Addendum to this book devoted to eigenfunctions associated with chaotic regions, which, in my opinion, makes for a more complete exposition of the subject.

I am grateful to Springer-Verlag for their great patience. I thank many people (it is impossible to mention all of them here) who helped me directly or indirectly during the preparation of the manuscript. Without their help the book could not be finished.

Barcelona, 11 May, 1993 V.F. Lazutkin

Table of Contents

Introduction

We shall study the asymptotic behaviour of eigenvalues and eigenfunctions of a certain class of linear differential self-adjoint operators. A typical example is the Schrödinger equation of quantum mechanics:

$$-\frac{\hbar^2}{2}\Delta\psi + V(x)\psi = E\psi. \tag{0.1}$$

Here $\Delta = \sum_{i=1}^{n} \partial^2/\partial x_i^2$ is the Laplace operator in $\mathbb{R}^n$, $\hbar$ is a small number, the Planck constant (in CGSE units $\hbar = 1.0544 \times 10^{-27}$ erg s), the potential V is a real-valued function which reflects the physical properties of the system, and E is the spectral parameter (energy). One is looking for a function $\psi: \mathbb{R}^n \longrightarrow \mathbb{C}$, a wave function, which satisfies the equation (0.1) and behaves regularly at infinity. The values of E such that the equation (0.1) has a nonzero square-integrable solution, the eigenvalues, form the discrete spectrum $\{E_k\}$ of the Schrödinger operator. Corresponding nonzero and decreasing at infinity solutions of (0.1) and called eigenfunctions, or bound states of the quantum system. (Here "nonzero" means that the eigenfunction is not identically zero, but it is allowed to be zero on some subsets of $\mathbb{R}^n$.) Each chemical substance has an equation of the form (0.1) which determines the structure of its molecule, and one can observe its eigenvalues as dark thin strips in the real physical spectrum of the light passing through the substance.

A similar problem is to find the frequencies of the sound vibrations in a bounded domain $Q \subset \mathbb{R}^3$. These frequencies $\{\omega_k\}$ can be determined as the square roots of the eigenvalues of the wave operator $-c^2\Delta$ where Δ is the Laplace operator in Q and c is the velocity of sound in Q, the corresponding wave functions ψ_k are the nonzero solutions of the equation $-c^2\Delta\psi_k = \omega_k^2\psi_k$, satisfying certain homogeneous conditions at the boundary of Q which reflect the physical conditions there. A typical boundary condition is the Dirichlet one: $\psi|_{\partial Q} = 0$, which assumes that the density of the matter is very high outside Q.

The discrete spectrum of (0.1) is wholly determined by the potential V, as the frequency spectrum of the domain Q is determined by its shape and boundary conditions. How does one compute the spectrum? There exist very elaborate variational methods which enable us to calculate the eigenvalues consecutively, starting with the lowest one. Unfortunately these methods require a large amount of calculation, even with the help of modern computers. They become practically unrealizible if, in (0.1), $\hbar$ is small, $n \geqq 2$, and the potential is sufficiently complicated. The same is true for the problem of finding the high frequencies of a general domain Q.

For the reasons mentioned one is forced to use other approximate methods. Ariadne's clue to find them is the idea that *the quantum mechanics solutions turn into the classical ones as* $\hbar$ *tends to zero.* This means that one can construct approximate solution to (0.1) using as a starting point, or as a skeleton, the solutions to the corresponding equations of the classical mechanics case, Newton's ones. The latter are ordinary differential equations, which may be written in the Hamiltonian form for the variables $x=(x_1, x_2, \ldots, x_n)$ and $p=(p_1, p_2, \ldots, p_n)$:

$$\dot{x}_i = \frac{\partial H(x,p)}{\partial p_i}, \quad \dot{p}_i = -\frac{\partial H(x,p)}{\partial x_i}, 1 \leqq i \leqq n, \tag{0.2}$$

where the Hamiltonian function is

$$H(x,p) = \tfrac{1}{2}|p|^2 + V(x). \tag{0.3}$$

One expects to handle the equation (0.2) easier than the original partial differential equation (0.1). The analogue of the classical counterpart to the problem of eigenfrequencies of a bounded domain Q is the billiards in Q, the mechanical system which describes the motion of a material point (a billiard ball) in Q with elastic reflections at the boundary. Because of their origin the said methods were called semiclassical ones. They were used by physicists mostly for the one-dimensional Schrödinger equations, and for those which could be reduced to the above equations. Nevertheless, the class of problems solved in this way was large enough to provide explanations to many physical phenomena, and the semiclassical methods accompany quantum mechanics at all stages of its development starting with the initial steps when N. Bohr used the planetary atom model to obtain the formula for the spectrum of the hydrogen atom.

Serious obstacles arise when we attempt to extend the semiclassical methods to a general multidimensional case. One may naturally suppose that *the eigenfunctions (bound states) have bounded motions in the corresponding classical system as in their quantum counterparts.* It means that there must be a definite way to construct an approximation to an eigenfunction and eigenvalue which starts with a suitable bounded invariant set in the phase space of the classical system, and that, using this method, one can approximate all eigenfunctions if one considers all such invariant subsets. This approach suggests the investigation of invariant sets and their properties. So what do the invariant sets of a general multidimensional mechanical system look like? The reader can get some idea of the appearance of the partition of the phase space into invariant sets by looking at Fig. 1 which presents some trajectories of the well known standard map, which is considered to be a model for a general Hamiltonian system with two degrees of freedom. Such maps arise as Poincaré maps of cross sections in real mechanical systems. All the features of genuine, general systems are presented in this phase portrait.

We see that some trajectories are organized in regular curves, these being concentrated near periodic points. These are the so called KAM curves. Such curves, in a Poincaré section of a Hamiltonian system of differential equations, are the traces of KAM tori which bear quasiperiodic motions. The abbreviation KAM consists of the first letters of the names of the men who have discovered

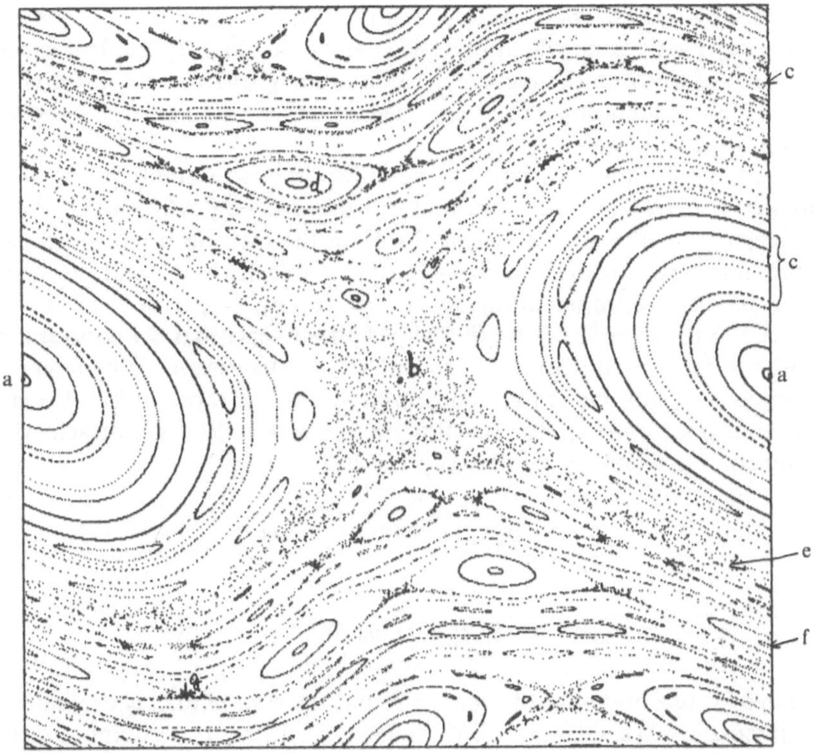

Fig. 1. This is the phase portrait of the standard Taylor–Chirikov–Greene mapping [(Example 8.8 and Eqs. (17.4)] for $k = 0.9$. Several trajectories are plotted on the unit square $0 \leqq \phi \leqq 1$, $-1/2 \leqq I \leqq 1/2$. The following invariant sets are marked: (a) elliptic fixed point, (b) hyperbolic fixed point, (c) KAM curves, (d) island of period 3, (e) stochastic layers, (f) elliptic periodic point of period 3, (g) hyperbolic periodic point of period 3

and proved the existence of the invariant tori: A. N. Kolmogorov, V. I. Arnol'd, and J. Moser. The first part of this book is devoted to the study of these tori.

The KAM curves intermit with filled zones which grow away from hyperbolic periodic trajectories and look like stretched ovals joined in a sort of wreath. These wreaths have nonzero width, and they bear a stochastic motion. Such wreaths are called stochastic layers. They pervade the phase portrait densely as well as the KAM curves and periodic trajectories. There does exist another type of invariant set which is not seen in this picture; Cantori which are ruined KAM curves. The reader can look at the portrait of a cantorus in Fig. 44.

There is strong evidence that the characteristic feature of the whole picture is its self-similarity: an arbitrarily chosen small piece contains another smaller piece which resembles the initial picture provided one magnifies it with appropriate coefficients in the appropriate directions. This is a typical phase portrait for general Hamiltonian systems with two degrees of freedom. The two extreme degenerate cases of this picture are firstly, an integrable one, when KAM curves form smooth families and fill all phase space, and, secondly, a wholly chaotic one, when there is one large stochastic zone which occupies the whole phase space. In the case of higher degrees-of-freedom the picture is similar, but KAM tori do not divide the phase space, and stochastic layers are tied together in one large everywhere penetrating, stochastic web.

How can one use the invariant sets described above to build approximations to eigenfunctions? The most suitable bricks turn out to be KAM tori, i.e. the most regular elements of the picture. Let us dwell on the method of constructing the said approximations attached to a KAM torus. Recall that a state in classical mechanics involves a momentum vector $p=(p_1,p_2,\ldots,p_n)$ as well as space coordinates $x=(x_1,x_2,\ldots,x_n)$, so the whole phase space has $2n$ coordinates. A KAM torus $\mathcal{T}$ is an n-dimensional torus in $\mathbb{R}^{2n}$. In the case $n=2$ its intersection with a Poincaré section looks like a closed curve such as those in Fig. 1. Let π be the projection map onto space coordinates: $\pi(x,p)=x$. We will attach to $\mathcal{T}$ a smooth function ψ whose value at x is given by

$$\psi(x)=\sum_l A_l(x)\exp\left\{\frac{i}{\hbar}S_l(x)\right\}, \tag{0.4}$$

where

$$S_l(x)=\int_{x_0}^{x} p_l(x')dx', \tag{0.5}$$

where each $p_l(x)$ is a preimage of x under π, and the sum in (0.4) is evaluated over all such preimages. We set $\psi(x)=0$ if $\pi^{-1}(x)=\varnothing$. In (0.5) the point $x_0\in\pi(\mathcal{T})$ is arbitrary, the integral does not depend on the shape of a path joining x_0 and x, and depends only on its homological class. The invariant torus $\mathcal{T}$ contains a dense trajectory, hence it lies wholly in some energy hypersurface $H(x,p)=E$, where $H(x,p)$ is the classical Hamiltonian function (0.3). The value E is an approximation to an eigenvalue E_k which the eigenfunction, approximated by ψ, corresponds to. Substituting (0.4) in (0.1), one obtains that the terms of order 1 cancel, as do the terms of order $\hbar$, provided each $A_l(x)$ obeys some transport equation. The latter determine, in the final analysis, the system of functions A_l uniquely up to a common constant multiplier. The functions A_l thus obtained have singularities at images under π of singular points on $\mathcal{T}$ with respect to the projection π. Such images form a submanifold of $\mathbb{R}^n$ called the *caustic*. The approximation (0.4) fails near the caustic since $A_l(x)$ tends to infinity as x approaches the corresponding component of the caustic. One way to repair the asymptotics there is to perform the Fourier transform in the direction transversal to the caustic to obtain a regular asymptotic expression for the eigenfunction in the mixed coordinate–momentum representation, and then to apply the inverse Fourier transform. Such a procedure suggests certain rules for the gluing together, at a point x^*, of the caustic (see Fig. 2) the terms of (0.4) corresponding to the sheets of $\mathcal{T}$ which meet at a preimage of x^*. Taking into account these rules, and requiring the univalency of (0.4), results in some quantisation rules, the typical one having the form

$$(2\pi\hbar)^{-1}\oint_{\gamma_j} p\,dx = k_j+\tfrac{1}{8}m(\gamma_j). \tag{0.6}$$

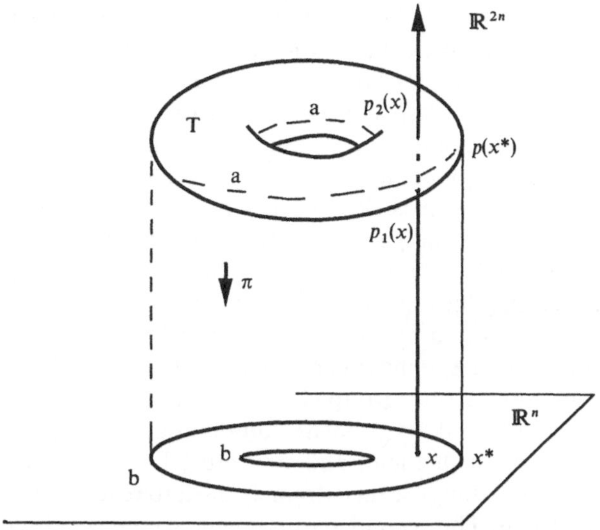

Fig. 2. An invariant torus $\mathscr{T}$ in the phase space $\mathbb{R}^{2n}$ and its projection onto the coordinate space $\mathbb{R}^n$. Marked: (a) singular submanifolds with respect to the projection, (b) caustics. A point x belonging to the interior of the torus projection has two pre-images with momenta $p_1(x)$ and $p_2(x)$. That belonging to the caustic, x^*, has only one pre-image under projection

Here the domains of integration γ_j are closed curves on $\mathscr{T}$ which represent elements of a base of the 1-dimensional homology group $H_1(\mathscr{T}, \mathbb{Z})$, the symbols k_j are integers, the *quantum numbers*, $m(\gamma)$ is an integer valued function of a closed curve γ, the *Maslov index* (roughly speaking $m(\gamma)$ is equal to the algebraic number of intersections of γ with the cycle C of singular points with respect to the projection π). Quantum rules need not be obeyed exactly but to an error of the order of $\hbar^{-\alpha}$, $0 < \alpha < 1$. They may be considered as the equations for finding an appropriate KAM torus (look again at Fig. 1 and take into account the discontinuous character of the set of KAM curves). Once $\mathscr{T}$ is found, it determines the value of $E = H(x, p)|_{(x,p)\in\mathscr{T}}$. In such a manner we obtain the approximate eigenvalue $E_k, k = (k_1, k_2, \ldots, k_n)$, as a function of quantum numbers, the latter varying over the subset $\Lambda \subset \mathbb{Z}^n$ consisting of all those k for which an appropriate torus exists.

Thus we have briefly described the procedure for building the first approximation to eigenvalues and eigenfunctions which uses KAM tori as a skeleton. Of course one may specify this approximation by writing out asymptotic series in the powers of $\hbar$. This programme has been realized in detail (for the case of more general operators on a manifold) in Chapters VI and VII of this book. This can be considered as our central result. Other chapters contain preparatory or accompanying material.

Numerous questions arise in connection with the construction described. We shall now outline the main ones.

Are the constructed "approximate" expressions for eigenfunctions and eigenvalues the true approximations to the actual eigenfunctions and eigenvalues, and, if so, to

what extent are they accurate? As a matter of fact we construct pairs (ψ_k, E_k), ψ_k a function, E_k a number, which satisfy the equation (0.1) approximately, i.e. they give a small discrepancy of order $\hbar^N$ (N is some fixed large number) when substituted into (0.1) instead of ψ and E, respectively. V. I. Arnol'd suggested that we call such functions ψ_k quasimodes, and found that quasimodes were not appropriate approximations to the original eigenfunctions, while the functions E_k were really close to the genuine eigenvalues. We discuss the corresponding mathematics in Chapter V, here we give some general comments. The nature of the mentioned phenomenon is concealed in the occurrence of the 'asymptotic degeneration of the spectrum'. Many eigenvalues are concentrated in close groups, the eigenvalues coinciding or being separated by small spaces of order $\hbar^N$ within such a group. A quasimode ψ_k approximates a linear combination of genuine eigenfunctions which correspond to eigenvalues of a group. Conversely, one can hope to construct a genuine eigenfunction by taking a linear combination of quasimodes. Of course one must possess enough of them, and one must know how to make the said construction. Both these things seem extremely hard to reach, at least at the present state of our knowledge. The second requirement is connected with the 'effect of tunnelling' and will be discussed in Notes to Chapter V. The first one generates the question about the completeness of the set of quasimodes which in turn yields a whole sheaf of questions, which we shall partially touch further.

The next problem is *to estimate the quota which on the approximately found eigenvalues form among all the eigenvalues of the Schrödinger operator*. The answer is simple: this quota is proportional to the invariant Liouville volume, $\mathrm{Vol}(\mathscr{K})$, of the set $\mathscr{K}$, occupied by KAM tori, which serve as a skeleton for our building. Moreover, we have obtained the following estimate for the number $N(\mathscr{K})$ of such approximated eigenvalues:

$$N(\mathscr{K}) \approx \frac{\mathrm{Vol}(\mathscr{K})}{(2\pi\hbar)^n}. \tag{0.7}$$

The precise statement and the proof of this result require some technical steps, the most important being to establish the so called transversal smoothness of KAM tori. This means that KAM tori form a sort of family which one may supplement with other tori (not necessarily invariant) so that the completed family becomes a smooth n-parametric family of tori filling an open domain in the $2n$-dimensional phase space. The original set of invariant tori is locally diffeomorphic to the product of a torus and a Cantor set. The consequent development of notions linked with such a structure results in the concept of a KAM set which has been developed in Chapter II. The formula (0.7) is nicely in agreement with the naive concept of physisists about a 'quantum cell' of volume $(2\pi\hbar)^n$ being occupied by a 'single quantum state'.

The problem that arise straight away is to estimate the whole volume (measure of the set of all KAM tori. Does this volume equal that of all the phase space? If it does, then the preceding construction must yield a complete collection of quasimodes, the genuine eigenfunctions being constructed as linear combinations of members of this collection. Alternatively, one may ask if there are other types of invariant sets (e.g. stochastic layers) which occupy a positive volume in the phase space. If so, how does one construct the corresponding quasimodes?

The possibility of answering these questions depends strongly on our knowledge of the structure of invariant sets of a general Hamiltonian system. It is a surprising fact for nonspecialists that such an ancient discipline as classical mechanism turns out to be incomplete and makes its first steps nowadays whereas nonrelativistic quantum mechanics which only started in the twentieth century has acquired its mathematical apparatus entirely and now is a very elaborate and transparent branch of mathematical physics. We know very well what the spectrum of Schrödinger operator (0.1) is and how it depends on the potential V, but, in contrast, very little is known about the intricate ingredients of the picture drawn by computer represented in Fig. 1. It must be confessed that at present our knowledge is insufficient to answer the above posed questions.

However, there are some rigorous results on the last question concerning the structure of quasimodes corresponding to chaotic zones. They are placed in the Addendum written by A. I. Shnirelman. The most interesting result by Shnirelman concerns the case when the corresponding classical dynamical system is ergodic, i.e. transitive on each submanifold of constant energy. He has proved that the local Fourier transform of an eigenfunction results in an invariant density, uniformly distributed upon the energy hypersurface. This can be interpreted figuratively as follows. Let us consider an eigenfunction $\psi(x)$ as the light wave and let the intensity of light vibrations at a point x be proportional to $|\psi(x)|$. An observer, placed at the point x of a large universe where such an 'ergodic' eigenfunction is excited, sees the dome of the sky shining uniformly, while he sees a finite number of stars provided an eigenfunction of the form (0.4) is excited.

We conclude this Introduction by describing the contents and the general structure of the book. It is divided into two parts. The first part, 'KAM theory', contains general facts about the symplectic structure and dynamics (Chapter I), theory of KAM tori (Chapter II), and discussions on phenomena occurring beyond the tori: Arnol'd diffusion, stochastic layers, cantori, etc. (Chapter III). The central notion on which the first part is devoted to is that of a KAM set. The main theorem which asserts the persistence of a KAM set under small perturbations of an integrable Hamiltonian is proved in Chapter IV. Thus all the questions on the classical dynamics are considered in the first part.

In contrast, the second part 'Eigenfunctions asymptotics' concentrates wholly on the questions of $\hbar$-tends-to-zero asymptotics of eigenfunctions of differential operators such as (0.1). The first two chapters contain auxiliary material on self-adjoint operators in a Hilbert space (Chapter V) and Maslov's operator attached to an invariant Lagrangian submanifold in the phase space (Chapter VI). Some topological questions concerning the Maslov index which enters the quantum rules, and the quantum rules themselves, are also discussed here. The Maslov operator serves as a tool for constructing quasimodes in the next Chapter (VII). The latter contains our main construction which assignes to each KAM set $\mathscr{K}$ a set of quasimodes such that the formula (0.7) holds.

The basic text, which consists of the seven described Chapters united in two parts, is accompanied by an Addendum by A. I. Shnirelman and three technical Appendices.

For the convenience of the reader we place the List of general mathematical notations before the main text.

List of General Mathematical Notations

Set theory

$x \in X$	x belongs to a set X
$x \notin X$	x does not belong to a set X
$A \subset B$	a set A is a subset of a set B (this does not exclude the case $A = B$)
$A \cup B, A \cap B, A \backslash B$	the union, intersection, difference of two sets A and B
$\{x \in A: P(x)\}$	the set of all $x \in A$ such that the condition $P(x)$ is true
$f: X \to Y$	f is a map defined on X (the domain of definition of f) with values in Y
$f(x)$	the value of a map f at an element x
$x \longmapsto f(x)$	a map f carries x to $f(x)$
$f\|_A$	the restriction of a map f onto a subset A of the domain of definition of f
$f \circ g$	the superposition of maps g and f, i.e. $f \circ g(x) = f(g(x))$ for all x belonging to the domain of definition of g such that $g(x)$ belongs to the domain of definition of f
$f^{-1}(A)$	the preimage of a set A under a map f, i.e. the set of all x such that $f(x) \in A$
$\varnothing$	the empty set
$\{a, b, c, \dots\}$	the set consisting of the elements $a, b, c, \dots$
$\mathrm{id}: X \longrightarrow X$	the identity map which assignes x to itself for each $x \in X$
id_X	the same as the previous notation if we wish to indicate the underlying set X

In the case of X being a vector space we will also denote the identity map as 1 (or 1_X in order to indicate the space), the identity matrix will be denoted also by I.

Topology

$\bar{A}$	the closure of a subset A in a topological space
∂X	the boundary of a topological manifold X
$\mathbb{Z}$	the set of all integer numbers
$\mathbb{R}$	the set of all real numbers
$\mathbb{C}$	the set of all complex numbers
$\mathbb{Z}_+$	$= \{n \in \mathbb{Z}: n > 0\}$

$\mathbb{R}_+$	$=\{x\in\mathbb{R}: x\geqq 0\}$
$\mathbb{N}$	$=\{n\in\mathbb{Z}: n>0\}$
$\mathbb{R}^n$	n-dimensional Euclidean space, its elements are of the form $x=(x^1,x^2,\ldots,x^n)$, where $x\in\mathbb{R}$, $1\leqq i\leqq n$
$\mathbb{C}^n$	n-dimensional complex space with elements of the form $z=(z^1,z^2,\ldots,z^n)$, $z_i\in\mathbb{C}$, $1\leqq i\leqq n$
ort_i	$=(0,0,\ldots,1,0,\ldots 0)$, 1 standing at ith place and 0 at the remaining ones, the ith ort of $\mathbb{R}^n$ (or $\mathbb{C}^n$)
$\langle x,y\rangle$	$=\sum_{i=1}^n x^i y^i$, *standard inner product* of $x,y\in\mathbb{R}^n$
$\lvert x\rvert$	$=\langle x,x\rangle$, *standard norm* of $x\in\mathbb{R}^n$
$\mathrm{dist}\,(x,y)$	$=\lvert x-y\rvert$, the *distance* between $x,y\in\mathbb{R}^n$
$\mathrm{dist}\,(x,A)$	$=\inf_{y\in A}\mathrm{dist}\,(x,y)$, the *distance* between a point $x\in\mathbb{R}^n$ and a subset $A\subset\mathbb{R}^n$
$\mathrm{diam}\,A$	$=\sup_{x,y\in A}\mathrm{dist}\,(x,y)$, the *diameter* of a set $A\subset\mathbb{R}^n$

If $A\subset\mathbb{R}^n$ and δ is a positive number, we define

$[A]^\delta$	$=\{x\in\mathbb{R}^n: \mathrm{dist}\,(x,A)<\delta\}$, *$\delta$-dilation* of a set A
$[A]_\delta$	$=\{x\in A: \mathrm{dist}\,(x,\mathbb{R}^n\backslash A)\geqq\delta\}$, *$\delta$-contraction* of A
$\mathcal{M}es\,A$	the Lebesgue measure of $A\subset\mathbb{R}^n$
$B_\delta(x)$	$=\{y\in\mathbb{R}^n: \mathrm{dist}\,(y,x)<\delta\}$, the open ball with the centre x and the radius δ

$\mathbb{R}^n$ possesses the *standard topology* with a base consisting of all open balls. If $f:X\longrightarrow\mathbb{R}^n$ is a map with a topological space X as the domain, we define

$\mathrm{supp}f$	$=\overline{\{x\in X: f(x)\neq 0\}}$, i.e. the closure of the set where $f(x)$ does not vanish, the *support* of f.

Linear Algebra

If V is a vector space over $\mathbb{R}$ or ($\mathbb{C}$), then

V^*	denotes its conjugate, or dual space, i.e. the space of all linear maps $l: V\longrightarrow\mathbb{R}$ (or $\mathbb{C}$)
$\dim V$	the dimension of V.

if $f: V\longrightarrow W$ is a linear map between vector spaces V and W, then

$\ker f$	$=\{x\in V: f(x)=0\}$, the kernel of f,
$\mathrm{im}\,f=f(V)$	$=\{y\in W: \exists x\in V, f(x)=y\}$, the image of f.

If V is a vector space, and W, W' are vector subspaces of V, then we write

$V=W\oplus W'$	if each vector $v\in V$ has unique representation $v=w+w'$ where $w\in W$ and $w'\in W'$. In this case we call W' an algebraic complement of W, and say W and W' are transversal, or W' is transversal to W. The decomposition $V=W\oplus W'$ is called a splitting

	of a vector space V, and V with a fixed splitting is called a splitted space. If W and W' are transversal subspaces of V, there is uniquely defined linear projection $P: V \longrightarrow V$ with $\ker P = W$ and $\operatorname{im} P = W'$. By definition $Pv = w'$ where w' is the member of the decomposition $v = w + w'$, $w \in W$, $w' \in W'$. We call a projection P with $\ker P = W$ a projection along W;
$\det A$	the determinant of a square matrix A; if $f: V \longrightarrow V$ is a linear map of a vector space into itself (endomorphism), then
$\det f$	the determinant of f, is equal to the determinant of the matrix of f in arbitrary chosen vector base in V, it is independent of the choice of the base;
$\operatorname{sign} M$	the signature of a symmetric square matrix M, i.e. the difference between the numbers of positive and negative eigenvalues.

Function Spaces

If X is a topological space then

$C(X, \mathbb{R})$	is the space of all continuous real valued functions defined on X;
$C(X, \mathbb{C})$	is the space of continuous complex valued functions defined on X;
$C(X)$	denotes one of the previous two spaces if it is clear what space is meant, the same abbreviation will be used for other spaces.

If X is a smooth manifold then

$C^\infty(X)$	is the space of all smooth functions defined on X. The lower index c at the symbol of a space denotes the subspace of functions with compact support, e.g.
$C_c(X)$	is the space of all continuous functions $\varphi = X \longrightarrow \mathbb{R}$ or (C) with compact support.

Let $\rho: X \longrightarrow X$ be a map. The lower index ρ at the symbol of a function space of functions defined on X denotes the subspace of functions φ which take equal values at point of X with common image under ρ, i.e. $\varphi(x) = \varphi(y)$ if $\rho(x) = \rho(y)\ \forall x, y \in X$. Only the letter ρ will be used for this purpose.

If X is a manifold with a boundary ∂X, then the lower index 0 denotes the subspace of functions vanishing in a neighbourhood of ∂X in X.

Let $D \subset \mathbb{R}^n$ be an open set, and a be a positive number. The lower index a at the symbol of a function space of functions defined on $\mathbb{R}^n \times D$ denotes the subspace consisting of functions whose supports are contained in $\mathbb{R}^n \times [D]_a$. The letters c, ρ, and 0 are not to be used for this purpose.

The subscript *per* at the symbol of a function space of functions defined on $\mathbb{R}^n = D$ denotes the subspace consisting of functions $\varphi(x, y)$, $x \in \mathbb{R}^n$, $y \in D$, which are periodic in the x-variable with period 1, i.e. $\varphi(x + \operatorname{ort}_i, y) = \varphi(x, y)\ \forall_i \in \{1, 2, \ldots, n\}$, $x \in \mathbb{R}^n$, $y \in D$.

All above indices may be used simultaneously, e.g.

$C^{\infty}_{\mathrm{per},a}(\mathbb{R}^n \times D)$ denotes the space of all smooth, periodic with period 1 with respect to the variable x, functions $\varphi(x, y), x \in \mathbb{R}^n, y \in D$, which vanishes in the a-neighbourhood of the boundary of D.

The concept of a space being "connected" in this book is equivalent to the more usual concept of a space being "path connected" or "pathwise connected" which is preferred by many other authors, though the former is the term used in Fuks and Rokhlin.

The norms in the spaces of functions defined on $\mathbb{R}^n \times D$ are defined in §8. The notations for the sets of diffeomorphisms are also defined §8.

Part I. KAM Theory

Chapter I. Symplectic Dynamical Systems

§1. Symplectic Vector Spaces

In this section, we shall work out the necessary parts of linear algebra. The notations and the statements will be used in the next sections for developing the symplectic geometry and the theory of Hamiltonian systems. The following definitions can be found summarized in "list of General Mathematical Notations" earlier.

The term vector space below will denote, unless otherwise stated, a finite-dimensional vector space over the field $\mathbb{R}$ of real numbers. The symbol $\dim V$ denotes the dimension of a vector space V. If $f: V \longrightarrow W$ is a linear map between vector spaces, $\ker f$ denotes the kernel of f, i.e. the set of all $v \in V$ such that $f(v) = 0$; $\operatorname{im} f$ denotes the image of f, i.e. $f(V)$. We write $V = W \oplus W'$ where W and W' are vector subspaces of V if each vector $v \in V$ has a unique representation $v = w + w'$ with $w \in W$ and $w' \in W'$. In this case, we call W' an algebraic complement of W, and say W and W' are transversal. The decomposition $V = W \oplus W'$ is called a splitting of the vector space V, and V with a fixed splitting is called a split space. If W and W' are transversal subspaces of V, there is a uniquely defined linear projection $P: V \longrightarrow V$ with $\ker P = W$ and $\operatorname{im} P = W'$. By definition, $Pv = w'$ where w' is a member of the decomposition $v = w + w'$, $w \in W$, $w' \in W$. We call a projection P with $\ker P = W$ a projection along W.

The dual space V^* of a vector space V is the space of all linear functions $V \longrightarrow \mathbb{R}$.

Nonsingular Bilinear Forms

A bilinear form on a vector space V is a map $\Omega: V \times V \longrightarrow \mathbb{R}$ which is linear in each of its two variables when the other one is fixed.

Two notions are linked with an arbitrary bilinear form. The first is orthogonality. A vector u is said to be orthogonal to a vector v if $\Omega(u, v) = 0$. This is written as $u \perp v$. If it is necessary to mark the dependence of this notion on Ω, we say u is orthogonal to v with respect to Ω, or Ω-orthogonal, and write $\perp(\Omega)$ instead of $\perp$. A vector u is said to be orthogonal to a set $M \subset V$ if u is orthogonal to each of the vectors belonging to M. The set of all vectors which are orthogonal to M is called the orthogonal complement of a set M, and is denoted as $M^{\perp}$. It is always a vector subspace of V.

The second notion is the "flat" map $\flat: V \longrightarrow V^*$ from V into its dual space V^*. The value of $u^\flat$ of the map $\flat$ at a vector $u \in V$ is the linear function whose value at $v \in V$ is $u^\flat(v) = \Omega(u, v)$. It follows from bilinearity of Ω that the map $\flat$ is linear. Again, we write $\flat(\Omega)$ if there is necessity to emphasize the dependence on Ω.

A bilinear form Ω is called nonsingular if the only vector which is Ω-orthogonal to the whole space V is the zero one. Since the kernel of the "flat" map coincides with $V^{\perp(\Omega)}$, we may assert that a form Ω is nonsingular if and only if the "flat" map $\flat(\Omega)$ is a linear isomorphism between V and V^*. In this case there exists an inverse map of $\flat(\Omega)$ which we denote by $\#$ and call the "sharp" map corresponding to the nonsingular bilinear form Ω. To indicate the dependence on Ω, if necessary, we shall sometimes write $\#(\Omega)$. The value of the sharp map on a linear form $l \in V^*$ will be denoted by $l^\#$.

Symplectic Linear Structures

A bilinear form Ω on a vector space V is called symplectic if it is nonsingular and skew-symmetric. The latter means $\Omega(u, v) = -\Omega(v, u)$ for any two vectors u and v. A pair (V, Ω) where V is a vector space and Ω is a symplectic form on V is called a symplectic vector space. We shall often write only V for (V, Ω), the symplectic form Ω being understood. If $V = (V, \Omega)$ is a symplectic vector space, then form Ω is called its symplectic structure.

Example 1.1 (*Standard 2n-dimensional symplectic vector space*). Let $\mathbb{R}^{2n}$ be a $2n$-dimensional Euclidean space. Take in this space the standard orthonormal vector basis $\{\mathrm{ort}_i, 1 \leqq i \leqq 2n\}$ (ort_i has 1 in its ith place and 0 in all others) and define the form Ω_{st} by its values on this basis as follows:

1.1a. $\Omega_{st}(\mathrm{ort}_i, \mathrm{ort}_{n+i}) = -\Omega_{st}(\mathrm{ort}_{n+i}, \mathrm{ort}_i) = 1$ if $1 \leqq i \leqq n$, and $\Omega_{st}(\mathrm{ort}_i, \mathrm{ort}_j) = 0$ for all other values of i, j.

By bilinearity Ω_{st} uniquely extends to the whole set $\mathbb{R}^{2n} \times \mathbb{R}^{2n}$ and is manifestly symplectic.

Let (V, Ω) and (W, Σ) be symplectic vector spaces. A linear map $f\colon V \longrightarrow W$ is called symplectic if $f^*\Sigma = \Omega$. We recall that $f^*\Sigma(u, v) = \Sigma(f(u), f(v))$.

Proposition 1.2. *A linear symplectic map is injective.*

Proof. Let $x \in \ker f$. Then, for an arbitrary $y \in V$, $\Omega(x, y) = \Sigma(f(x), f(y)) = \Sigma(0, f(y)) = 0$. □

A symplectic linear map is called a symplectic linear isomorphism, or simply an isomorphism, if it is a linear isomorphism, i.e. it has a linear two-sided inverse. Two symplectic spaces V and W are called isomorphic if there exists a symplectic linear isomorphism between them. The following theorem classifies all symplectic vector spaces up to isomorphism.

Theorem 1.3 (Linear part of Darboux theorem). *Any symplectic vector space has even dimension. For each integer $n \geqq 1$ there exists a symplectic vector space of* *dimension 2n, namely* $(\mathbb{R}^{2n}, \Omega_{st})$. *Any two symplectic vector spaces of the same dimension are isomorphic.*

The part concerning the existence of a symplectic vector space of prescribed dimension has already been considered in Example 1.1. We shall prove the rest of this theorem in the following subsection. It is further convenient to regard, by definition, the point (=0-dimensional vector space) as a symplectic vector space of dimension zero.

Symplectic Subspaces

If W is a vector subspace of a vector space V, and Ω a symplectic form on V, we may consider the restriction $\Omega|_W$ of the form Ω on the subspace W. This is the restriction of the map $\Omega: V \times V \longrightarrow \mathbb{R}$ to the subset $W \times W \subset V \times V$. In other terms $\Omega|_W = i^*\Omega$ where $i: W \longrightarrow V$ is the inclusion mapping. The form $\Omega|_W$ on the vector space W is bilinear and skew-symmetric. If it is nonsingular, the pair $(W, \Omega|_W)$ is a symplectic vector space which is called a symplectic subspace of a symplectic vector space (V, Ω). From the above, it follows that the inclusion $i: W \longrightarrow V$ of a symplectic subspace is a symplectic map. Since the symplectic structure on W is uniquely defined by that of V, we shall also refer to W as a symplectic subspace of a symplectic vector space V.

Example 1.4 (*Coordinate symplectic subspaces of the standard symplectic space*). Let $\alpha \subset \{1, \ldots, n\}$ be an arbitrary subset. Denote by S_α the subspace on $\mathbb{R}^{2n}$ which is generated by the vectors $\{\mathrm{ort}_i, \mathrm{ort}_{n+i}, i \in \alpha\}$. Then S_α is a symplectic subspace of $(\mathbb{R}^{2n}, \Omega_{st})$ of dimension $2\,\mathrm{card}\,\alpha$.

Proposition 1.5. *The orthogonal complement* $W^\perp$ *of a symplectic subspace* W *is also a symplectic subspace. It is an algebraic complement of* W.

Proof. If $v \in W$ and simultaneously $v \perp W$, then v must be zero by the nonsingularity of $\Omega|_W$. This proves the property $W \cap W^\perp = 0$. Since $\dim W + \dim W^\perp = \dim V$, which is due to the nonsingularity of Ω, it yields $V = W + W^\perp$. If $v \in W^\perp$ and $v \perp W^\perp$, then v is orthogonal to the whole space V, from which it follows $v = 0$. This proves the nonsingularity of $\Omega|_{W^\perp}$. □

Corollary. *If* W *is a symplectic subspace, then* $W^{\perp\perp} = W$.

An orthogonal splitting of a symplectic vector space V is its representation in the form $V = W_1 \oplus W_2$ where $W_i, i = 1, 2$, are its mutually orthogonal symplectic subspaces. The word "symplectic" in this definition is unnecessary:

> if $V = W_1 \oplus W_2$ and the $W_i, i = 1, 2$, are mutually orthogonal, then they are necessarily symplectic subspaces.

We may now consider orthogonal splittings of V into more than two symplectic subspaces.

This notion is very useful. Now we are going to use it to prove Theorem 1.3. Let V be a symplectic vector space. We may assume that V is not a single point. Take an arbitrary nonzero vector $g_1 \in V$. Then there exists a vector $h_1 \in V$ such that $\Omega(g_1, h_1) \neq 0$. This is due to the nonsingularity of Ω. We may assume $\Omega(g_1, h_1) = 1$, otherwise multiply g_1 by a suitable number. Consider the two-dimensional subspace W_1 generated by the vectors g_1 and h_1. It is obviously symplectic. Form the orthogonal splitting $V = W_1 \oplus W_1^\perp$. Then, we may apply the same construction to $W_1^\perp$ in place of V, provided $\dim W_1^\perp > 0$, because it is, by 1.5, itself a symplectic space. Acting in such a manner, we get the splitting

$$V = W_1 \oplus W_2 \oplus \cdots \oplus W_n$$

of our symplectic space V into an orthogonal sum of a finite number of two-dimensional symplectic subspaces $W_i, i = 1, 2, \ldots, n$. This proves the first assertion of Theorem 1.3. In each subspace W_i there are two (chosen in our construction) vectors g_i and h_i such that $\Omega(g_i, h_i) = 1$. Since $\{g_i, h_i\}$ is a basis of W_i for each i, the set $\{g_i, h_i, 1 \leqq i \leqq n\}$ is a basis of V. Form a linear map $f: V \longrightarrow \mathbb{R}^{2n}$ by assigning $f(g_i) = \mathrm{ort}_i, f(h_i) = \mathrm{ort}_{n+i}, 1 \leqq i \leqq n$ (see Example 1.1). It follows from 1.1a that f^* carries Ω_{st} to Ω. So, $f: (V, \Omega) \longrightarrow (\mathbb{R}^{2n}, \Omega_{st})$ is symplectic. Clearly f is an isomorphism. This completes the proof of Theorem 1.3.

Let us return to the notion of an orthogonal splitting. If W is a symplectic subspace of V and $V = W \oplus W^\perp$ the corresponding orthogonal splitting, we may define the orthogonal projection $P: V \longrightarrow V$ with $\operatorname{im} P = W$ and $\operatorname{ket} P = W^\perp$. One can easily prove the following proposition which characterizes the orthogonal projections among all linear maps of V into itself.

Proposition 1.6. *A linear map $P: V \longrightarrow V$ of a symplectic vector space V to itself is an orthogonal projection if and only if $P^2 = P$ and $P^* = P$.*

Here P^* is the adjoint linear map of P with respect to the symplectic structure Ω of V, i.e. $\Omega(Pu, v) = \Omega(u, P^*v) \forall u, v \in V$. The nonsingularity of Ω yields the existence and uniqueness of such a map.

Let $(V_i, \Omega_i), i = 1, 2$, be symplectic vector spaces. Consider the direct sum $V = V_1 \oplus V_2$ of their underlying vector spaces. The space V may be uniquely supplied with a symplectic structure Ω so that the preceding splitting becomes an orthogonal splitting, and (V_i, Ω_i) become symplectic subspaces. The form Ω is defined by the formula $\Omega(u_1 \oplus u_2, v_1 \oplus v_2) = \Omega_1(u_1, v_1) + \Omega_2(u_2, v_2)$. We shall write $\Omega = \Omega_1 \oplus \Omega_2$ in this case and call the symplectic vector space (V, Ω) the *direct sum of the spaces* (V_i, Ω_i), and the form Ω the *direct sum of the forms* Ω_i. Analogous definitions, notations and terminology hold for an arbitrary finite collection of spaces.

Lagrangian Subspaces

There is another very useful type of subspaces of a symplectic vector space, namely Lagrangian ones.

A vector subspace W of a symplectic vector space V is called *isotropic* if the symplectic structure Ω of V is zero on vectors belonging to W. This can be expressed equivalently as $\Omega|_{W\times W}=0$, or as $W^\perp\supset W$. By virtue of the skew-symmetry property of Ω, each one-dimensional subspace is isotropic. The nonsingularity of Ω yields the whole space V not to be isotropic.

A subspace is called *Lagrangian* if it is isotropic and maximal with respect to this property, that is, a subspace W is Lagrangian means (1) $W^\perp\supset W$, (2) if W' is any subspace of V such that $W'\supset W$ and $(W')^\perp\supset W'$ then $W'=W$. It follows from this definition that a subspace W is Lagrangian if and only if $W^\perp=W$. One can easily prove that any isotropic subspace is contained in some Lagrangian subspace.

Proposition 1.7. *Consider a splitting $V=W\oplus W'$. If both W and W' are isotropic, then they are Lagrangian.*

Proof. Let $v\in W'$ and $v=w+w'$ be its decomposition corresponding to the splitting. Since W is isotropic, $w\perp W$, as is w'. Since W' is isotropic, $w'\perp W'$. We have that w' is orthogonal to both W and W', hence it is zero, and $v=w\in W$. We have proved $W^\perp=W$, i.e. W is Lagrangian. Apply the same proof for W'. □

Theorem 1.8. *All Lagrangian subspaces of a symplectic vector space V have the same dimension of $\frac{1}{2}\dim V$. Any Lagrangian subspace has an algebraic complement which is a Lagrangian subspace.*

Proof. Let us start with proving the second statement. Let L be a Lagrangian subspace. Consider all isotropic subspaces M such that $M\cap L=\{0\}$. Among them there exists a maximal subspace L'. Since $L\cap L'=\{0\}$, we have $V=L^\perp+(L')^\perp=L+(L')^\perp$. Also $(L')^\perp\subset L\oplus L'$. Otherwise there exists a vector $v\perp L'$ and $v\notin L\oplus L'$. The subspace M' generated by L' and v is isotropic, $M'\cap L=\{0\}$, and $M'\supsetneqq L'$, which contradicts the maximality of L'. Hence $V=L+(L')^\perp\subset L\oplus L'$, i.e. $V=L\oplus L'$. It follows from 1.7 that L' is Lagrangian.

In order to prove the first statement, let us define a linear map $A\colon L\longrightarrow(L')^*$ by the formula $(Av)(v')=\Omega(v,v')$ for $v\in L, v'\in L'$. If $v\in\ker A$ then $\Omega(v,v')=0$ for all $v'\in L'$. Since L is isotropic, $\Omega(v,v'')=0$ for all $v''\in L$. Hence $\Omega(v,w)=0$ for all $w\in V$, and we have $v=0$ due to the nonsingularity of Ω. We have obtained that A is an isomorphism. So $\dim L=\dim(L')^*=\dim L'=\frac{1}{2}\dim V$. □

Corollaries 1.9. (a) *An isotropic subspace is Lagrangian if and only if its dimension is equal to half of that of the whole space.*

(b) *Two Lagrangian subspaces W and W' are transversal if and only if $W\cap W'=\{0\}$.*

Lagrangially Split Spaces

Let W and W' be transversal Lagrangian subspaces. We call the splitting $V=W\oplus W'$ a *Lagrangian splitting*, and W' a *Lagrangian complement* of W. A symplectic vector space V with fixed Lagrangian splitting $V = W \oplus W'$ is called a *Lagrangially split space*. A projection π with both Lagrangian $\ker \pi$ and $\operatorname{im} \pi$ is called a *Lagrangian projection*.

Let $f: W \oplus W' \longrightarrow R \oplus R'$ be a symplectic map between two Lagrangially split spaces. We say f *preserves splittings* provided $f(W) \subset R$ and $f(W') \subset R'$. Two Lagrangially split spaces are called *isomorphic* if there exists an isomorphism between them which preserves splittings.

Now, we are going to construct a canonical model for a Lagrangially split symplectic vector space. Take an arbitrary vector space W, form the dual space W^* consisting of all linear functions on W, and let

1.10. $V_c = W \oplus W^*$, *and for the vectors* $u \oplus u^*$, $v \oplus v^* \in V_c$

$$\Omega_c(u \oplus u^*, v \oplus v^*) = v^*(u) - u^*(v).$$

One can easily verify (V_c, Ω_c) to be a symplectic vectors space, W and W^* to be Lagrangian subspaces.

Proposition 1.11. *Any Lagrangially split symplectic vector space $V = W \oplus W'$ is naturally isomorphic to the canonical model 1.10 (the W is the same) by an isomorphism of the form $1_W \oplus h$ where $h: W' \longrightarrow W^*$ is the uniquely defined linear isomorphism.*

Proof. Let there exist $h: W' \longrightarrow W^*$ with the required properties. Then $(1_W \oplus h)^* \Omega_c = \Omega$, Ω being the symplectic structure of V. This means that for $u, v \in W$ and $u', v' \in W'$

$$(*) \qquad \Omega(u + u', v + v') = h(v')(u) - h(u')(v).$$

Taking $u = 0, v' = 0$ one gets

$$(**) \qquad (h(u')(v) = -\Omega(u', v), \quad u' \in W', v \in W.$$

This proves the uniqueness of h. Define $h: W' \longrightarrow W^*$ by the formula $(**)$. Obviously, h is linear. Let $u' \in \ker h$. Then $\Omega(u', v) = 0$ for all $v \in W$. On the other hand, $\Omega(u', v') = 0$ for all $v' W'$ since W' is isotropic. Hence $u' = 0$. We have proved that $\ker h = \{0\}$. Since $\dim W' = \dim W = \dim W^*$, h is a linear isomorphism. Now, we need only verify $1_W \oplus h$ to be a symplectic map. This is equivalent to the validity of $(*)$. Using (1) the bilinearity of Ω, (2) W and W' being isotropic, (3) the skew-symmetry of Ω and the definition $(**)$, one can easily get, in the notations of $(*)$, $\Omega(u + u', v + v') = \Omega(u, v') + \Omega(u', v) = -\Omega(v', u) + \Omega(u', v) = h(v')(u) = h(u')(v)$. □

The following proposition says that any two Lagrangially split spaces of the same dimension are isomorphic; moreover, an isomorphism is uniquely defined by its restriction to the first summands of the splittings.

Proposition 1.12. *Let $W \oplus W'$ and $U \oplus U'$ be Lagrangially split symplectic vector spaces, and $f: W \longrightarrow U$ a linear isomorphism. Then f can be uniquely extended to a symplectic linear isomorphism $F: W \oplus W' \longrightarrow U \oplus U'$ preserving the splittings.*

Proof. By Proposition 1.11 we may reduce our problem to that for the spaces of the form $W \oplus W^*$ and $U \oplus U^*$ endowed with canonical symplectic structures 1.10. In this case, we take $F = f \oplus (f^{-1})^*$ where $(f^{-1})^*: W^* \longrightarrow U^*$ is algebraically adjoint to the linear map $f^{-1}: U \longrightarrow W$. For $u, v \in W$, $u^*, v^* \in W^*$ we have

$$\begin{aligned}(F^*\Omega_c)(u \oplus u^*, v \oplus v^*) &= \Omega_c(f(u) \oplus (f^{-1})^*(u^*), f(v) \oplus (f^{-1})^*(v^*)) \\ &= (f^{-1})^*(v^*)(f(u)) - (f^{-1})^*(u^*)(f(v)) \\ &= v^*(f^{-1}f(u)) - u^*(f^{-1}f(v)) \\ &= \Omega_c(u \oplus u^*, v \oplus v^*),\end{aligned}$$

i.e. $f \oplus (f^{-1})^*$ is symplectic. If there is another such symplectic isomorphism which preserves the splittings and extends f, namely F_1, then the superposition $F_1^{-1} \circ F: W \oplus W^* \longrightarrow W \oplus W^*$ is an identical map when restricted to W, so, by 1.11, it is an identical map on the whole space $W \oplus W^*$. □

Example 1.13 (*Coordinate subspaces of* $\mathbb{R}^{2n}$). Consider the standard symplectic vector space $(\mathbb{R}^n, \Omega_{st})$ (see Example 1.1). Let $\mathfrak{a}$ and $\mathfrak{b}$ be two arbitrary subsets of the set $\{1, 2, \ldots, n\}$. Denote by $L^{\mathfrak{a}}_{\mathfrak{b}}$ the subspace of $\mathbb{R}^{2n}$ generated by the vectors $\{\text{ort}_i, i \in \mathfrak{a}, \text{ort}_{n+j}, j \in \mathfrak{b}\}$. One can easily be convinced that

$$\begin{aligned}&L^{\mathfrak{a}}_{\mathfrak{b}} \quad \text{is isotropic if and only if } \mathfrak{a} \cap \mathfrak{b} = \emptyset, \\ &L^{\mathfrak{a}}_{\mathfrak{b}} \quad \text{is Lagrangian if and only if } \mathfrak{b} = \{1, 2, \ldots, n\} \backslash \mathfrak{a}.\end{aligned}$$

We shall denote the complement of $\mathfrak{a}$ in $\{1, 2, \ldots, n\}$ by $\bar{\mathfrak{a}}$. Obviously $L^{\mathfrak{a}}_{\bar{\mathfrak{a}}} \cap L^{\bar{\mathfrak{a}}}_{\mathfrak{a}} = \{0\}$. The Lagrangian subspaces of the form $L^{\mathfrak{a}}_{\bar{\mathfrak{a}}}$ will be called *coordinate Lagrangian subspaces*, and the Lagrangian splitting

(1.13a) $$\mathbb{R}^{2n} = L^{\mathfrak{a}}_{\bar{\mathfrak{a}}} \oplus L^{\bar{\mathfrak{a}}}_{\mathfrak{a}}$$

a *coordinate Lagrangian splitting* of $(\mathbb{R}^{2n}, \Omega_{st})$. There are precisely $2n$ such splittings. The Lagrangian projection onto $L^{\mathfrak{a}}_{\bar{\mathfrak{a}}}$ with kernel $L^{\bar{\mathfrak{a}}}_{\mathfrak{a}}$ will be denoted $\pi^{\mathfrak{a}}_{\bar{\mathfrak{a}}}$.

The following property of the coordinate Lagrangian subspaces will be useful in our future constructions.

Proposition 1.14. *Let W be an arbitrary Lagrangian subspace of $(\mathbb{R}^{2n}, \Omega_{st})$. Then there exists a subset $\alpha \subset \{1, 2, \ldots, n\}$ such that $L^{\mathfrak{a}}_{\bar{\mathfrak{a}}}$ is a Lagrangian complement of W.*

Firstly we prove an auxiliary Lemma. By coordinate subspace of $\mathbb{R}^n$ we mean a subspace generated by some of the vectors ort_i, where, as above, ort_i is the ith standard coordinate unit vector in $\mathbb{R}^n$.

Lemma. *Let $M \subset \mathbb{R}^n$ be a vector subspace. Then $\mathbb{R}^n$ has a coordinate subspace which is a complement of M.*

Proof. Let K be any coordinate subspace in $\mathbb{R}^n$ such that

(+) $$M \cap K = \{0\}.$$

We may form $M \oplus K$. If $M \oplus K \neq \mathbb{R}^n$, there exists a coordinate vector $\mathrm{ort}_i \notin M \oplus K$. Let L be the 1-dimensional coordinate subspace generated by ort_i. Then $(K \oplus L) \cap M = \{0\}$ by the associativity of $\oplus$. In such a manner we can build up a coordinate subspace preserving the property $(+)$ up to the situation $M \oplus K = \mathbb{R}^n$. □

Proof of 1.14. Consider $M = W \cap (\mathbb{R}^n \oplus 0)$. Here $\mathbb{R}^n \oplus 0$ is the first summand of the splitting $\mathbb{R}^{2n} = \mathbb{R}^n \oplus \mathbb{R}^n$ (which is 1.13a with $\mathfrak{a} = \{1, 2, \dots, n\}$). Take a coordinate subspace K of $\mathbb{R}^n \oplus 0 = \mathbb{R}^n$ which is a complement of M in $\mathbb{R}^n \oplus 0$ (apply Lemma above). The coordinate subspace K is generated by some collection of coordinate vectors, say $\{\mathrm{ort}_i, i \in \mathfrak{a}\}$. This $\mathfrak{a}$ is that which we look for. Indeed, if $u \in L^{\mathfrak{a}}_{\bar{\mathfrak{a}}} \cap W$, then $u \perp L^{\mathfrak{a}}_{\bar{\mathfrak{a}}}$ and $u \perp W$ because both subspaces are Lagrangian. Since $K \subset L^{\mathfrak{a}}_{\bar{\mathfrak{a}}}, u \perp K$. Since $M \subset W, u \perp M$. Hence $u \perp K \oplus M = \mathbb{R}^n \oplus 0$, and $u \in \mathbb{R}^n \oplus 0$ because $\mathbb{R}^n \oplus 0$ is also Lagrangian. We have $u \in L^{\mathfrak{a}}_{\bar{\mathfrak{a}}} \cap W \cap (\mathbb{R}^n \oplus 0) = (L^{\mathfrak{a}}_{\bar{\mathfrak{a}}} \cap (\mathbb{R}^n \oplus 0)) \cap (W \cap (\mathbb{R}^n \oplus 0)) = K \cap M = \{0\}$. Apply 1.9(b). □

Construction 1.15 (*Symplectic Basic*). Let $V = W \oplus W'$ be a Lagrangially split space. Take a basis $\{e_i, 1 \leqq i \leqq n\}$ in W. The correspondence $e_i \longmapsto \mathrm{ort}_i$ defines, by 1.12, the unique isomorphism between Lagrangially split spaces $V = W \oplus W'$ and $\mathbb{R}^{2n} = \mathbb{R}^n \oplus \mathbb{R}^n$. The latter is (1.13a) with $\mathfrak{a} = \{1, 2, \dots, n\}$. Then the preimage $\{e_i, n+1 \leqq i \leqq 2n\}$ of the system $\{\mathrm{ort}_i, n+1 \leqq i \leqq 2n\}$ is a basis of W'. By (1.1a) we have the following relations:

(1.15a)
$$\Omega(e_i, e_{n+i}) = -\Omega(e_{n+i}, e_i) = 1, \quad 1 \leqq i \leqq n,$$
$$\Omega(e_i, e_j) = 0 \quad \text{for all other values of } i, j.$$

We call such a basis a *symplectic basis* in a (Lagrangially split) symplectic vector space V.

Subspaces in a Lagrangially Split Space

Let $V = W \oplus W'$ be a Lagrangially split space, and Ω be its symplectic structure. Denote by π and π' the Lagrangian projections which correspond to this splitting: $1_V = \pi + \pi'$ $\mathrm{im}\,\pi = \ker \pi' = W$, $\mathrm{im}\,\pi' = \ker \pi = W'$. Consider the bilinear form

(1.16)
$$b(u, v) = \Omega(\pi u, v) = \Omega(u, \pi' v).$$

Let $L \subset V$ be a subspace. The restriction b_L of the form b to $L \times L \subset V \times V$ is called the *bilinear form* of L.

Proposition 1.17.

(i) *L is isotropic* $\Leftrightarrow$ *b_L is symmetric.*
(ii) *Let L be Lagrangian, then*

L is transversal to both W and W' $\Leftrightarrow$ *b_L is nonsingular.*

Proof. Part (i) follows from the equality $b(u, v) - b(v, u) = \Omega(u, v)$ which is an immediate consequence of (1.16) and the skew symmetry of Ω.

Let L be Lagrangian. Consider L as a vector space supplied with a symmetric bilinear form b_L, and let $L^{\perp(b_L)}$ be the orthogonal complement of L in L with respect to b_L. The nonsingularity of b_L may be equivalently expressed as $L^{\perp(b_L)}=\{0\}$. The assertion (ii) follows now from the equality

$$L^{\perp(b_L)}=(L\cap W)\oplus(L\cap W'). \tag{1.17a}$$

To prove (1.17), we use the expression

$$b_L(u,v)=\Omega(\pi u,v)=-\Omega(\pi' u,v) \tag{1.17b}$$

which is (1.16), the symmetry being taken into account. Let $u\in L^{\perp(b_L)}$. Then (1.17b) yields $\pi u\perp^{\Omega}L$ and $\pi' u\perp^{\Omega}L$. Since L is Lagrangian, it follows that $\pi u\in L$ and $\pi' u\in L$. We have $u=\pi u+\pi' u\in(L\cap W)\oplus(L\cap W')$. Conversely, each $u\in L\cap W$ satisfies $0=\Omega(u,l)=\Omega(\pi u,l)=b_L(u,l)$ for all $l\in L$, and hence belongs to $L^{\perp(b_L)}$. Analogous considerations result in $L\cap W'\subset L^{\perp(b_L)}$. □

Projections along Lagrangian Subspace

Let $W\subset V$ be a Lagrangian subspace. Here we investigate projections along W, i.e. those whose kernels coincide with W. The set of all such projections is in one-to-one correspondence with the set of all n-dimensional subspaces $L\subset V$, $\dim V=2n$, which are transversal to W, the correspondence attaches to each L the projection π_L along W onto L.

Fix another Lagrangian subspace W' transversal to W and take a symplectic vector base $\{e_i, 1\leqq i\leqq 2n\}$ for the Lagrangially split space $V=W\oplus W'$. Let π and π' be the Lagrangian projections of this splitting as in the previous subsection. If a subspace L is transversal to W, then $\pi'|_L$ maps L bijectively onto W', and the vectors

$$f_i=(\pi'|_L)^{-1}e_{n+i},\quad 1\leqq i\leqq n, \tag{1.18}$$

form a basis of L.

Proposition 1.19.

$$f_i=\sum_{j=1}^{n} b_{ij}e_j+e_{n+i},\quad 1\leqq i\leqq n, \tag{1.19a}$$

where

$$b_{ij}=b_L(f_i,f_j) \tag{1.19b}$$

is the matrix of the bilinear form of L in the Lagrangially split space $V=W\oplus W'$ with symplectic basis $\{e_i\}$.

Proof. Setting $f_i=\sum_{j=1}^{2n} b_{ij}e_j$, $1\leqq i\leqq n$, applying $\Omega(e_k,\cdot)$, and taking into account (1.15a), we obtain $\Omega(e_k,f_i)=b_{i,n+k}$, $\Omega(e_{n+k},f_i)=-b_{ik}$, $1\leqq i,k\leqq n$. Since $f_i=\pi f_i+\pi' f_i$ and $e_{n+k}=\pi' f_k$, we have $b_{ik}=-\Omega(\pi' f_k,f_i)=b_L(f_i,f_k)$, $1\leqq i,k\leqq n$, and $b_{i,n+k}=\Omega(e_k,\pi' f_i)=\Omega(e_k,e_{n+i})=\delta_{ik}$. □

Proposition 1.20. *The matrix of π_L, L transversal to W, in the basis $\{e_k\}$, described above, has the form*

$$\left(\begin{array}{c|c} 0 & B^t \\ \hline 0 & I \end{array}\right), \tag{1.20a}$$

where $B^t = \{b_{ji}\}$ is the transposed matrix of the bilinear form b_L of L in the base $\{f_k\}$, I is the $n \times n$ unit matrix. The subspace L is Lagrangian if and only if B is symmetric. If L is Lagrangian then L is transversal also to W' if and only if B is invertible.

Conversely, every matrix of the form (1.20a) is the matrix of a projection along W.

Proof. Denote the coefficients of the matrix in question by a_{ij}, $1 \leqq i, j \leqq 2n$. We have

$$\pi_L e_i = \sum_{j=1}^{2n} a_{ji} e_j, \quad 1 \leqq i \leqq 2n.$$

Using (1.15a), we obtain

$$\begin{aligned} a_{ki} &= \Omega(\pi_L e_i, e_{n+k}), \\ a_{n+k,i} &= -\Omega(\pi_L e_i, e_k), \quad 1 \leqq k \leqq n,\ 1 \leqq i \leqq 2n. \end{aligned} \tag{1.20b}$$

Since $\ker \pi_L = W$ and $e_i \in W$, $1 \leqq i \leqq n$, $a_{ji} = 0$ if $1 \leqq i \leqq n$. It follows from (1.19a) that

$$\pi_L e_{n+i} = \pi_L f_i - \sum_{j=1}^{n} b_{ij} \pi_L e_j = f_i, \quad 1 \leqq i \leqq n. \tag{1.20c}$$

Substituting (1.20c) into (1.20b), we obtain $a_{k,n+i} = \Omega(f_i, e_{n+k}) =$ (use (1.19a) again) $\Omega(f_i, \pi' f_k) = b(f_i, f_k) = b_{ik}$, $a_{n+k,n+i} = -\Omega(f_i, e_k) =$ (use (1.19a)) $- \Omega(e_{n+i}, e_k) = \delta_{ik}$. The converse assertion of the Proposition is obvious. □

Corollary 1.21. *Given a pair W_i, $i = 1, 2$, of transversal Lagrangian subspaces, there is a Lagrangian subspace W_3 transversal to both of the W_i, $i = 1, 2$.*

The requirement for the spaces W_i to be transversal is unnecessary here.

Proposition 1.22. *Given a pair W_i, $i = 1, 2$, of Lagrangian subspaces of a symplectic vector space V, there is a Lagrangian subspace W_3 of V which is transversal to both of the W_i, $i = 1, 2$.*

Proof. We shall prove the proposition by using induction on $\dim V$. Let the proposition be true if $\dim V < 2n$. Now let $\dim V = 2n$. Consider the isotropic subspace $M = W_1 \cap W_2$. If $\dim M = 0$ then the assertion is nothing more than that of Corollary 1.21. Let $k = \dim M > 0$. Then there is another isotropic subspace M' of dimension k such that $M' \cap M = \{0\}$ and $K = M \oplus M'$ is symplectic. To prove this take a Lagrangian complement W'_1 to W_1 and a symplectic basis $\{e_j, 1 \leqq j \leqq 2n\}$ in $V = W_1 \oplus W'_1$ such that $\{e_j, 1 \leqq j \leqq k\}$ form a basis of M (see Construction 1.15). Then $\{e_{n+j}, 1 \leqq j \leqq k\}$ spans a subspace M' with the desired property.

We claim that

(a) $$W_i \cap M' = \{0\}, \quad i = 1, 2.$$

Otherwise there would be a v such that $0 \neq v \in W_i \cap M'$ then there would exist $m \in M$ such that $\Omega(v, m) = 1$. This would contradict the fact that W_i is isotropic; $v, m \in W_i$.

(b) $$K \cap W_i = M, \quad i = 1, 2.$$

The inclusion $M \subset K \cap W_i$ is obvious. If $v \in K \cap W_i$, then $v = m + m'$, $m \in M$ and $m' \in M'$. Since both v and m belong to M_i, we have $m' \in W_i$. Then (a) yields $m' = 0$.

Define the subspaces $U_i = W_i \cap K^{\perp}$. We assert that

(c) $$W_i = M \oplus U_i, \quad i = 1, 2.$$

The inclusions $M \oplus U_i \subset W_i$ and the equality $M \cap U_i = \{0\}$ are obvious. Let $v \in W_i$ and $v = k + k_1$ be its orthogonal expansion corresponding to the orthogonal splitting $V = K \oplus K^{\perp}$. Then $k = m + m'$, $m \in M$, $m' \in M'$. The vector $m' \in K$ is orthogonal to M since each term of the expansion $m' = v - m - k_1$. Hence $m' = 0$ since M is Lagrangian in K. We have $v = m + k_1$, and $k_1 = v - m \in W_i \cap K^{\perp} = U_i$. This proves the inverse inclusion and hence (c) itself.

(d) $$U_i \text{ is Lagrangian in } K^{\perp}, \quad i = 1, 2.$$

Otherwise there would exist $\tilde{U}_i \supsetneqq U_i$ isotropic in $K^{\perp}$, and $M \oplus \tilde{U}_i$ would be an isotropic subspace in V larger than W_i (see (c)).

By the induction hypothesis there is a Lagrangian subspace U_3 in $K^{\perp}$ which is transversal to both of the spaces U_i, $i = 1, 2$, in $K^{\perp}$. The subspace $W_3 = U_3 \oplus M'$ is Lagrangian and transversal to both of the spaces $W_i, i = 1, 2$. □

The Manifold of Lagrangian Subspaces

Denote by $\mathscr{L}(V)$ the set of all Lagrangian subspaces of a symplectic vector space V, $\dim V = 2n$. $\mathscr{L}(V)$ may be regarded as a subset of the Grassmanian manifold $\mathscr{G}_n(V)$ of all n-dimensional subspaces of V. It is well known that $\mathscr{G}_n(V)$ is a closed n^2-dimensional manifold (see Fuks and Rokhlin [1], Chap. 3, §2.2).

Proposition 1.23. *$\mathscr{L}(V)$ is a smooth closed $n(n+1)/2$-dimensional connected submanifold of $\mathscr{G}_n(V)$.*

Proof. We shall construct a collection of special charts which cover $\mathscr{L}(V)$. Choose a Lagrangian subspace W and consider the set $\mathscr{L}_W \subset \mathscr{L}(V)$ which consists of all the Lagrangian complements to W. Obviously, $\mathscr{L}_W \subset \mathscr{G}_W$, where $\mathscr{G}_W$ is the set of all algebraic complements to W. The set $\mathscr{G}_W$ is in one-to-one correspondence with the set of all projections along W, and $\mathscr{L}_W$ with those of Lagrangian ones. Fix $W' \in \mathscr{L}_W$ and take a symplectic basis $e = \{e_i, 1 \leqq i \leqq 2n\}$ of the Lagrangian splitting $V = W \oplus W'$. Define the chart

(1.23a) $$\Psi_{W,W',e}: \mathscr{G}_W \longrightarrow \mathbb{R}^{n^2}$$

where $\Psi_{W,W',e}$ assigns to each algebraic complement L of W the lexicographically ordered set of elements of the matrix B which arises in (1.20a) (the matrix of the bilinear form of L in the base f_k given by (1.18)). The set $\mathscr{G}_W$ is open in $\mathscr{G}_n(V)$ and the map (1.23a) is a diffeomorphism. As a matter of fact, the smooth structure of $\mathscr{G}_n(V)$ may be introduced by taking all such maps as charts, W and W' running over all mutually transversal pairs of Lagrangian subspaces, e running over all symplectic bases in $V = W \oplus W'$. Thereby the map (1.23a) is a smooth chart in $\mathscr{G}_n(V)$ with the property (due to Proposition 1.20) that the intersection of $\mathscr{L}(V)$ with the domain of (1.23a) is carried by the map onto a linear $n(n+1)/2$-dimensional subspace of $\mathbb{R}^{n^2}$. Theorem 1.8 implies that the sets $\mathscr{L}_W = \mathscr{L}(V) \cap \mathscr{G}_W$ cover $\mathscr{L}(V)$. Furthermore, as follows from 1.14, one may choose a finite number, namely 2^n, of such charts whose domains cover $\mathscr{L}(V)$. This proves that $\mathscr{L}(V)$ is a $n(n+1)/2$-dimensional smooth submanifold of $\mathscr{G}_n(V)$. Obviously $\mathscr{L}(V)$ has no boundary. If $L \in \mathscr{G}_n(V) \setminus \mathscr{L}(V)$, then, due to 1.20, there is an open neighbourhood of L is $\mathscr{G}_n(V)$ which has no intersection with $\mathscr{L}(V)$. This proves that $\mathscr{L}(V)$ is closed.

Let W_1 and W_2 be two Lagrangian subspaces of V. Proposition 1.22 yields the existence of $W \in \mathscr{L}(V)$ transversal to both of these $W_i, i = 1, 2$. Hence $W_i \in \mathscr{L}_W$. Since $\mathscr{L}_W$ is diffeomorphic to $\mathbb{R}^{n(n+1)/2}$, there exists a smooth path joining W_1 and W_2 and lying in $\mathscr{L}_W \subset \mathscr{L}(V)$. This proves that of $\mathscr{L}(V)$ is connected. □

We shall obtain another use of our construction by defining the barycentric map. Let m be a positive integer. Denote by Δ_{m-1} the $(m-1)$-dimensional standard symplex in $\mathbb{R}^m$ which consists of all points $t = (t_1, t_2, \ldots, t_m) \in \mathbb{R}^m$ such that $t_i \geqq 0$, $1 \leqq i \leqq m$, $\sum_{i=1}^m t_i = 1$. The formula 1.19a and the assertion of Proposition 1.19, despite being stated "in coordinates", show us that if $\{P_i, 1 \leqq i \leqq m\}$ is a collection of Lagrangian projections along M, the sum $\sum_{i=1}^m t_i P_i$ is again a Lagrangian projection along M. Identifying Lagrangian complements of M with Lagrangian projections onto them along M, we introduce this operation in $\mathscr{L}_M$. We have the C^a-map

$$\Delta_{m+1} \times \underbrace{\mathscr{L}_M \times \cdots \times \mathscr{L}_M}_{m \text{ times}} \longrightarrow \mathscr{L}_M$$

which acts by the formula

$$(t_1, t_2, \ldots, t_m, W_1, W_2, \ldots, W_m) \longmapsto \sum_{i=1}^m t_i W_i.$$

We shall refer to this map as the barycentric map.

Subspaces of a Symplectic Vector Space

Symplectic and Lagrangian subspaces which were our interest in previous subsections are only particular kinds of subspaces in a symplectic vector space. Here we shall investigate the geometry of an arbitrary vector subspace. The following proposition plays a key role in this investigation.

Proposition 1.24 (the linear part of Givental's theorem). *Let Ω_0 and Ω_1 be two symplectic forms on a vector space V, and M be a vector subspace of V. If Ω_0 and Ω_1 coincide on the vectors lying in M, i.e. $\Omega_0|M = \Omega_1|M$, then there exists a linear isomorphism $f: V \longrightarrow V$ such that $f|M = 1_M$ and $f^*\Omega_1 = \Omega_0$.*

Proof. Consider the form $\omega = \Omega_0|M = \Omega_1|M$ in the vector space M. It does not need to be symplectic, but we may split M into $M_0 \oplus M_1$ where M_0 is an orthogonal complement of M in M in the sense of ω, and M_1 is an arbitrary algebraic complement of M_0 in M. Then, the pair $(M_1, \omega|M_1)$ is a symplectic vector space, and so are the orthogonal complements V_i of M_1 in symplectic vector spaces (V, Ω_i), $i = 0, 1$. Consider the symplectic vector spaces $(V_i, \Omega_i|V_i)$. They both contain M_0 as an isotropic subspace. Extend M_0 to Lagrangian subspaces W_i in these symplectic vector spaces, take an arbitrary linear isomorphism between the spaces W_i (note that W_1 has the same dimension, as W_2) which is an identity on M_0, and apply Proposition 1.12. We obtain an isomorphism $g: (V_0, \Omega_0|V_0) \longrightarrow (V_1, \Omega_1|V_1)$ such that $g|M_0 = 1_{M_0}$. Put $f = g \oplus 1_{M_1}$. □

Two pairs (V_i, M_i), $i = 1, 2$, where V_i are symplectic vector spaces, and M_i are vector subspaces of V_i, are called *isomorphic* if there exists an isomorphism $f: V_1 \longrightarrow V_2$ of symplectic vector spaces which takes M_1 to M_2. Such an isomorphism f is called an isomorphism of pairs. Proposition 1.20 reduces the problem of classification of pairs up to an isomorphism to that of classification of pairs having the form (M, ω), where M is a vector space and ω is a skew-symmetric bilinear form in M which may be degenerate, and to such pairs which occur as vector subspaces of a symplectic vector space with induced bilinear form. As consequences, we obtain

1.25. *Any two pairs (V_i, M_i), $i = 1, 2$, with V_i symplectic vector spaces of the same dimension, and M_i their isotropic subspaces of the same dimension, are isomorphic.*

1.26. *Any two pairs (V_i, M_i), $i = 1, 2$, with V_i symplectic vector spaces of the same dimension, and M_i their hyperplanes (subspaces of codimension 1), are isomorphic.*

Let us consider the latter case in more detail, namely the case of hyperplanes. If $M \subset V$ is a hyperplane, its orthogonal complement $L = M^\perp$ is one-dimensional and lies in M (otherwise, M would be a symplectic subspace of V which contradicts the odd dimension of M). We call this one-dimensional subspace L the zero-line of a hyperplane M. Take an arbitrary algebraic complement R of L in M. We have $M = R \oplus L$ and R is a symplectic subspace of V. If $M_i = R_i \oplus L_i$, $i = 1, 2$, are two hyperplanes in symplectic spaces of the same dimension decomposed in such a manner, we may construct a linear isomorphism $f: M_1 \longrightarrow M_2$ of the form $f = g \oplus h$, $g: R_1 \longrightarrow R_2$ being a symplectic isomorphism, $h: L_1 \longrightarrow L_2$ an arbitrary linear isomorphism, proving the validity of 1.26.

Endomorphisms of a Symplectic Vector Space

We will study endomorphisms of a symplectic vector space (V, Ω) or in the other words we are considering linear symplectic self-maps $f: V \longrightarrow V$. Due to 1.2, such mappings are invertible, so they form a group, The *symplectic group* denoted by $\mathrm{Sp}(V)$.

$\mathrm{Sp}(V)$ is a subgroup of $\mathrm{GL}(V)$ the group of all linear automorphisms of V. $\mathrm{GL}(V)$ is a Lie group, one can embed it into $\mathbb{R}^{4n^2}$ by fixing a basis $\{e_i, 1 \leqq i \leqq 2n\}$ via the map Ψ which assignes the matrix f_{ij} to a map $f \in \mathrm{GL}(V)$ where

$$f(e_i) = \sum_{j=1}^{2n} f_{ji} e_j. \tag{1.27}$$

The elements of the matrix $\{f_{ij}\}$ are ordered lexicographically to become points of $\mathbb{R}^{4n^2}$. The image $\Psi(\mathrm{GL}(V))$ consists of nonsingular matrices, so is an open dense subset of $\mathbb{R}^{4n^2}$. The following proposition will be useful in further studying Maslov's index.

Proposition 1.28. $\mathrm{Sp}(V)$ *is a Lie subgroup of* $\mathrm{GL}(V)$, *and* $\dim \mathrm{Sp}(V) = n(2n+1)$.

Proof. Take a symplectic basis $\{e_i, 1 \leqq i \leqq 2n\}$ as in 1.15. The above map Ψ, associated with this base, assigns the matrix $\{f_{ij}\}$ to $f \in \mathrm{Sp}(V)$ where $\{f_{ij}\}$ satisfies the equations

$$F_{ik} \stackrel{\mathrm{def}}{=} \sum_{j=1}^{n} (f_{ji} f_{n+j,k} - f_{n+j,i} f_{jk}) = \Omega(e_i, e_k), \quad 1 \leqq i < k \leqq 2n, \tag{a}$$

conveying the conservation of Ω. None of the columns of the matrix $\{f_{ij}\}$ is zero to ensure invertability. The gradients of the left-hand sides of (a) are linearly independent. Indeed, by representing the said gradients as a matrix $\{\partial F_{ik}/\partial f_{i'k'}, 1 \leqq i' < k' \leqq 2n\}$, one finds this matrix that possess two nonzero columns, namely: the ith one, $(f_{n+1,k}, f_{n+2,k}, \ldots, -f_{1k}, -f_{2k}, \ldots, -f_{nk})^t$, and the kth one, $(-f_{n+1,i}, -f_{n+2,i}, \ldots, -f_{2n,i}, f_{1i}, f_{2i}, \ldots, f_{ni})^t$, all other columns are equal to zero. So, we may apply the theorem of AI.53 to the map $\{f_{ik}\} \longrightarrow \{F_{ik}\}$. We obtain that $\mathrm{Sp}(V)$ is a submanifold of $\mathrm{GL}(V)$, $\operatorname{codim} \mathrm{Sp}(V) =$ the number of equations (a) $= n(2n-1)$, and $\dim \mathrm{Sp}(V) = 4n^2 - n(2n-1) = n(2n+1)$. □

Two elements $f_k \in \mathrm{Sp}(V_k)$, $k = 1, 2$, V_k being two symplectic vector spaces of the same dimension, are *conjugated* if $f_1 = h^{-1} \circ f_2 \circ h$ where $h: V_1 \longrightarrow V_2$ is a symplectic linear isomorphism. The problem of classification of symplectic automorphisms up to conjugacy, which we shall merely touch on here, looks more pliable for the complex case. So we describe the procedure of complexification of an underlying space.

Define the *complexification* $V \otimes \mathbb{C}$ of V as the product $V \times V$ together with addition $(x, y) + (x', y') = (x + x', y + y')$ and multiplication by a complex number $c = a + ib : c \cdot (x, y) = (ax - by, ay + bx)$. Obviously $V \otimes \mathbb{C}$ is a vector space over the field $\mathbb{C}$. The initial real space V is embedded in $V \otimes \mathbb{C}$ by setting $x = (x, 0)$ for $x \in V$.

In view of that identification, each vector $z=(x,y)\in V\otimes\mathbb{C}$ may be uniquely written as $z=x+iy$. The vectors x and y will be referred to as the *real* and *complex parts* of z respectively. The symplectic structure Ω has a natural extension onto $V\otimes\mathbb{C}$ given by the formula $\Omega(x+iy,u+iv)=\Omega(x,u)-\Omega(y,v)+i\Omega(x,v)+i\Omega(y,u)$, $x,y,u,v\in V$.

The extended Ω is bilinear in the sense of the complex linear structure of $V\otimes\mathbb{C}$. Another useful operation defined on $V\otimes\mathbb{C}$ is *complex conjugation* $x+iy\longmapsto x-iy$ which will usually be denoted by the upper line: $\bar{z}$ is the complex conjugate of z. Trivally $\overline{\Omega(z,w)}=\Omega(\bar{z},\bar{w})$.

Define the *complexification* ${}^{\mathbb{C}}f\colon V\otimes\mathbb{C}\longrightarrow V\otimes\mathbb{C}$ of f by setting

$$ {}^{\mathbb{C}}f(x+iy)=f(x)+if(y),\quad f\in\mathrm{Sp}(V),\ x,y\in V. $$

The transformation ${}^{\mathbb{C}}f$ preserves the features of linearity and symplecticity, but in the new "complex" sense. We have

$$ {}^{\mathbb{C}}f(\bar{z})=\overline{{}^{\mathbb{C}}f(z)}\quad\text{for all } f\in\mathrm{Sp}(V),\ z\in V\otimes\mathbb{C}. $$

A complex number λ is called an *eigenvalue* of $f\in\mathrm{Sp}(V)$ if there is a nonzero vector $v\in V\otimes\mathbb{C}$ satisfying the equation ${}^{\mathbb{C}}f(v)=\lambda v$. This v is called an *eigenvector* (of f) corresponding to λ. Eigenvectors corresponding to λ span the *eigenspace* $\ker({}^{\mathbb{C}}f-\lambda 1)$. The latter is contained in the *root space* $V_\lambda=\ker(({}^{\mathbb{C}}f-\lambda 1)^{2n})$. If $v\in V_\lambda$ then there exists a chain of nonzero *principal* vectors v_k, $1\leqq k\leqq m$, such that $f(v_k)=\lambda v_k+v_{k+1}$, $1\leqq k\leqq m-1$, $v_1=v$, and v_m is an eigenvector for λ. The direct sum of root spaces corresponding to all eigenvalues is equal to $V\otimes\mathbb{C}$. An eigenvalue is called *simple* if the (complex) dimension of its root space is 1. The above definitions and the aforementioned fact of completeness of root spaces hold for an arbitrary linear map $f\colon V\otimes\mathbb{C}\longrightarrow V\otimes\mathbb{C}$, but in the case $f\in\mathrm{Sp}(V)$ there are some particular properties which we now summarize in the following

Proposition 1.29. *Let $f\in\mathrm{Sp}(V)$ and let λ be one of its eigenvalues. Then* (1) $\lambda\neq 0$; (2) λ^{-1} *is also an eigenvalue of f*; (3) *so is $\bar{\lambda}$, and, if v is an eigenvector corresponding to λ then $\bar{v}$ is an eigenvector corresponding to $\bar{\lambda}$*; (4) *if λ' is another eigenvalue of f such that $\lambda\lambda'\neq 1$, then the root spaces of λ and λ' are mutually orthogonal*; (5) *each of two numbers:* 1 *and* -1, *cannot be a simple eigenvalue of f.*

Proof. (1) follows from the existence of f^{-1}. To prove (2) notice that an eigenvector v corresponding to λ is orthogonal to the image of ${}^{\mathbb{C}}f-\lambda^{-1}1$. Indeed, if $w={}^{\mathbb{C}}f(u)-\lambda^{-1}u$, then $\Omega(v,w)=\Omega(v,{}^{\mathbb{C}}f(u)-\lambda^{-1}u)=\Omega({}^{\mathbb{C}}f^{-1}(v)-\lambda^{-1}v,u)=\Omega(0,u)=0$. Hence $\mathrm{im}({}^{\mathbb{C}}f-\lambda^{-1}1)\neq V\otimes\mathbb{C}$, and consequently $\ker({}^{\mathbb{C}}f-\lambda^{-1}1)\neq\{0\}$. One obtains the assertion (3) by immediately applying conjugation to the eigenvector equation. Now let $v\in V_\lambda$ and $v'\in V_{\lambda'}$, $\lambda\lambda'\neq 1$. Then there are chains of nonzero principal vectors v_k, $1\leqq k\leqq m$, v'_j, $1\leqq j\leqq m'$. The calculation

$$ \Omega(v_k,v'_j)=\Omega({}^{\mathbb{C}}f(v_k),{}^{\mathbb{C}}f(v'_j))=\lambda\lambda'\Omega(v_k,v'_j)+\lambda\Omega(v_k,v'_{j+1})+\lambda'\Omega(v_{k+1},v'_j) $$

enables us to prove by induction the orthogonality of v_k and v'_j proving (4). Let 1, or -1, be a simple eigenvalue, and w its eigenvector. Then (4) implies w is orthogonal to all root spaces corresponding to all the other eigenvalues. It is also orthogonal to itself. Since V_λ spans $V \otimes \mathbb{C}$, it is a contradiction to the nonnullity of w. □

Let us study two special classes of symplectic endomorphisms. We say a map $f \in \mathrm{Sp}(V)$ is *simple elliptic* if all its eigenvalues are simple and lie in the unit circle $S = \{z \in \mathbb{C}: |z| = 1\}$. The structure of simple elliptic endomorphisms is described entirely by the following proposition.

Proposition 1.30. *If f is simple elliptic then there is a symplectic basis $\{e_k, f_k, 1 \leqq k \leqq n\}$ in V (see the definition in 1.15) such that the matrix of f in that basis has a quasi-diagonal form, namely it is the direct sum of 2×2-blocks of the form*

$$\begin{pmatrix} \cos \alpha_k & -\sin \alpha_k \\ \sin \alpha_k & \cos \alpha_k \end{pmatrix}$$

corresponding to each pair e_k, f_k. The numbers $\alpha_k \in]0, \pi[\cup] -\pi, 0[$ are pointwise distinct. The set $\{\alpha_k\}$ is determined uniquely, it is invariant under conjugacy and determines f up to conjugacy. The eigenvalues of f are $\exp(\pm i\alpha_k)$, $1 \leqq k \leqq n$.

Proof. Let $\exp(i\beta_k)$, $\beta_k \in]0, \pi[$, be eigenvalues of f lying in the upper halfplane, and $w_k = u_k + iv_k$ be their eigenvectors, $1 \leqq k \leqq n$. Then, due to Proposition 1.29, $\exp(-i\beta_k)$ are also eigenvalues with eigenvectors $\bar{w}_k = u_k - iv_k$. One can easily deduce that $\Omega(w_k, \bar{w}_k)$ is nonzero (apply 1.29(4)), purely imaginary, and $\varepsilon_k = \mathrm{sign}(i\Omega(w_k, \bar{w}_k))$ does not depend on the choice of w_k in the eigenspace corresponding to $\exp(i\beta_k)$. Let us normalize w_k so that $i\Omega(w_k, \bar{w}_k) = 2\Omega(u_k, v_k) = 2\varepsilon_k$. If $\varepsilon_k = +1$ we set $e_k = u_k$, $f_k = v_k$, and $\alpha_k = -\beta_k$. If $\varepsilon_k = -1$, our choice is $e_k = v_k$, $f_k = u_k$, and $\alpha_k = \beta_k$. The eigenvalue equation ${}^{\mathbb{C}}f(w_k) = \exp(i\beta_k) w_k$ yields $f(u_k) = u_k \cos \beta_k - v_k \sin \beta_k$, $f(v_k) = u_k \sin \beta_k + v_k \cos \beta_k$. These give in both of the above cases the desired form of the matrix of f in the 2-dimensional symplectic subspaces of V generated by u_k, v_k. All such subspaces are invariant, and their sum is V. □

A map $f \in \mathrm{Sp}(V)$ is called *hyperbolic* if the spectrum of ${}^{\mathbb{C}}f$ has no intersection with the unit circle in $\mathbb{C}$. A linear self map $g: L \longrightarrow L$ of a finite dimensional vector space L is called a *contraction* if $g^n(x) \longrightarrow 0$ as $n \longrightarrow +\infty$ for all $x \in L$. A necessary and sufficient condition for g to be a contraction is that the spectrum of its complexification lies strictly within of the unit circle.

Proposition 1.31. *Let $f: V \longrightarrow V$ be a hyperbolic linear symplectic map. Then there is a Lagrangian splitting $V = V^s + V^u$ invariant with respect to f and such that $f|_{V^s}$ and $f^{-1}|_{V^u}$ are contractions.*

Proof. Consider the resolvent $R_\lambda = ({}^{\mathbb{C}}f - \lambda 1)^{-1}$ of ${}^{\mathbb{C}}f$. This is a linear map which commutes with ${}^{\mathbb{C}}f$ and which is a rational function of the variable $\lambda \in \mathbb{C}$, the poles being at points of the spectrum of ${}^{\mathbb{C}}f$. Define the operators

$$P_s = \frac{1}{2\pi i} \oint R_\lambda \, d\lambda \quad \text{and} \quad P_u = 1 - P_s,$$

the integration spreading over the unit circle in the complex plane. It is well known that P_s and P_u are both projections, i.e. satisfies the equation $P^2 = P$. There is a splitting $V \otimes \mathbb{C} = V'_s + V'_u$ where $V'_{s,u} = \{x \in V \otimes \mathbb{C}\colon P_{s,u}x = x\}$. This splitting is invariant with respect to ${}^{\mathbb{C}}f$, and ${}^{\mathbb{C}}f|_{V'_s}$, $({}^{\mathbb{C}}f)^{-1}|_{V'_u}$ are contractions. Now consider $V \otimes \mathbb{C}$ as a real vector space of double dimension and split it into two subspaces $V \otimes \mathbb{C} = V + iV$, where V is the original space, i is multiplication by the imaginary unit. Since the integration path in the integral defining P_s is symmetric with respect to the real axis (the reflection must be completed by reversing the direction of integration), P_s and P_u leave this second splitting invariant. (Take a real basis in V and consider it as a (complex) basis of $V \otimes \mathbb{C}$. Then the matrix of ${}^{\mathbb{C}}f$ in this basis is real and those of R_λ and $R_{\bar{\lambda}}$ consist of mutually conjugate entries. The imaginary part of the matrix of P_s cancels in the process of integrating. Set $V^s = V \cap V'_s$ and $V^u = V \cap V'_u$. These are invariant under ${}^{\mathbb{C}}f$ and split the original real space: $V = V^s \oplus V^u$. Since ${}^{\mathbb{C}}f|_V = f$, we have that $f|_{V^s}$ and $f^{-1}|_{V^u}$ are contractions. Let $x, y \in V^s$, and Ω be the symplectic structure of V. Then $\Omega(x,y) = \Omega(f^n(x), f^n(y)) \longrightarrow 0$ as $n \longrightarrow +\infty$. Hence V^s is isotropic. Similar arguments show V^u is isotropic also. Applying 1.7 yields V^s and V^u are Lagrangian. □

Lagrangian subspaces V^s and V^u of the above splitting are called *stable* and *unstable subspaces* of f.

Exterior Algebra of a Vector Space

The aim of this final subsection is to indicate the position of a symplectic form in the exterior algebra of a vector space V where the exterior algebra, is the direct sum

$$\Lambda(V) = \sum_{p=0}^{\dim V} \oplus \Lambda^p(V), \tag{1.32}$$

and where $\Lambda^p(V)$ is the space of all skew-symmetric real valued p-linear forms

$$\alpha\colon \underbrace{V \times V \times \cdots \times V}_{p \text{ times}} \longrightarrow \mathbb{R}$$

p-linearity means that α must be linear in each of its p variables separately, skew-symmetry means that permutation of the values of any two variables changes the sign of the value of α. The dimension of $\Lambda^p(V)$ equals $\binom{\dim V}{p}$ (binomial coefficient). In (1.32) $\Lambda^0(V) = \mathbb{R}$ and $\Lambda^1(V) = V^*$ (the dual space). Elements of $\Lambda^p(V)$ are called homogeneous to the power of p.

Let $\mathcal{S}^p$ denote the group of all p-permutations. An element of $\mathcal{S}^p$ is a bijection $\varepsilon\colon \{1, 2, \ldots, p\} \longrightarrow \{1, 2, \ldots, p\}$, the group multiplication is the superposition of bijections. Define the action of $\mathcal{S}^p$ on functions of p variables by the formula

$$\varepsilon f(x_1, x_2, \ldots, x_p) = f(x_{\varepsilon(1)}, x_{\varepsilon(2)}, \ldots, x_{\varepsilon(p)})$$

and define the antisymmetrisation operator $\mathfrak{A}$ on real valued functions of p

variables as follows

$$\mathfrak{A}f = \sum_{\varepsilon \in \mathcal{S}^p} \sigma(\varepsilon)\varepsilon f,$$

where $\sigma(\varepsilon) = \pm 1$ is the sign of a permutation ε. The exterior product of two homogeneous forms $\alpha \in \Lambda^p(V)$ and $\beta \in \Lambda^q(V)$ is the form $\alpha \wedge \beta \in \Lambda^{p+q}(V)$ defined by the formula

(1.33) $$\alpha \wedge \beta = \frac{1}{p!q!} \mathfrak{A}\alpha \otimes \beta,$$

where the tensor product of forms is

(×) $$\alpha \otimes \beta(x_1, x_2, \ldots, x_p, x_{p+1}, \ldots, x_{p+q}) = \alpha(x_1, x_2, \ldots, x_p) \cdot \beta(x_{p+1}, \ldots, x_{p+q}).$$

If $p + q > \dim V$, formula (1.33) yields us $\alpha \wedge \beta = 0$, so there is no contradiction with (1.32).

One can prove the following properties of $\wedge$-multiplication:

($\wedge$ 1) bilinearity, i.e. linearity in each of its multipliers,
($\wedge$ 2) associativity: $(\alpha \wedge \beta) \wedge \gamma = \alpha \wedge (\beta \wedge \gamma)$
($\wedge$ 3) commutation rule: $\alpha \wedge \beta = (-1)^{pq} \beta \wedge \alpha$ if $\alpha \in \Lambda^p(V)$, $\beta \in \Lambda^q(V)$,
($\wedge$ 4) if $\{f_i, 1 \leqq i \leqq \dim V\}$ is a basis of $\Lambda^1(V) = V^*$, then the set $\{f_{i_1} \wedge f_{i_2} \wedge \cdots \wedge f_{i_p}, i_1 < i_2 < \cdots < i_p, 1 \leqq i_s \leqq \dim V, 1 \leqq s \leqq p\}$ is a basis of $\Lambda^p(V)$,
($\wedge$ 5) if $\lambda \in \Lambda^0(V) = \mathbb{R}$, $\alpha \in \Lambda^p(V)$, then $\lambda \wedge \alpha$ coincides with the usual product $\lambda\alpha$ of a number λ and a form α.

Also, the following formula for the product of more then two multipliers $\alpha \in \Lambda^p(V)$, $\beta \in \Lambda^q(V), \ldots, \gamma \in \Lambda^r(V)$ is valid

(1.34) $$\alpha \wedge \beta \wedge \cdots \wedge \gamma = \frac{1}{p!q! \cdots r!} \mathfrak{A}(\alpha \otimes \beta \otimes \cdots \otimes \gamma),$$

the tensor product $\alpha \otimes \beta \otimes \cdots \otimes \gamma$ being defined inductively by taking into account the associativity of the binary operation $\otimes$ (see formula (×) above).

We extend the multiplication $\wedge$ onto the whole vector space $\Lambda(V)$ demanding the validity of ($\wedge$ 1) for the extended operation. Such an extension exists and is unique. The property ($\wedge$ 2) also remains valid. Thus, we get a graded algebra $\Lambda(V)$, called the exterior algebra of a vector space V. We use the name exterior form for an element of $\Lambda(V)$, and exterior p-form for that of $\Lambda^p(V)$.

If $f: V \longrightarrow V'$ is a linear map between two vector spaces, then it induces the homomorphism $f^*: \Lambda(V') \longrightarrow \Lambda(V)$ which is defined on homogeneous forms $\alpha \in \Lambda^p(V)$ by the formula

(1.35) $$f^*\alpha(v_1, v_2, \ldots, v_p) = \alpha(f(v_1), f(v_2), \ldots, f(v_p)).$$

The map f^* preserves linear operations, grading and multiplication. We may resume our constructions above by saying that Λ is a contravariant functor from the category of real vector spaces into the category of real associative graded algebras.

A symplectic form Ω in a vector space V is a member of $\Lambda^2(V)$. The following proposition answers the question of when a member of $\Lambda^2(V)$ is a symplectic form, the answer being in terms of the algebraic structure of $\Lambda(V)$.

Proposition 1.36. *Let V be a vector space of even dimension $2n$. A form $\Omega \in \Lambda^2(V)$ is nonsingular if and only if its nth power $\Omega^n = \Omega \wedge \Omega \wedge \cdots \wedge \Omega$ (n times) is nonzero.*

Proof. Let $\{e_i, 1 \leqq i \leqq 2n\}$ be a basis in V. Then by (1.34) we have

$$(*) \qquad \Omega^n(e_1, \ldots, e_{2n}) = \frac{1}{2^n} \sum_{\varepsilon \in \mathscr{S}_{2n}} \prod_{i=1}^{n} \Omega(e_{\varepsilon(2i-1)}, e_{\varepsilon(2i)})$$

If Ω is singular then there exists a nonzero vector which is Ω-orthogonal to the whole space V. We can take this vector as a member of the basis above. Then the right-hand side of $(*)$ is zero. This implies (by virtue of $(\wedge 4)$) that $\Omega^n = 0$. If Ω is nonsingular then we take a basis with the property 1.18a and substitute it into $(*)$. Immediate calculation shows that the right-hand side of $(*)$ equals $n!$ in this basis.

We conclude the section by the definition of the inner product of a vector and a homogeneous exterior form. Let $v \in V$ and $\alpha \in \Lambda^p(V)$, $p \geqq 1$. Then their inner product $v \lrcorner \alpha$ is the form in $\Lambda^{p-1}(V)$ whose value on vectors $v_1, v_2, \ldots, v_{p-1}$ is

$$(1.37) \qquad v \lrcorner \alpha(v_1, v_2, \ldots, v_{p-1}) = \alpha(v, v_1, v_2, \ldots, v_{p-1}).$$

In particular, if $\Omega \in \Lambda^2(V)$, $v \lrcorner \Omega = v^{\flat(\Omega)}$.

§2. Symplectic Manifolds

The study of Hamiltonian systems requires a detailed investigation of the underlying phase space with its intrinsic geometric structures. The main such structure is the symplectic one. In this section we give the definition of a symplectic manifold and deduce its properties. The basic facts about smooth manifolds, which we shall refer to, are collected in Appendix I.

Symplectic Structure

By a *symplectic form* on a smooth manifold X we mean a nonsingular closed 2-form, that is, a smooth section $\Omega: X \longrightarrow \Lambda^2 \mathrm{TX}$ which assignes to each point $x \in X$ a bilinear skew-symmetric form Ω_x on the tangent space $T_x X$ so that

(i) Ω_x is nonsingular for all $x \in X$, and thereby the pair $(T_x X, \Omega_x)$ becomes a symplectic vector space,

(ii) $d\Omega = 0$ (A form Ω satisfying condition (ii) is called closed.)

A pair (X, Ω), where X is a smooth manifold and Ω is a symplectic form on X, is called a *symplectic manifold*. We shall often only write X for (X, Ω), the letter X designating simultaneously both a symplectic manifold and the underlying

manifold. If (X,Ω) is a symplectic manifold, then Ω is called the *symplectic structure* of a symplectic manifold X.
Theorem 1.3 and Proposition 1.36 immediately yield:

2.1. *The dimension of a symplectic manifold is even, every symplectic manifold is orientable.*

We may take the $\frac{1}{2}\dim X$ – power of the symplectic structure of X for the orientation form on a manifold X. This power defines also a measure on X. We call it the *Liouville measure*, or *Liouville volume* and denote Vol(.).

Example 2.2. $\mathbb{R}^{2n}$ has a form equal to the standard symplectic form Ω_{st} of §1 (Example 1.1) at each point if one identifies $T_x\mathbb{R}^{2n}$ with the vector space $\mathbb{R}^{2n}$ in the usual way. Again we denote this symplectic form by Ω_{st}. It will be convenient to adopt the following notations for the coordinates of a point $x\in\mathbb{R}^{2n}$: $x=(q^1,q^2,\ldots,q^n,p_1,p_2,\ldots,p_n)$. In this notation the form Ω_{st} is expressed by

$$\Omega_{st}=\sum_{i=1}^{n} dq^1\wedge dp_i. \tag{2.2a}$$

This form is nonsingular and obviously closed. The symplectic manifold $(\mathbb{R}^{2n},\Omega_{st})$ will be called the *standard* (2*n*-dimensional) *symplectic manifold*.

A somewhat more general example is a symplectic vector space (V,Ω). If we consider V as a manifold and identify T_xV with V at any point $x\in V$, the constant form, which equals Ω at each point, it symplectic, and the pair (V,Ω) may be considered as a symplectic manifold.

A smooth map $f:X\longrightarrow Y$ between the underlying manifolds of two symplectic manifolds (X,Ω) and (Y,Σ) is called *symplectic* if f preserves the symplectic structures, i.e. $\Omega=f^*\Sigma$. It follows from Proposition 1.2 that

2.3. *Every symplectic map is an immersion.*

A *symplectic submanifold* S of a symplectic manifold (X,Ω) is a smooth submanifold $S\subset X$ such that the tangent space T_xS is a symplectic vector subspace of T_xX at each point $x\in S$. Let $i:S\subset X$ be the inclusion map. Then the form $i^*\Omega$ is symplectic in S, and the pair $(S,i^*\Omega)$ is a symplectic manifold. We call $i^*\Omega$ the *induced structure* and always regard S as a symplectic manifold with this structure. The inclusion map i is obviously symplectic. One particular case of a symplectic submanifold is that of a non-empty open subset. As new examples of symplectic manifolds we obtain

Example 2.3. (a) any non-empty open subset of $\mathbb{R}^{2n}$ is a symplectic submanifold of $(\mathbb{R}^{2n},\Omega_{st})$;

(b) the half-space $\mathbb{R}^{2n}_+=\{x=(q^1,q^2,\ldots,q^n,p_1,p_2,\ldots,p_n)\in\mathbb{R}^{2n}\colon\ q^1\geqq 0\}$ is a submanifold of $(\mathbb{R}^{2n},\Omega_{st})$, we shall refer to this example as the *standard halfspace*;

(c) the coordinate symplectic subspaces defined in Example 1.4 become symplectic submanifolds of $(\mathbb{R}^{2n},\Omega_{st})$ if the latter is regarded as a symplectic manifold.

Given a finite or countable collection (X_k, Ω_k), $k \in K$, of symplectic manifolds of the same dimension, the disconnected sum $\coprod_{k\in K} X_k$ of their underlying manifolds has the unique symplectic structure, $\coprod_{k\in K}\Omega_k$, such that all the inclusions $i_r: X_r \longrightarrow \coprod_{k\in K} X_k$, $r \in K$, are symplectic. The symplectic manifold $(\coprod_{k\in K} X_k, \coprod_{k\in K}\Omega_k)$ is called the *disconnected sum* of a family (X_k, Ω_k), $k \in K$. The members X_k of a family can be regarded as symplectic submanifolds of $\coprod_{k\in K} X_k$.

Given a finite collection (X_k, Ω_k), $k \in K$ of symplectic manifolds, at most one of the spaces X_k having non-empty boundary, we supply the product $\prod_{k\in K} X_k$ of underlying manifolds with a symplectic structure in the following manner. The tangent space of $X = \prod_{k\in K} X_k$ at the point $x = (x_k, k \in K) \in X$ has the form $T_x X = \sum_{k\in K} \oplus T_{x_k} X_k$ (see AI.47). Define the linear symplectic structure Ω_x on this space as the direct sum of those of X_k at points x_k (see the definition on p. 18). Ω_x is nonsingular at each point $x \in X$, and the 2-form $x \longmapsto \Omega_x$ on X is closed. This structure is denoted by $\sum_{k\in K} \oplus \Omega_k$, it converts $\prod_{k\in K} X_k$ into a symplectic manifold, the product of a family (X_k, Ω_k), $k \in K$. Note that the projection maps $p_i: \prod_{k\in K} \longrightarrow X_i$, $i \in K$, are not symplectic unless X is a single point.

Let us now turn to the concept of a *quotient* space. Consider a smooth map $p: X \longrightarrow Y$, X being a symplectic manifold, Y a smooth one, and p being onto. Does there exist a symplectic structure on Y such that p is a symplectic map? If the answer is yes, such a structure is unique. The manifold Y endowed with this structure is called the quotient symplectic manifold, and the said structure is called the *quotient* symplectic *structure generated* (or *induced*) by a map p. It follows that the quotient symplectic manifold Y must be of the same dimension as the original manifold X. Here we shall consider in more detail one particular case of the concept of a quotient manifold, namely the case of *covering*. We recall that a map $p: X \longrightarrow Y$ between topological spaces X and Y is a covering if it is onto, X and Y are non-empty and connected, and each point $y \in Y$ has an open neighbourhood U such that the preimage $p^{-1}(U)$ is a disjoint union of open subsets V_i such that for each V_i the restriction $p|V_i$ is a homeomorphism of V_i onto U. A map $p: X \longrightarrow Y$ between smooth manifolds X and Y is called a *smooth covering* if it is (i) a covering, (ii) smooth, and (iii) a submersion. These condition yield that the maps $p|V_i$ of the above definition of a covering are diffeomorphisms of V_i onto U. Now let X be a symplectic manifold with the symplectic structure Ω, and let $p: X \longrightarrow Y$ be a smooth covering. Then Y has the quotient symplectic structure if and only if in the above notations

(+) *the symplectic forms* $((p|V_i)^{-1})^*\Omega$ *on* U *are the same for all* V_i.

These define the unique symplectic form on Y, the latter becoming the quotient symplectic manifold.

The preceding situation can be described in terms of group actions as follows. All smooth diffeomorphisms $f: X \longrightarrow X$ with the property $p \circ f = p$ form a group if one takes the superposition of diffeomorphisms as the group operation, the group $\operatorname{Aut} p$ of automorphisms of a covering $p: X \longrightarrow Y$. This group acts on X (1) freely and (2) in a totally disconnected manner. (1) means that any $f \in \operatorname{Aut} p$ has no fixed points unless f equals 1_X, and the (2) that any point $x \in X$ has a neighbourhood V such that $f(V)$ and $g(V)$ do not intersect if $f \neq g$, f, $g \in \operatorname{Aut} p$. The above condition $(+)$ for p to induce a quotient structure implies that

$(+')$ *all maps in* $\operatorname{Aut} p$ *are symplectic.*

Note that $(+')$ is weaker than $(+)$. They are equivalent only for the case of a normal covering. A covering $p: X \longrightarrow Y$ is called normal if for each point $x \in X$ the image of the fundamental group $\pi_1(X, x)$ under the induced map $p_*: \pi_1(X, X) \longrightarrow \pi_1(Y, p(x))$ is a normal subgroup of $\pi_1(Y, p(x))$. In the case of a normal covering $\operatorname{Aut} p$ is isomorphic to the quotient group $\pi_1(Y, p(x))/p_*(\pi_1(X, x))$. We note two special cases of a normal covering: the *universal covering* which is the covering with a simply connected space X, i.e. with $\pi_1(X, x) = 0$, and the other case is a covering over the space Y whose fundamental group $\pi_1(Y, y)$ is commutative.

There is also the inverse construction. Let Γ be a group which acts on a symplectic manifold (X, Ω) effectively (i.e. only identity element of Γ acts as 1_X), freely and in a totally disconnected manner by symplectic diffeomorphisms. Then we may construct the quotient manifold X/Γ, points of X/Γ being orbits $\{f(x), f \in \Gamma\}$ of points $x \in X$, with X/Γ having the unique structure of a smooth manifold so that the canonical projection $p: X \longrightarrow X/\Gamma$, which assignes to any point $x \in X$ its orbit, is a covering. The symplecticity of the action yields the existence of the unique symplectic structure Σ on X/Γ such that $p^*\Sigma = \Omega$. The obtained smooth covering is normal. The group $\operatorname{Aut} p$ coincides with Γ in this construction if we identify Γ with the corresponding subgroup of transformations of X. We refer the reader to Fuks and Rokhlin (1984), Chapter V, §6.2, or to Wolf (1972), §1.8 for a detailed account of the theory of coverings.

Example 2.4 (*Symplectic Tori*). Consider a lattice in the standard symplectic manifold $(\mathbb{R}^{2n}, \Omega_{st})$, i.e. a discrete subgroup Γ_e of the additive group $\mathbb{R}^{2n}$ generated by the basis $e = \{e_1, e_2, \ldots, e_{2n}\}$ of $\mathbb{R}^{2n}$. The group Γ_e consists of all vectors $f \in \mathbb{R}^{2n}$ of the form $f = \sum_{i=1}^{2n} m_i e_i$, $(m_1, m_2, \ldots, m_{2n}) \in \mathbb{Z}^{2n}$. We regard Γ_e as a group of transformations $f: \mathbb{R}^{2n} \longrightarrow \mathbb{R}^{2n}$ acting as shifts $f(x) = x + f$. This action of Γ_e is symplectic, effective, free, and totally disconnected. Hence, we have the symplectic quotient manifold $\mathbb{T}_e^{2n} = \mathbb{R}^{2n}/\Gamma_e$, called the *symplectic torus*, a closed symplectic manifold. The *standard symplectic torus* $\mathbb{T}_{e_0}^{2n}$ is one which is obtained if the standard basis $e_0 = \{\mathrm{ort}_i, 1 \leqq i \leqq 2n\}$ is taken for e.

Example 2.5 (*Symplectic Annulus*). Let Γ be an additive subgroup of $\mathbb{R}^{2n}$ generated by $\{\mathrm{ort}_i, 1 \leqq i \leqq n\}$, and acting on $\mathbb{R}^{2n}$ by shifts. Let D be a non-empty open subset in $\mathbb{R}^n$. Consider the stripe $\mathbb{R}^n \times D \subset \mathbb{R}^n \times \mathbb{R}^n = \mathbb{R}^{2n}$. The action of Γ leaves this stripe invariant. Supply $\mathbb{R}^n \times D$ with the restriction of the standard symplectic structure Ω_{st} of $\mathbb{R}^{2n}$ and restrict the action to $\mathbb{R}^n \times D$, using the same notation Γ for this restricted action. Since shifts preserve Ω_{st}, we obtain the quotient symplectic manifold $A_D = (\mathbb{R}^n \times D)/\Gamma$, the *symplectic annulus*. Taking $D = \mathbb{R}^n$, we obtain the *standard symplectic annuals* $A_{\mathbb{R}^n}$. The underlying manifold of A_D is naturally diffeomorphic to $\mathbb{T}^n \times D$, where $\mathbb{T}^n = \mathbb{R}^n/\mathbb{Z}^n$ is the standard n-torus. We fix the following notations: $\tilde{p}_2$ for the second projection of the product $\mathbb{R}^n \times D$, p_2 for that of $A_D = \mathbb{T}^n \times D$, and p for the canonical projection onto the quotient space. Our maps are shown in the following commuting diagram

$$\begin{array}{ccc} \mathbb{R}^n \times D & \xrightarrow{\tilde{p}_2} & D \\ \downarrow p & \nearrow p_2 & \\ A_D = \mathbb{T}^n \times D & & \end{array} \tag{2.5'}$$

We also use the notation $(\varphi^1, \varphi^2, \ldots, \varphi^n, I_1, I_2, \ldots, I_n) = (\varphi, I)$ for the Cartesian coordinates on the covering space $\mathbb{R}^n \times D$ of the annulus. These coordinates in view of future utilisations will be referred to as the *angle-action coordinates.* Consider the differentials $d\varphi^i \in \mathscr{E}^1(\mathbb{R}^n \times D)$, $1 \leqq i \leqq n$, of the angle coördinates $\varphi^i : \mathbb{R}^n \times D \longrightarrow \mathbb{R}$. They are invariant under the shift transformations, in particular under those of Γ, so there are uniquely defined 1-forms on A_D whose images under $p^* : \mathscr{E}^1(A_D) \longrightarrow \mathscr{E}^1(\mathbb{R}^n \times D)$ are equal to the forms $d\varphi^i$. It is convenient to denote these 1-forms in $\mathscr{E}^1(A_D)$ again by $d\varphi^i$.

For future applications we now explicitly compute the cohomology group $H^1(A_D; \mathbb{R})$ provided that D is simply connected. Recall that a topological space is *simply connected* if it is connected and its fundamental group vanishes. In our case D is open in $\mathbb{R}^n$ and the first condition means that every two points of D may be joined by a smooth path lying in D. The second condition means that every two paths joining two points $x_0, x_1 \in D$ and lying in D may be deformed one to the other in a continuous manner, the ends x_0, x_1 being fixed and the paths staying in D during the deformation. We identify the real cohomology groups and the corresponding de Rham cohomology groups via de Rham theorem (see AI.61) and consider them as vector spaces over the real numbers. Each closed 1-form $\lambda \in \mathscr{E}^1(A_D)$ belongs to a cohomology class in $H^1(A_D; \mathbb{R})$ which we denote by $[\lambda]$.

Proposition 2.5a. *Let D be simply connected. Then the classes $[d\varphi^i]$, $1 \leqq i \leqq n$, form the basis in $H^1(A_D; \mathbb{R})$. In particular $H^1(A_D; \mathbb{R})$ is isomorphic, as a vector space, to $\mathbb{R}^n$.*

Proof. Denote by S_j, $1 \leqq j \leqq n$, the circles in A_D which are images of the segments

$$\begin{aligned} l_j = \{&(\varphi^1, \varphi^2, \ldots, \varphi^n, I_1, I_2, \ldots, I_n) \in \mathbb{R}^n \times D : \\ &I_i = I_i^0, \quad 1 \leqq i \leqq n, \quad 0 \leqq \varphi^j \leqq 1, \quad \varphi^i = 0, \quad i \neq j\}, \end{aligned} \tag{2.5b}$$

$(I_1^0, I_2^0, \ldots, I_n^0) \in D$ being fixed, under the canonical projection p. Note that

$$\int_{S_j} \mathrm{d}\varphi^i = \delta_j^i. \tag{2.5c}$$

Let $c_i \in \mathbb{R}$, $1 \leqq i \leqq n$, be such that $\sum_{i=1}^n c_i [d\varphi^i] = 0$. This means that there is a function $\varphi : A_D \longrightarrow \mathbb{R}$ such that $\sum_{i=1}^n c_i\, d\varphi^i = d\varphi$. Integrating the latter equality over S_j, we obtain by (2.5c) that $c_j = 0$, $1 \leqq j \leqq n$. This proves the linear independence of the forms $d\varphi^j$. To prove the completeness consider an arbitrary closed 1-form λ on A_D and let

$$c_i = \int_{S_i} \lambda, \quad 0 \leqq i \leqq n. \tag{2.5d}$$

Consider the form $\mu = \lambda - \sum_{i=1}^{n} c_i \mathrm{d}\varphi^i \in \mathscr{E}^1(A_D)$. It follows that

(2.5e)
$$\int_{l_j} p^*\mu = \int_{S_j} \mu = 0, \quad 1 \leqq j \leqq n.$$

Since $\mathbb{R}^n \times D$ is also simply connected, there is a function Φ such that $p^*\mu = \mathrm{d}\Phi$, and (2.5e) shows that Φ is periodic with period 1 in each angle variable φ^i. The latter property yields the existence of a function $\Phi_1 \in C^\infty(A_D)$ such that $\Phi = p^*\Phi_1$. We have $\mu = d\Phi_1$, i.e. $[\mu] = 0$, and

(2.5f)
$$[\lambda] = \sum_{i=1}^{n} c_i [\mathrm{d}\varphi^i]. \qquad \square$$

Proposition 2.5a enables us to identify $H^1(A_D{:}\mathbb{R})$ with $\mathbb{R}^n$ via the isomorphism which carries $[\mathrm{d}\varphi^i]$ to $\mathrm{ort}_i \in \mathbb{R}^n$, $1 \leqq i \leqq n$, provided D is simply connected. The preceding proof yields the following explicit formula for the map [ ] which assignes to each closed 1-form $\lambda \in \mathscr{E}^1(A_D)$ its cohomology class $[\lambda] \in H^1(A_D{:}\mathbb{R}) = \mathbb{R}^n$. The lift of λ onto $\mathbb{R}^n \times D$ has the form

$$p^*\lambda = \sum_{i=1}^{n} (a_i(\varphi, I)\,\mathrm{d}\varphi^i + b^i(\varphi, I)\,dI_i),$$

where $a_i(\varphi, I)$ and $b^i(\varphi, I)$ are periodic in each angle variable $\varphi^i, 1 \leqq i \leqq n$, with period 1. We have

$$[\lambda] = ([\lambda]_1, [\lambda]_2, \ldots, [\lambda]_n) \in \mathbb{R}^n,$$

where

(2.5g)
$$[\lambda]_i = \int_{l_i} a_i(\varphi, I)\,\mathrm{d}\varphi^i, \quad 1 \leqq i \leqq n,$$

and the segment l_i is given by (2.5b) (see the formulae (2.5f) and (2.5d)). Note that the right-hand side of (2.5g) is independent of I. This follows since λ is closed.

Remark 2.5h. The map [ ]: is linear and continuous, the space $\mathscr{E}^1(A_D)$ being endowed with a compact-open topology (see AI-17), the subspace of closed forms with the induced one.

2.5j. *A special symplectic diffeomorphism between two annuli.*

Let $\mathrm{GL}(n, \mathbb{Z})$ denote the group of square $n \times n$ matrices with integer coefficients and with determinant equal to ± 1, the group operation being multiplication of matrices. We also identify matrices with linear maps in $\mathbb{R}^n$ in the obvious way and use the same letter for denoting a matrix and its corresponding map. Given a matrix $M \in \mathrm{GL}(n, \mathbb{Z})$ consider the map $g_M{:}A_D \longrightarrow A_{D_1}, D_1 = M(D)$, whose left $\tilde{g}_M{:}\mathbb{R}^n \times D \longrightarrow \mathbb{R}^n \times D_1$ is defined as $(\varphi, I) \longmapsto ((M^t)^{-1}\varphi, MI)$. Here the upper index t denotes the operation of transposition of a matrix: $(M^t)_{ik} = M_{ki}$. The map g_M is symplectic. It preserves the fibration of annuli into tori $I = \mathrm{const}$. As we shall see in 7.13, a general symplectic map between annuli with such a property differs nonessentially from g_M. This map also will be useful later.

Cotangent Bundle

Let Q be a manifold of dimension $n \geqq 1$. The total space T^*Q of its contangent bundle $\pi: T^*Q \longrightarrow Q$ itself is a manifold of dimension $2n$. It turns out that this manifold bears a natural symplectic structure. This provides us with an important class of examples of symplectic manifolds. In order to describe the above structure let us consider the following commutative diagram

(2.6)
$$\begin{array}{ccc} TT^*Q & \longrightarrow & T^*Q \\ {\scriptstyle T\pi}\downarrow & & \downarrow{\scriptstyle \pi} \\ TQ & \longrightarrow & Q \end{array}$$

where horizontal arrows are tangent bundle projections, and $T\pi$ is the tangent map of the cotangent bundle projection π. Take $q \in Q$, $p \in T_q^*Q$ and $v \in T_pT^*Q$. Then $T\pi(v)$ is a tangent vector which belongs, by the commutativity of diagram (2.6), to T_qQ. We may apply the cotangent vector $p \in T_q^*Q$ which we started with to $T_p\pi(v) \in T_qQ$ and get, as a result, the real number $p(T\pi(v))$. All the operations involved are linear, so the map $p \circ T_p\pi: T_pT^*Q \longrightarrow \mathbb{R}$ is a linear function on the vector space T_pT^*Q and defines the element $\Theta_Q(p) \in T_p^*T^*Q$. By definition

(2.7)
$$\Theta_Q(p)(v) = p \circ T_p\pi(v), \quad v \in T_pT^*q.$$

We have obtained the smooth section $p \longmapsto \Theta_Q(p)$ of the vector bundle $T^*T^*Q \longrightarrow T^*Q$. The 1-form Θ_Q on T^*Q which arises is called the *Liouville form*. We claim that the 2-form

(2.8)
$$\Omega_Q = -d\Theta_Q$$

is a symplectic one. It is obvious that the form Ω_Q is closed, moreover, it is exact. It is only necessary to prove nonsingularity. To do this it is sufficient to make local considerations. Let (U, α) be a chart in Q and $q^i = \mathrm{pr}_i \circ \alpha$, $1 \leqq i \leqq n$, the corresponding coordinate functions. Consider the associated chart $(T_UQ, \tilde{\alpha})$ in T^*Q (see AI.60), denoting the coordinate function by $(q^1, q^2, \ldots, q^n, p_1, p_2, \ldots, p_n)$. A point $p \in T_q^*Q$ has the expansion

(2.9)
$$p = \sum_{i=1}^{n} p_i \, dq^i|_q.$$

A vector $v \in T_pT^*Q$ expands in analogous way as

(2.10)
$$v = \sum_{i=1}^{n} \left(v_i^i \frac{\partial}{\partial q^i}\bigg|_p + w_i \frac{\partial}{\partial p_i}\bigg|_0 \right).$$

The cotangent bundle projection $\pi: T^*Q \longrightarrow Q$ has the following local representation in the chart $(T_U Q, \alpha)$:

$$(q^1, q^2, \ldots, q^n, p_1, p_2, \ldots, p_n) \longmapsto (q^1, q^2, \ldots, q^n). \text{ So}$$

(2.11)
$$T\pi(v) = \sum_{i=1}^{n} v^i \frac{\partial}{\partial q^i},$$

where v is given by (2.10). Applying (2.9) to (2.11) and taking into account (AI.60a) we obtain (see 2.7)):

(2.12)
$$\Theta_Q(p) = \sum_{i=1}^{n} p_i \, dq^i|_p.$$

and applying the d-operation:

(2.8a)
$$\Omega_Q | T_U^* Q = \sum_{i=1}^{n} dq^i \wedge dp_i.$$

The expression (2.8a) yields the desired nonsingularity of Ω_Q. We shall refer to the form Ω_Q as the *canonical symplectic structure* on the cotangent bundle of a manifold Q.

Example 2.9. The cotangent bundle of $\mathbb{R}^n$ may be identified with $\mathbb{R}^{2n}$ (cf. AI.46). The associated chart to the global chart $(\mathbb{R}^n, 1_{\mathbb{R}^n})$ yields the isomorphism between $T^*\mathbb{R}^n$ and $\mathbb{R}^{2n}$ (see AI.60). In particular, if the coordinates in $\mathbb{R}^{2n}$ are denoted $(q^1, q^2, \ldots, q^n, p_1, p_2, \ldots, p_n)$, the Liouville form has the form (2.12), and $T^*\mathbb{R}^n$ is isomorphic, as a symplectic manifold, to $(\mathbb{R}^{2n}, \Omega_{st})$ of Example 2.2 via the said identification.

Let us investigate the Liouville form of a manifold in more detail. It may be characterized by the following interesting property.

Proposition 2.10. *A 1-form θ defined on T^*Q equals the Liouville form Θ_Q if and only if the equality*

(2.10a)
$$\sigma^*\theta = \sigma$$

holds for all smooth sections $\sigma: Q \longrightarrow T^*Q$ of the cotangent bundle.

Proof. We first prove that Θ_Q satifiies (2.10a). Take $u \in T_q Q$ and compute $\sigma^*\Theta_Q$ at the vector u. We have

$$\sigma^*\Theta_Q(u) = \Theta_Q(\sigma_q)(T_q\sigma u) = \sigma_q(T_{\sigma q}\pi \circ T_q\sigma(u)) = \sigma_q(u) \quad \text{(by 2.6)}$$

since $\pi \circ \sigma = 1_Q$. Conversely, let θ be a 1-form on T^*Q which satisfies (2.10a) for all σ. We have $\sigma^*\theta = \sigma^*\Theta_Q$ and $\theta = \Theta_Q$. This follows since the vectors of the form $T_q\sigma u$ span $T_p T^*Q$ provided σ runs over all sections of the cotangent bundle with fixed value $\sigma_q = p$, $q \in Q$, and u runs over $T_q Q$. □

Now we shall deduce two remarkable properties of the Liouville form. The first one concerns its behaviour under the action of smooth mappings. Let $f: Q \longrightarrow R$ be an immersion between manifolds Q and R of the same dimension. Define the contragredient map $T^{*-1}f: T^*Q \longrightarrow T^*R$ as follows: if $p \in T_q^*Q$ then

$$T^{*-1}f(p) = [(T_q f)^*]^{-1}(p). \tag{2.11}$$

The contragredient map is an immersion. It is a morphism of bundles. If f is a diffeomorphism, so is $T^{*-1}f$.

Proposition 2.12. (Functional property of the Liouville form). *If $f: Q \longrightarrow R$ is a diffeomorphism between manifolds Q and R then*

$$(T^{*-1}f)^*\Theta_R = \Theta_Q.$$

In particular, the contragredient map of a diffeomorphism preserves the symplectic structures of the cotangent bundles.

Proof. Denote $\theta' = (T^{*-1}f)\Theta_R$. Let $\sigma: Q \longrightarrow T^*Q$ be a smooth section. Denote $\sigma' = T^{*-1}f \circ \sigma \circ f^{-1}$. We have

$$\sigma^*\theta' = \sigma^*(T^{*-1}f)\Theta_R = (T^{*-1}f \circ \sigma)^*\Theta_R = (\sigma' \circ f)^*\Theta_R = f^*\sigma'^*\Theta_R = f^*\sigma' \tag{2.12a}$$

(due to 2.10).

To compute $f^*\sigma'$, take $v \in T_qQ$, $q \in Q$, and denote $v' = T_qf(v) \in T_{f(q)}R$. Then $f^*\sigma'(v) = \sigma'(T_qf(v)) = \sigma'(v') = ((T^{*-1}f)_q \cdot \sigma_q)(v') = [(T_qf)^{-1}]^*\sigma_q(v') = \sigma_q((T_qf)^{-1}v') = \sigma_q(v)$. We have obtained $f^*\sigma' = \sigma$, and then (2.12a) shows that θ' satisfies the test of Proposition 2.10. Hence $\theta' = \Theta_Q$. □

Proposition 2.13. (Locality property of the Liouville form). *If $U \subset Q$ is an open non-emptry set then $\Theta_U = \Theta_Q|_{T_U^*Q}$ (we identify $T_U^*Q \subset T^*Q$ with T^*U in an obvious manner).*

Proof. Apply 2.12 to the inclusion $i: U \hookrightarrow Q$. □

There is a useful description of the cotangent bundle to the quotient manifold. Let Q_0 be a manifold, and Γ_0 be a group which acts on Q_0 effectively, freely and in a totally disconnected manner. Consider the canonical projection $\rho_0: Q_0 \longrightarrow Q_0/\Gamma_0$. Since the tangent map $T_q\rho_0: T_qQ_0 \longrightarrow T_{\rho_0(q)}(Q_0/\Gamma_0)$, $q \in Q_0$, is an isomorphism of vector spaces, the formula (2.11) defines the contragredient map $T^{*-1}\rho_0: T^*Q_0 \longrightarrow T^*(Q_0/\Gamma_0)$ which is also a covering, and additionally is a bundle map. The contragredient maps $T^{*-1}\gamma: T^*Q_0 \longrightarrow T^*Q_0$, $\gamma \in \Gamma_0$, form the action of Γ_0 on T^*Q_0. This is also effective, free, and a totally disconnected action. We denote it by Γ. It follows from 2.12 that Γ consists of symplectic transformations. Hence T^*Q_0/Γ obtains the symplectic structure. One can easily deduce that $T^{*-1}\rho_0$ is invariant under Γ and induces a symplectic isomorphism $(T^*Q_0)/\Gamma \approx T(Q_0/\Gamma_0)$ (due to 2.12). We identify these two symplectic manifolds.

Example 2.14. Cotangent bundle to the standard n-torus. Let $\mathbb{T}^n = \mathbb{R}^n/\mathbb{Z}^n$ be the standard n-torus. Take the action of $\mathbb{Z}^n$ on $\mathbb{R}^n$ by shifts as Γ_0 in the preceding construction. We obtain, using Examples 2.5 and 2.9, the following equalities of symplectic manifolds:

$$T^*\mathbb{T}^n = (T^*\mathbb{R}^n)/\Gamma = \mathbb{R}^{2n}/\Gamma = A_{\mathbb{R}^n}.$$

Here Γ is generated by $\{\text{ort}_i, 1 \leqq i \leqq n\}$. That is, the cotangent bundle to the standard n-torus coincides with the standard symplectic annulus $A_{\mathbb{R}^n}$.

Consider a matrix $M \in \mathrm{GL}(n, \mathbb{Z})$. It defines a linear diffeomorphism $M: \mathbb{T}^n \longrightarrow \mathbb{T}^n$ (we use the same letter, since there is no abuse). The contragredient map $M^{*-1}: T^*\mathbb{T}^n \longrightarrow T^*\mathbb{T}^n$ coincides with the fibre-preserving map $g_M: A_{\mathbb{R}^n} \longrightarrow A_{\mathbb{R}^n}$ defined in 2.5j.

Global and Local Properties of Symplectic Manifolds

Recall that two symplectic manifolds are called isomorphic if they are diffeomorphic via a symplectic diffeomorphism. It is a difficult problem to classify symplectic manifolds up to isomorphism. However, it has an exhaustive solution in two-dimensional case.

Information 2.15. *Any two-dimensional orientable manifold admits a symplectic structure.* A symplectic form Ω on a surface X is a volume form. The set $\{\int_{X_i}\Omega\}$ of total areas of the connected components X_i of X is invariant under symplectic diffeomorphisms. As a matter of fact, it is the unique symplectic invariant:

Any two homeomorphic connected symplectic surfaces of the same area are isomorphic.

Let us now turn to the local variant of the isomorphism problem. Let $X_i, i = 1, 2$, be symplectic manifolds, and $x_i \in X_i$. We say that the X_i are locally isomorphic at the points x_i if there are open neighbourhoods U_i of x_i in X_i and a symplectic isomorphism $f: U_1 \longrightarrow U_2$ such that $f(x_1) = x_2$, U_i being endowed with the induced symplectic structures. The following theorem gives the complete solution of the problem of local isomorphism in points.

Theorem 2.16. (Darboux). *Any two symplectic manifolds of the same dimension are locally isomorphic at all of their interior points. The same statement is valid with respect to boundary points.*

We shall obtain the proof of Darboux' theorem later as a consequence of the more general Weinstein's and (or) Givental's theorems. We shall now discuss the meaning of the assertions of 2.16.

Note that a map which realizes a local isomorphism also does this in the sense of differential and topological structures. Thereby, the equality of dimensions and distinguishing the cases of interior and boundary points are the necessary conditions which arise from topological properties of manifolds and do not bear any relation to the specificity of the symplectic structure. I. M. Gelfand expressed this by saying: "the symplectic manifold (locally) has no features". The natural question that arises is whether the symplectic geometry exists at all. Locally it does not, globally it is ruled by differential topological invariants, the latter having the limitations which are connected with the presence of a nonsingular closed two-form. The assertions of 2.15 may be interpreted as the nonexistence of symplectic geometry in dimension two. There is almost nothing left for the symplectic structure to have a part to play. The situation changes in higher dimensions. Shall now we formulate the remarkable result by M. Gromov in this direction. Consider the following two open subsets of $\mathbb{R}^{2n}$: the ball $B_r^{2n} = \{x = (q^1, q^2, \ldots, q^n, p_1, p_2, \ldots, p_n): \sum_{i=1}^{n}((q^i)^2 + (p_i)2) < r^2\}$ and the slice $A_\varepsilon^1 \times \mathbb{R}^{2n-2} = \{x = (q^1, q^2, \ldots, q^n, p_1, p_2, \ldots, p_n): (q^1)^2 + (p_1)^2 < \varepsilon\}$, and supply them with the inherited symplectic structure from $(\mathbb{R}^{2n}, \Omega_{st})$. The symplectic map is called a symplectic embedding if it is homeomorphism onto its image.

Theorem (M. Gromov). *If there is a symplectic embedding $B_r^{2n} \longrightarrow A_\varepsilon^1 \times \mathbb{R}^{2n-2}$ then $r \leqq \varepsilon$.*

In other words one cannot rumple a symplectic domain symplectically in an arbitrary manner, there are some intrinsic symplectic obstructions more comprehensive than the difference of the volumes.

Darboux' theorem implies that any symplectic manifold of dimension $2n$ at any of its points is locally isomorphic to $(\mathbb{R}^{2n}, \Omega_{st}) \sqcup (\mathbb{R}_+^{2n}, \Omega_{st})$. A symplectic chart is a local symplectic diffeomorphism $\alpha: U \longrightarrow \mathbb{R}^{2n} \sqcup \mathbb{R}_+^{2n}$ where U is open. We may introduce a symplectic atlas which consists of symplectic charts whose domains cover the symplectic manifold in question, 2.16 asserting the existence of such an atlas. This enables us to give an alternative definition of a symplectic structure on a manifold. The set of all local symplectic diffeomorphisms of $(\mathbb{R}^{2n}, \Omega_{st}) \sqcup (\mathbb{R}_+^{2n}, \Omega_{st})$ form a pseudogroup (see Golubitsky and Guillemin (1973) for the definition). Let X be a smooth $2n$-dimensional manifold. A smooth atlas on X with values in $\mathbb{R}^{2n} \sqcup \mathbb{R}_+^{2n}$ is called symplectic if for any two of its charts $\alpha_i: U_i \longrightarrow \mathbb{R}_+^{2n}, i = 1, 2$, the superposition $\alpha_1 \circ \alpha_2^{-1}$ belongs to the said pseudogroup. Two symplectic atlases are regarded as equivalent if their union is again a symplectic atlas. An equivalency class of symplectic atlases is called a symplectic structure on X. If a manifold X is endowed with a symplectic structure in the latter sense, one may construct the symplectic form Ω on X such that $\Omega|_U = \alpha^* \Omega_{st}$ for each chart (U, α) which belongs to some symplectic atlas of this structure. Thus, the two definitions of a symplectic structure on a manifold are equivalent.

For practical reasons it is convenient to slightly change the definition of a symplectic chart. Further by "a symplectic chart in a symplectic manifold X of dimension $2n$" we mean a pair (U, α) where U is an open set in X and α is a symplectic diffeomorphism of U onto some open set in $(\mathbb{R}^{2n}, \Omega_{st})$ or $(\mathbb{R}_+^{2n}, \Omega_{st})$. Let us collect together the main objects connected with a symplectic chart and relations between them in the following list.

2.17. List (What is present in a symplectic chart (U, α)).

2.17a. The coordinate functions

$$q^i = \mathrm{pr}_i \circ \alpha : U \longrightarrow \mathbb{R},$$
$$p_i = \mathrm{pr}_{n+i} \circ \alpha : U \longrightarrow \mathbb{R}, \qquad 1 \leqq i \leqq n.$$

These coordinates $(q^1, q^2, \ldots, p_n, p_2, \ldots, p_n)$ also are called *Darboux' coordinates.*

2.17b. Their differentials dq^i, dp_i, $1 \leqq i \leqq n$, which are smooth sections of $T^*_U X$, their values form a basis in $T^*_x X$ for each $x \in U$.

2.17c. Vector fields

$$\frac{\partial}{\partial q^i}, \quad \frac{\partial}{\partial p_i}, \qquad 1 \leqq i \leqq n, \text{ are smooth sections of } T_U X,$$

their values form a basis in $T_X U$ for each $x \in U$.

2.17d. A symplectic structures Ω has the local expression

$$\Omega|_U = \sum_{i=1}^{n} dq^i \wedge dp_i.$$

This is due to the equality $\Omega|_U = \alpha^* \Omega_{st}$ (cf. (2.2a)). We have the following values of Ω on the basic vector fields:

$$\text{(2.17e)} \quad \Omega|_U\left(\frac{\partial}{\partial q^i}, \frac{\partial}{\partial q^j}\right) = 0, \qquad \Omega|_U\left(\frac{\partial}{\partial p_i}, \frac{\partial}{\partial p_j}\right) = 0, \qquad \Omega|_U\left(\frac{\partial}{\partial q^i}, \frac{\partial}{\partial p_j}\right) = \delta^i_j,$$
$$1 \leqq i, j \leqq n.$$

2.17f. The "sharp" map # linked up with Ω (see §1) has the following values at the basic differentials 2.17b:

$$(dq^i)^{\#} = -\frac{\partial}{\partial p_i},$$
$$(dp_i)^{\#} = \frac{\partial}{\partial q^i}, \qquad 1 \leqq i \leqq n.$$

One can easily deduce this formulae from 2.17e and (AI.60a).

Weinstein's and Givental's Theorems

The theorems in the title are basic for the study of the geometry of submanifolds in a symplectic manifold. The term submanifold will be used, unless otherwise stated, for denoting a submanifold without boundary and without intersections with the boundary of the surrounding manifold.

Let X be a smooth manifold and $S \subset X$ a submanifold. If Ω is an exterior form defined in some neighbourhood of S, we may consider the restrictions of Ω onto S

in two different ways: firstly, for any $x \in S$ we consider $\Omega_x(v_1, v_2, \ldots, v_p)$, p being the degree of Ω, only on vectors v_i, $1 \leqq i \leqq p$, which belong to $T_x S$. We denote such a restriction by $\Omega|_{TS}$. This is the section of a bundle $\Lambda^p TS \longrightarrow S$. It can be obtained as $i^*\Omega$ where $i: S \longrightarrow X$ is the inclusion map. In the second way, we consider $\Omega_x(v_1, v_2, \ldots, v_p)$ for $x \in S$ on all vectors $v_i \in T_x X$, $1 \leqq i \leqq p$, and obtain, in such a manner, the section of the bundle $i^! \Lambda^p TX \longrightarrow S$. We denote this restriction by $\Omega|_S$.

Theorem 2.18. (A. Weinstein). *Let Ω_0 and Ω_1 be two symplectic forms defined in some neighbourhood of a submanifold S of a manifold X, and let*

$$\text{(W)} \qquad \Omega_0|_S = \Omega_1|_S.$$

Then there is a diffeomorphism $f: U \longrightarrow V \subset X$ defined in some neighbourhood U of S in X such that $f|_S = 1_S$ and $f^*\Omega_1|_V = \Omega_0|_U$.

Darboux' theorem 2.16 for interior points follows from 2.18. Indeed, let (X, Ω) and (Y, Σ) be two symplectic manifolds of the same dimension, and let $x_0 \in X$, $y_0 \in Y$ be their interior points. We may choose a local diffeomorphism g between some neighbourhood of these points so that (1) $g(x_0) = y_0$, (2) $(g^*\Sigma)|_{x_0} = \Omega_{x_0}$. Such a choice is possible due to Theorem 1.3. Then apply Theorem 2.18 taking the single-point set $\{x_0\}$ for S, and the forms Ω and g for Ω_0 and Ω_1.

In order to prove Theorem 2.18 we first make the construction of the homotopy operation.

Construction 2.19. Let $\pi: U \longrightarrow S$ be a smooth vector bundle. Define the family $h_t: U \longrightarrow U$, $t \in [0, 1]$, of smooth mappings, by the formula $h_t(v) = tv$, $v \in U$. If α is a p-form defined in U, we have the differentiation formula (see AI.66):

$$\frac{d}{dt} h_t^* \alpha = h_t^* \lrcorner\, d\alpha + d(h_t^* \lrcorner\, \alpha).$$

Integrating with respect to t over $[0, 1]$, we obtain

$$(*) \qquad h_1^*\alpha - h_0^*\alpha = \int_0^1 h_t^* \lrcorner\, d\alpha\, dt + d\left(\int_0^1 h_t^* \alpha\, dt\right).$$

We introduce the *homotopy operator* $I: \mathscr{E}^p(U) \longrightarrow \mathscr{E}^{p-1}(U)$ by the formula

$$\text{(2.19a)} \qquad I\alpha = \int_0^1 h_t^* \lrcorner\, \alpha\, dt.$$

Since $h_1 = 1_U$, we have $h_1^*\alpha = \alpha$, and $(*)$ yields

$$\text{(2.19b)} \qquad \alpha - h_0^*\alpha = I \circ d\alpha + d \circ I\alpha.$$

Let us identify S with its image in U under the zero-section. Then $h_0^*\alpha = 0$ if and only if $\alpha|_{TS} = 0$, since h_0 assignes to each vector vU its projection $\pi(v)$. As a consequence of (2.19b), we have

$$\text{(2.19c)} \qquad \alpha|_{TS} = 0 \quad \text{and} \quad d\alpha = 0 \;\Rightarrow\; \alpha = d(I\alpha).$$

Another property of operator I is that

(2.19d) $$I\alpha|_S = 0.$$

This is because $h_t|_S = 1_S$, hence $dh_t/dt|_S = 0$.

Corollary 2.19e (Poincaré's Lemma). *Let $U \subset X$ be an open subset diffeomorphic to the ball $B^n \doteq \{x \in \mathbb{R}^n : |x| < 1$, $n = \dim X$. Then there is a linear map $I: \mathscr{E}^p(U) \longrightarrow \mathscr{E}^{p-1}(U)$, $1 \leqq p \leqq n$, such that*

$$I \circ d + d \circ I = 1|_{\mathscr{E}^p(U)}.$$

In particularly, any closed form $\alpha \in \mathscr{E}^p(U)$, i.e. having the property $d\alpha = 0$, is an exact one: there is a form $\beta \in \mathscr{E}^{p-1}(U)$ such that $\alpha = d\beta$.

Proof. Supply U with the vector bundle structure over the single point as the base space via a diffeomorphism $f: B^n \longrightarrow U$. □

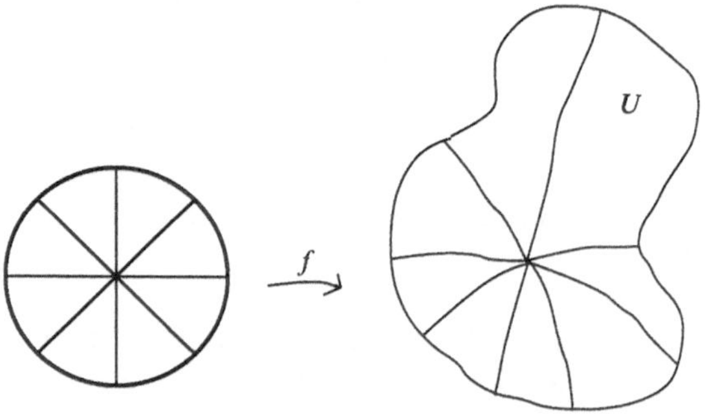

Fig. 3

Proof of Theorem 2.18. Consider the one-parameter family of forms

$$\Omega_t = (1-t)\Omega_0 + t\Omega_1 = \Omega_0 + t\omega,$$

where $\omega = \Omega_1 - \Omega_0$, $t \in [0, 1]$. We have

(+) $$\omega|_{TS} = 0 \quad \text{and} \quad s\omega = 0,$$

the latter being because both of the forms Ω_i, $i = 0, 1$, are closed. We may restrict our considerations to a tubular neighbourhood U of S in X (see AI.54), so there exists the homotopy operator I defined by (2.19a) exists. Let $\Phi = I\omega$. Since $\Omega_t|_S = \Omega_0|_S$, the forms Ω_t are nonsingular in some sufficiently small neighbourhood of S, $t \in [0, 1]$. In this neighbourhood we may use the isomorphism $\#(\Omega_t)$. Define the family of vector fields ξ_t, $t \in [0, 1]$ by the formula

$$\xi_t = -\Phi^{\#(\Omega_t)}.$$

It follows from (2.19d) that $\xi_t|_S = 0$. By the well-known theorems of differential equations theory (see AI.67) this family of vector fields may be integrated. Indeed, it is easily shown that ξ_t may be integrated on S: the solution is the family of identities. Then the assertion follows from the theorems of openness of the domain of definition and a smooth dependence on initial data of the solution of a system of differential equations. There exists the family f_t, $t \in [0, 1]$ of local diffeomorphisms defined in some open neighbourhood of S such that f_0 is the inclusion, $f_t|_S = 1_S$, $t \in [0, 1]$, and $df_t/dt = \xi_t \circ f_t$. Applying the differentiation formula (AI.68) and taking into account (+) and 2.19c, we obtain

$$\frac{d}{dt}(f_t^* \Omega_t) = f_t^* \frac{d\Omega_t}{dt} + f_t^*(d(\xi_t \lrcorner \Omega_t)) = f_t^*(\omega - d\Phi) = 0.$$

Since $f_0^* \Omega_0 = \Omega_0$, it follows that $f_t^* \Omega_t = \Omega_0$ for each $t \in [0, 1]$. The diffeomorphism f_1 has the required properties. □

The natural question that arises is whether one may relax the condition (W) of Theorem 2.18, demanding instead of (W) the weaker condition

(G) $$\Omega_0|_{TS} = \Omega_1|_{TS}.$$

It is only in the nonsingularity of $\Omega_t|_S$ that the condition (W) is fully satisfied. The condition (G) does not yield such a nonsingularity. There is another formulation which is satisfied with (G) alone but loses its global properties in S in exchange. One may, by applying a suitable linear transformation in local coordinates, make the forms Ω_0 and Ω_1 equal at a single point (the Proposition 1.24 and (G)). Then Ω_t is nonsingular at this point since it is equal to Ω_0, and the same is true in some neighbourhood of this point. The conclusion is that we may repeat the above considerations in a sufficiently small neighbourhood of an arbitrary point of S. Thereby we have the following theorem.

Theorem 2.20 (Givental). *Let condition* (W) *in* 2.18 *be replaced by condition* (G). *Then for each point $x \in S$, there exists a neighbourhood U of x and a local diffeomorphism $f: U \longrightarrow X$ such that $f|_{U \cap S} = 1_{U \cap S}$ and $f^* \Omega_1 = \Omega_0|_U$.*

Givental's theorem shows that the local properties of a submanifold in a symplectic manifold are completely defined by the restriction of the symplectic structure on vectors tangent to the submanifold. In other words, the intrinsic geometry of a submanifold determines its exterior geometry. This can be formulated as follows. We say that the submanifolds S_i, $i = 1, 2$, of symplectic manifolds X_i are *locally isomorphic at points* $x_i \in S_i$ if there is a symplectic diffeomorphism $f: U_1 \longrightarrow U_2$, U_i being open neighbourhoods of x_i in X_i, such that $f(x_1) = x_2$ and $f(U_1 \cap S_1) = U_2 \cap S_2$.

Corollary 2.21. *Two submanifolds S_i, $i = 1, 2$, of the same dimension in symplectic manifolds $(X_i$, fix $\Omega_i)$ of the same dimension are locally isomorphic at points $x_i \in S_i$ if and only if there is a diffeomorphism $f: U_1 \longrightarrow U_2$, U_i being open neighbourhoods of x_i in S_i, such that $f(x_1) = x_2$ and*

$$f^*((\Omega_2|_{TS_2})|_{U_2}) = (\Omega_1|_{TS_1})|_{U_1}.$$

We have in particular that *symplectic* submanifolds of the same dimension and codimension are all locally isomorphic in all their (interior) points.

Hypersurfaces and the Boundary

Let $S \subset X$ be a hypersurface of a symplectic manifold X, i.e. a submanifold of codimension one. Then its tangent hyperplanes T_xS contains the zero-line L_x (see §1 at the end of subsection "Subspaces of a Symplectic Space"). We have the *zero-line bundle* LS over S which is a subbundle of TS. The bundle LS may be integrated to obtain the one-dimensional *zero-foliation* of S.

Theorem 2.22. *Any two hypersurfaces in symplectic manifolds of the same dimension are locally isomorphic.*

Recall that we are considering submanifolds without boundary and having no intersection with the boundary of surrounding manifold. Otherwise the statement of 2.22 is false, and the problem of local classification becomes rather difficult (see Suzuki (1985)).

Proof of Theorem 2.22. By 2.22 it is sufficient to prove that any two closed 2-forms Ω_i, $i = 1, 2$, defined on $(2n\text{-}1)$-dimensional manifolds, such that the Ω_i-orthogonal complement of the whole tangent spaces have dimension one at each point, are locally isomorphic. By taking suitable charts one can reduce this problems to that of comparing two closed 2-forms Ω_i, $i = 1, 2$, defined in the cylinder $]-\delta, \delta[\times B^{2n-2}$ where B^{2n-2} is a $(2n-2)$-dimensional open ball, the zero-leaves of Ω_i being the straight line $]-\delta, \delta[\times \{x\}$, $x \in B^{2n-2}$. Note that the restrictions $\Omega_i|_{T(\{z\} \times B^{2n-2})}$ are symplectic forms on the manifolds $\{z\} \times B^{2n-2}$, $z \in]-\delta, \delta[$. We may assume that

$$\Omega_1|_{\{0\} \times B^{2n-2}} = \Omega_2|_{\{0\} \times B^{2n-2}}. \tag{0}$$

Otherwise apply the first part of Darboux' Theorem 2.16 to the slice $\{0\} \times B^{2n-2}$ and take a deffeomorphism, not depending on z, which carries one of our forms to another on this slice, and then perhaps this construction can be performed in a smaller ball. Consider the difference $\omega = \Omega_2 - \Omega_1$. We are going to prove $\omega = 0$. In coordinates $(z, x^1, \ldots, x^{2n-2}) \in]-\delta, \delta[\times B^{2n-2}$ it has the form

$$\omega = dz \wedge \gamma + \alpha,$$

where

$$\alpha = \sum_{i<k} \alpha_{ik}(z, x^1, \ldots, x^{2n-2}) dx^i \wedge dx^k,$$

$$\gamma = \sum_k \gamma_k(z, x^1, \ldots, x^{2n-2}) dx^k.$$

Since $\partial/\partial z$ is the vector field tangent to the zero-foliation, we have $\omega(\partial/\partial z, \partial/\partial x^k) = 0$, $1 \leqq k \leqq 2n-2$. Using the above coordinate expression for ω, we obtain that $\gamma_k = 0$, $1 \leqq k \leqq 2n-2$.

We may regard our cylinder $]-\delta,\delta[\times B^{2n-2}$ as a line bundle over $\{0\}\times B^{2n-2}$ in an obvious way. Put $\theta = dz + I\omega$ when I is the homotopy operator (2.19a) constructed for this bundle. We have

(a) $$\omega = \sum_{i<k} \alpha_{ik}(z, x^1, \ldots, x^{2n-2}) dx^i \wedge dx^k,$$

(b) $$\omega = d\theta \quad \text{(by 2.19c and (0))},$$

(c) $$\theta = \theta_0(z, x^1, \ldots, x^{2n-2}) + \sum_{k=1}^{2n-2} \theta_k(z, x^1, \ldots, x^{2n-2}) dx^k,$$

(d) $$\theta_0(0, x^1, \ldots, x^{2n-2}) = 1,$$

(e) $$\theta_k(0, x^1, \ldots, x^{2n-2}) = 0,\ 1 \leqq k \leqq 2n-2$$

$\Big\}\ \forall (x^1, \ldots, x^{2n-2}) \in B^{2n-2}$, (due to (2.19d).

Take the new coordinates $(z', x'_1, \ldots, x'_{2n-2})$ defined by the formulas $z' = \int_0^z \theta_0(z, x^1, \ldots, x^{2n-2})\, dz$, $x'_k = x^k$, $1 \leqq k \leqq 2n-2$. The equality (d) yields that these new coordinates are defined in some cylinder $]-\delta', \delta'[\times B^{2n-2}$, $0 < \delta' < \delta$. Therefore we may assume

(d') $$\theta_0(z, x^1, \ldots, x^{2n-2}) = 1 \text{ identically.}$$

The equations (a), (b), (c), (d') give us that $\partial\theta_k/\partial z = 0$, $1 \leqq k \leqq 2n-2$. Hence

(e') $$\theta_k = 0 \text{ identically.}$$

Equations (b), (c), (d') and (e') imply that $\omega = 0$. □

All our considerations in this subsection are still valid if we replace our submanifold S by the boundary ∂X of a manifold X. The above proof then yields the validity of the remaining assertion of Theorem 2.16 concerning the boundary points.

Lagrangian Submanifolds

A submanifold L of a symplectic manifold X is called *isotropic* if the tangent space to L at each of its points is an isotropic subspace of the tangent space to X at this point. A *Lagrangian* submanifold is an isotropic one of dimension equal to half of that of X, or equivalently: a submanifold with Lagrangian tangent spaces. Corollary 2.21 yields

2.23. *Any two isotropic submanifolds of the same dimension and codimension are locally isomorphic. In particular, any two Lagrangian submanifolds of the same dimension are locally isomorphic.*

Recall that our submanifolds have no boundaries and no intersection points with the boundary of a manifold.

Example 2.24. Any Lagrangian subspace of a symplectic vector space considered as a symplectic manifold is a Lagrangian submanifold. In particular, the coordinate

Lagrangian subspaces of $(\mathbb{R}^{2n}, \Omega_{st})$ (see Example 1.16) are Lagrangian submanifolds.

Example 2.25. The image of the zero-section of a cotangent bundle is a Lagrangian submanifold in the canonical symplectic structure. One obtains that this manifold is Lagrangian by a particular case of a more general proposition which describes the Lagrangian submanifolds of the cotangent bundle provided that they are images of smooth sections.

Proposition 2.26. *The image of a smooth section $\sigma: Q \longrightarrow T^*Q$ is a Lagrangian submanifold of the cotangent bundle if and only if σ is closed*, i.e. $d\sigma = 0$.

Proof. Obviously, $\operatorname{im}\sigma$ has the required dimension. Corollary 1.9a shows that it is sufficient to examine when the equation $\sigma^*\Omega_Q = 0$ holds (see (2.7) and (2.8) for the definition of Ω_Q). By use of Proposition 2.14 we have

$$\sigma^*\Omega_Q = \sigma^* d\Theta_Q = d\sigma^*\Theta_Q = d\sigma. \qquad \square$$

Let us identify a manifold with its image under the zero-section of its cotangent bundle. By "canonical neighbourhood of a Lagrangian submanifold L in X" we mean a pair (U, ψ) where U is an open neighbourhood of L in T^*L, V is an open neighbourhood of L in X, and $\psi: U \longrightarrow X$ is a symplectic local diffeomorphism onto some open neighbourhood of L in X where the diffeomorphism leaves points of L fixed.

Theorem 2.27. *Any Lagrangian submanifold has a canonical neighbourhood.*

A canonical neighbourhood is a useful tool for the investigation of Lagrangian submanifolds. Theorem 2.27 shows that a Lagrangian submanifold cannot be situated in the surrounding manifold in an arbitrary manner. A Lagrangian submanifold itself models its neighbourhood by intrinsic means, the geometry of the manifold playing no role.

Proof of Theorem 2.27. Let L be a Lagrangian submanifold of X and $\dim X = 2n$. By 2.23 any point of L has a neighbourhood W in X and symplectic diffeomorphism $\varphi: W \longrightarrow B^n \times B^n \subset \mathbb{R}^{2n}$, $\mathbb{R}^{2n}$ being supplied with the standard symplectic structure, B^n being a ball in $\mathbb{R}^n$, such that $\varphi(W \cap L) = B^n \times \{0\}$. For $x \in W$ define $M_x^{(W)} = T_x\varphi^{-1}(\{\varphi_1(x)\} \times B^n)$. Here $\varphi_1(x)$ is the projection of $\varphi(x)$ on the first multiplier of the product $B^n \times B^n$. It is apparent that $M_x^{(W)}$ is a Lagrangian complement of T_xL in T_xX. Cover L by such neighbourhoods $\{W_i\}$ and take a partition of unity $\{\eta_i\}$ which is subordinate to this cover. We may construct the space $M_x = \sum_i \eta_i(x) M_x^{(W_i)}$, where $\operatorname{supp}\eta_i \subset W_i$, by applying the baricentric map defined in §1, subsection "The Manifold of Lagrangian Subspaces". The subspace M_x is a Lagrangian complement of T_xL in T_xX and depends smoothly on x. We have a Lagrangian splitting of the restriction to L of the tangent bundle:

$$T_LX = TL \oplus M, \tag{2.27a}$$

where $M = \bigcup_{x\in L} M_x$ is a bundle over L, $\oplus$ is the Whitney sum. Proposition 1.12 gives us the bundle isomorphism

$$TL \oplus M \xleftarrow{1_{TL}\oplus h} TL \oplus T^*L,$$

which preserves the symplectic linear structures on the fibres. These are: Ω_x in $T_xX = T_xL \oplus M_x$ and $\Omega_{L,x}$ in $T_xT^*L = T_xT \oplus T_x^*L$, $x\in L$, which coincide with that of the canonical model (1.13). Applying the tubular neighbourhood theorem AI.54 to the submanifold L taking M as a normal bundle we obtain a diffeomorphism $g: M \longrightarrow V_1$ onto some open neighbourhood V_1 of L in X such that $g_L = 1_L$ and $T_L g = 1_{TL} \oplus 1_M$ (We have $L \subset M$ as the zero-section, and $T_xM = T_xL \oplus M_x$ at each $x\in L$ (see AI.48).). Consider the map $k = g\circ h: T^*L \longrightarrow V_1$. It is a diffeomorphism and its linearisation on $L\subset T^*L$ is $T_Lk = T_Lg\circ T_Lh = (1_{TL}\oplus 1_M)\circ(1_{TL}\oplus h) = 1_{TL}\oplus h$. By the above consideration k preserves the restrictions onto L of the symplectic structures of T^*L and X, i.e. $k^*\Omega|_L = \Omega_L|_L$. Applying Theorem 2.18 we get a diffeomorphism $l: U \longrightarrow V_2$ between two neighbourhoods of L in T^*L such that $l^*\circ k^*\Omega|_U = \Omega_L|_U$. Then the pair $(U, k\circ l)$ is a canonical neighbourhood of L. □

Remark 2.28. The preceding construction yields the required symplectic diffeomorphism with the additional property that its inverse carries the prescribed distribution M of Lagrangian subspaces in T_LX, M satisfying (2.27a), to $T^*L \subset TT^*L$. We say that a distribution M is *transversal* to TL if (2.27a) holds.

The above definition of a canonical neighbourhood of a Lagrangian submanifold does not require that the neighbourhood bears (or admits) the structure of a tubular neighbourhood in the sense of differential topology (see AI.54). If this is the case then we call such a canonical neighbourhood the canonical tubular neighbourhood. Any canonical neighbourhood contains a canonical tubular neighbourhood.

Example 2.29. Canonical neighbourhood of a Lagrangian n-torus. A *Lagrangian n-torus* $\mathcal{T}$ in a $2n$-dimensional symplectic manifold X is an embedding $\chi: \mathbb{T}^n \longrightarrow X$ whose image $\chi(\mathbb{T}^n)$ is a Lagrangian submanifold of X. Here $\mathbb{T}^n = \mathbb{R}^n/\mathbb{Z}^n$ is the standard n-torus. Combining 2.27 and 2.14 one may assert that there is a ball

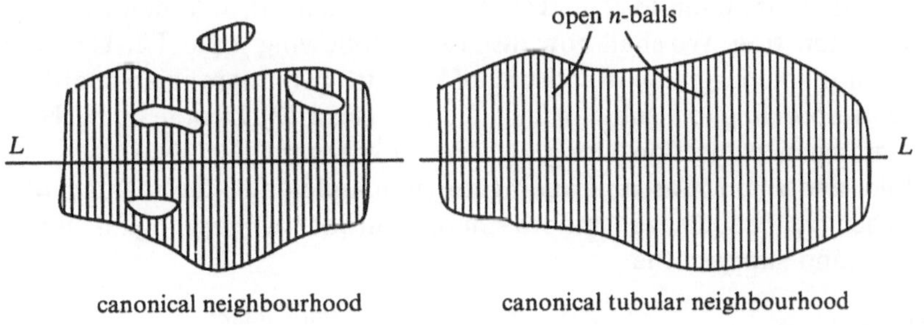

Fig. 4

$B=\{I\in\mathbb{R}^n: \|I\|<D'\}$ and a symplectic local diffeomorphism $\psi: A_B=\mathbb{T}^n\times B\longrightarrow X$ such that $\psi|_{\mathbb{T}^n\times\{0\}}=\chi$ and (A_B,ψ) is a canonical tubular neighbourhood of $\chi(\mathbb{T}^n)$. Here, for simplicity we identify $\mathbb{T}^n\times\{0\}$, $\mathbb{T}^n$, and its image in X under χ. In other words there is a canonical neighbourhood of a Lagrangian torus with the structure of a symplectic annulus, the torus being situated in the annulus like $\mathbb{T}^n\times\{\text{point}\}$.

Graphs of Symplectic Maps

Let $f: X\longrightarrow Y$ be a smooth map between two symplectic manifolds (X,Ω) and (Y,Σ). The *graph of f* is the set

$$\operatorname{graph} f=\{(x,y)\in X\times Y: y=f(x)\}.$$

It is a submanifold of $X\times Y$, and its dimension equals that of X. Supply $X\times Y$ with the symplectic structure $\Pi=\Omega\oplus(-\Sigma)$ (we change the sign of the structure of the second manifold!).

Proposition 2.30. *A map f is symplectic if and only if graph f is an isotropic submanifold of* $(X\times Y,\Pi)$.

Proof. Consider the vectors $w_i=(u_i,v_i)\in T_{(x,y)}X\times Y$, $i=1,2$, where $u_i\in T_xX$ and $v_i\in T_yY$ (see AI.47). Let $(x,y)\in\operatorname{graph} f$. Then $w_i\in T_{(x,y)}\operatorname{graph} f$ if and only if $v_i=T_xf(u_i)$. Computing

$$\Pi(w_1,w_2)=\Omega(u_1,u_2)-\Sigma(v_1,v_2)=\Omega(u_1,u_2)-f^*\Sigma(u_1,u_2),$$

we have that $\Pi(w_1,w_2)=0$ for all $w_1,w_2\in T_{(x,y)}\operatorname{graph} f$ and $(x,y)\in\operatorname{graph} f$ if and only if f preserves the symplectic structures. □

In particular, a local diffeomorphism is symplectic if and only if its graph is a Lagrangian submanifold with an appropriate symplectic structure on the product of its source and target.

Generating Function of a Lagrangian Submanifold in $\mathbb{R}^{2n}$

A Lagrangian submanifold $\mathscr{L}$ in $(\mathbb{R}^{2n},\Omega_{st})$ can be defined, at least locally, by means of a scalar function. We shall now discuss the following topic: Let $\mathbb{R}^{2n}=L^{co}_{\mathfrak{a}}+L^{co}_{\bar{\mathfrak{a}}}$ be the coordinate Lagrangian splitting of $\mathbb{R}^{2n}$ into two Lagrangian coordinate subspace corresponding to a subset $\mathfrak{a}\subset\{1,2,\dots,n\}$ (see 1.13). Denote by $\pi_{\mathfrak{a}}:\mathbb{R}^{2n}\longrightarrow L^{co}_{\mathfrak{a}}$ the Lagrangian projection along $L^{co}_{\bar{\mathfrak{a}}}$. We now regard $(\mathbb{R}^{2n},\Omega_{st})$ as a symplectic manifold, and $L^{co}_{\mathfrak{a}}$, $L^{co}_{\bar{\mathfrak{a}}}$ as its Lagrangian submanifolds in the manner of 2.2 and 2.24. Then $\pi_{\mathfrak{a}}$ becomes a smooth map. Consider a Lagrangian submanifold $\mathscr{L}\subset\mathbb{R}^{2n}$, and suppose that

(2.31) $\qquad \pi_{\mathfrak{a}}|_{\mathscr{L}}$ *is a local diffeomorphism*

Let U be its image in $L_{\mathfrak{a}}^{co}$. It is convenient to arrange our objects in the following commutative diagram

$$\begin{array}{ccccc} \mathscr{L} & \overset{i}{\hookrightarrow} & \mathbb{R}^{2n} & \overset{\pi_{\mathfrak{a}}}{\longrightarrow} & L_{\mathfrak{a}}^{co} \\ & & & \nwarrow^{h} & \cup \\ & & & & U \end{array}$$

where i denotes the inclusion map of $\mathscr{L}$ into $\mathbb{R}^{2n}$, $h = i\circ(\rho\circ i)^{-1}$, $\rho = \pi_{\mathfrak{a}}$. Obviously $\mathscr{L} = h(U)$. Consider the 1-form defined on $\mathbb{R}^{2n}$:

$$\theta = \sum_{i\in\mathfrak{a}} p_i dq^i - \sum_{j\in\bar{\mathfrak{a}}} q^j dp_j.$$

The form $\lambda = h^*\theta \in \mathscr{E}^1(U)$ is closed. Indeed, $d\lambda = h^* d\theta = -h^*\Omega_{st} = -[(\rho\circ i)^{-1}]^* i^*\Omega_{st} = 0$ since $i^*\Omega_{st} = 0$ due to $\mathscr{L}$ being Lagrangian: Suppose in addition that

(2.32) *$\mathscr{L}$ is deffeomorphic to the n-ball.*

Then Poincaré's Lemma 2.19e yields the existence of a function $S \in C^\infty(U)$ such that $\lambda = dS$. The equation of $\mathscr{L}$ may be written as $p_i = h_i(x)$, $q^j = h_{n+j}(x)$, $i\in\mathfrak{a}$, $j\in\bar{\mathfrak{a}}$, $x\in U$. Take Cartesian coordinates $(q^i, p_j,\ i\in\mathfrak{a},\ j\in\bar{\mathfrak{a}})$ in $U \subset L_{\mathfrak{a}}^{co}$ induced by those of $\mathbb{R}^{2n}$. Since

$$\lambda = \sum_{i\in\mathfrak{a}} h_i dq^i - \sum_{j\in\bar{\mathfrak{a}}} h_{n+j} dp_j,$$

the equations of $\mathscr{L}$ read

$$p_i = \frac{\partial S}{\partial q^i}, \qquad q^j = -\frac{\partial S}{\partial p_j}, \qquad i\in\mathfrak{a},\ j\in\bar{\mathfrak{a}}. \tag{2.33}$$

The function $S = S(q^i, p_j,\ i\in\mathfrak{a},\ j\in\bar{\mathfrak{a}})$ is called the *generating function* of $\mathscr{L}$, the coordinates $(q^i, p_j,\ i\in\mathfrak{a},\ j\in\bar{\mathfrak{a}})$ are called the *focal coordinates*.

In general, a Lagrangian submanifold $\mathscr{L}$ of $(\mathbb{R}^{2n}, \Omega_{st})$ does not satisfy the conditions (2.31) and (2.32). However, it follows from 1.14 that the said conditions are fulfilled locally: each point $z\in\mathscr{L}$ has a neighbourhood $\mathscr{L}'$ in $\mathscr{L}$ which can be defined by means of a generating function with some $\mathfrak{a}$, the latter depending on z.

Application 2.34. Consider a symplectic map $f: X \longrightarrow \mathbb{R}^{2n}$, where X is an open subset in $(\mathbb{R}^{2n}, \Omega_{st})$ supplied with the induced symplectic structure. Then, due to 2.28, graph f is a Lagrangian submanifold of $(\mathbb{R}^{2n}\times\mathbb{R}^{2n}, \Omega_{st}\oplus(-\Omega_{st}))$. Hence, at least locally, it may be described in terms of a generating function. There are many possibilities for the choice of focal coordinates. If graph f has nondegenerate projection to $L_{\mathfrak{a}}^{co}\times L_{\mathfrak{b}}^{co} \subset \mathbb{R}^{2n}\times\mathbb{R}^{2n}$, where $\mathfrak{a}$ and $\mathfrak{b}$ are subsets of $\{1, 2, \ldots, n\}$, and X is diffeomorphic to a $2n$-ball, then the map

$$f: (q^1, q^2, \ldots, q^n, p_1, p_2, \ldots, p_n) \longmapsto (\bar{q}^1, \bar{q}^2, \ldots, \bar{q}^n, \bar{p}_1, \bar{p}_2, \ldots, \bar{p}_n)$$

can be written in terms of the generating function S depending on the variables

$q^i, i\in \mathfrak{a}$, $p_j, j\in \bar{\mathfrak{a}}$, $\bar{q}^k$, $k\in \mathfrak{b}$, $\bar{p}_1, 1\in \bar{\mathfrak{b}}$:

$$p_i = \frac{\partial S}{\partial q^i}, \quad i\in\mathfrak{a}, \qquad q^j = -\frac{\partial S}{\partial p_j}, \quad j\in\bar{\mathfrak{a}},$$
(2.35)
$$\bar{p}_k = -\frac{\partial S}{\partial \bar{q}^k}, \quad k\in\mathfrak{b}, \qquad \bar{q}^l = \frac{\partial S}{\partial \bar{p}_l}, \quad l\in\bar{\mathfrak{b}}.$$

The formulae (2.35) follow from (2.33) by use of the elementary isomorphism $(\mathbb{R}^{2n}\times\mathbb{R}^{2n}, \Omega_{st} + (-\Omega_{st})) \longrightarrow (\mathbb{R}^{2n}, \Omega_{st})$ given by

$$(q^1, q^2, \ldots, q^n, p_1, p_2, \ldots, p_n, \bar{q}^1, \bar{q}^2, \ldots, \bar{q}^n, \bar{p}_1, \bar{p}_2, \ldots, \bar{p}_n)$$
$$\longmapsto (q^1, q^2, \ldots, q^n, \bar{p}_1, \bar{p}_2, \ldots, \bar{p}_n, p_1, p_2, \ldots, p_n, \bar{q}^1, \bar{q}^2, \ldots, \bar{q}^n).$$

Example 2.36. The generating function of a contragredient map in Euclidean space. Let U and V be non-empty open sets in $\mathbb{R}^n$ and $f: U \longrightarrow V$ be a diffeomorphism. Identifying T^*U and T^*V with $U\times\mathbb{R}^n$ and $V\times\mathbb{R}^n$ as in 2.9, the contragredient map $T^{*-1}f: T^*U \longrightarrow T^*V$ is transformed to the following expression in Cartesian coordinates: $T^{*-1}f(q^1, q^2, \ldots, q^n, p_1, p_2, \ldots, p_n) = (\tilde{q}^1, \tilde{q}^2, \ldots, \tilde{q}^n, \tilde{p}_1, \tilde{p}_2, \ldots, \tilde{p}_n)$ where $\tilde{q}^i = pr_i f(q)$, $q = (q^1, q^2, \ldots, q^n)\in U$, and

(2.36a)
$$\tilde{p}_j = \sum_{i=1}^n p_i \frac{\partial q^i(\tilde{q})}{\partial \tilde{q}^j}, \quad q^i(\tilde{q}) = pr_i f^{-1}(\tilde{q}),$$
$$\tilde{q} = (\tilde{q}^1, \tilde{q}^2, \ldots, \tilde{q}^n)\in V.$$

Thus, the transformation $T^{*-1}f$ may be written by means of the generating function as

(2.36b)
$$R(p_1, p_2, \ldots, p_n, \tilde{q}^1, \tilde{q}^2, \ldots, \tilde{q}^n) = -\sum_{i=1}^n p_i q^i(\tilde{q}).$$

Example 2.37. Rotation by the angle α of the symplectic plane $(\mathbb{R}^2, \Omega_{st}): (q,p) \longmapsto (\bar{q}, \bar{p})$, where

$$\bar{q} = q\cos\alpha - p\sin\alpha,$$
$$\bar{p} = q\sin\alpha + p\cos\alpha$$

may be given by the following generating functions

(2.37a)
$$R_1(q, \bar{p}) = \frac{1}{\cos\alpha} q\bar{p} - \frac{1}{2}\tan\alpha(q^2 + (\bar{p})^2)$$

provided $\alpha \neq (k + 1/2)\pi$, $k\in\mathbb{Z}$, and

(2.37b)
$$R_2(q, \bar{q}) = -\frac{1}{\cos\alpha} q\bar{q} + \frac{1}{2}\cot\alpha(q^2 + (\bar{q})^2)$$

provided $\alpha \neq k\pi$, $k\in\mathbb{Z}$.

Straightening of a 2-dimensional Symplectic Annulus

Later we need the solution of one particular case of a global isomorphism problem for symplectic manifolds. The manifold in question is the two-dimensional annulus $\mathbb{T}^1 \times [y_1, y_2]$ where $\mathbb{T}^1 = \mathbb{R}^1/\mathbb{Z}^1$ is the one-dimensional torus (= circle). Denote the coordinates in the covering space $\mathbb{R} \times [y_1, y_2]$ of $\mathbb{T}^1 \times [y_1, y_2]$ by (x, y). Supply $\mathbb{T}^1 \times [y_1, y_2]$ with a symplectic structure Ω which is a perturbation of the standard structure Ω_{st}. In the coordinates introduced in the covering space the lifts of these structures read $\tilde{\Omega}_{st} = dx \wedge dy$ and $\tilde{\Omega} = (1 + \beta(x, y))dx \wedge dy$ respectively. Here β is a smooth function defined on $\mathbb{R} \times [y_1, y_2]$ which satisfies the following conditions:

(i) β is periodic in its x-variable with period 1;

(ii) $\operatorname{supp} \beta \subset \mathbb{R} \times [y_1 + \gamma, y_2 - \gamma]$, γ being a positive constant;

(iii)
$$\int_0^1 \int_{y_1}^{y_2} \beta(x, y)dx \wedge dy = 0.$$

Proposition 2.38. *Given a positive constant C_1 and a positive integer r_0 there are positive constants $C = C(r_0, C_1)$ and $\varepsilon = \varepsilon(r_0, C_1)$ such that the following is true. If β satisfies* (i) – (iii) *and in addition*

(iv)
$$y_2 - y_1 \leqq C_1, \ \min(1 + \beta(x, y)) \geqq C_1^{-1},$$

(v)
$$\| \beta \|_{r_0 + 1} \leqq \min\{\varepsilon, \varepsilon\gamma\}$$

then there is a diffeomorphism $g: \mathbb{T}^1 \times [y_1, y_2] \longrightarrow \mathbb{T}^1 \times [y_1, y_2]$ which carries Ω to the standard structure $\Omega_{st} = dx \wedge dy$:

(a)
$$g^*\Omega = \Omega_{st},$$

and satisfies the inclusion and the estimate:

(b)
$$\operatorname{supp}(g - \mathrm{id}) \subset \mathbb{T}^1 \times [y_1 + \gamma, y_2 - \gamma],$$

(c)
$$\| g - \mathrm{id} \|_r \leqq C \| \beta \|_r, \quad 1 \leqq r \leqq r_0.$$

Proof. In the course of the proof "const" will denote a positive constant only depending on r_0 and C_1. Set $\beta_0(y) = \int_0^1 \beta(x, y)dx$ and define a 1-form

$$\alpha(x, y) = -\left(\int_{y_1}^{y} \beta_0(y')\,dy'\right)dx + \left(\int_0^x (\beta(x', y) - \beta_0(y))dx'\right)dy.$$

It is obvious that $d\alpha = \beta\,dx \wedge dy$ and α is periodic in the variable x, and vanishes outside the strip $\mathbb{R} \times [y_1 + \gamma, y_2 - \gamma]$. The subsequent proof follows the lines of that of Theorem 2.18. We consider the family of vector fields

(d)
$$\xi_t = -\frac{\int_0^x (\beta(x', y) - \beta_0(y))\,dx'}{1 + t\beta(x, y)} \frac{\partial}{\partial x} - \frac{\int_{y_1}^y \beta_0(y')\,dy'}{1 + t\beta(x, y)} \frac{\partial}{\partial y}, \quad t \in [0, 1],$$

which satisfies the equation $\alpha = -\tilde{\Omega}_t(\xi_t, \cdot)$, $\tilde{\Omega}_t = \tilde{\Omega}_{st} + t\beta\,dx \wedge dy$, and define the family of diffeomorphisms $\tilde{g}_t: \mathbb{R} \times [y_1, y_2] \longrightarrow \mathbb{R} \times [y_1, y_2]$ as a solution to the problem

$$\text{(e)} \qquad \frac{d\tilde{g}_t}{dt} = \xi \circ \tilde{g}_t, \qquad \tilde{g}_0 = \text{id}.$$

The problem (e) is equivalent to the integral equation

$$\text{(f)} \qquad \tilde{g}_t(x, y) = (x, y) + \int_0^t c(\tilde{g}_t(x, y), t')\, dt',$$

where $c(z, t)$ is a vector valued function of the variable $z = (x, y)$ and $t \in [0, 1]$, whose components are those of (d). We will solve (f) by the usual iteration method by setting

$$\text{(g)} \qquad \begin{gathered} g_t(x, y) = g(x, y, t), \qquad g_0(x, y, t) = (x, y), \\ g_{k+1}(x, y, t) = (x, y) + \int_0^t c(g_k(x, y, t'), t')\, dt'. \end{gathered}$$

Note that $\operatorname{supp}(g_k - \text{id}) \subset \mathbb{R} \times [y_1 + \gamma, y_2 - \gamma]$. One immediately obtains the estimate

$$\text{(h)} \qquad \| c(\cdot, t) \|_r \leqq M = \text{const} \, \| \beta \|_r.$$

We require the right-hand side of (h) to be less than $\varepsilon' = \min\{1/2, \gamma/\sqrt{2}, 1/2C_1(2)\}$, where $C_1(2)$ is the constant in AII.26(c). Let us inductively suppose that

$$\text{(i)} \qquad \left\| \int_0^t c(g_s(\cdot, t'), t')\, dt' \right\|_r \leqq \left(1 - \frac{1}{2^{s+1}}\right)\varepsilon', \quad s \leqq k.$$

Then g_{k+1} is a diffeomorphism (Lemma AII.21) and

$$\text{(j)} \qquad \max\{ \| g_{s+1}(\cdot, t) \|_r, \| g_{g+1}^{-1}(\cdot, t) \|_r \} \leqq \text{const}, \quad s \leqq k$$

(use equation (g) and Lemma AII.26). Applying Lemma AII.16 yields that

$$\text{(k)} \qquad \| c(g_{s+1}, t) - c(g_s, t) \|_r \leqq L \| g_{s+1}(\cdot, t) - g_s(\cdot, t) \|_r,$$

where $L = \text{const} \, \| \beta \|_{r+1}$. We suppose that $L \leqq 1$. Usual induction applied to equation (g) results in

$$\text{(l)} \qquad \| g_{s+1}(\cdot, t) - g_s(\cdot, t) \|_r \leqq \frac{ML^s}{(s+1)!} t^{s+1} \leqq \frac{1}{(s+1)!}, \quad s \leqq k.$$

Then one finds that the estimate (i) for $s = k + 1$ is a consequence of (k) provided $\| \beta \|_{r+1}$ is sufficiently small.

The estimate (l) shows that $g_k(\cdot, t)$ converges to a map $g(z, t) = g_t(z)$ which satisfies (f). The familiar theorem of the theory of differential equations ensures that g_t is C^∞. The estimate (l) also results in $\| g_t - \text{id} \|_r \leqq \text{const} \, \| \beta \|_r$ which shows that g_t is a diffeomorphism provided $\| \beta \|_r$ is sufficiently small. The same arguments as in the proof of Theorem 2.18 show that $\tilde{g}_t^+ \tilde{\Omega}_t = \tilde{\Omega}_{st}$ and, in particular, $\tilde{g} = \tilde{g}_1$ is the lift of a diffeomorphism g satisfying (a), (b) and (c). □

§3. Symplectic Dynamical Systems

The goal of this section is to discuss the possible kinds of the concept of a symplectic dynamical system, to fix the terminology, and to investigate the simplest properties of the objects that arise. It is convenient, at first, to introduce the cases of discrete and continuous times separately.

Symplectic Dynamical Systems with Discrete Time

Let X be a symplectic manifold. A *local symplectic diffeomorphism* in X is a smooth injective map $f: U \longrightarrow X$, where U is an open subset in X, with f preserving the symplectic structure. A *symplectic dynamical system* (with discrete time) is a pair (X, f) where X is a symplectic manifold and f is a local symplectic diffeomorphism in X. We refer to X as the *phase space* and to f as the *generator* of the system. The domain U of the generator is referred to as the *domain* of a system. The *number of degrees of freedom* is, by definition, $n+1$ where $n = \frac{1}{2}\dim X$.

Starting with f we may define the *local symplectic cascade* as follows: put $f^0 = i_U$ (the inclusion map of U in X), $f^1 = f$, and let f^{-1} be the inverse of f restricted to the set $U \cap f(U)$. Define f^k, $k > 0$ an integer, as $f \circ f \circ \cdots \circ f$ (k times) considered on the maximal open subset of U where this expression remains sensible, and define f^k for $k < 0$ integer as $f^{-1} \circ f^{-1} \circ \cdots \circ f^{-1}$ ($|k|$ times), using the same agreement for the domain as in the case $k > 0$. Some of the obtained maps may be void. We also call the local symplectic cascade $\{f^t : t \in \mathbb{Z}\}$ the *dynamics* of the system.

The *orbit*, or *trajectory*, of a point $x \in U$ is the set $\{f^t(x): t \in \mathbb{Z}$ and $f^t(x)$ exists$\}$.

If the domain U of a system coincides with the phase space X, we say the dynamics are *complete*, and refer to the system as a *system with complete dynamics* or *symplectic cascade*. In other words, a complete dynamics is a family of symplectic diffeomorphisms $f^t: X \longrightarrow X$, $t \in \mathbb{Z}$, such that $f^0 = \mathrm{id}$ and $f^{t_1 + t_2} = f^{t_1} \circ f^{t_2}$ for all $t_1, t_2 \in \mathbb{Z}$, i.e. it is a representation of $\mathbb{Z}$ into the group of symplectic diffeomorphisms of the objects that arise. It is convenient, at first, to introduce the cases of discrete and continuous times separately.

Symplectic Dynamics Systems with Continuous Time

Let X be a symplectic manifold with symplectic structure Ω. A smooth vector field ξ on X is said to be *locally Hamiltonian* if the form $\xi^\flat$ is closed, i.e. $d\xi^\flat = 0$. Here $\flat = \flat(\Omega)$: $\mathscr{H}(X) \longrightarrow \mathscr{E}^1(X)$ is the "flat" operation generated by Ω (see §1). A *symplectic dynamical system* (with continuous time) is a pair (X, ξ) where X is a symplectic manifold and ξ is a locally Hamiltonian vector field on X. We refer to X as the *phase space* and to ξ as the *generator* of the system. The *number of degrees of freedom* is, by definition, $n = \frac{1}{2}\dim X$.

It follows from the definition that each point $x \in X$ has a neighbourhood U in X such that $\xi^{\flat}|_U = dH$ where $H: U \longrightarrow \mathbb{R}$ is a smooth function, called a *local Hamiltonian.* Conversely, this property characterizes locally Hamiltonian vector fields. It is not difficult, using AI.60(c) and 2.17f, to deduce the following explicit expression for a locally Hamiltonian vector field in Darboux's coordinates $(q^1, q^2, \ldots, q^n, p_1, p_2, \ldots, p_n)$, $2n = \dim X$, it terms of its local Hamiltonian H:

$$\xi = \sum_{i=1}^{n} \left(\frac{\partial H}{\partial p_i} \frac{\partial}{\partial q^i} - \frac{\partial H}{\partial q^i} \frac{\partial}{\partial p_i} \right). \tag{3.1}$$

If there is a global function $H: X \longrightarrow \mathbb{R}$ such that $\xi^{\flat} = dH$, then the field ξ is called *Hamiltonian* and such an H is called a *Hamiltonian of* ξ, *or a Hamiltonian function* of ξ. A symplectic dynamical system with Hamiltonian generator is called a *Hamiltonian dynamical system.* The Hamiltonian of its generator will also be referred to as the *Hamiltonian* (*Hamiltonian function*) of the system.

The *orbit*, or *trajectory*, of a point $x \in X$ is the union of all integral curves of a constant family $\xi_t = \xi$, $t \in \mathbb{R}$, (see AI.67) passing through x. The latter are solutions to the equation

$$\frac{df^t}{dt} = \xi \circ f^t \tag{3.2}$$

of the form $f^t: * \longrightarrow X$, $t \in \Delta \subset \mathbb{R}$ ($*$ being a point), such that $f^{t_0}(*) = x$ for some $t_0 \in \Delta$. The orbit itself is the image of the maximal integral curve starting at x (i.e., $f^0(*) = x$ and $\Delta \ni 0$ is the maximal interval such that a solution to (3.2) with this initial condition exists). Note that (3.1) implies that the equation (3.2) in Darboux coordinates reads

$$\frac{dq^i}{dt} = \frac{\partial H}{\partial p_i}, \qquad \frac{dp_i}{dt} = -\frac{\partial H}{\partial q^i}, \qquad 1 \leqq i \leqq n. \tag{3.3}$$

These are the familiar *Hamiltonian equations.*

A *normed local flow* of a vector field ξ is a solution $f^t: U \longrightarrow X$, $t \in \Delta$, to (3.2), such that U is a non-empty open subset of X, $0 \in \Delta$, $f^0 = i_U$ (the inclusion map of U into X), and all of the f^t are local diffeomorphisms (see AI.67, 69, 70). The set U is called the *domain* of the local flow f^t. As follows from Proposition 3.4 below, a normed local flow of a locally Hamiltonian vector field consists of symplectic local diffeomorphisms. For further use, we will formulate this assertion in a more general setting.

Proposition 3.4. *Let* ξ_t, $t \in \Gamma$, *be a smooth family of locally Hamiltonian vector fields in* X, Γ *being an open interval in* $\mathbb{R}$. *Consider the differential equation*

$$\frac{\mathrm{d}f_t}{\mathrm{d}t} = \xi_t \circ f_t. \tag{3.4a}$$

Let $f_t: U \longrightarrow X$, $t \in \Delta$, *be a general integral of* (3.4a) (see AI.67), U *being a nonempty open subset of* X, $\Delta \subseteq \Gamma$ *an open interval. If* f_{t_0} *is symplectic for some* $t_0 \in \Delta$, *then* f_t *are also symplectic for all* $t \in \Delta$.

Proof. Applying differentiation formula AI.68, we have

$$\frac{d}{dt}(f_t)^*\Omega = df_t^*(\xi_t \lrcorner \Omega) = df_t^* \xi_t^\flat = f_t^* d(\xi_t^\flat) = 0.$$

Integrating by t over the interval with ends t_0 and t we obtain $f_t^*\Omega = f_{t_0}^*\Omega = \Omega$. □

Remain 3.4b. The above calculation leads to the following converse assertion. Let $\xi_t, t \in \Gamma \ni 0$, be a family of vector fields on a symplectic manifold X, and let the assertion of Proposition 3.3 be valid for this family, that is, any general integral $f_t: U \longrightarrow X, t \in \Delta \subset \Gamma, U \subset X$, containing one symplectic map, consists of symplectic maps. Then ξ_t, $t \in \Gamma$, is locally Hamiltonian. To prove this it is sufficient to restrict ourselves to general integrals with the property $f_{t_0} = i_U$ for some $t_0 \in \Delta$, to reverse the proof of 3.3, and to take into account that X may be covered by domains U of such integrals.

The *dynamics* of a symplectic dynamical system (X, ξ) is a collection of normed local flows of ξ whose domains cover the interior of the manifold. Dynamics always exist (AI.67).

If ξ is complete (i.e. each its integral curve may be extended to an infinite interval $\mathbb{R}$ of the time variable, AI.71), then the dynamics may be given by the global flow $f^t = \exp(t\xi): X \longrightarrow X$, $t \in \mathbb{R}$. We call such dynamics *complete* and refer to the system (X, ξ) as a *system with complete dynamics*, or a *symplectic flow*. If the system is Hamiltonian we also use the term *Hamiltonian flow*. In the case of complete dynamics we also refer to the family f^t, $t \in \mathbb{R}$, as a flow, or a system itself.

Proposition 3.5. *If f^t, $t \in \mathbb{R}$, is a Hamiltonian flow with a Hamiltonian function H, then H is an integral of the system* i.e. *$H \circ f^t$ is independent of t.*

Proof. Let ξ be the generator of f^t. By differentiation formula AI.68

$$\frac{d}{dt} H \circ f^t = \frac{d}{dt}(f^t)^* H = (f^t)^* \xi \lrcorner dH = (f^t)^* \xi \lrcorner \xi^\flat = (f^t)^* \Omega(\xi, \xi) = 0.$$ □

Generally speaking, a vectorfield on a manifold does not generate a flow. There are two reasons for a vector field not to be complete. The first reason occurs due to the presence of the boundary: a vector field ξ may turn transversely on the boundary ∂X of a manifold X as is shown in Fig. 5. Integral curves of such vector field passing through points of ∂X cannot be extended to the whole real axis of the time variable. Another source of noncompleteness is the noncompactness of a supporting manifold: integral curves may reach infinity in finite time. Example 3.8 illustrates this phenomenon.

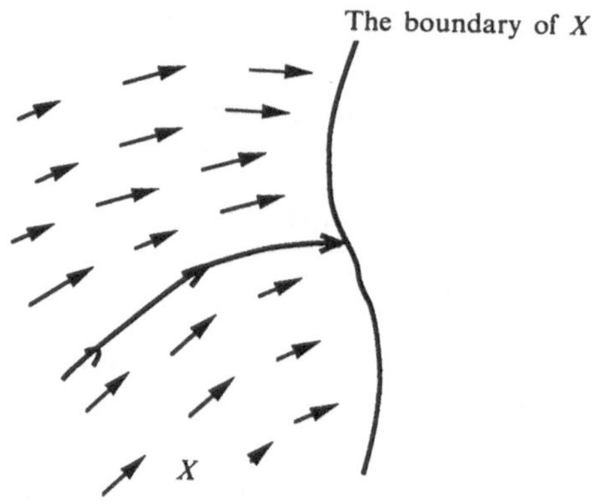

Fig. 5. Vector field ξ may meet transversely the boundary of the phase space

The connection between various types of symplectic dynamical systems with continuous time may be illustrated in the following diagram

(3.6)

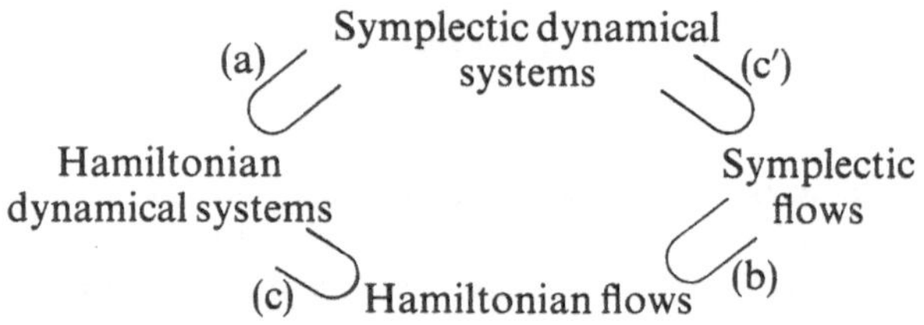

In general, none of these inclusions is an equality. We may claim the following assertions concerning the cases when the inclusions in the above diagram become equalities for a given phase space X.

3.6a. *The inclusion* (a) *is an equality if and only if the first cohomology group* $H^1(X;\mathbb{R})$ *is zero.*

3.6b. *The inclusion* (b) *is an equality if* $H^1(X;\mathbb{R})=0$.

3.6c. *Both the inclusions* (c) *and* (c′) *are equalities if* X *is a closed manifold* (i.e. *compact and without boundary*).

Example 3.7. Fix $(\omega_1,\omega_2)\in\mathbb{R}^2$ and consider the symplectic flow $f^t:(\mathbb{R}^2,\Omega_{st})\longrightarrow(\mathbb{R}^2,\Omega_{st})$, defined by

$$f^t(q,p)=(q+\omega_1 t, p+\omega_2 t),\quad t\in\mathbb{R}.$$

This flow commutes with the action of $\mathbb{Z}^2\subset\mathbb{R}^2$ by shifts and thereby defines the symplectic flow $g^t:\mathbb{T}^2\longrightarrow\mathbb{T}^2$ of the symplectic standard 2-torus (see Example 2.4). The vector field of the flow g^t is given by

$$\xi=\omega_1\frac{\partial}{\partial q}+\omega_2\frac{\partial}{\partial p}.$$

If both ω_1 and ω_2 are nonzero and ω_1/ω_2 is an irrational number then the flow g^t is not Hamiltonian. Otherwise a Hamiltonian function H could be constant on a trajectory of the flow by 3.5. The trajectory is dense in $\mathbb{T}^2$, so $H = \text{const}$ which contradicts the nonnullity of ξ.

This example shows that the inclusion (b), and therefore (a), is not generally an equality.

Example 3.8. Consider the Hamiltonian vector field ξ on $(\mathbb{R}^2, \Omega_{st})$ with Hamiltonian function $H(q, p) = qp^2$, i.e.

$$\xi = 2pq\frac{\partial}{\partial q} - p^2\frac{\partial}{\partial p}.$$

The corresponding differential equations (3.2) have the general integral $p(t) = (t - c_1)^{-1}$, $q(t) = c_2(t - c_1)^2$, and the singular integral curves $p_0(t) = 0$, $q_0(t) = c$. It is evident that there is no solution defined on the whole real axis R of the variable t which passes through the point (q, p) with $p \neq 0$. This example shows that the inclusion (c), and therefore the inclusion (c′), is not generally an equality.

We conclude this subsection by a definition applicable to both of the cases of discrete and continuous times. An *isomorphism* between two symplectic dynamical systems of the same kind of time is a symplectic diffeomorphism $h: X_1 \longrightarrow X_2$ between their phase spaces X_i, $i = 1, 2$, which conjugates the maps f_i^t, $i = 1, 2$, of their dynamics, i.e. $h(\text{domain of } f_1^t) = \text{domain of } f_2^t$, and the diagram

$$\begin{array}{ccc} X_1 \supset \text{domain of } f_1^t & \xrightarrow{f_1^t} & X_1 \\ \downarrow h & & \downarrow h \\ X_2 \supset \text{domain of } f_2^t & \xrightarrow{f_2^t} & X_2 \end{array} \tag{3.9}$$

computes for all possible t. In the case of noncomplete symplectic dynamical systems with continuous time the same must be true for all possible local flows in X_1 and X_2 corresponding to f_1^t and f_2^t.

Two symplectic dynamical systems are said to be *isomorphic* if there is an isomorphism between them.

Invariant Sets with Complete Dynamics

The situations described below often arise when one is interested in the behaviour of trajectories of a dynamical system in a neighbourhood of an invariant set. Let (X, ξ) be a symplectic dynamical system with continuous time. A set $M \subset X$ is called *invariant* if for each $x \in M$ there is an integral curve $s: \Delta \longrightarrow X$, Δ being an open interval in $\mathbb{R}$, of ξ such that s passes through x, i.e. $s(t_0) = x$ for some $t_0 \in \Delta$, and $s(\Delta) \subset M$. An invariant set M is said to have *complete dynamics, or be completely invariant* if, in the previous definition, one may take $\Delta = \mathbb{R}$ for each point $x \in M$. If M is an invariant set with complete dynamics then there is the flow $f^t: M \longrightarrow M$ generated by ξ on M. This is a family of homeomorphisms of M onto itself having the properties: (1) $f^0 = \text{id}_M$, (2) $f^{t+\tau} = f^t \circ f^\tau$ $\forall t, \tau \in \mathbb{R}$, (3) the map $F: M \times \mathbb{R} \longrightarrow M$ defined as $F(x, t) = f^t(x)$ is continuous, (4) each point $(x_0, t_0) \in M \times \mathbb{R}$ has a neighbourhood in $X \times \mathbb{R}$ of the form $U \times \Delta$, U open in X and Δ an open interval in $\mathbb{R}$, such that there is a local flow $\tilde{f}^t : U \longrightarrow X$, $t \in \Delta$, of ξ, consisting of symplectic maps whose restrictions on $U \cap M$ coincide with those of the functions f^t.

Example 3.10. The *zero-set* $Z(\xi) = \{x \in X : \xi_x = 0\}$ of a locally Hamiltonian vector field ξ is invariant and has complete dynamics. The flow generated by ξ on $Z(\xi)$ consists of identity maps for all $t \in \mathbb{R}$. Any subset of $Z(\xi)$ is an invariant set with complete dynamics.

Example 3.11. A *periodic orbit* (or *periodic trajectory*) of a system (X, ξ) is an integral curve $s : \mathbb{R} \longrightarrow X$ of ξ with the property that there is a positive number T, called the *period*, such that $s(t+T) = s(t)$ for all $t \in \mathbb{R}$ and $s(t+\tau) \neq s(t)$ if $0 < \tau < T$. The range $s(\mathbb{R})$ of a periodic orbit s is an invariant set with complete dynamics. It is homeomorphic to the circle, and the system in question is Hamiltonian, being restricted to a tubular neighbourhood of $s(\mathbb{R})$.

The corresponding situation in the case of discrete time may be presented as follows. Consider a symplectic dynamical system (X, f) with discrete time. Let U be its domain. A subset $K \subset X$ is called *invariant* if $f(K \cap U) \subset K$. We say that an invariant set K *has complete dynamics*, or is *completely invariant*, if $K \subset U$ and $f(K) = K$. In this case there is a cascade $f^t : K \longrightarrow K$, $t \in \mathbb{Z}$, generated by f. It actually coincides with the restriction on K of the local symplectic cascade generated by f. The use of the same notation, f^t, does not involve ambiguity.

Example 3.12. A *periodic orbit* (or *periodic trajectory*) of a system (X, f) with domain U is a finite set $\{x_0, x_1, \ldots, x_{N-1}\}$ contained in U such that $f(x_k) = x_{k+1}$, $0 \leqq k \leqq N-2$, and $f(x_{N-1}) = x_0$. Points of a periodic orbit are called *periodic points*. The number N is called the *period* of a periodic orbit (and of a periodic point). A periodic orbit is obviously a completely invariant set. If $N = 1$, we have a *fixed point*.

If $\{x_i,\ 0 \leqq i \leqq N-1\}$ is a periodic orbit of (X, f), the eigenvalue of $Tf^N(x_i) : T_{x_i}X \longrightarrow T_{x_i}X$ do not depend on i and serve as symplectic invariants of the system in the neighbourhood of the periodic orbit. In particular we may classify periodic orbits according to the type of $Tf^N(x_i)$ (see §1 Endomorphisms of Symplectic Vector Space). We say the periodic orbit is *simple elliptic* or *hyperbolic* if $Tf^N(x_i)$ is also.

The Calabi Invariant

It is useful to introduce a quantity which measures the deviation from being Hamiltonian. The invariant in the title plays the role of such a quantity. Firstly consider the case of continuous time, namely that of a locally Hamiltonian vector field ξ defined on a symplectic manifold (X, Ω). Due to the definition the 1-form $-\xi^\flat = -\xi \lrcorner \Omega$ is closed and therefore defines the cohomology class $\mathrm{Ca}(\xi) = [-\xi^\flat] \in H^1(X; \mathbb{R})$, the *Calabi invariant* of the system. A locally Hamiltonian vector field is Hamiltonian if and only if its calabi invariant zero.

In the case of discrete time we shall only define the analogous notion for the case when the symplectic structure Ω is *exact*. This means that there is a 1-form Θ, the *Liouville form*, such that $\Omega = -d\Theta$. Note that the Liouville form is not uniquely defined by Ω; Liouville forms of the same structure differ from each other by the addition of a closed form. There are many important examples of symplectic manifolds with exact symplectic structures. Among these we have the contangent bundle with its canonical symplectic structure (see (2.8)) and the symplectic annulus (Example 2.5). The latter example has

$$\Theta = -\sum_{i=1}^{n} I_i d\varphi^i \tag{3.13}$$

and its Liouville form. Note that a closed manifold never has an exact symplectic structure.

Let X be a symplectic manifold with exact symplectic structure and Θ be its Liouville form. Consider a local diffeomorphism $f\colon U \longrightarrow X$, U being open in X. Since f is symplectic the form $f^*\Theta - \Theta|_U$ is forced to be closed. The *Calabi invariant* of f is by definition the cohomology class

$$\mathrm{Ca}(f) = [f^*\Theta - \Theta|_U] \in H^1(U; \mathbb{R}). \tag{3.14}$$

In general this definition depends on the choice of a Liouville form. But this is not so provided f is smoothly homotopic to zero, that is, there is a family $f_t\colon U \longrightarrow X$, $t \in [Q, 1]$, of smooth maps, such that f_0 is the inclusion of U into X and $f_1 = f$. Indeed, let Θ_i, $i = 1, 2$, be two Liouville forms of the same structure Ω. Then $\lambda = \Theta_1 - \Theta_2$ is closed. We have (see AI.66):

$$\frac{d}{dt} f_t^* \lambda = f_t^* \lrcorner\, d\lambda (= 0) + d(f_t^* \lrcorner\, \lambda),$$

$$f_0^* \lambda = \lambda|_U,$$

$$f^*\lambda - \lambda|_U = \int_0^1 d(f_t^* \lrcorner\, \lambda)\, dt = d \int_0^1 f_t^* \lrcorner\, \lambda\, dt,$$

i.e. $f^*\Theta_i - \Theta_i|_U$ differ by the differential of a function. We obtain that the Calabi invariant is correctly defined for local symplectic diffeomorphisms smoothly homotopic to zero, defined in a symplectic manifold with exact symplectic structure.

To make the geometric meaning of the Calabi invariant clear let us consider the following simple example.

Example 3.15 (Drift map). Let $D \subset \mathbb{R}^n$ be an open set. Fix a vector valued function $\psi\colon \mathbb{R}^1 \longrightarrow \mathbb{R}^n$ periodic with period 1 and define a local diffeomorphism of the symplectic annulus 2.5 $f\colon A_D \longrightarrow A_{\mathbb{R}^n}$ by the formula for the lift $\tilde{f}\colon \mathbb{R}^n \times D \longrightarrow \mathbb{R}^n \times \mathbb{R}^n$ of f:

$$\tilde{f}(\varphi, I) = (\varphi^i, I_i + \psi_i(\varphi^i),\ 1 \leqq i \leqq n).$$

We use the angle action coordinates of Example 2.5. Obviously f is symplectic and smoothly homotopic to zero. One can easily calculate its Calabi invariant

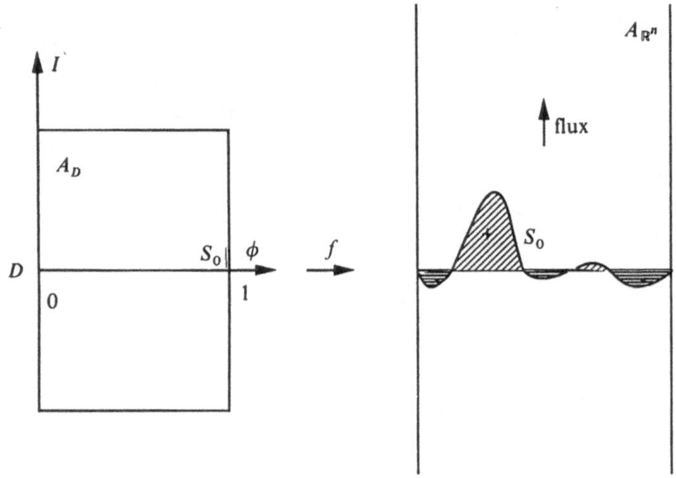

Fig. 6

using the definitions (3.11) and (2.5g):

$$\mathrm{Ca}(f) = \int_0^1 \psi(\varphi)\, d\varphi.$$

The case $n=1$ is shown in Fig. 6. The value of the Calabi invariant is equal to the (algebraic) area of the domain between S_0 and $f(S_0)$ where S_0 is a circle with $I=0$. $\mathrm{Ca}(f)$ measures the area flux in the I-direction under iterations of f. In the multi-dimensional case this flux is a vectorial quantity.

There is a connection between the definitions of the Calabi invariant for the cases of discrete and continuous time. To investigate such a connection we restrict ourselves the case $X = A_{\mathbb{R}^n}$, $U = A_D$, D being open, simply connected in $\mathbb{R}^n$ with compact closure $\bar{D}$. Consider a locally Hamiltonian vector field ξ defined on the set $\tilde{u}$ of the form $\tilde{u} = A_{\tilde{D}}$, $\tilde{D}$ being an open set containing $\bar{D}$. Then there is a well defined local flow $f^t: U \longrightarrow A_{\mathbb{R}^n}$, $|t| \leqq \delta$ with some $\delta > 0$ consisting of symplectic local diffeomorphisms smoothly homotopic to the identity.

Proposition 3.16. *Under the above conditions*

$$\mathrm{Ca}(f^t) = t\,\mathrm{Ca}(\xi|_U).$$

Proof. By AI.68

$$\frac{d}{dt}(f^{t*}\Theta) = f^{t*}(\xi \lrcorner\, d\Theta) + d f^{t*}(\xi \lrcorner\, \Theta).$$

Integrating over the interval $[0, t]$ we obtain

$$f^{t*}\Theta - \Theta|_U = -\int_0^t (f^t)^* \xi^b\, dt + d\int_0^t f^{t*}(\xi \lrcorner\, \Theta)\, dt.$$

Applying the map [ ] which assigns a closed form to its cohomology class and taking into account Remark 2.5f (which yields the commutativity of [ ]) and integrating over t, we have

(a)
$$\mathrm{Ca}(f^t) = -\int_0^t [(f^t)^*\xi^b]\,dt.$$

Again using the differentiation formula:

$$\frac{d}{dt}(f^{t*}\xi^b) = f^{t*} \lrcorner\, d\xi^b + d(f^{t*} \lrcorner\, \xi^b) = d(f^{t*} \lrcorner\, \xi^b),$$

after integration we obtain that

(b)
$$[f^{t*}\xi^b] = [\xi^b] = -\mathrm{Ca}(\xi).$$

The conclusion of the Proposition follows from (a) and (b). □

Note that this proof only makes use of the finiteness of the dimension of $H^1(U,\mathbb{R})$ and continuity of the map [ ].

Reversible Systems

In studying dynamical systems one frequently meets a special kind of symmetry, called reversibility. Let $f_t^t: X \longrightarrow X$ be a flow or cascade (topological, smooth, or symplectic). A *reversor* to f^t is a map $\sigma: X \longrightarrow X$ which (i) is idempotent: $\sigma\circ\sigma = \mathrm{id}$, (ii) reverses the direction of time:

(3.17)
$$\sigma\circ f^t\circ\sigma = f^{-t} \quad \forall t.$$

A system is called *reversible* if it has a reversor. If the system in question does not possess global dynamics, but there are local dynamics f^t defined on part of the phase space, then (3.17) can be reformulated in an obvious manner.

A curious fact is that smooth reversible systems share many properties with Hamiltonian ones. There is an advanced parallel theory (Arnol'd and Sevrjuk (1986)) which copies the main results of KAM theory, but we shall not touch on it here. In many cases both the symplectic structure and reversibility are present, the reversor as a rule being a smooth antisymplectic map: $\sigma^*\Omega = -\Omega$. In this book we shall meet such situations in the following examples of generalized billiards (§6, 14) and the standard map (§8).

Warning. All the systems considered in this book are reversible in the sense that the maps f^t describing dynamics have opposite ones f^{-t}. Do not mix up this general notion of reversibility with the special case just described.

§4. Symplectic Gluing

Here we return to developing symplectic geometry. We use symplectic dynamical systems as a tool for making the gluing operation. This is a well-known method in differential topology (see, for example, Hirsch (1976) or Fuks and Rokhlin (1984)) but now we must take into account the symplectic structure.

General Gluing

In the most general situation the symplectic gluing problem may be stated as follows. Let X be a symplectic manifold and C_i, $i=1,2$, subsets of the boundary ∂X of X such that (1) each C_i consists of whole connected components of ∂X, (2) C_1 and C_2 do not intersect. Consider a diffeomorphism $g: C_1 \longrightarrow C_2$. We shall try to glue X along the subsets C_i by means of g in order to again obtain a symplectic manifold. Form the *quotient topological space X_g glued by g*. The *set X_g* is obtained by identifying each point $x \in C_1$ with its image $g(x) \in C_2$. Denote the projection map which assigns to $x \notin C_1 \cup C_2$ the point x itself, to $x \in C_1$ the class $\{x, g(x)\}$, and to $x \in C_2$ the class $\{g^{-1}(x), x\}$ by $p: x \longrightarrow x_g$. The *topological structure* on X_g is defined by the natural requirement that the projection p to should continuous and for the topology to be the strongest among those satisfying the first requirement. The set $U \subset X_g$ is open in this topology if and only if $p^{-1}(U)$ is open in X. Obviously X_g is a topological manifold. The problem is to supply X_g with smooth and then with symplectic structures so that the map p becomes symplectic. Note that the symplectic structure on X_g is uniquely defined if it exists provided the smooth structure is fixed. If this problem has positive solution, i.e. X_g admits a smooth structure such that there exists a (necessarily unique) symplectic form Ω_g on X_g with the property $p^*\Omega_g = \Omega$, Ω being the symplectic structure of X, then the map g is called a *glueing map*, and the symplectic space (X_g, Ω_g) the *space glued by g*.

We may also consider another setting of this problem by considering two (or more) symplectic spaces X_i, $i=1,2$, with specific sets C_i of ∂X_i consisting of whole components and given a diffeomorphism $g: C_1 \longrightarrow C_2$, and trying to supply

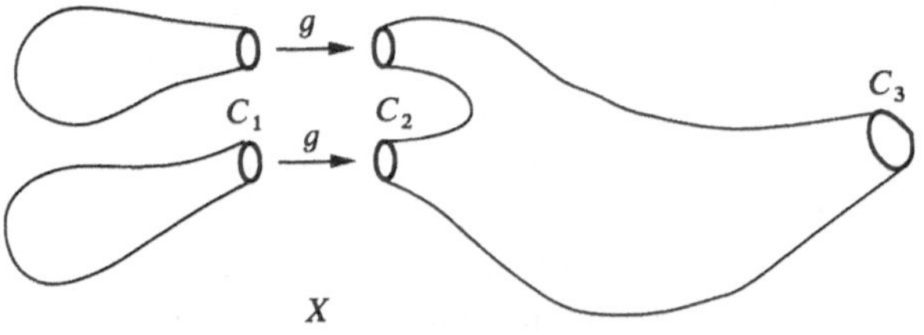

Fig. 7. Before gluing. The component C_3 of ∂X does not take part in gluing

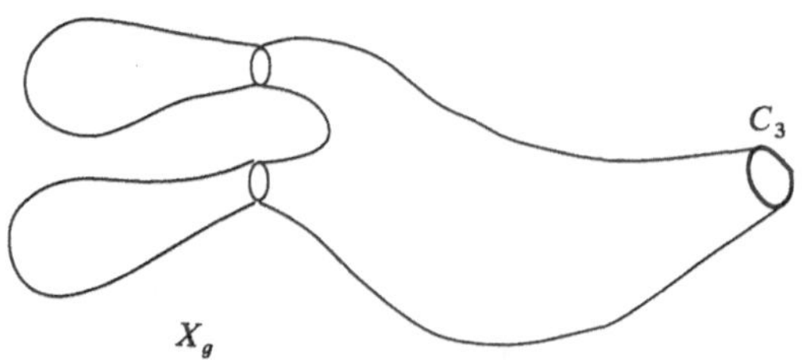

Fig. 8. After gluing

$X_1 \bigcup_g X_2$ (the topological manifold glued by g) with the appropriate symplectic structure. This problem reduces to the preceding one if one substitutes $X = X_1 \cup X_2$.

What conditions must a map $g: C_1 \longrightarrow C_2$ satisfy to be a glueing map? Obviously it must preserve the restrictions on the sets C_i of the symplectic structure of X. But this property by itself is insufficient as the following example shows.

Example 4.1. Let X_i, $i = 1, 2$, be two identical copies of the unit ball $B \subset \mathbb{R}^4$ supplied with the inherited symplectic structure $\Omega_{st}|_B$, and let $g: \partial X_1 \longrightarrow \partial X_2$ be an arbitrary diffeomorphism (the identity map for example). Then $X_1 \bigcup_g X_2$ cannot be supplied with a symplectic structure having the desired properties, i.e. the gluing problem has no solution in this case. Indeed, the glued space $X_1 \bigcup_g X_2$ is homeomorphism to the four-dimensional sphere S^4 which cannot possess any symplectic structure due to the nullity of its two-dimensional cohomology space $H^2(S^4; \mathbb{R})$. (I am obliged to Ya.M. Eliaschberg who directed me to this example as well as to Theorem 4.2 and also Example 4.11).

The following theorem completely solves the problem of gluing as stated above.

Theorem 4.2 (Eliaschberg). *A diffeomorphism $g: C_1 \longrightarrow C_2$ between two nonintersecting sections of the boundary ∂X of a symplectic manifold X, C_i consisting of whole components, is a gluing map if and only if* (1) *g preserves the restrictions to C_i of the symplectic structure of X,* (2) *g reverses the orientations induced by that of X onto C_i.*

We omit the proof of this theorem since further on we only deal with the special case of gluing when the glued parts of boundaries are equipped with locally Hamiltonian vector fields, these fields being transversal to those parts. Such a type of gluing is called *locally Hamiltonian gluing.* The main tool for carrying out this gluing is the *symplectic collar.*

The Symplectic Collar

Before stating the definition of a symplectic collar we small introduce some preliminary notions. Let M be a smooth odd-dimensional manifold without boundary. *Symplectic collar data* on M is a pair (ω, θ) where

(i) ω is a closed 2-form on M of maximal rank, that is, its zero-subspace $Z_x = \{v \in T_x M \colon \omega(v, w) = 0 \ \forall w \in T_x M\}$ is one-dimensional for each $x \in M$;

(ii) θ is a closed 1-form on M which is non-zero on zero-subspaces of ω, that is, $\theta(v) \neq 0$ if $0 \neq v \in Z_x \ \forall x \in M$.

Note that a 2-form on an odd-dimensional manifold cannot be nonsingular in the sense of §1 since the nonsingularity property requires a manifold to have even dimension. Therefore, (i) induces the maximal possible nondegeneracy of ω.

Construction 4.3 (Model collars). Given an odd dimensional manifold M without boundary and symplectic collar data (ω, θ) on M, consider the manifold $\mathbb{R} \times M$

and supply it with the 2-form

(4.3a) $$\Omega_{co} = p_1^* dt \wedge p_2^* \theta + p_2^* \omega.$$

Here p_i, $i = 1, 2$, are the projections of $\mathbb{R} \times M$ onto the first and the second factors respectively, dt is the standard 1-form on $\mathbb{R}$, being the differential of the identity map $\mathbb{R} \longrightarrow \mathbb{R}$. Consider the vector field $\partial/\partial t = p_1^! d/dt$ on $\mathbb{R} \times M$ where d/dt is the standard vector field on $\mathbb{R}$ (see AI.46, 47). The value of the field $\partial/\partial t$ at a point $(t, x) \in \mathbb{R} \times M$ generates the first summand in the decomposition

(4.36b) $$T_{(t,x)}\mathbb{R} \times M = T_t\mathbb{R} \oplus T_x M.$$

The second summand being decomposed in turn as

(4.3c) $$T_x M = \ker \theta_x \oplus Z_x$$

as follows from (i) and (ii). Immediate calculation yields the following formulae for the values of (4.3a):

$$\Omega_{co}\left(\frac{\partial}{\partial t}, w\right) = \theta(w), \quad w \in T_x M,$$

(4.3d) $$\Omega_{co}(v, w) = \omega(v, w), \quad v, \quad w \in T_x M.$$

Conversely (4.3d) implies (4.3a).

Proposition 4.3e. *Ω_{co} is symplectic.*

Proof. One needs to check that Ω_{co} is closed and nonsingular (see above text). The former is obvious because the differential commutes with p_i^* and the forms θ and ω are closed. The rule of differentiating $\wedge$ (see AI.58 (iii)) must be applied. To see the nonsingularity, take a vector $v \in T_{(t,x)}\mathbb{R} \times M$, $(t, x) \in \mathbb{R} \times M$, and suppose that

(+) $$\Omega_{co}(v, w) = 0 \quad \text{for all } w \in T_{(t,x)}\mathbb{R} \times M.$$

Decompose v and w in accordance with (4.3b):

$$v = v_1 \frac{\partial}{\partial t} + v_2, \quad w = w_1 \frac{\partial}{\partial t} + w_2, \quad v_1, w_1 \in \mathbb{R}, \; v_2, w_2 \in T_x M.$$

The formulae (4.3d) yield

(4.3f) $$\Omega_{co}(v, w) = v_1 \theta(w_2) - w_1 \theta(v_2) + \omega(v_2, w_2).$$

Taking $w_1 = 0$ and $0 \neq w_2 \in Z_x$, one obtains by virtue of $\theta(w_2) \neq 0$ that $(+) \Rightarrow v_1 = 0$. Taking $w_1 = 1$ and $w_2 = 0$, one obtains that $(+) \Rightarrow v_2 \in \ker \theta_x$. So, $(+)$ becomes $\omega(v_2, w_2) = 0$, and we have $v_2 \in Z_x$. Finally (4.3c) yields that $v_2 = 0$. □

The symplectic manifold $(\mathbb{R} \times M, \Omega_{co})$ is called the *model double collar* of M with data (ω, θ). The symplectic submanifolds $\mathbb{R}_+ \times M$ and $\mathbb{R}_- \times M$ of $\mathbb{R} \times M$ are called the *model right* and model *left collars* of M with data (ω, θ) respectively. Here $\mathbb{R}_+ = [0, +\infty[$, $\mathbb{R}_- =]-\infty, 0]$. The modal collars are equipped with the vector field $\partial/\partial t$ defined above (we use the same notation $\partial/\partial t$ for the above defined vector field on $\mathbb{R} \times M$ and for its restrictions to the right and left collars). The following equality will be useful in later considerations.

$$(p_2^*\theta)^{\#(\Omega_{co})} = \frac{\partial}{\partial t}. \tag{4.3g}$$

Proof. Take $(t,x) \in \mathbb{R} \times M$ and write in accordance with (4.3b) $(p_2^*\theta)^{\#}|_{(t,x)} = \xi = \xi_1 \,\partial/\partial t + \xi_2$, $\xi_1 \in \mathbb{R}, \xi_2 \in T_x M, \# = \#(\Omega_{co})$. Let $w = w_1 \,\partial/\partial t + w_2 \in T_{(t,x)}\mathbb{R} \times M, w_1 \in \mathbb{R}$, $w_2 \in T_x M$, be an arbitrary vector. The definition of the sharp map (see §1) yields

$$\Omega_{co}(\xi, w) = (p_2^*\theta)(w) = \theta(w_2). \tag{α}$$

On the other hand (4.3f) yields

$$\Omega_{co}(\xi, w) = \xi_1 \theta(w_2) - w_1 \theta(\xi_2) + \omega(\xi_2, w_2). \tag{β}$$

Comparing (α) and (β), we have the equality

$$\theta(w_2) = \xi_1 \theta(w_2) - w_1 \theta(\xi_2) + \omega(\xi_2, w_2), \tag{γ}$$

which holds identically with respect to w_1 and w_2.

Taking in $w_1 = 1$ and $w_2 = 0$ (γ): we have: (δ) $\theta(\xi_2) = 0$, again taking $w_2 \in Z_x$ and $\theta(w_2) = 1$ in (γ) we have that $\xi_1 = 1$ (the desired result). The equality (γ) becomes (ε) $\dot{\omega}(\xi_2, w_2) = 0$. (4.3c), ($\delta$) and ($\varepsilon$) yield that $\xi_2 = 0$. □

Let X be a symplectic manifold with non-empty boundary ∂X, and C be a subset of ∂X consisting of whole components. A *right (left) symplectic collar* in X is a pair (K, k) where K is an open neighbourhood of C in X with $\partial K = C$, and k is a symplectic diffeomorphism of K onto some neighbourhood of C in the right (left) model collar of C, such that $K|_C$ coincides with the inclusion map $x \longmapsto (0, x)$. We refer to K as the *collar domain* and to k as the *collar map*. The general name for right and left symplectic collars is a *symplectic collar*, or, simply, a *collar*, we agree to sometimes omit the word "symplectic" if there is no danger of confusion.

Note that a symplectic collar may be considered, by forgetting the symplectic structure, as the more usual collar in the sense of differential topology (the requirement for a collar domain to be diffeomorphic to $\mathbb{R}_+ \times C$, which one usually poses, is unnecessary).

It follows from the definition of a symplectic collar that the first member ω of the collar data (ω, θ) on C, which enters into the definition of the model collars, is equal to $\Omega|_{T\partial X}$, where Ω is the symplectic structure of X. The second member, θ, remains free. We call it the *boundary form* of a collar.

Remark 4.4. The right collar may be turned to become the left one (and vice versa) by taking the superposition of the collar map with the map $(t, x) \longmapsto (-t, x)$ of the model right collar onto the left one while simultaneously changing the collar boundary form θ to $-\theta$ in order to make the above map symplectic.

Gluing by Collars

Construction 4.5. Let (X,Ω) be a symplectic manifold with non-empty boundary ∂X, and $C_i, i=1,2$, be two non-empty non-intersecting subsets of ∂X consisting of whole components. *Gluing data* are:

$$\text{(D)}\begin{cases}\text{a left collar } (K_1,k_1) \text{ with } \partial K_1 = C_1,\\ \text{a right collar } (K_2,k_2) \text{ with } \partial K_2 = C_2, K_1 \cap K_2 = \varnothing,\\ \text{a diffeomorphism } g: C_1 \longrightarrow C_2, \text{ the } \textit{gluing map},\end{cases}$$

preserving the collar data, i.e. $g^*\omega_2 = \omega_1, g^*\theta_2 = \theta_1$, (ω_i, θ_i) being the collar data of (K_i, k_i).

Given gluing data, we firstly form the topological manifold X_g glued by g in the same way as at the beginning of the section. Our goal is to supply X_g with smooth and symplectic structures with the property that the canonical projection $p: X \longrightarrow X_g$ is symplectic. In order to do this we cover X_g by two open sets, $p(K_1 \cup K_2)$ and $p(X \backslash (C_1 \cup C_2))$, and supply these sets with appropriate smooth and symplectic structures so that they agree on this intersection.

The building of a symplectic structure on $p(K_1 \cup K_2)$. Define two model double collars $\mathbb{R} \times C_i, i=1,2$, by supplying $\mathbb{R} \times C_i$ with a symplectic structure of form (4.3a) with ω_i, θ_i standing for ω, θ. Consider the map $\tilde{g}: \mathbb{R} \times C_1 \longrightarrow \mathbb{R} \times C_2$ defined as $\tilde{g}(t,x) = (t, g(x))$. Since g preserves the collar data, it immediately follows that the map $\tilde{g}$ is a symplectic diffeomorphism. Define the maps $\tilde{k}_i: K_i \longrightarrow \mathbb{R} \times C_1$ (the first double collar!), $i=1,2$, by the formulas $\tilde{k}_1 = j_1 \circ k_1$, where $j_1: k_1(K_1) \hookrightarrow \mathbb{R} \times C_1$ is the inclusion, and $\tilde{k}_2 = \tilde{g}^{-1} \circ k_2$. Both $\tilde{k}_i, i=1,2$, are symplectic local diffeomorphisms. There is the uniquely defined map

$$(+) \qquad \tilde{k}_1 \bigcup_g \tilde{k}_2 : p(K_1 \cup K_2) \longrightarrow \mathbb{R} \times C_1$$

whose restrictions on $p(K_i)$ coincide with $\tilde{k}_i \circ (p|_{K_i})^{-1}$ (note that each $p|_{K_i}: K_i \longrightarrow p(K_i)$ is a homeomorphism). Indeed, the maps $\tilde{k}_i$ agree with gluing being restricted to glued

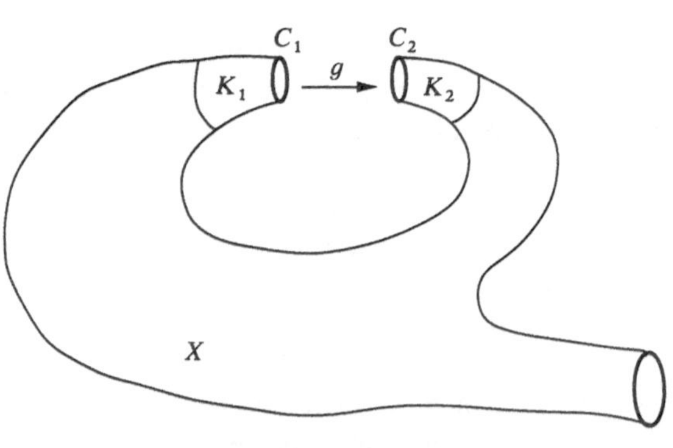

Fig. 9. Before symplectic gluing. The components C_1, C_2 are ready for gluing: they are supplied with symplectic collars K_1, K_2

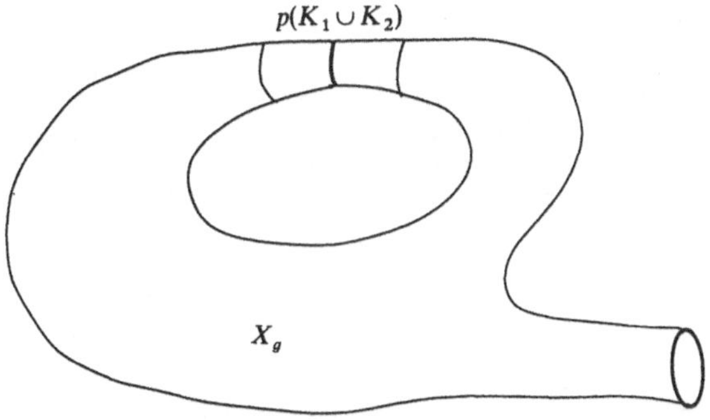

Fig. 10. After symplectic gluing

parts, C_i. Hence the maps $\tilde{k}_i \circ (p|_{K_i})^{-1}$ agree on $p(C_1) = p(C_2)$. The map (+) is a homeomorphism onto its image, the latter being an open neighbourhood of $\{0\} \times C_1$. We transfer the smooth and symplectic structures from $\mathbb{R} \times C_1$ to $p(K_1 \cup K_2)$ via the map (+). Since $p|_{K_1} = (\tilde{k}_1 \bigcup_g \tilde{k}_2)^{-1} \circ \tilde{k}_1, p|_{K_2} = (\tilde{k}_1 \bigcup_g \tilde{k}_2)^{-1} \circ \tilde{k}_2 \circ \tilde{g}$ (all maps are symplectic!), and K_1 and K_2 are nonintersecting open sets, $p|_{K_1 \cup K_2}$ is symplectic.

The symplectic structure on $p(X \backslash (C_1 \cup C_2))$ is transferred from X via p. Such a definition is possible because the restriction of p on $X \backslash (C_1 \cup C_2)$ is a homeomorphism onto $p(X \backslash (C_1 \cup C_2))$.

The agreement onto the intersection is due to the symplecticity of p restricted to the intersection $(X \backslash (C_1 \cup C_2)) \cap (K_1 \cup K_2)$, the image being endowed with either of the structures. Note that the said restriction is again a homeomorphism onto its image.

Now the symplectic atlas for X_g may be obtained as the union of those of $p(X \backslash (C_1 \cup C_2))$ and $p(K_1 \cup K_2)$. We call X_g with the obtained symplectic structure Ω_g the *symplectic manifold glued by gluing data* (D).

Existence of a Collar

The above construction shows us the use of an object thing as a symplectic collar. How does one obtain it? To consider this question, let us introduce some additional notions. The *form of a collar* (K, k), or the *collar form*, is the 1-form λ defined on the set K by the formula

$$\lambda = K^* p_2^* \theta, \tag{4.6}$$

θ being the collar boundary form. The vector field of a collar (K, k), or the collar field, is given by

$$\xi = \lambda^{\#(\Omega)} = (Tk^{-1}) \circ \frac{\partial}{\partial t} \circ k, \tag{4.7}$$

Ω being the symplectic structure. The equality of the two expressions for ξ is a consequence of (4.3g) and (4.6). These are both *closed* forms and *locally Hamiltonian*

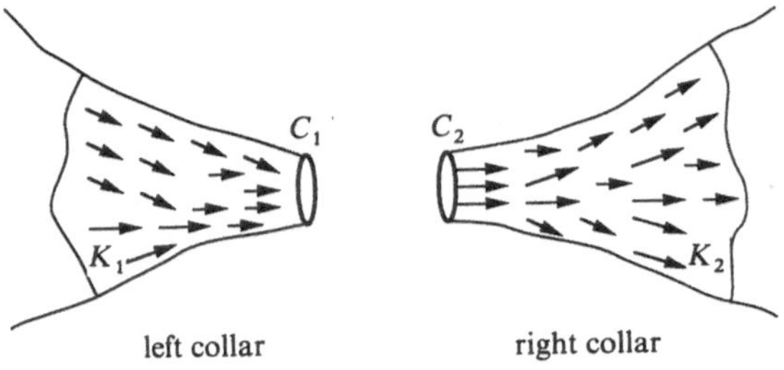

Fig. 11

vector fields. Both are also everywhere nonzero. The vector field ξ is transversal to the boundary, ∂K. The latter assertion follows from (4.3b) and (4.7). The formula (4.7) also enables us to clarify the geometric maning of the difference between right and left collars in terms of their fields as shown in the following picture: i.e. the field of the left collar *looks out of* a manifold at points of the boundary, that of the right collar looks *into* the manifold.

It is an immediate consequence of Construction 4.5 that

4.8. In the gluing operation, collar forms λ_i, and collar vector fields $\xi_i, i = 1, 2$, become the restrictions to K_i of the smooth closed form λ and the smooth locally Hamiltonian vector field ξ defined on $p(K_1 K_2)$. We call these form and field the *glued* ones.

The collar vector field, or equivalently the collar form, is the most significant characteristic of a collar. As a matter of fact, the collar is uniquely determined by its vector field. The following theorem provides us with a collar if we have a suitable vector field. In the following X denotes a symplectic manifold with non-empty boundary $\partial X, C \subset \partial X$ a subset consisting of whole components.

Theorem 4.9. *Given a locally Hamiltonian vector field ξ defined in some neighbourhood of C in X, transversal to C, and looking into X at points of C, there is the right symplectic collar (K, k) in X with $\partial K = C$ and $\xi|_K$ being its vector field. A collar is uniquely restored by its vector field.*

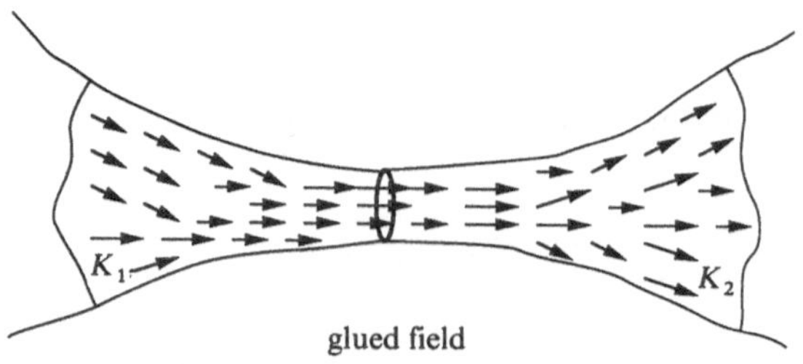

Fig. 12

Proof. For each point $x \in C$ there is a uniquely defined solution $y(t)$, $0 \leqslant t < \tau(x)$, $\tau(x) > 0$, to the initial value problem

$$\text{(a)} \qquad \frac{dy}{dt} = \xi \circ y, \quad y(0) = x.$$

The set $V = \{(t, x) \in \mathbb{R}_+ \times C : 0 \leqslant t < \tau(x)\}$ is open in $\mathbb{R}_+ \times C$, and the map $f: V \longrightarrow X$ defined by $(t, x) \longmapsto y(t)$ is smooth. These facts are obvious consequences of the theorems of the theory of ordinary differential equations (see AI.67). We clim that

(A) f is a local diffeomorphism, when restricted to some smaller neighbourhood $V_1 \subset V$ of C in $\mathbb{R}_+ \times C$,

(B) $f|_{V_1}$ is symplectic provided $\mathbb{R}_+ \times C$ is supplied with the symplectic structure (4.3a) where M is changed to C, $\omega = \Omega|_{TC}$, $\theta = \xi^\flat|_{TC}$, Ω being the symplectic structure of X, $\flat = \flat(\Omega)$.

If this is so, then $(f(V_1), (f|_{V_1})^{-1})$ is the required collar.

In order to prove (A) we shall calculate $T_{(0,x)}f$, $x \in C$. It follows from the definition of f that

$$\text{(b)} \qquad Tf\left(\frac{\partial}{\partial t}\right) = \xi \circ f,$$

and that $f|_{\{0\} \times C}$ coincides with the inclusion map $C \hookrightarrow X$. Therefore $T_{(0,x)}f w = w$ for all $w \in T_x C$. Since both $\partial/\partial t$ and ξ are transversal to C, we conclude that $T_{(0,x)}f$ is an isomorhpism. Statement (A) then follows as a consequence of the inverse function theorem.

When proving (B), we may carry out our considerations in a local manner. It is sufficient to prove that the restrictions of f onto sets of the form $[0, T[\times S \subset V_1$, $T > 0$, S being open in C, are symplectic. Consider the family of maps $g_t: S \longrightarrow X$, $t \in [0, T[$, defined by the formula $g_t = f \circ i_t$ where $i_t: S \longrightarrow \mathbb{R}_+ \times C$ is the inclusion $x \longmapsto (t, x)$. It follows from the definition of f that g_t satisfies the differential equation $dg_t/dt = \xi \circ g_t$ (take into account (b)). Applying the differentiation formula of AI.68 to the family g_t, one easily obtains

$$\frac{d}{dt} g_t^* \Omega = g_t(\xi \lrcorner\, d\Omega) + g_t^* d(\xi \lrcorner\, \Omega) = 0,$$

since Ω is closed, ξ is locally Hamiltonian and

$$\frac{d}{dt} g_t^* \lambda = g_t(\xi \lrcorner\, d\lambda) + g_t^* d(\xi \lrcorner\, \lambda) = 0,$$

where $\lambda = \xi^\flat$, $\flat = \flat(\Omega)$. To prove the last equality notice that λ is again closed, so the first term in the middle of the last formula is zero. We also have

$$\xi \lrcorner\, \lambda = \xi \lrcorner\, \xi^\flat = \Omega(\xi, \xi) = 0.$$

therefore, the forms $g_t^* \Omega$ and $g_t^* \lambda$ are independent of t. Taking into account that $g_0 = f \circ i_o: S \longrightarrow X$ is the inclusion map and using the definition of ω and θ given in

(B), we have

(c) $g_t^*\Omega = \omega|_S$ and $g_t^*\lambda = \theta|_S$ for all $t \in [0, T[$.

Take $(t, x) \in S \times [0, T[$ and calculate $f^*\Omega$ on the tangent space $T_{(t,x)}\mathbb{R}_+ \times C = T_t\mathbb{R}_+ \oplus T_xC$ as follows: for $v, w \in T_xC$ $f^*\Omega\left(\frac{\partial}{\partial t}, w\right) = \Omega\left(T_{(t,x)}f\frac{\partial}{\partial t}, T_{(t,x)}fw\right) =$ (use (b) and the definition of g) $= \Omega(\xi, T_xg_tw) = \lambda(T_xg_tw) = g_t^*\lambda(w) =$ (use (c)) $= \theta(w)$, $f^*\Omega(v, w) = \Omega(T_{(t,x)}fv, T_{(t,x)}fw) = \Omega(T_xg_tv, T_xg_tw) = g_t^*\Omega(v, w) =$ (use (c)) $= \omega(v, w)$.

The values obtained are the same as in (4.3d) (with C substituted for M), therefore, $f^*\Omega = \Omega_{co}$ as required.

The last assertion of the theorem is a consequence of the uniqueness theorem for the solution of a system of ordinary differential equations. Indeed, if (K, k) is a collar with collar vector field ξ then the map $t \longmapsto k^{-1}(t, x)$ is the solution to the initial value problem (a). □

Remark 4.10. Given a closed 1-form θ on the set $C \subset \partial X$ consisting of whole components of the boundary ∂X of a symplectic manifold X with the property $\theta|_{(TC)^\perp} \neq 0$ everywhere, there exists a symplectic collar with θ being its boundary form. One can easily deduce this from the collar theorem of differential topology AI.55 and Theorem 4.9.

Thus, the existence of a symplectic collar is equivalent to that of a form θ. The following example shows us that not every symplectic manifold with boundary admits a symplectic collar (although the usual collar in the sense of AI.55 always exists).

Example 4.11. Let $B \subset (\mathbb{R}^4, \Omega_{st})$ be the unit ball with inherited symplectic structure. The zero-foliation induced by Ω_{st} on its boundary $\partial B = S^3$ coincides with the famous Hopf fibration whose fibres are circles. Since S^3 is simply connected, the required form θ would be exact, i.e. it would be the differential of a function. The nonnullity of θ on $(T\partial B)^\perp$ means that this function has no critical points, being restricted to a fibre, which is impossible.

On the other hand, if such a form θ exists, there are many possibilities of extending it to a closed form λ onto a neighbourhood of C. Weinstein's theorem 2.18 yields that given a section v of T_CX transversal to TC, there exists a closed extension λ of θ with the property that $\lambda(v)$ is nowhere zero. Taking different v, we obtain different collars. This enables us to construct examples of symplectic gluing with the same θ and different resulting smooth structures, and consequently different symplectic ones.

As Example 4.11 shows, symplectic gluing is possible without any symplectic collar: $\mathbb{R}^4 \setminus B$ and B are symplectically glued to form $(\mathbb{R}^4, \Omega_{st})$.

Hamiltonian Gluing

Let us now return to gluing by collars. It is useful to single out the case of Hamiltonian vector field ξ in Theorem 4.9. We call a collar obtained by such a field a *Hamiltonian collar*.

4.12. Let (X,Ω) be a symplectic manifold, $C_i \subset \partial X, i=1,2$, two nonintersecting non-empty subsets consisting of whole components, and $H_i, i=1,2$, smooth functions defined in some nonintersecting neighbourhoods of C_i such that $(dH_i)^{\#}|_{C_i}$ are transversal to C_i, $(dH_1)^{\#}$ looking out of X, $(dH_2)^{\#}$ looking into X. Let $g: C_1 \longrightarrow C_2$ be a diffeomorphism with the properties that

(a) $\qquad g^*H_2|_{C_2} = H_1|_{C_1}$ and (b) $\qquad g^*\Omega_2|_{TC_2} = \Omega_1|_{TC_1}$.

Then Theorem 4.9 and Construction 4.3 yield the unique gluing of manifold X into the symplectic manifold X_g, and there is also a smooth function H defined in some neighbourhood of $p(C_1 \cup C_2)$ in X_g which coincides with $H_i \circ (p|_{K_i})^{-1}$ being restricted to $p(K_i)$, $i=1,2$. Here $p: X \longrightarrow X_g$ is the canonical projection onto the glued space. The spaces $K_i, i=1,2$, are the corresponding collar domains.

The gluing described in 4.12 is called *Hamiltonian gluing.*

§5. Cross-sections

In this section we establish links between symplectic dynamical systems with continuous time and those with discrete time. We describe the special constructions which in some cases enable us to pass from the study of Hamiltonian vector fields to that of (local) symplectic cascades of lower dimension and vice versa. This approach is most useful if one is interested in discovering the behaviour of a system in the neighbourhood of an invariant set. In the latter case we shall deal with an invariant set with complete dynamics (see § 1.3).

Cross-sections and Poincaré Maps

We start with some simple definitions which are concerned with the case of an arbitrary vector field. Let $A \subset X$ be a subset of a manifold X and ξ a smooth vector field on X. Fix $x \in A$ and let $s(t), t \in \Delta$, the maximal integral curve of ξ starting at x, i.e. $s(t), t \in \Delta$, is an integral curve of ξ (see AI.67), $0 \in \Delta$, $s(0) = x$, and Δ is as large as possible. A priori there may be three possibilities: (a) $x \in Z(\xi)$ (see Example 3.11), the set of stationary points of ξ, then $s(t) = x$ for all $t \in \Delta = \mathbb{R}$, (b) $x \notin Z(\xi)$ and $s(\Delta \cap]0,\infty[)$ does not intersect A (in particularly, x is of type (b) if $x \in A \cap \partial X$ and $\xi_x \neq 0$ looks out of X), (c) there is a $\tau > 0$ such that $s(\tau) \in A$ and $s(\tau') \notin A$ if $0 < \tau' < \tau$. We define the *return-time function* $\tau: A \longrightarrow \mathbb{R}_+ \cup \{+\infty\}$ of a set A with respect to a field ξ as follows:

$$\tau(x) = \begin{cases} 0 & \text{if } x \text{ is of type (a)} \\ +\infty & \text{if } x \text{ is of type (b)} \\ \tau & \text{if } x \text{ is of type (c), } \tau \text{ being defined in (c).} \end{cases}$$

Now let X be a symplectic manifold and ξ a Hamiltonian vector field on X with Hamiltonian function H. The *energy surface*, with value of energy E is, by definition, the set $H^{-1}(E) = \{x \in X: H(x) = E\}$. Consider a subset M of X with the following

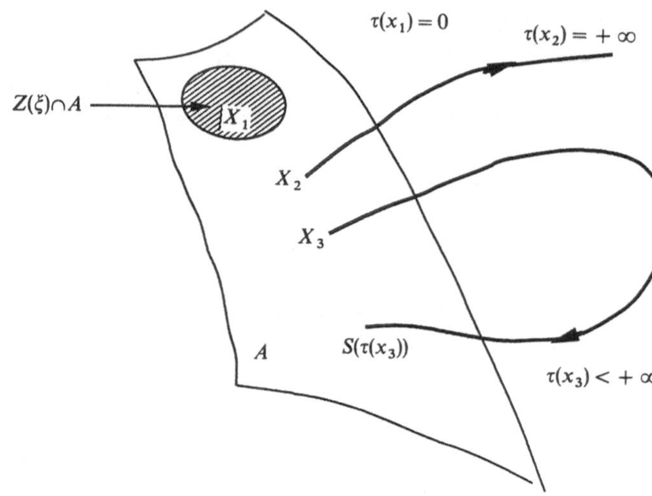

Fig. 13. The returning of a trajectory to the set A. The return-time function $\tau(x)$

properties:

(i) $M \subset H^{-1}(E)$ for some $E \in \mathbb{R}$.
(ii) M does not intersect $Z(\xi)$, the zero-set of ξ.
(iii) M is invariant with respect to ξ and has complete dynamics.

Note that the empty set $\varnothing$ satisfies conditions.

A symplectic submanifold $S \subset X$ of dimension $2n-2$ is called a *cross-section* for M satisfying (i)–(iii) if

(iv) $\partial S = \varnothing, S \cap \partial X = \varnothing, S \cap Z(\xi) = \varnothing$.
(v) $S \subset H^{-1}(E)$.
(vi) There is an open neighbourhood U of the set $K = S \cap M$ in S such that the return-time function τ of U is finite, positive, and smooth.
(vii) The field ξ is transversal to S in $H^{-1}(E)$. This means that $T_x H^{-1}(E) = T_x S \oplus L_x$, for all $x \in S$, where L_x is the one-dimensional subspace of $T_x X$ generated by ξ_x. Note that $H^{-1}(E)$ need not be a submanifold but $H^{-1}(E) \setminus Z(\xi)$ is, so $T_x H^{-1}(E)$ makers sense for $x \in S$.
(viii) Any point $x \in M$ lies in the positive half-orbit $0_y^+ = \{s(t): t \geqslant 0, s(t), T \in \Delta \in 0,$ being the maximal integral curve of ξ with $s(0) = y\}$ of some point $y \in K = M \cap S$.

Let ξ be a Hamiltonian vector field on X with Hamiltonian function H, let M be a set satisfying (i)–(iii), and let S be its cross-section. Then each point $x \in U$, where U is the set defined in (vi), has a neighbourhood W in X such that there is a local flow $f^t: W \longrightarrow X$ of ξ defined on the time interval $\tau(x) - \varepsilon < t < \tau(x) + \varepsilon$ for some $\varepsilon > 0, \tau$: $U \longrightarrow \mathbb{R}$ being the return-time function of set U. We use the same notation f^t for all such flows since they coincide on the intersections of their domains. We define the *Poincaré map* $g: U \longrightarrow S$ by the formula

(5.1) $$g(x) = f^{\tau(x)}(x), \quad x \in U.$$

It is obvious that the set $K = S \cap M$ is invariant under the map g.

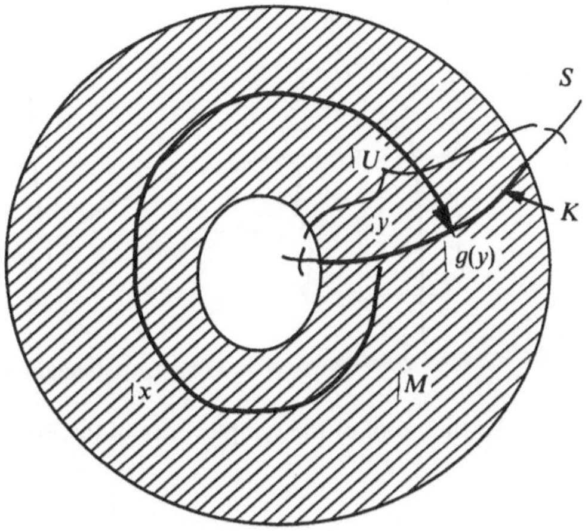

Fig. 14. To the definition of a cross-section

Proposition 5.2. *The Poincaré map is symplectic.*

Proof. Some complexity of this proof arises from the fact that the dependence on x enters twice into the expression (5.1) for g. Let $f^t: W \longrightarrow X$ be some local flows of ξ which we have considered when defining g. Denote $W' = W \cap S$. We are going to prove that $g|_{W'}$ preserves the symplectic structure. At first we consider the case when the symplectic structure Ω of X is exact, i.e. $\Omega = d\Theta$, Θ being a 1-form. We have by AI.68

(a) $$\frac{d}{dt}(f^{t*}\Theta) = d((\xi \lrcorner \Theta)\circ f^t) + f^{t*}\, dH.$$

Denote by j_E the inclusion map of the energy surface $H^{-1}(E)$ in X. Since $H^{-1}(E)$ is invariant under f^t,

(b) $$(j_E|_W)^* \circ f^{t*} = (f^t|_{\tilde{W}})^* \circ j_E^*,$$

where $\tilde{W} = W \cap H^{-1}(E)$, and

(c) $$j_E^*\, dH = 0.$$

Applying the operation j_E^* to (a) and using (b), (c), one obtains

(d) $$\frac{d}{dt}(f^{t*}\Theta_1 - \Theta_1) = dj_E^*((\xi \lrcorner \Theta)\circ f^t).$$

Here $\Theta_1 = j_E^*\Theta$. Our following considerations will be in $H^{-1}(E)$. We define some functions on the set $\tilde{W} = W \cap H^{-1}(E)$: $\tilde{\tau}(x)$ is the minimal positive t such that $f^t(x) \in S$ (if $x \in U$ $\tilde{\tau}(x) = \tau(x)$); $h(x) = \int_0^{\tilde{\tau}(x)} (\xi \lrcorner \Theta)\circ f^t(x)\, dt$, a real valued function; $k(x) = \int_0^{\tilde{\tau}(x)} d(\xi \lrcorner \Theta)\circ f^t(x)\, dt$, a 1-form; $\tilde{g}(x) = f^{\tilde{\tau}(x)}(x)$, a smooth map of $\tilde{W}$ into S (obviously $\tilde{g}|_{W'} = g|_{W'}$).

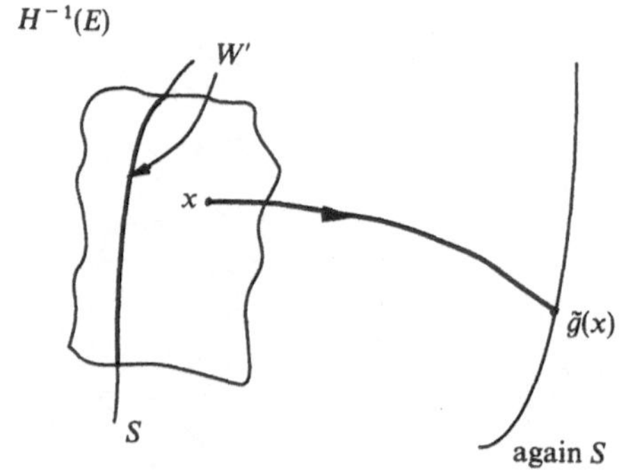

Fig. 15. To the proof of the symplectness of the Poincaré map

By usual formula for differentiating the integral, we have

(e) $$dh = ((\xi \lrcorner \Theta)\circ \tilde{g})\, d\tilde{\tau} + k.$$

Define also the linear map $\mathscr{F}$, which carries 1-forms from $H^{-1}(E)$ to those on $\tilde{W}$, by the formula

$$\mathscr{F}\alpha(x) = (f^{\tilde{\tau}(x)*}\alpha)(x), \alpha \in \mathscr{E}^1(H^{-1}(E)), x \in \tilde{W}.$$

For any 1-form α on $H^{-1}(E)$

(f) $$\tilde{g}^*\alpha = \mathscr{F}\alpha + ((\xi \lrcorner \alpha)\circ \tilde{g})\, d\tilde{\tau}.$$

Indeed, there is the following formula for differentiation of the compound map $\tilde{g}(x) = f^{\tilde{\tau}(x)}(x)$: if $x \in \tilde{W}$, $v \in T_x H^{-1}(E)$

$$T_x \tilde{g}(v) = T_x f^t(v)|_{t=\tilde{\tau}(x)} + d\tilde{\tau}(v)\cdot \xi \circ g(x).$$

Applying α to both sides of the latter formula, one obtains (f).

After these preparations we return to considering (d). Integrating (d) at each point $x \in \tilde{W}$ over $t \in [0, \tilde{\tau}(x)]$, we get

(g) $$\mathscr{F}\Theta_1 - \Theta_1 = k.$$

The formulae (e)–(g) yield

$$\tilde{g}^*\Theta_1 - \Theta_1 = dh.$$

Applying the d-operator, we obtain that $\tilde{g}$ preserves $j_E^*\Omega$. This includes that $\tilde{g}|_{W'} = g|_{W'}$ preserves $\Omega|_{TS}$.

Suppose that Ω is not exact. Then we may restrict our considerations to the tubular neighbourhood of the segment $\{f^t(x_0)\colon -\delta \leqq t \leqq \tau(x_0) + \delta\}, \delta > 0, x_0 \in S$, if $g(x_0) \neq x_0$, or to that of the circle $\{f^t(x_0); 0 \leqq t \leqq \tau(x_0)\}$ if $g(x_0) = x_0$. In the former case the neighbourhood is simply connected, so the symplectic form is exact when restricted to it. In the latter case we may pass to the universal covering manifold of the tubular neighbourhood of the circle. □

Poincaré Map of a Periodic Orbit. Classification of Periodic Orbits

We will apply the notions just introduced to classify periodic orbits of a symplectic dynamical system with continuous time.

Consider a periodic orbit M of (X, ξ) (see Example 3.11). As has already been mentioned, ξ is Hamiltonian when restricted to some tubular neighbourhood of M. Let H be its Hamiltonian. Since H is constant on trajectories, 3.5, there is as $E \in \mathbb{R}$ such that $M \subset H^{-1}(E)$. Fix an arbitrary $x_0 \in M$ and take a $(2n-2)$-dimensional symplectic submanifold S which (1) lies in $H^{-1}(E)$, (2) intersects M in only one point x_0, (3) is transversal to the field ξ in $H^{-1}(E)$. Such an S is a cross-section. In this case the set K reduces to the one-point set $\{x_0\}$. To prove the existence of S with the above properties let us take a symplectic chart which carries x_0 into the origin and the part of M contained in the domain of the chart onto the q^1-axis. This is possible due to 2.21. We have in chosen local coordinates

$$\dot{q}^j = \left.\frac{\partial H(q,p)}{\partial p_j}\right|_M = 0 \quad \text{if } j \geqq 2 \quad \text{and} \quad \dot{p}_j = -\left.\frac{\partial H(q,p)}{\partial q^j}\right|_M = 0, \quad 1 \leqslant j \leqslant n.$$

Hence the Taylor expansion of H in the neighbourhood of M is

$$H(q^1, q^2, \ldots, q^n, p_1, p_2, \ldots, p_n) = E + \frac{\partial H}{\partial p_1}(q^1, 0, \ldots) p_1 + \text{higher-order terms.}$$

Note that $dH(x) \neq 0$ if $x \in M$, otherwise x would be a fixed point. Hence $\partial H/\partial p_1(q^1, 0, \ldots) \neq 0$. From this it follows that the tangent space to the energy surface $H^{-1}(E)$ at x_0 is generated by vectors $\partial/\partial q^i, 1 \leqslant i \leqslant n, \partial/\partial p_j, 2 \leqslant j \leqslant n$. Take any $(2n-2)$-dimensional submanifold $\tilde{S}$ of $H^{-1}(E)$ containing x_0 with $T_{x_0}\tilde{S}$ generated by $\partial/\partial q^j, \partial/\partial p_j, 2 \leqslant j \leqslant n$. There is an open neighbourhood S of x_0 in $\tilde{S}$ which is a symplectic submanifold and has x_0 as the unique point of intersection with M. We have $T_{x_0}H^{-1}(E) = T_{x_0}M \oplus T_{x_0}S$, so the transversality condition is fulfilled at x_0. Take S sufficiently small to be transversal to ξ at all its points. Obviously, S is a desired cross-section for M. Thus the study of the behaviour of the system near a periodic trajectory is reduced to that of the behaviour of a local diffeomorphism $g: U \to S$ near a fixed point x_0.

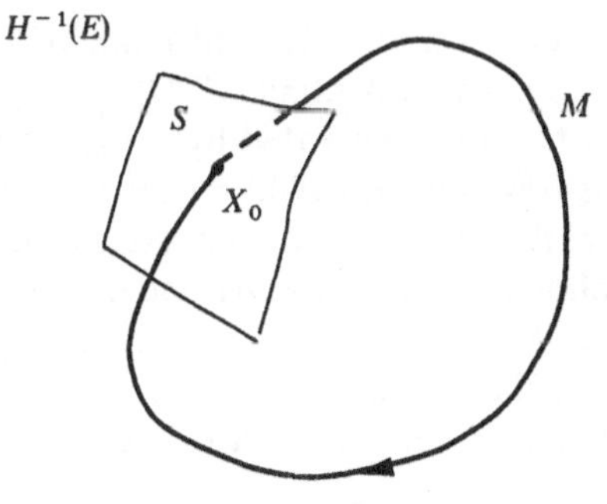

Fig. 16. Cross-section to a periodic orbit

Consider the tangent map $T_{x_0}g: T_{x_0}S \longrightarrow T_{x_0}S$. It belongs to $\mathrm{Sp}(T_{x_0}S)$. We say M is *simple elliptic* or *hyperbolic* whenever so is $T_{x_0}g$. This definition does not depend on the choice of g.

Remark 5.3. It is not difficult to prove, using AI.53, that, in the case of continuous time, periodic trajectories of the both above types form in fact a smooth one-parameter family, each member of that family lying in one energy level $H^{-1}(E)$, E ranging some interval.

Remark 5.4. Analogously, trajectories of above types (one parameter families for flows) persist under small perturbations of the system.

Suspended Hamiltonian Systems

There is an inverse construction to that of the previous subsection. Here we restrict ourselves to only considering the case of diffeomorphisms defined globally. The starting data are

(i) a symplectic manifold (S, Ω) of dimension $2n-2, n \geqslant 2$, S having no boundary;
(ii) a symplectic diffeomorphism $g: S \longrightarrow S$;
(iii) a smooth positive function τ defined on $S \times \Delta$ where $\Delta \subset \mathbb{R}$ is an open interval containing zero. In addition we suppose that
(iiia) τ is separated from zero by a positive constant;
(iiib) the following Cauchy problem for the first order differential equation

$$\frac{d}{dp} g_p = -Tg_p(d_x\tau_p)^{\#}, \quad g_p|_{p=0} = g \tag{5.5}$$

has a solution defined on the whole interval Δ of the variable p, each member of the family $\{g_p, p \in \Delta\}$ being a diffeomorphism of S onto itself.

In (5.5) we used the notation d_x for the differential with respect to the variable $x \in S$. $\# = \#(\Omega)$, and $\tau_p, p \in \Delta$, represents the functions obtained from $\tau(x, p)$ by freezing the variable p.

Our goal is to construct a complete Hamiltonian vector field ξ on a symplectic manifold X of dimension $2n$, which contains (S, Ω) as a symplectic submanifold, so that S is cross-section for the flow generated by ξ, $\tau|_S$ the return-time function and g the Poincaré map. This flow will be called the *suspended Hamiltonian flow*.

A remarkable fact is that to fulfil this construction one needs the extended return-time function τ defined on a neighbourhood of the domain S of g in X instead of only knowing it on S. As a matter of fact the extended return-time function together with the diffeomorphism g define the suspended Hamiltonian flow uniquely (up to isomorphism), and the former as may be chosen in objects arbitrary manner.

Proposition 5.6. *Each g_p defined by* (5.5) *is symplectic.*

Proof. Consider the family of two-forms $\hat{\Omega}_p = g_p^*\Omega$, $p \in \Delta$. Applying the differentiation formula AI.66 yields

$$(\alpha) \qquad \frac{d}{dp}\hat{\Omega}_p = d_x(g_p^* \lrcorner\, \Omega).$$

Using the definition of $* \lrcorner$-operation AI.66b,c, and equation (5.5), one calculates the expression under the sign of the d_x-operation in (α) on a vector $v \in T_xS$ as follows

$$\begin{aligned}(g_p^* \lrcorner\, \Omega)(v) &= \Omega\left(\frac{dg_p}{dp}, Tg_pv\right) \\ &= -\Omega(Tg_p(d_x\tau_p)^{\#}, Tg_pv) \\ &= -\hat{\Omega}((d_x\tau_p)^{\#}, v) = -((d_x\tau_p)^{\#} \lrcorner\, \hat{\Omega}_p)(v).\end{aligned}$$

Inserting this in (α) and taking into account that g is symplectic and the initial condition of (5.5), one obtains that $\{\hat{\Omega}_p\}$ is a solution to the following Cauchy problem

$$(\beta) \qquad \frac{d}{dp}\hat{\Omega}_p = -d_x((d_x\tau_p)^{\#} \lrcorner\, \hat{\Omega}_p), \quad \hat{\Omega}_p|_{p=0} = \Omega.$$

Since (β) has only one solution, it is sufficient to notice that the constant family $\hat{\Omega}_p = \Omega$, $p \in \Delta$, satisfies (β). □

Let $(S \times \mathbb{R}^2, \tilde{\Omega})$ be the product of symplectic manifolds (S, Ω) and $(\mathbb{R}^2, \Omega_{st})$. The coordinates in $\mathbb{R}^2$ will be denoted by (q, p). Define the subset $Z \subset S \times \mathbb{R}^2$ as follows: $(x, q, p) \in Z$ if and only if $p \in \Delta$ and $0 \leqq q \leqq \tau(x, p)$. Obviously Z is a submanifold of $S \times \mathbb{R}^2$ of the same dimension $2n$. Its boundary ∂Z may be divided into two sets C_i, $i = 0, 1$, consisting of the whole components: $C_0 = S \times \{0\} \times \Delta$ and $C_1 = \{(x, q, p) \in Z \colon q = \tau(x, p)\}$. Define the map $F \colon C_1 \longrightarrow C_0$ by the formula

$$(5.7) \qquad F(x, q, p) = (g_p(x), 0, p).$$

Obviously, F is a diffeomorphism.

Proposition 5.8. *F preserves the restrictions $\tilde{\Omega}|_{TC_i}$, $i = 0, 1$, of the symplectic structure.*

Proof. At first, we present an explicit expression for the product-structure $\tilde{\Omega}$. Any vector $w \in T_{(x,q,p)}(S \times \mathbb{R}^2)$ can be expressed as

$$(a) \qquad w = v + \alpha\frac{\partial}{\partial q} + \beta\frac{\partial}{\partial p},$$

where $v \in T_xS$, $\alpha, \beta \in \mathbb{R}$, $\partial/\partial q$, $\partial/\partial p$ are obtained by applying the map $p_2^!$ to the coordinate vector fields in $\mathbb{R}^2$, $p_2 \colon S \times \mathbb{R}^2 \longrightarrow \mathbb{R}^2$ being the second projection (see AI.46, 47). The definitions of the direct sum of symplectic vector spaces (§1.1 Symplectic Subspaces) and the product of symplectic manifolds (§1.2 Symplectic

Structure) yield the formula

(b) $$\tilde{\Omega}(w_1, w_2) = \Omega(v_1, v_2) + \alpha_1\beta_2 - \alpha_2\beta_1,$$

where v_i, α_i, β_i are the components of w_i in the expansion of the form (a). We must now prove the equality

(c) $$\tilde{\Omega}(TFw_1, TFw_2) = \tilde{\Omega}(w_1, w_2)$$

for each pair of vectors $w_i \in T_{(x,q,p)}C_1 \subset T_{(x,q,p)}S \times \mathbb{R}^2$, $i = 1, 2$, $(x, q, p) \in C_1$. The equation $q = \tau(x, p)$ of C_1 implies that $dq(w_i) = d_x\tau(w_i) + \partial\tau/\partial p dp(w_i)$. Therefore the components v_i, α_i, β_i of the decomposition (a) for w_i are linked by the relations

$$\alpha_i = d_x\tau_p(v_i) + \frac{\partial\tau}{\partial p}\beta_i, \quad i = 1, 2.$$

The right-hand side of (c) reads

(d) $$\begin{aligned}\tilde{\Omega}(w_1, w_2) &= \Omega(v_1, v_2) + d_x\tau_p(v_1)\beta_2 - d_x\tau_p(v_2)\beta_1\\ &= \Omega(v_1, v_2) - \beta_2\Omega(v_1, (d_x\tau_p)^{\#}) - \beta_1\Omega((d_x\tau_p)^{\#}, v_2).\end{aligned}$$

The components containing $\partial\tau/\partial p$ cancel. To calculate $T_{(x,q,p)}Fw_i$ let us take a local function φ defined on an open neighbourhood of $F(x, q, p)$ in C_0. Since $q = 0$ on C_0, φ may be considered as a function of variables $(x', p') \in S \times \Delta$. We have

$$\begin{aligned}(T_{(x,q,p)}F(w_i))\varphi &= w_i(\varphi \circ F)\\ &= w_i\varphi(g_p(x), p)\\ &= \left(v_i + \alpha_i\frac{\partial}{\partial q} + \beta_i\frac{\partial}{\partial p}\right)\varphi(g_p(x), p)\\ &= T_xg_p(v_i)\varphi + \beta_i\left(d_x\varphi\left(\frac{dg_p}{dp}\right) + \left.\frac{\partial\varphi}{\partial p}\right|_{(g_p(x),p)}\right).\end{aligned}$$

The last calculation shows that

$$T_{(x,q,p)}F(w_i) = T_xg_p(v_i) + \beta_i\left(\frac{dg_p}{dp} + \frac{\partial}{\partial p}\right),$$

and we have the following expression for the left-hand side of (c) (use (b) and take into account that the summands containing $\partial/\partial q$ vanish, i.e. $\alpha_i = 0$, $i = 1, 2$):

(e) $$\begin{aligned}\tilde{\Omega}(TFw_1, TFw_2) &= \Omega(Tg_p(v_1), Tg_p(v_2))\\ &\quad + \beta_1\Omega\left(\frac{dg_p}{dp}, Tg_p(v_2)\right) + \beta_2\Omega\left(Tg_p(v_1), \frac{dg_p}{dp}\right)\\ &= \Omega(v_1, v_2) + \beta_1\Omega\left((Tg_p)^{-1}\frac{dg_p}{dp}, v_2\right)\\ &\quad + \beta_2\Omega\left(v_1, (Tg_p)^{-1}\frac{dg_p}{dp}\right).\end{aligned}$$

Here we used the fact that g_p is symplectic. The right-hand sides of (d) and (e) coincide due to the equation (5.5). □

We have obtained the following objects: (1) a symplectic manifold $(Z, \tilde{\Omega}_Z)$, its boundary ∂Z being divided into two subsets C_0 and C_1 consisting of whole components; (2) a Hamiltonian vector field $\partial/\partial q|_Z$ on Z with Hamiltonian function p; (3) a diffeomorphism $F: C_1 \longrightarrow C_0$, which preserves the restrictions of the symplectic structure on C_i, $i = 1, 2$, and the restrictions of the Hamiltonian, $p|_{C_i}$, $i = 1, 2$ (see formula (5.7)). Applying 4.12, we glue Z in a Hamiltonian manner by using the said data. We obtain the glued symplectic manifold $X = Z_F$ and the glued Hamiltonian vector field ξ on X, the image of $\partial/\partial q|_Z$ under the canonical map $Z \longrightarrow X$. The field turns out to be complete: trajectories of its flow obtained by gluing together the segments $x = \text{const}$, $p = \text{const}$ in Z under the aforementioned canonical map, the condition (iiia) ensuring the possibility of infinite continuation of the trajectory with respect to the time variable. Let us embed (S, Ω) in X as the image of $S \times \{(0,0)\}$ under the canonical map. Obviously S becomes a cross-section for the flow generated by ξ, the Poincare′ map coinciding with our initial g.

§6. Generalized Geodesic Flows

In this section we introduce the large class of Hamiltonian systems known as generalized geodesic flows. This notion is important due to numerous applications of it to geometry, physics and classical mechanics. It enables us to treat a large number of different systems in a similar manner. Particular cases of these systems are given by billiards, geodesic flows and mechanics systems. We shall also meet generalized geodesic flows in part II of this book when considering the Laplace–Beltrami–Schrödinger operator. It is necessary to note that the term "flow" here is used in a somewhat conventional sense since the arising vector fields need not be complete (and often are not).

General Construction

The starting data needed for the construction of the system in question are a *Riemannian manifold* (Q, g) and a smooth function $V: Q \longrightarrow \mathbb{R}$ called a *potential*. A Riemannian manifold (Q, g) consists of a manifold Q and a Riemannian metric g on Q. The latter is a smooth section $g: Q \longrightarrow S^2 T^* Q$ of the bundle of symmetric bilinear forms on Q, the values of g being positive definite forms. In the local coordinates $(q^1, \ldots, q^n)$ of some smooth chart $\alpha: U \longrightarrow \mathbb{R}^n_{(+)}$ in Q, $n = \dim Q$, this section has the expression

(6.1) $$g|_U = \sum_{i,k=1}^{n} g_{ik}\, dq^i \odot dq^k.$$

Here the symbol $\odot$ denotes the symmetrized product

$$dq^i \odot dq^k = \tfrac{1}{2}(dq^i \otimes dq^k + dq^k \otimes dq^i)$$

(see the definition of $\otimes$-product of forms on p. 32). The coefficients in (6.1) are smooth real functions $g_{ik}: U \longrightarrow \mathbb{R}$, symmetric with respect to indices the i, k: $g_{ik} = g_{ki}$. They are referred to as the *components of the metric tensor* in a chart (U, α).

The Riemannian metric g defines, for each $q \in Q$, the isomorphism

(6.2) $$\#(g): T_q^*Q \longrightarrow T_qQ$$

between cotangent and tangent spaces (the "sharp" map of g, see §1) and the *inner product* $\langle \cdot, \cdot \rangle$ in T_q^*Q

(6.3) $$\langle p, h \rangle = g_q(p^{\#(g)}, h^{\#(g)}), \quad p, h \in T_q^*Q.$$

The norm of a covector p, corresponding to this inner product, will be denoted by $|p|$.

Our preliminary notions are the *pre-Hamiltonian* which is the function $\tilde{H}: T^*Q \longrightarrow \mathbb{R}$ whose value at a point $p \in T_q^*Q$, $q \in Q$, is given by the formula

(6.4) $$\tilde{H}(p) = \tfrac{1}{2}|p|^2 + V(q)$$

and the *pre-Hamiltonian vector field*

(6.5) $$\tilde{\xi} = (d\tilde{H})^{\#(\Omega)}.$$

Here $\Omega = \Omega_Q$, the canonical symplectic 2-form on T^*Q (see §2).

If Q has no boundary, the above notions are those we need. The perfix "pre" and the sign $\sim$ must be omitted in this case. On the contrary, if ∂Q is non-empty, complications arise, the preliminary objects $\tilde{H}$ and $\tilde{\xi}$ are insufficients for our purposes, and we need an additional gluing to obtain more satisfactory ones.

Firstly, we obtain a useful formula which relates p to $\tilde{\xi}$. Let $\pi: T^*Q \longrightarrow Q$ be the contangent bundle projection.

Proposition 6.6. *For each $p \in T^*Q$*

$$T_p\pi\tilde{\xi} = p^{\#(g)}.$$

Proof. We use the local coordinates. Take a chart (U, α) in Q, and let $(q^1, q^2, \ldots, q^n, p_1, p_2, \ldots, p_n)$, $q = (q^1, q^2, \ldots, q^n) \in Q$, be the coordinates in the associated chart $(\tilde{U} = T_U^*Q, \tilde{\alpha})$ in T^*Q (see AI60). We have the following formulae which express the quantities in question in the said coordinates:

(6.6a) $$p = \sum_{i=1}^{n} p_i\, dq^i,$$

(6.6b) $$p^{\#(g)} = \sum_{i,k} g^{ik}(\pi(p))\, p_i \frac{\partial}{\partial q^k},$$

g^{ik} being the elements of the inverse matrix of $\{g_{ik}\}$,

(6.6c) $$dH|_{\tilde{U}} = \sum_{i,k} g^{ik} p_i\, dp_k + \frac{1}{2}\sum_{i,k,j}\left(\frac{\partial g^{ik}}{\partial q^j} p_i p_k\, dq^j\right) + \sum_j \frac{\partial V}{\partial q^j} dq^j,$$

$$(6.6d)\qquad (d\tilde{H})^{\#(\Omega)}|_{\tilde{v}} = \sum_{i,k} g^{ik} p_i \frac{\partial}{\partial p^k} - \sum_j \left(\frac{1}{2}\sum_{i,k} \frac{\partial g^{ik}}{\partial q^j} p_i p_k + \frac{\partial V}{\partial q^j}\right)\frac{\partial}{\partial p_j},$$

$$(6.6c)\qquad T_p\pi(dH)^{\#(\Omega)} = \sum_{i,k} g^{ik}(\pi(p))p_i \frac{\partial}{\partial q^k}.$$

Comparing (6.6b) and (6.6c) yields the desired result. □

Corollary 6.7. *Let $p \in \partial T^*Q$ and $q = \pi(p)$. A vector $\tilde{\xi}_p$ is transversal to ∂T^*Q if and only if the vector $p^{\#(g)}|_q$ is transversal to ∂Q.*

Proof. This is an immediate consequence of the formula $\pi(\partial T^*Q) = \partial Q$ and Proposition 6.6. □

Denote by Σ the set of all $p \in T^*Q$ such that $\tilde{\xi}_p$ fails to be transversal to the boundary, i.e. such that $\tilde{\xi}_p \in T_p\partial T^*Q$. The above Corollary may then be expressed by writing $\Sigma = \#(g)^{-1}(T\partial Q)$, the set of all covectors whose tangent images are tangent to the boundary of Q.

We now continue our construction to ensure that the suitable object is a phase space. Consider the manifold $T^*Q \setminus \Sigma$. Its boundary may be divided into two nonintersecting parts C_{out} and C_{in} containing points where the field $\tilde{\xi}$ looks out, or correspondingly into the manifold. The subsets C_{out} and C_{in} consist of whole components of the boundary. Define the reflection map $r: C_{\text{out}} \longrightarrow C_{\text{in}}$ by the formula

$$(6.8)\qquad r(p) = p - 2\langle p, n\rangle n,$$

where n is the unit cotangent vector in T_q^*Q, $q = \pi(p) \in \partial Q$, which is orthogonal to the boundary: $|n| = 1$ and $n(v) = 0$ for all $v \in T_q\partial Q$.

It will be useful to obtain the coordinate expression for the reflection map. Take a chart (V, α) in Q with $V \cap \partial Q \neq \varnothing$ and consider the associated chart $(T_V Q, T^{*-1}\alpha)$ in T^*Q (see AI.60). We use the notation $(q^1, q^2, \ldots, q^n, p_1, p_2, \ldots, p_n)$ for the coordinates in the image of the coordinate map $T^{*-1}\alpha$. Recall that this image lies in $\mathbb{R}_+^{2n}$, the first coordinate $q^1 > 0$ at interior points, and $q^1 = 0$ at boundary points of T^*Q.

Proposition 6.9. *The reflection map has the following expression in the coordinates of the associated chart $\tilde{\alpha} = T^{*-1}\alpha$: $\tilde{\alpha} \circ r \circ \tilde{\alpha}^{-1}(0, q^2, \ldots, q^n, p_1, p_2, \ldots, p_n) = (0, q^{2\prime}, \ldots, q^{n\prime}, p_1', p_2', \ldots, p_n')$ where*

$$(6.9a)\qquad \begin{aligned} &q^{i\prime} = q^i, \qquad p_i' = p_i, \quad 2 \leqq i \leqq n, \\ &p_1' = p_1 - 2\left(\sum_{j=1}^n g^{j1} p_j\right)(g^{11})^{-1}. \end{aligned}$$

Proof. Let $p = \sum p_i\, dq^i \in T_q^*Q$, $q \in \partial Q$. Then $q = (0, q^2, \ldots, q^n)$ and $p' = r(p) \in T_q^*Q$ by definition. Hence $q' = (q^{1\prime}, q^{2\prime}, \ldots, q^{n\prime}) = q$, i.e. the part of (6.9a) involving q is valid. Consider the covector

$$(6.9b)\qquad n = (g^{11}(q))^{-1/2}\, dq^1 \in T_q^*Q.$$

As is easy to verify, it is of unit norm, and is orthogonal to the boundary. The calculation of the inner product with p yields

$$(6.9c) \qquad \langle p, n \rangle = g(p^{\#(g)}, (dq^1)^{\#(g)})(g^{11})^{-1/2} = \sum_{i,k,j} g^{ik} g^{1j} p_i (g^{11})^{-1/2} g\left(\frac{\partial}{\partial q^k}, \frac{\partial}{\partial q^j}\right)$$

$$= \sum_{i,k,j} g^{ik} g^{1j} g_{kj} (g^{11})^{-1/2} = \sum_i g^{i1}(g^{11})^{-1/2}.$$

Substituting (6.9b), (6.9c) into (6.8) we obtain the remaining part of (6.9a).   □

Corollary 6.10. *The reflection map preserves the restrictions of the symplectic structure.*

Proof. Since $dq^1|_{C_{in}} = dq^1|_{C_{out}} = 0$,

$$r^*\left(\sum_{i=1}^n p_i \, dq^i|_{C_{in}}\right) = r^*\left(\sum_{i=2}^n p_i \, dq^i\right) = \sum_{i=2}^n p_i' \, dq^{i'}$$

(use (6.9a))

$$\sum_{i=2}^n p_i \, dq^i = \sum_{i=1}^n p_i \, dq^i|_{C_{out}}.$$

We obtain that the reflection map even preserves the restrictions of the Liouville form.   □

Corollary 6.11. *The matrix R of the tangent map*

$$(T_{r(p)}\tilde{\alpha}|_{C_{out}}) \circ T_p r \circ (T_p \tilde{\alpha}|_{C_{in}})^{-1} : \mathbb{R}^{2n-1} \longrightarrow \mathbb{R}^{2n-1},$$

$p \in C_{in}$, *has the form*

$$R = \begin{array}{|c|c|} \hline I_{n-1} & 0 \\ \hline \diagdown\!\!\!\!\!\diagdown\; -1\; \diagdown\!\!\!\!\!\diagdown & \\ \hline 0 & I_{n-1} \\ \hline \end{array} \quad \longleftarrow n\text{th row}$$

where I_{n-1} is the $(n-1) \times (n-1)$ unit matrix.

Proof. The corollary follows immediately from (6.9a).   □

We finish our construction by gluing $T^*Q \backslash \Sigma$ via the reflection map r in a Hamiltonian manner, the $\tilde{H}$ playing the role of the gluing Hamiltonian (see 4.12). What we obtain is the glued symplectic manifold Z with Hamiltonian function H. The latter defines a Hamiltonian dynamical system on Z, the generalized geodesic flow. The corresponding vector field on Z will be denoted by ξ. We collect the notations for the maps we shall use further in the following commutative diagram

(6.12)

$$T^*Q \subset T^*Q \backslash \Sigma \xrightarrow{\;\rho\;} Z$$

$$\pi \downarrow \qquad \qquad \swarrow \pi'$$

$$Q \longleftarrow$$

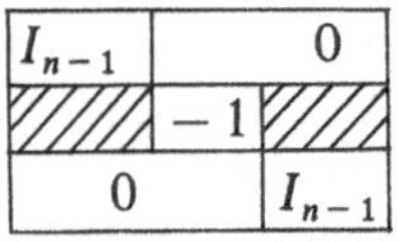

Here π is the cotangent bundle projection, ρ is the natural projection onto the quotient space, π' is induced by π. The image of $C_{\text{in}} \cup C_{\text{out}}$ under the map ρ will be denoted by Π. It is a smooth submanifold (without boundary) of dimension $2n-1$, the maps $\rho|_{C_{\text{in}}}: C_{\text{in}} \longrightarrow \Pi$, $\rho|_{C_{\text{out}}}: C_{\text{out}} \longrightarrow \Pi$, and $\rho|_{T^*Q \backslash (\Sigma \cup C_{\text{in}} \cup C_{\text{out}})}: T^*Q \backslash (\Sigma \cup C_{\text{in}} \cup C_{\text{out}}) \longrightarrow Z \backslash \Pi$ being diffeomorphisms.

The following 1-form Θ induced on Z by the Liouville form Θ_Q will play an important role later in the construction of Maslov's operator (see Chapter VI). We define the post-Liouville form Θ by the equation

$$\rho^*\Theta = \Theta_Q|_{T^*Q \backslash \Sigma} \tag{6.13}$$

In view of the noninvertability of ρ at points of Π, the form Θ is only uniquely defined at points of $Z \backslash \Pi$. At a point $z \in \Pi \Theta$ has exactly two values $\Theta_{\text{in}}|_z$ and $\Theta_{\text{out}}|_z$ where

$$\Theta_\iota = ((\rho|_{C_\iota})^{-1})^* \Theta_Q, \quad \iota \in \{\text{in}, \text{out}\}. \tag{6.14}$$

So, the post-Liouville form is neither continuous nor single valued if $\partial Q \neq \emptyset$, these problems occurring on Π. However

$$\Theta_{\text{in}}|_{T\Pi} = \Theta_{\text{out}}|_{T\Pi}, \tag{6.15}$$

as follows from the coordinate expressions (2.7a) and (6.9a).

The dynamical system just constructed is reversible in the sense explained at the end of §3. One obtains a reversor by factorizing the map $\hat{\sigma}: T^*Q \longrightarrow T^*Q$ which changes the sign of moment: $\hat{\sigma}(q, p) = (q, -p)$. This map preserves Σ (Corollary 6.7) and commutes with the reflection map, when restricted onto ∂T^*Q. Hence it induces a smooth idempotent map $\sigma: Z \longrightarrow Z$ which reverses the direction of time on trajectories of ξ. We call it the *momenta-reversor* of the generalized geodesic flow.

Particular Cases

Here we briefly outline the diversity of possibilities that our general scheme yields.

6.10. The *system of classical mechanics* arises if one uses the Euclidean space $(\mathbb{R}^n, g_{st})$ as the coordinate Riemannian manifold. Here

$$g_{st} = \sum_{i=1}^{n} dq^i \odot dq^i, \tag{6.10a}$$

the standard Euclidean metric tensor.

6.11. *Geodesic flow* is the system with zero potential. $V(q) = 0$. A characteristic feature of such a system is the homogeneity of its Hamiltonian with respect to the p-variables: $\forall \tau \in \mathbb{R}$

$$H(\tau p) = \tau^2 H(p). \tag{6.11a}$$

This leads to the behaviour of the system being completely determined by that of its restriction onto any hypersurface of constant energy $\mathscr{E}_C = \{z \in Z: H(z) = C\}$, $C > 0$. Indeed, the reflection map commutes with the family of homotheties $p \longmapsto \tau p$, $\tau > 0$. Therefore they induce the family of diffeomorphisms $\psi_\tau: Z \longrightarrow Z$ of a glued space Z. We have $\psi_\tau(\mathscr{E}_{1/2}) = \mathscr{E}_{(1/2)\tau^2}$. Since $\psi_\tau^* H = \tau^2 H$ (see (6.11a)), the tangent map $T_{\mathscr{E}_{1/2}}\psi_\tau$ carries the field $\xi|_{\mathscr{E}_{1/2}}$ to the field $\tau\xi|_{\mathscr{E}_{1/2\tau^2}}$. So ψ_τ conjugates the system restricted to $\mathscr{E}_{1/2}$ with that on $\mathscr{E}_{1/2\tau^2}$ provided the change of the time variable $t \longmapsto \tau t$ is made for the latter case. As for the zero-energy hypersurface $\mathscr{E}_0 = \{z: H(z) = 0\}$, the system is trivial there.

We distinguish geodesic flows *with reflections* (*billiards*) which occur when $\partial Q \neq \varnothing$, and geodesic flows *without reflections* (the case $\partial Q = \varnothing$). In the latter case the hypersurface $\mathscr{E}_{1/2}$ is nothing more than $S_1^* Q = \{p \in T^*Q: |p| = 1\}$, the unit sphere subbundle of the cotangent bundle. It is worthwhile here to mention two remarkable classes of examples of geodesic flows without reflections. They demonstrate quite different ergodic properties.

6.12. *Riemannian torus.* Consider the Riemannian space $(\mathbb{R}^n, \tilde{g})$ with $\tilde{g}$ invariant under the action of a lattice Γ in $\mathbb{R}^n$. Then $\tilde{g}$ induces he Riemannian metric g on the n-torus $\mathbb{R}^n = \mathbb{R}^n/\Gamma$. Geodesic flow on $(\mathbb{T}^n, g)$ has a very simple structure when $\tilde{g} = \text{const}$. Small perturbations of the constant Riemannian metric on a torus will be investigated within the framework of KAM theory.

6.13. *Surface of constant negative curvature.* Firstly, we discuss Lobachevski's plane (Q_0, g_0) in Poincaré representation: $Q_0 = \{z = x + iy \in \mathbb{C}: y > 0\}$, the upper half-plane, and $g_0 = y^{-2}(dx \odot dx + dy \odot dy)$. There is a convenient description of the reduced phase space $S_1^* Q_0$ as the real projective special linear group in dimension two OSL(2, $\mathbb{R}$). Any element of this group is a pair $(m, -m)$ where m is a 2×2-matrix with real coefficients and determinant 1. The sharp map $\#(g_0)$ establishes the isomorphism between $S_1^* Q_0$ and the bundle of unit spheres $S_1 Q_0$ of the tangent bundle. Denote $v_0 = \partial/\partial y|_i \in S_1 Q_0$. Then for any $v \in S_1 Q_0$ there is a unique, up to simultaneous change of the sign of coefficients, 2×2-matrix

$$m = \begin{pmatrix} a & b \\ c & d \end{pmatrix}$$

with real coefficients and determinant 1 such that m transforms v_0 into v. In more detail, if $v \in T_z Q_0$ then

$$m(i) = \frac{ai + b}{ci + d} = z$$

and $T_i m(v_0) = v$. Thus we have a one-to-one correspondance, smooth in both directions, between $S_1^* Q_0$ and PSL(2, $\mathbb{R}$). We identify these two manifolds. Note that under this identifications, the unit element of PSL(2, $\mathbb{R}$) corresponds to $v_0^\flat$. Consider the trajectory $m_0^t, t \in \mathbb{R}$, of geodesic flow passing through $v_0^\flat$ at time 0. In the canonical coordinates (x, y, p_x, p_y) associated with coordinates (x, y) the equations of the trajectory read

$$\dot{x} = y^2 p_x, \qquad \dot{y} = y^2 p_y, \qquad \dot{p}_x = 0, \qquad \dot{p}_y = -y(p_x^2 + p_y^2).$$

They have a solution $x(t) = 0, y(t) = e^t, p_x = 0, p_y = e^{-t}$. In matrix representation our trajectory looks like

$$m_0^t = \begin{pmatrix} e^{t/2} & 0 \\ 0 & e^{-t/2} \end{pmatrix}.$$

One can easily deduce that the trajectory through $m \in \mathrm{PSL}(2, \mathbb{R})$ is $m \cdot m_0^t$ (here $\cdot$ denotes matrix multiplication).

A great amount of interesting examples can be obtained by taking the left quotient, with respect to a discrete subgroup Γ, of $\mathrm{PSL}(2, \mathbb{R})$. The right action $m \longmapsto m \cdot m_0^t$ of the subgroup $\{m_0^t, t \in \mathbb{R}\}$, which yields the geodesic flow on $S_1^* Q_0 \cong \mathrm{PSL}(2, \mathbb{R})$, commutes with the left action of Γ. So it induces a flow on $\Gamma \backslash \mathrm{PSL}(2, \mathbb{R})$ which may be interpreted as the geodesic flow on $S_1^* Q$ where $Q = Q_0/\Gamma$, the Riemannian surface obtained by identification of the points of Q_0 which may be carried one to another by Γ. It is well known that any connected surface with Riemannian matrix of constant negative curvature -1 is realizable as Q_0/Γ with suitable discrete subgroup Γ. (See Wolf (1972), Ahlfors and Sario (1960) and Berdon (1983)) If Q_0/Γ has finite Riemannian volume, the geodesic flow on $S_1(Q_0/\Gamma)$ is ergodic and mixing (Kornfeld, Sinai and Fomin 1980).

The systems described in 6.12 and 6.13 have complete dynamics. In the former case the compactness argument works, in the latter one the dynamics is written explicitly.

Let us now consider billiards. We shall mainly consider billiards with constant Euclidean metric of the form (6.10a).

6.14. *Sinai's billiards.* Take a strictly convex open domain $D \subset \mathbb{R}^2$ with smooth boundary such that D lies entirely within the open unit square in $\mathbb{R}^2$ (see Fig. 17). The strictness of the convexity means that there is a positive constant detatching the curvature of the boundary from zero. Let Γ_0 be the discrete transformation group of $\mathbb{R}^2$ generated by ort_i, $i = 1, 2$. Consider the torus with a 'hole' $Q = (\mathbb{R}^2 \backslash \bigcup_{\gamma = \Gamma_0} \gamma D)/\Gamma_0$ with induced Euclidean metric. The billiard in Q was extensively investigated by Ya. Sinai in Sinai (1970). It was proved to be ergodic on $S_1^* Q$. Note that this system does not possess complete dynamics, but an "almost complete" one: i.e. the dynamics is defined almost everywhere with respect to Lebesgue measure (see Kornfeld, Sinai and Fomin (1980) for the proof).

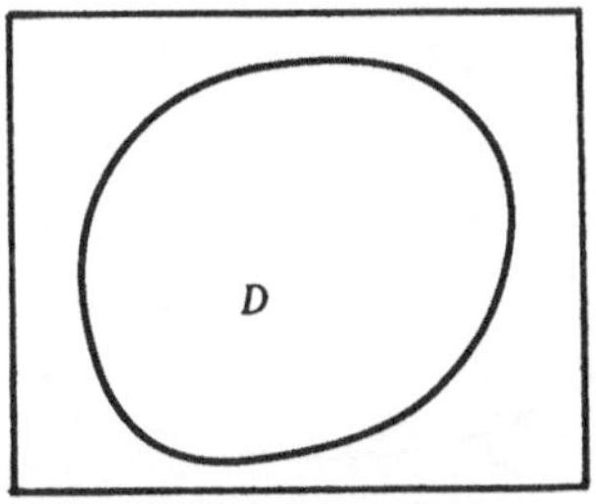

Fig. 17. Sinai's billiards

6.15. *Convex plane billiard.* Here Q is a compact domain in $\mathbb{R}^2$ with smooth boundary. We suppose that Q is strictly convex in the sense of 6.14. Supplying Q with the Euclidean metric (6.10a), we obtain the coordinate space of the convex plane billiard. There are two features of the corresponding dynamical system which we want to underline. Firstly, it turns out to have complete dynamics. This is not obvious. There is a counterexample with boundary of class C^2, but smoothness of at least class C^3 is sufficient (Halpern [1]). We shall obtain the aimed completeness later as a consequence of the KAM theorem. Secondly, the submanifold $\Pi = \rho(\partial S_1^*Q)$, the image under the gluing of reflecting unit covectors, turns out to be a cross section. This is obvious since any $p \in \partial S_1^*Q$ has an immediate predecessor p' and an immediate successor p'' on its trajectory in the (noncomplete) geodesic flow on T^*Q generated by the Euclidean metric (see Fig. 18). Topologically Π is an open annulus. The Poincaré map $f: \Pi \longrightarrow \Pi$, which assignes $\rho(p'')$ to $\rho(p')$, is a diffeomorphism. Let us introduce the coordinates s and ϑ in Π. If $p' \in T_q^*Q$ with $\rho(p') \in \Pi$, then the corresponding coordinate s is the (multivalued) length of the arc joining a chosen point $q_0 \in \partial Q$ with q in the anticlockwise direction, and ϑ is the angle between the unit vector $\tau(s)$ *tangent to* ∂Q at the point q, $\tau(s)$ pointing in the direction of increasing s, and $(p')^{\#(g_0)}$, $0 < \vartheta < \pi$ (see Fig. 18). Let $\gamma(s)$ be the angle between $\tau(s)$ and ort_1. Then we have the following formulae for the coordinates (q^1, q^2, p_1, p_2) of a point in Π in terms of (s, ϑ):

$$p_1 = \cos(\vartheta + \gamma(s)), \qquad p_2 = \sin(\vartheta + \gamma(s)),$$

$$\frac{dq^1}{ds} = \cos\gamma(s), \qquad \frac{dq^2}{ds} = \sin\gamma(s).$$

Immediate calculation yields the formula for induced symplectic structure on Π:

$$\Omega|_{T\Pi} = \sin\vartheta\, d\vartheta \wedge ds. \tag{6.21a}$$

The submanifold Π is invariant under the momenta-reversor σ. Denote $\hat{\sigma} = \sigma|_{\Pi}$. The cascade generated by the Poincaré map f, defined above, is reversible with $\hat{\sigma}$

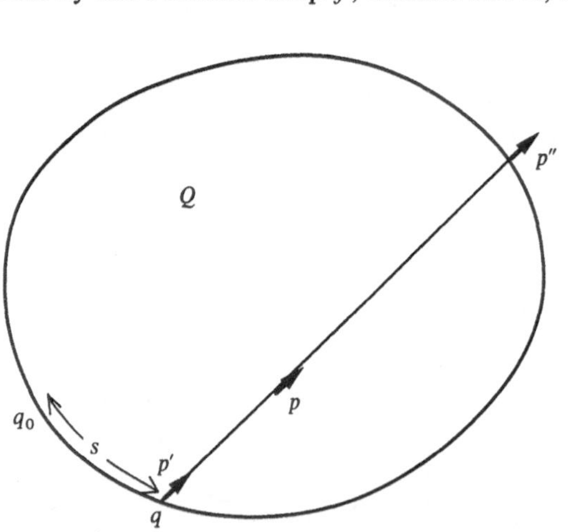

Fig. 18. The motion inside a planar billiard.

Fig. 19. Stadium

as a reversor. The expression for $\hat{\sigma}$ in the coordinate (s, ϑ) is $\hat{\sigma}:(s, \vartheta) \longmapsto (s, \pi - \vartheta)$. Further properties of convex planar billiards will be discussed in §14.

6.22. The case of a *partially smooth boundary*. Our construction may be applied to domains whose boundaries have singularities. One need only remove these singularities, i.e. consider the domain $Q \backslash S$, where S is the set of all singular points of ∂Q, as a Riemannian manifold to which the preceeding scheme is applied to. Thus we may consider billiards in polygons and in domains such as the stadium: having interesting ergodic properties (see the Notes to this chapter for the references).

§7. Completely Integrable Hamiltonian Systems

There is a remarkable class of Hamiltonian systems, namely integrable ones, possessing the most simple character of motion. Unfortunately they seldom occur and are destroyed under general small perturbations. Nevertheless it is necessary to study them since in many cases we deal with systems which only slightly differ from integrable ones. We begin our study with the investigation of the geometric structure which is closely connected to the systems in question. All manifolds in this section are supposed to have no boundary.

Lagrangian Fibrations

A smooth map $p: X \longrightarrow B$, X and B being manifolds, is called a *fibration* if (i) p is onto, (ii) p is a submersion. We regard fibration as a particular case of a bundle (see Appendix I). The manifold B is called the *base* of the fibration $p: X \longrightarrow B$, or one can say: "a fibration *over B*". The *fibre* $p^{-1}(\{b\})$ over a point $b \in B$ is a submanifold of X, and $\operatorname{codim} p^{-1}(\{b\}) = \dim B$. The *restriction* of a fibration $p: X \longrightarrow B$ onto a submanifold $C \subset B$ is again a fibration, $p|_{p^{-1}(C)}: p^{-1}(C) \longrightarrow C$.

A fibration $p: X \longrightarrow B$ is called a *Lagrangian fibration* if (iii) X is a symplectic manifold, (iv) the fibres are Lagrangian submanifolds of X. It follows that $\dim X = 2 \dim B$ for the Lagrangian fibration $p: X \longrightarrow B$. The restriction of a Lagrangian fibration over B onto a submanifold $C \subset B$ is again a Lagrangián fibration provided $\dim C = \dim B$ and $p^{-1}(C)$ is supplied with the induced symplectic structure.

An *isomorphism* from a Lagrangian fibration $p: X \longrightarrow B$ to a Lagrangian fibration $q: Y \longrightarrow G$ is a pair (f, g) where $f: X \longrightarrow Y$ is a symplectic diffeomorphism and

$g: B \longrightarrow C$ is a diffeomorphism such that the diagram

$$\begin{array}{ccc} X & \xrightarrow{\ f\ } & Y \\ {\scriptstyle p}\downarrow & & \downarrow{\scriptstyle q} \\ B & \xrightarrow{\ g\ } & C \end{array}$$

commutes. From this it follows that the restriction of f onto the fibre $p^{-1}(\{b\})$ maps the latter diffeomorphically onto the fibre $q^{-1}(\{g(b)\})$. Two Lagrangian fibrations are said to be isomorphic if there is an isomorphism from one to the other. Two Lagrangian fibrations $p: X \longrightarrow B$ and $q: Y \longrightarrow C$ are *locally isomorphic* at points $b \in B$ and $c \in C$ if there are open neighbourhoods U of b in B and V of c in C such that the restrictions of our fibrations onto U and V respectively are isomorphic via an isomorphism carrying b onto c.

7.1. Example. Let A_D be the symplectic annulus of Example 2.5 with base D being an open set in $\mathbb{R}^n$. We may represent A_D by the product $A_D = \mathbb{T}^n \times D$ where $\mathbb{T}^n = \mathbb{R}^n/\mathbb{Z}^n$ is the n-dimensional torus. The second projection $p_2: A_D \longrightarrow D$ of this product is a Lagrangian fibration over D. The fibres of p_2 are n-dimensional Lagrangian tori $\mathbb{T}^n \times \{d\}$, $d \in D$. We shall refer to the fibration $p_2: A_D \longrightarrow D$ as the *standard toroidal Lagrangian* fibration over D, or with base D. All standard toroidal Lagrangian fibrations of the same dimension are locally isomorphic.

Theorem 7.2. *If the fibres of a Lagrangian fibration are connected and compact then it is locally isomorphic at each point of its base to the standard toroidal Lagrangian fibration.*

We shall prove this theorem later in this section. To do this we need some additional notions which are also of independent interest.

Poisson Brackets

The *Poisson brackets* on a symplectic manifold (X, Ω) is the binary operation defined on the space $C^\infty(X)$ of real-valued smooth functions by the formula

$$\{\varphi, \psi\} = \Omega((d\varphi)^\#, (d\varphi)^\#), \quad \varphi, \psi \in C^\infty(X). \tag{7.3}$$

Here $\# = \#(\Omega)$, the "sharp" map of Ω.

In Darboux' coordinates $(q^1, q^2, \ldots, q^n, p_1, p_2, \ldots, p_n)$, $n = (\dim X)/2$, we have, after simple calculations using formulae 2.17:

$$\{\varphi, \psi\} = \sum_{i=1}^{n} \left(\frac{\partial\varphi}{\partial q^i}\frac{\partial\psi}{\partial p_i} - \frac{\partial\varphi}{\partial p_i}\frac{\partial\psi}{\partial q^i} \right). \tag{7.4}$$

The Poisson brackets are bilinear, skew-symmetric and satisfy the Lie identity

$$\{\{\varphi_1, \varphi_2\}, \varphi_3\} + \{\{\varphi_2, \varphi_3\}, \varphi_1\} + \{\{\varphi_3, \varphi_1\}, \varphi_2\} = 0 \tag{7.5}$$

for all $\varphi_i \in C^\infty(X)$, $i = 1, 2, 3$. So it converts $C^\infty(X)$ into a Lie algebra. One can easily verify the following proposition by using the coordinate expression (7.4).

Proposition 7.6. *The map*

(7.6a) $$\varphi \longmapsto (d\varphi)^{\#}$$

is a homomorphism between the Lie algebra $C^\infty(X)$ *and the Poisson brackets as the Lie product onto the Lie algebra of Hamiltonian vector fields on* X, *the latter being supplied with the Lie brackets* $[\cdot,\cdot]$ *(see* AI.45) *as the Lie product. The kernel of the homomorphism* (7.6a) *coincides with locally constant functions on* X.

Recall that a homomorphism of Lie algebras is a map which is linear and preserves Lie products.

We say that a set $\{\varphi_i, i\in I\}$ of real valued smooth functions on a symplectic manifold is *in involution* (or the functions φ_i are *in involution*), if their Poisson brackets vanish, i.e. $\{\varphi_i, \varphi_j\} = 0\, \forall i, j\in I$. Vector fields $\xi_i, i\in I$, are said to commute if their brackets vanish, i.e. $[\xi_i, \xi_j] = 0\ \forall_{i,j}\in I$.

Corollary 7.7 *Hamiltonian vector fields commute if and only if their Hamiltonian functions are in involution.*

The following proposition reveals the dynamical meaning of Poisson brackets.

Proposition 7.8. *Let* φ *and* ψ *be smooth functions on a symplectic manifold* X, *and let* $g^t: U \longrightarrow X$, $t\in\Delta\ni 0$, *be a normed local flow of the vector field* $(d\psi)^{\#}$ (see AI.69). *Then*

(7.8a) $$\{\varphi, \psi\}|_U = \frac{d}{dt}(g^t)^*\varphi|_{t=0}.$$

Proof. By the differentiation formula AI.68 we have

$$\frac{d}{dt}(g^t)^*\varphi = (g^t)^*((d\varphi)^* \lrcorner\, d\varphi) = (g^t)^* d\varphi((d\varphi)^{\#})$$

$$= (g^t)^*\Omega((d\varphi)^{\#}, (d\psi)^{\#}) = (g^t)^*\{\varphi, \psi\}.$$

Put $t = 0$. □

Proof of Theorem 7.2. Let $p: X \longrightarrow B$ be a Lagrangian fibration. In virtue of the local character of the assertion of the theorem with respect to the base, we may assume that B is an open set in $\mathbb{R}^n$.

Proposition 7.9. *A fibration* $p: X \longrightarrow B$ *over an open set* $B \subset \mathbb{R}^n$, X *being a symplectic manifold of dimension* $2n$, *is a Lagrangian fibration if and only if its components* $J_i = \mathrm{pr}_i \circ p: X \longrightarrow \mathbb{R}^1$, $1 \leqslant i \leqslant n$, *are functions in involution. Here* $\mathrm{pr}_i: \mathbb{R}^n \longrightarrow \mathbb{R}^1$ *is the ith coordinate projection.*

Proof. Consider the fibre $L_b = p^{-1}(\{b\})$ over a point $b\in B$, and take a point $x\in L_b$. A vector $v\in T_xX$ belongs to T_xL_b if and only if $T_xp(v) = 0$ which is equivalent to $T_xJ_i(v) = 0$, $1 \leqslant i \leqslant n$. The latter equations may be written as $T_xJ_i(v) = dJ_i(v) =$

$\Omega((dJ_i)^{\#}, v) = 0$ or

$$(+) \qquad v \perp (dJ_i)^{\#}, \quad 1 \leqslant i \leqslant n.$$

It follows from the definition of fibration that the vectors $(dJ_i)^{\#}, 1 \leqslant i \leqslant n$, are linearly independent in T_xX. Let T_xL_b be Lagrangian. Then $(T_xL_b)^{\perp} = T_xL_b$, so $(+)$ implies $(dJ_i)^{\#} \in T_xL_b$. In this case $(dJ_i)^{\#}, 1 \leqslant i \leqslant n$, form a basis of T_xL_b, and the equalities $\{J_i, J_k\} = \Omega((dJ_i)^{\#}, (dJ_k)^{\#}) = 0$ follow since T_xL_b is Langrangian. Conversely, let J_i, $1 \leqslant i \leqslant n$, be in involution. Then $(dJ_i)^{\#} \perp (dJ_k)^{\#}, 1 \leqslant i, k \leqslant n$. Hence $(dJ_i)^{\#}$ satisfies the test $(+)$ for a vector to belong to T_xL_b. The vector fields $(dJ_i)^{\#}$ again form the basis of T_xL_b (use the equality of dim T_xL_b and the number of the $(dJ_i)^{\#}$). The property that the components J_i are in involution implies that $v \perp w \, \forall v, w \in T_xL_b$, i.e. T_xL_b is Lagrangian. □

Consider the vector fields $(dJ_i)^{\#}, 1 \leqslant i \leqslant n$, where each J_i is defined in the formulation of Proposition 7.9. Since our fibration $p: X \longrightarrow B$ is Lagrangian, it follows that these fields

(a) *are Hamiltonian,*
(b) *commute,*
(c) *are tangent to the fibres, and their values at a point $x \in L_b$ form a basis of T_xL_b,*
(d) *are complete.*

The property (a) is obvious, (b) is a consequence of 7.7, (c) has been obtained in the process of proving 7.9. To prove (d) let us note that each integral curve of the field $(dJ_i)^{\#}$ lies wholly in some fibre. Since fibres are supposed to be compact (and to have to boundary) the restriction of $(dJ_i)^{\#}$ onto each fibre is a complete vector field. We conclude that each integral curve of $(dJ_i)^{\#}$ may be extended to the integral curve defined on the whole real axis of the time variable.

Consider the flows $g_i^t = \exp t(dJ_i)^{\#}, t \in \mathbb{R}, 1 \leqslant i \leqslant n$. Their properties are:

(e) *$g_i^t: X \longrightarrow X$ are symplectic diffeomorphisms,*
(f) *they preserve the fibres: $g_i^t(L_b) = L_b$,*
(g) *they commute: $g_i^t \circ g_j^\tau = g_j^\tau \circ g_i^t$.*

The property (e) follows from 3.3, (f) has just been proved. Let us now prove (g). Locally, the fields $(dJ_i)^{\#}$ may be simultaneously straightened (see AI.45), i.e. there is a chart (V, β) in X, V being a neighbourhood of a given point, the chart having coordinates $(u_1, u_2, \ldots, u_{2n})$ such that $(dJ_i)^{\#} = \partial/\partial u_i$, $1 \leqslant i \leqslant n$. In these coordinates g_i^t reads as $u_j = u_j^0, j \neq i$, $u_i = u_i^0 + t$ provided t is sufficiently small and $u^0 = (u_1^0, u_2^0, \ldots, u_{2n}^0)$ ranges over a compact subset of $\beta(V) \subset \mathbb{R}^{2n}$. Obviously, (g) holds for such values t, τ, and $u^0 = \beta(x)$. If we restrict our considerations to a fixed fibre L_b, we conclude that L_b may be covered by a finite number of domains of these described charts. So there is a $t_0 > 0$ such that (g) holds for $|t|, |\tau| \leqslant t_0$, and $x \in L_b$. Since (f) holds, it follows that (g) may be extended to all values of t, τ from those lying in the interval $|t| \leqslant t_0$ by the flow property $g_i^{t+\tau} = g_i^t \circ g_i^\tau$.

We now may define the action of the additive group $\mathbb{R}^n$ by the formula $\mathbb{R}^n \ni t = (t_1, t_2, \ldots, t_n) \longmapsto g_t: X \longrightarrow X$, where

$$g_t = g_1^{t_1} \circ g_2^{t_2} \circ \cdots \circ g_n^{t_n}.$$

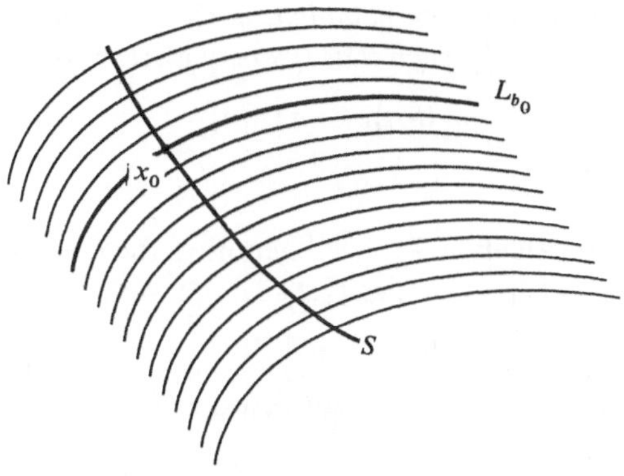

Fig. 20

The commutativity property (g) yields that this formula really does define an action, i.e. $g_{t+s} = g_t \circ g_s$, $t, s \in \mathbb{R}^n$. The maps g_t are symplectic diffeomorphisms and preserve the fibres.

The next step in our construction is the fixing of a Lagrangian transversal slice. Fix $b_0 \in B$, $x_0 \in L_{b_0}$, and introduce a symplectic chart (U, α) where $U \ni x_0$. The vector subspace $V = T_{x_0} \alpha(T_{x_0} L_{b_0})$ is a Lagrangian subspace of $T_{\alpha(x_0)}\mathbb{R}^{2n} = \mathbb{R}^{2n}$, the latter being supplied with Ω_{st} as the symplectic structure. Take some Lagrangian complement W to V in $T_{\alpha(x_0)}\mathbb{R}^{2n}$, consider W as a Lagrangian submanifold of $\mathbb{R}^{2n}$, and take the preimage $S = \alpha^{-1}(W)$. This is a Lagrangian submanifold of X transversal to L_{b_0} at the point x_0, i.e.

(h) $$T_{x_0} X = T_{x_0} S \oplus T_{x_0} L_{b_0}.$$

We may assume that S meets any fibre L_b at most at one point and transversally. Otherwise it is sufficient to replace S by a smaller slice $S_1 \subset S$ with the desired property. The said property yields that the fibration projection p maps S diffeomorphically onto some open set $U_1 \subset B$ so that S becomes the image of a smooth section $s: U_1 \longrightarrow X$ of the restriction of our fibration onto U_1.

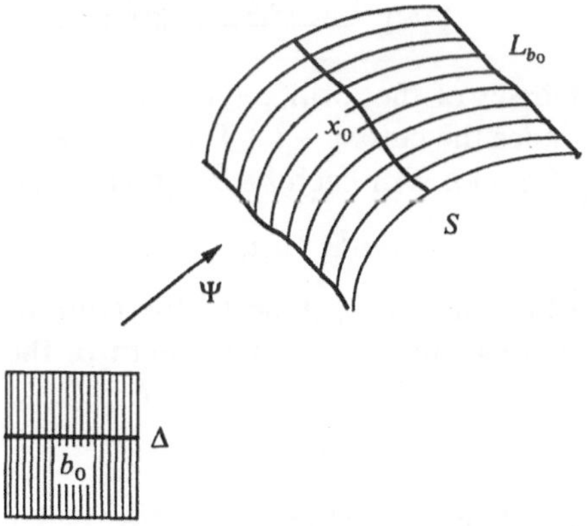

Fig. 21

Lemma 7.10. *There are Δ, an open neighbourhood of the origin in $\mathbb{R}^n$, and U_2, an open neighbourhood of b_0 in U_1, such that the map $\psi: U_2 \times \Delta \longrightarrow X$ defined by*

$$(b, t) \longmapsto g_t(s(b))$$

is a local diffeomorphism.

Proof. It is sufficient to calculate $T_{(b_0,0)}\psi$. Denoting the coordinates in $U_1 \times \mathbb{R}^n \subset \mathbb{R}^n \times \mathbb{R}^n = \mathbb{R}^{2n}$ as $(b, t) = (y_1, y_2, \ldots, y_n, t_1, t_2, \ldots, t_n)$ we have

$$\left.\frac{\partial \psi}{\partial y_i}\right|_{(b,t)} = Tg_t(s(b))\frac{\partial s(b)}{\partial y_i}, \qquad \left.\frac{\partial \psi}{\partial t_i}\right|_{(b,t)} = (dJ_i)^{\#}|_{s(b)}, \quad 1 \leqq i \leqq n.$$

Putting $t = 0$ and $b = b_0$, we have that the values $T_{(b_0,0)}\psi(\text{ort}_i)$ equal

$$\left\{\frac{\partial s(b_0)}{\partial y_1}, \ldots, \frac{\partial s(b_0)}{\partial y_n}, (dJ_1)^{\#}|_{x_0}, \ldots, (dJ_n)^{\#}|_{x_0}\right\}.$$

They form a basis of $T_{x_0}X$ due to (c) and (h). Therefore, ψ is a local diffeomorphism if $U_2 \times \Delta$ is sufficiently small. □

It follows from Lemma 7.10 that $L_{b_0} \cap \psi(U_2 \times \Delta)$ is just $\psi(\{b_0\} \times \Delta)$, and it is an open neighbourhood of x_0 in L_{b_0}. Note that $\psi(\{b_0\} \times \Delta) = \{g_t(x_0): t \in \Delta\}$ is a part of the orbit $\mathcal{O}_{x_0} = \{g_t(x_0): t \in \mathbb{R}^n\}$ of the action g_t, passing through the point x_0. We have obtained that $\mathcal{O}_{x_0}$ is open in L_{b_0} consists of whole orbits, each such orbit is open in L_{b_0}, and since L_{b_0} is connected, we conclude that $\mathcal{O}_{x_0} = L_{b_0}$, i.e. L_{b_0} is an orbit.

It is well known (see Bourbaki [1] Chap. 1, §7 or Fuks and Rokhlin [1] Chap. 4, §2, subsection 3) that the restriction to the orbit action, $g_t|_{L_{b_0}}: L_{b_0} \longrightarrow L_{b_0}$, is isomorphic to the action of $\mathbb{R}^n$ on the homogeneous space $\mathbb{R}^n/\Gamma$, where Γ is a subgroup of $\mathbb{R}^n$, the action being induced by shifts. This means that there is a one-to-one map $h: \mathbb{R}^n/\Gamma \longrightarrow L_{b_0}$ such that the following diagram commutes

(k)
$$\begin{array}{ccc} L_{b_0} & \xrightarrow{g_t|L_{b_0}} & L_{b_0} \\ \uparrow{\scriptstyle h} & & \uparrow{\scriptstyle h} \\ \mathbb{R}^n/\Gamma & \xrightarrow{\sigma_t} & \mathbb{R}^n/\Gamma \end{array}.$$

Elements of $\mathbb{R}^n/\Gamma$ are classes of the form $\tau + \Gamma = \{\tau + \gamma: \gamma \in \Gamma\}$ where $\tau \in \mathbb{R}^n$. In (k) σ_t is the shift which assignes the class $\tau + t + \Gamma$ to the class $\tau + \Gamma \in \mathbb{R}^n/\Gamma$. There are explicit expressions for Γ and h. For each $b \in B$ define the subgroup

(l) $$\Gamma_b = \{\gamma \in \mathbb{R}^n: g_\gamma(x_1) = x_1\},$$

where x_1 is some chosen point in L_b. Due to the commutativity of $\mathbb{R}^n$ and the transitivity of $g_t|_{L_b}$, the subgroup Γ_b is independent of the choice of $x_1 \in L_b$. The subgroup Γ in (k) coincides with Γ_{b_0} and the map h in (k) is obtained as

(m) $$h(t + \Gamma) = g_t(x_0).$$

The point $g_t(x_0)$ does not change if one replaces t by $t + \gamma, \gamma \in \Gamma$. Therefore, h is well defined, it depends only on the choice of x_0.

Lemma 7.10 yields that a small neighbourhood of the origin in $\mathbb{R}^n$ does not contain the points of Γ expert 0 itself, i.e. is a discrete subgroup of $\mathbb{R}^n$. All discrete subgroups of $\mathbb{R}^n$ are completely enumerated as follows. Given a linearly independent set $E=\{e_1,e_2,\ldots,e_k\}$ of vectors in $\mathbb{R}^n$, let Γ^E denote the smallest subgroup generated by E. We have $\Gamma^E=\{m_1e_1+m_2e_2+\cdots+m_ke_k:(m_1,m_2,\ldots,m_k)\in\mathbb{Z}^k\}$. So Γ^E is isomorphic to $\mathbb{Z}^k$. Here $0\leq k\leq n$. Obviously Γ^E is discrete.

Lemma 7.11. *Any discrete subgroup of the additive group* $\mathbb{R}^n$ *coincides with some* Γ^E.

The reader can find the proof of this Lemma in Arnol'd (1978) §49 G or in Bourbaki (1963) Chapter 7, §1.1.

The space $\mathbb{R}^n/\Gamma^E$ is a manifold, and is diffeomorphic to $\underbrace{S^1\times S^1\cdots\times S^1}_{k\text{-times}}\times\mathbb{R}^{n-k}$, where S^1 denotes the circle. Another consequence of Lemma 7.10 is that the map h defined by (m) is a diffeomorphism. Since L_{b_0} is compact, we have that $k=n$, and L_{b_0} is diffeomorphic to the n-torus $\mathbb{T}^n=S^1\times S^1\times\cdots\times S^1$ (n-times).

The next step is to prove that the subgroup Γ_b defined by (l) depends smoothly on $b\in B$. Consider the equation

$$g_e(s(b))=s(b) \tag{n}$$

with respect to $e\in\mathbb{R}^n$ as an unknown quantity.

Let $E^0=(e_1^0,e_2^0,\ldots,e_n^0)$ be a basis of $\mathbb{R}^n$ such that $\Gamma_{b_0}=\Gamma^{E^0}$. The vectors $e_i^0, 1\leq i\leq n$, satisfy (n) if one puts $b=b_0$. Lemma 7.10 shows that there are neighbourhoods V_2 of e_i^0 in $\mathbb{R}^n$ and U_2 of b_0 in B such that (n) may be solved uniquely, the solution being a smooth map $e_i:U_2\longrightarrow V_2$ satisfying the initial condition $e_i(b_0)=e_i^0$. Indeed, consider the map $\psi_1=g_{e_i^0}\circ\psi\circ h_{e_i^0}$, where $h_e:B\times\mathbb{R}^n\longrightarrow B\times\mathbb{R}^n$ is defined by $h_e(b,t)=(b,t-e)$, and ψ is as in Lemma 7.10, ψ_1 is defined in a neighbourhood of (b_0,e_i^0) and $\psi_1(b_0,e_i^0)=s(b_0)$. One may take the preimage of S under the map ψ_1 as the graph of e_i. Since S meets each orbit of g_t transversally, e_i is correctly defined and smooth in some neighbourhood of b_0 (which can be smaller than that of Lemma 7.10). Thus we obtain a map e_i with the property $g_{e_i(h)}(s(b))\in S$. Since g_t preserves fibres and S meets the fibre L_b in one point $s(b)$, we have $g_{e_i(b)}(s(b))=s(b)$. Let $U_3\subset B$ be an open neighbourhood of b_0 such that the n-set $E(b)=\{e_i(b),1\leq i\leq n\}$ of smooth solutions of (n) is defined on U_3, the initial conditions $e_i(b_0)=e_i^0$ being satisfied. We may assume that $e_i(b)$, $1\leq i\leq n$, are linearly independent for each $b\in U_3$, otherwise take a smaller neighbourhood. It follows that $\Gamma_b=\Gamma^{E(b)}$. Note that such a representation is valid in a small neighbourhood of b_0 contrary to the fact that Γ_b depends smoothly on b on the whole space B.

Define the map $\tilde{\Phi}:\mathbb{R}^n\times U_3\longrightarrow X$ by the formula

$$\tilde{\Phi}(\varphi,b)=g_{t(\varphi,b)}(s(b)),$$

where

$$t(\varphi,b)=\sum_{i=1}^{n}\varphi^ie_i(b),\qquad \varphi=(\varphi^1,\varphi^2,\ldots,\varphi^n)\in\mathbb{R}^n.$$

This map is a covering due to Lemma 7.10, and because of (n) it may be factored through the $\mathbb{T}^n \times U_3$:

$$\begin{array}{ccc} \mathbb{R}^n U_3 & \xrightarrow{\tilde{\Phi}} & X \\ {\scriptstyle \pi \times 1_{U_3}} \downarrow & \nearrow {\scriptstyle \Phi} & \\ \mathbb{T}^n \times U_3 & & \end{array}$$

where $\pi: \mathbb{R}^n \longrightarrow \mathbb{T}^n = \mathbb{R}^n/\mathbb{Z}^n$ is the canonical projection and Φ is a local diffeomorphism. It follows from (k), (l), (m), and (n) that Φ maps the tori $\mathbb{T}^n \times \{b\}$ diffeomorphically onto the fibres L_b. Consider the pullback of the symplectic structure of X:

(o) $$\tilde{\Phi}^*\Omega = \sum_{i<k} \alpha_{ik}(\varphi, b) d\varphi^i \wedge d\varphi^k + \sum_{i<k} \beta_{ik}(\varphi, b)\, db_i \wedge db_k + \sum_{i,k} \gamma_{ik}(\varphi, b)\, d\varphi^i \wedge db_k,$$

$(b_1, b_2, \ldots, b_n)$ being the coordinates in U_3. Since the submanifolds $b = \text{const}$ in $\mathbb{R}^n \times U_3$ are Lagrangian (they are the preimages of fibres), we have $\alpha_{ik} = 0$. The same is true for submanifolds $\varphi = \text{const}$ since S and $g_t S$ are Lagrangian. Hence $\beta_{ik} = 0$. The coefficients γ_{ik} in the remaining terms of (o) do not depend on φ. Indeed, the action g_t leaves Ω invariant. It may be lifted onto $\mathbb{R}^n \times U_3$ via $\tilde{\Phi}$, the lift being the action of $\mathbb{R}^n \times 1_{U_3}$ on $\mathbb{R}^n \times U_3$ by shifts. The invariance of $\tilde{\Phi}^*\Omega$ under this action implies the desired independence of γ_{ik} to φ. We rewrite (o) as follows:

(p) $$\tilde{\Phi}^*\Omega = \sum_{i=1}^{n} d\varphi^i \wedge \theta_i,$$

where

$$\theta_i(b) = \sum_{k=1}^{n} \gamma_{ik}(b)\, db_k, \quad 1 \leqq i \leqq n,$$

which may be regarded as 1-forms defined on U_3. The set $\theta_i(b)$, $1 \leqq i \leqq n$, is linearly independent. Taking U_3 diffeomorphic to a ball $B^n \subset \mathbb{R}^n$ and applying the Poincaré Lemma 2.19e, we obtain functions I_i, $1 \leqq i \leqq n$, such that $\theta_i = dI_i$. The formula (p) now reads

$$\tilde{\Phi}^*\Omega = \sum_{i=1}^{n} d\varphi^i \wedge dI_i.$$

Since the forms θ_i are linearly independent, there is a neighbourhood $U_4 \subset U_3$ of b_0 with the property that the map $I = (I_1, I_2, \ldots, I_n): U_4 \longrightarrow \mathbb{R}^n$ is a local diffeomorphism. It follows that the pair of maps $((1_{\mathbb{T}^n} \times I) \circ \Phi^{-1}, I)$ is the desired isomorphism between the restriction of our Lagrangian fibration onto U_4 and the standard toroidal Lagrangian fibration $p_2: \mathbb{T}^n \times U_5 \longrightarrow U_5$, where $U_5 = I(U_4)$.

Local Structure of an Isomorphism of Lagrangian Fibrations with Compact Connected Fibres

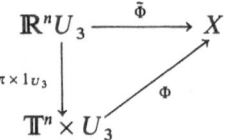

Theorem 7.2 reduces the problem of describing isomorphisms between Lagrangian fibrations with compact connected fibres to that of the standard toroidal fibrations

over open subsets of $\mathbb{R}^n$, the said problem being in a local sense with respect to points of base spaces. Let (f, g) be an isomorphism between two standard toroidal fibrations $p_2: \mathbb{T}^n \times U \longrightarrow U$ and $p_2: \mathbb{T}^n \times V \longrightarrow V$, U, V being non-empty open subsets of $\mathbb{R}^n$. We have the following commutative diagram

$$\begin{array}{ccc} \mathbb{R}^n \times U & \xrightarrow{\tilde{f}} & \mathbb{R}^n \times V \\ \downarrow 1_U & & \downarrow 1_V \\ \mathbb{T}^n \times U & \xrightarrow{f} & \mathbb{T}^n \times V \\ \downarrow p_2 & & \downarrow p_2 \\ U & \xrightarrow{g} & V \end{array} \tag{7.12}$$

where $\tilde{f}$ is the lift of f to the covering spaces. Obviously $\tilde{f}$ is a diffeomorphism. It is not uniquely defined. If $\tilde{f}_1$ and $\tilde{f}_2$ are two lifts of the same f then $\tilde{f}_1 \circ \tilde{f}_2^{-1}$ has the form $(\varphi, I) \longmapsto (\varphi + m, I)$ with $m \in \mathbb{Z}$. Up to the said uncertainty $\tilde{f}$ is well defined. Recall that f in (7.12) is a symplectic diffeomorphism, hence so is $\tilde{f}$. As usual, $\mathbb{R}^n \times U, \mathbb{R}^n \times V$ are supplied with symplectic structures induced from $(\mathbb{R}^n, \Omega_{st})$. Without loss of generality we may restrict ourselves to the case of U and (thereby) V diffeomorfic to a ball $B^n \subset \mathbb{R}^n$.

Proposition 7.13. *Let U in (7.12) be diffeomorphic to a ball. Then $\tilde{f}$ and g have the form*

$$\tilde{f}: (\varphi, I) \longmapsto (A(\varphi + \operatorname{grad} S(I)), A^{-1}I + b), \quad g: I \longmapsto A^{-1}I + b,$$

where $A = \{a_{ik}\}$ is a $n \times n$-matrix with integer coefficients, $\det A = \pm 1$, *S is a smooth function,* $\operatorname{grad} S(I) = (\partial S(I)/\partial I_1, \partial S(I)/\partial I_2, \ldots, \partial S(I)/\partial I_n)$ *and b is a constant vector.*

Proof. By the definition of an isomorphism of fibrations, the map $\tilde{f}$ has the form $(\varphi, I) \longmapsto (f^1(\varphi, I), \ldots, f^n(\varphi, I), g_1(I), \ldots, g_n(I))$. Since $\tilde{f}$ is symplectic we have

$$\sum_i d\varphi^i \wedge dI_i = \sum_{i,j,k} \frac{\partial f^i}{\partial \varphi^j} \frac{\partial g_i}{\partial I_k} d\varphi^j \wedge dI_k + \sum_{i, j<k} \left(\frac{\partial f^i}{\partial I_j} \frac{\partial g_i}{\partial I_k} - \frac{\partial f^i}{\partial I_k} \frac{\partial g_i}{\partial I_j} \right) dI_j \wedge dI_k.$$

Therefore

$$\sum_i \frac{\partial f^i}{\partial \varphi^j} \frac{\partial g_i}{\partial I_k} = \delta_j^k \quad \text{(the Kronecker symbol)}, \tag{a'}$$

$$\sum_i \left(\frac{\partial f^i}{\partial I_j} \frac{\partial g_i}{\partial I_k} - \frac{\partial f^i}{\partial I_k} \frac{\partial g_i}{\partial I_j} \right) = 0, \quad 1 \leqq k, j \leqq n. \tag{b'}$$

The equation (a′) yields that the matrix $\{\partial g_i / \partial I_k(I)\}$ is the inverse of the matrix $\{\partial f^i / \partial \varphi^j\}$. Since the former is independent of φ, so is the latter, and we have

$$f^i(\varphi, I) = \sum_{i=1}^n a_k^i(I) \varphi^k + h^i(I). \tag{c'}$$

It follows from the commutativity of diagram (7.12) that $\tilde{f}$ preserves the orbits of the actions of the group $\mathbb{Z}^n \times \{0\}$ on $\mathbb{R}^n \times U'$ and $\mathbb{R}^n \times V'$ by shifts. In particular

$$a_k^i(I) = f^i(\varphi + \mathrm{ort}_k, I) - f^i(\varphi, I) \in \mathbb{Z},$$

$1 \leqq i, k \leqq n$. Since U is connected a_k^i is independent of I. The same consideration, being applied to the inverse map $\tilde{f}^{-1}$, shows that A^{-1} also has integer coefficients. Hence $\det A = \pm 1$. We have obtained that the map g is linear with matrix A^{-1}. This proves the formula for g. Denote by c_k^i the coefficients of A^{-1}. In virtue of (c') the equations (b') now read

$$\sum_i \left(\frac{\partial h^i}{\partial I_j} c_i^k - \frac{\partial h_i}{\partial I_k} c_i^j \right) = 0, \quad 1 \leqq k, j \leqq n.$$

The last equalities are equivalent to the form $\sum_{j=1}^n \sum_{i=1}^n c_i^j h^i(I)\, dI_j$ being closed. By Poincaré Lemma 2.192 we have

$$\sum_{j=1}^n \sum_{i=1}^n c_i^j h^i(I)\, dI_j = dS(I) = \sum_{j=1}^n \frac{\partial S(I)}{\partial I_j} dI_j,$$

where S is a smooth function. This gives $h = A \operatorname{grad} S$. □

Fibre-Preserving Dynamical System

Let $p: X \longrightarrow B$ be a Langrangian fibration. A *fibre-preserving dynamical system* on this fibration is a symplectic flow (or cascade) $f^t: X \longrightarrow X, t \in \mathbb{R}$ (or $\mathbb{Z}$), which preserves the fibration, i.e. the diagram

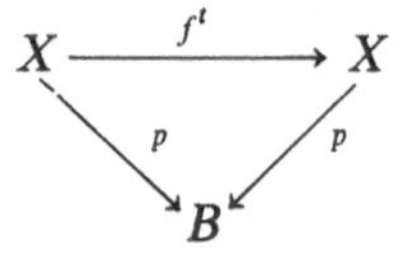

commutes for all t. This can also be expressed as $f^t(L_b) = L_b\, \forall_b \in B, \forall_t$, L_b denoting the fibre over a point b. Two fibre-preserving dynamical systems $f^t: X \longrightarrow X$ and $g^t: Y \longrightarrow Y$, (both of the same kind: flows or cascades) are said to be *isomorphic* if there is an *isomorphism* between them, this being an isomorphism between the underlying fibrations $p: X \longrightarrow B$, $q: Y \longrightarrow C$ which is consistent with the dynamics, i.e. such that the following diagram commutes for all t:

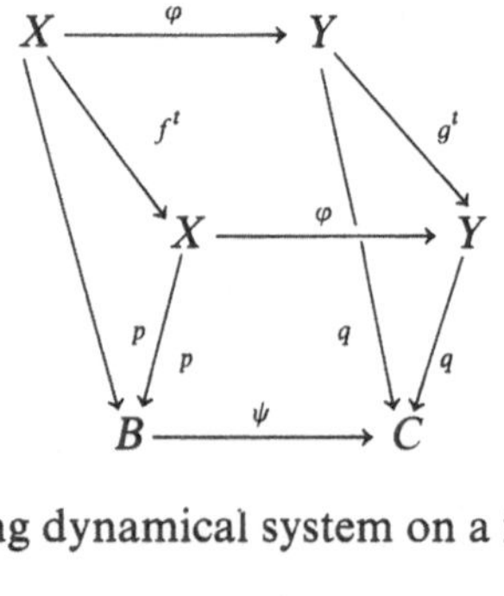

If $f^t: X \longrightarrow X$ is a fibre-preserving dynamical system on a fibration $p: X \longrightarrow B$, and

$U \subset B$ is a non-empty open set, then the *restriction* of f^t onto U is the system $f^t|_{p^{-1}(U)}: p^{-1}(U) \longrightarrow p^{-1}(U)$, $p^{-1}(U)$ being supplied with the restricted structure of the Lagrangian fibration. Two fibre-preserving dynamical systems are said to be *locally isomorphic* at points b and c of their bases B and C if there are neighbourhoods U of b in B and V of c in C such that the restrictions of the systems onto these neighbourhoods are isomorphic. The following example provides us with a local model for fibre-preserving dynamical systems.

Example 7.14 (*A standard integrable system in a symplectic annulus*). Let $p_2: A_D \longrightarrow D$ be the standard toroidal fibration over a non-empty open set $D \subset \mathbb{R}^n$ (see Example 7.1) and let $K: D \longrightarrow \mathbb{R}$ be a smooth function. Consider the covering $p: \mathbb{R}^n \times D \longrightarrow A_D$. Define the flow (cascade) $\tilde{f}^t: \mathbb{R}^n \times D \longrightarrow \mathbb{R}^n \times D$, $t \in \mathbb{R}$ (or $\mathbb{Z}$), by the formula

(7.14a) $$\tilde{f}^t(\varphi, I) = (\varphi + t\omega(I), I),$$

where

(7.14b) $$\omega(I) = \operatorname{grad} K(I) = \left(\frac{\partial K}{\partial I_1}, \frac{\partial K}{\partial I_2}, \ldots, \frac{\partial K}{\partial I_n}\right).$$

Straightforward calculation shows that $\tilde{f}^t$ preserves the symplectic structure on $\mathbb{R}^n \times D$ induced by the standard one. $\tilde{f}^t$ also commutes with the action of $\Gamma = \mathbb{Z}^n \times 1_D$ on $\mathbb{R}^n \times D$ by shifts. Here it induces a symplectic dynamics $\{f^t\}$ on A_D which preserves the fibres $\mathbb{T}^n \times \{d\}$, $d \in D$. The vector $\omega(I)$ defined by (7.14b) is called the *frequency-vector*, its components $\omega_i(I)$ are the *frequencies* of $\{f^t\}$. In the case of continuous time $\{f^t\}$ is Hamiltonian with Hamiltonian function $K \circ p_2$. The vector field of the system may be expressed in angle-action coordinates as

(7.14c) $$\xi = \sum_{i=1}^{n} \omega_i(I) \frac{\partial}{\partial \varphi^i}.$$

Theorem 7.15. *Any fibre-preserving symplectic dynamical system on a Lagrangian fibration with compact connected fibres is locally isomorphic at each point of its base to a standard integrable system of Example* 7.14.

Proof. In the case of discrete time this is an immediate consequence of Theorem 7.2 and Proposition 7.13. Let us now consider the case of continuous time. Theorem 7.2 and Proposition 7.13 give us the local isomorphism between our system and the system defined on the standard toridal fibration $p_2: A_D \longrightarrow D$, D being diffeomorphic to a ball, whose expression in angle-action coordinates on the covering space $\mathbb{R}^n \times D$ is given by

$$\tilde{f}^t: (\varphi, I) \longmapsto (\varphi + \operatorname{grad}_I S(I, t), I), \quad t \in \mathbb{R}.$$

The gradient operation grad_I acts only on the I-variables. For each $t \in \mathbb{R}$, the generating function $S(I, t)$ is defined uniquely up to a summand independent of I. Choose $I^0 \in D$ and norm S by the condition

(1) $$S(I^0, t) = 0 \qquad \forall t \in \mathbb{R}.$$

The flow identity $\tilde{f}^{t+\tau} = \tilde{f}^t \circ \tilde{f}^\tau$ yields

(2) $$\operatorname{grad}_I S(I, t+\tau) = \operatorname{grad}_I S(I, t) + \operatorname{grad}_I S(I, \tau).$$

(1) and (2) imply $S(I, t+\tau) = S(I, t) + S(I, \tau)$. Since $S(I, t)$ depends continuously on $t, S(I, t) = K(I)t$. Obviously $K(I)$ is smooth. □

Completely Integrable Hamiltonian Dynamical Systems

Let ξ be a locally Hamiltonian vector field on a symplectic manifold X. A smooth function $J: X \longrightarrow \mathbb{R}$ is called an *integral* of ξ if $\xi \cdot J = 0$ (see AI.45 for the definition of the application $\xi \cdot J$ of vector field ξ to a function J).

Proposition 7.16. *The following properties of a function J are equivalent:*

(A) *J is an integral of ξ,*

(B) *$J \circ f^t$ does not depend on t for any local flow f^t of ξ,*

(C) *$\{J|_U, H\} = 0$ for any local Hamiltonian $H: U \longrightarrow \mathbb{R}$ of ξ.*

Proof. Use of the differentiation formula AI.68 gives us

$$\left.\frac{dJ \circ f^t}{dt}\right|_{t=0,x} = \xi \lrcorner dJ|_x = \xi \cdot J|_x = (dH)^\# \lrcorner dJ|_x = \{H, J\}|_x.$$

The statements of 7.16 can be found by considering when the corresponding terms of this formula vanish. □

A locally Hamiltonian vector field on a $2n$-dimensional symplectic manifold X (a symplectic dynamical system with continuous time on a phase space X) is called *completely integrable* if there are smooth functions $J_i: X \longrightarrow \mathbb{R}$, $1 \leqq i \leqq n$, such that

(i′) J_i, $1 \leqq i \leqq n$, are integrals of ξ,

(ii′) they are in involution, i.e. $\{J_i, J_k\} = 0$, $1 \leqq i, k \leqq n$,

(iii′) their differentials $\{dJ_i|_x, 1 \leqq i \leqq n\}$ are linearly independent for almost all $x \in X$. Which means that the set of values x where this condition fails, is no where dense in X. The collection $\{J_i, 1 \leqq i \leqq n\}$ satisfying (i′)–(iii′) is called the *complete collection of integrals in involution.* For further use we shall often omit the word "completely" and say "an integrable system" instead.

Let $\{J_i, 1 \leqq i \leqq n\}$ be a complete collection of integrals in involution. Consider the map $J: X \longrightarrow \mathbb{R}^n$ defined by

$$J(x) = (J_1(x), J_2(x), \ldots, J_n(x)).$$

Denote by X_r the set of all $x \hat{\in} X$ for which the condition of (iii) is valid, the *regular set.* It is a dense open subset of X. Obviously, $B = J(X_r)$ is a non-empty open set in $\mathbb{R}^n$, and $J|_{X_r}: X_r \longrightarrow B$ is a Lagrangian fibration (take into account Proposition 7.9). We call it the *Lagrangian fibration generated by a system* $\{J_i, 1 \leqq i \leqq n\}$.

Analogous definitions and construction may be done for the case of discrete time. The sole change is in defining the integral: the condition $\xi \cdot J = 0$ must be replaced by

(B′) $J \circ f^t$ does not depend on t.

The property (B′) is clear when we deal with the global cascade $f^t: X \longrightarrow X$, $t \in \mathbb{Z}$. In the case of a local cascade generated by a local symplectic diffeomorphism $f: U \longrightarrow X$, $U \subset X$, when the domain of f^t depends on t, we interpret (B′) to be valid in a pointwise sense, i.e. $J \circ f^t(x) = J(x)$ whenever $f^t(x)$ is defined. We may also consider an integral only defined on U. The following example motivates the latter case of the above definition.

Example 7.17 (*Hyperbolic fixed point in the plane*). Consider a local symplectic diffeomorphism $f: U \longrightarrow \mathbb{R}^2$, $(0,0) \in U \subset \mathbb{R}^2$, $\mathbb{R}^2$ being supplied with the standard symplectic structure. Let f leave the origin $(0,0)$ fixed: $f(0,0) = (0,0)$. We call the fixed point $(0,0)$ *hyperbolic* if the tangent $T_{(0,0)}f: \mathbb{R}^2 \longrightarrow \mathbb{R}^2$ is a hyperbolic map, i.e. its spectrum has no intersection with the unit circle in the complex plane. In this case the eigenvalues λ_1 and λ_2 of $T_{(0,0)}f$ are real, and $\lambda_1 = \lambda_2^{-1} \neq 1$. Sternberg's theorem (see Hartman (1964)) asserts that there are smooth coordinates (X_1, X_2) with the origin at $(0,0)$ possibly defined in a smaller neighbourhood $(0,0) \in U_1 \subset U$, such that the map f in these coordinates read

$$(X_1, X_2) \longmapsto (\Lambda X_1, \Lambda^{-1} X_2), \tag{7.17a}$$

where $\Lambda = \lambda_1 + c(X_1 X_2)^m$, the constant $c \in \mathbb{R}$ and integer $m \geqq 0$ depending on f. The restriction $f|_{U_1}$ generates a local symplectic cascade which is integrable. The integral is $J = X_1 X_2$. The generated Lagrangian fibration in $(U_1)_r = U_1 \setminus \{(0;0)\}$ is shown in Fig. 22. There is no open non-empty domain where all maps $f^t, t \in \mathbb{Z}$, are defined simultaneously. Here the fibres are neither compact, nor connected.

Theorem 7.18. *Consider a completely integrable symplectic dynamical system on a symplectic manifold X, $\dim X = 2n$, and let $J_1, J_2, \ldots, J_n$ be its integrals in involution, $X_r \subset X$ be the regular set of the map $J = (J_1, J_2, \ldots, J_n): X \longrightarrow \mathbb{R}^n$. If the fibres $J^{-1}\{y\}, y \in J(X_r)$, are connected and compact then*

(1) *the restriction of our system to X_r has complete dynamics $f^t: X_r \longrightarrow X_r$, $t \in \mathbb{R}$ (or $\mathbb{Z}$ for discrete time) which preserves the Lagrangian fibration $J|_{X_r}: X_r \longrightarrow J(X_r)$,*

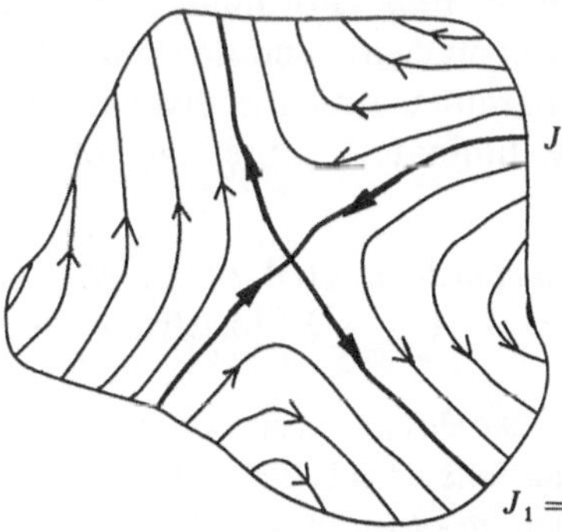

Fig. 22. The motion near a hyperbolic fixed point

(2) *each point of* $J(X_r)$ *has a neighbourhood* V *such that the restriction of* f^t *onto* $J^{-1}(V)$ *is isomorphic, as a fibre preserving dynamical system, to some standard integrable system on a symplectic annulus of Example* 7.14.

Proof. Is an immediate consequence of 7.9 and 7.15.

The Problem of Finding the Angle–Action Variables

In applications one often comes across the situation where we know a complete set $\{J_i, 1 \leqq i \leqq n\}$ of integrals in involution satisfying (i′)–(iii′), and we wish to find the angle–action coordinates (φ, I). Below we give the recipe for solving such a problem.

Since the J's are known, we can fined the invariant tori $\mathscr{T}_y$, $y = (y_1, y_2, \ldots, y_n)$, defined by the equations $J_i = y_i$, $1 \leqq i \leqq n$. Consider a set of such tori covering some invariant open set

(iv′) $$U = \bigcup_{y \in V} \mathscr{T}_y = J^{-1}(V) \subset X_r, \qquad V \subset \mathbb{R}^n \textit{ open}.$$

Recall that X_r is the set of "regular" points defined by condition (iii′). We are going to look for the unknown angle–action variables on the set U.

Theorem 7.18 ensures the existence of such variables provided U is sufficiently small. Then $U \subset \psi(A_D)$ where $\psi: A_D \longrightarrow X$ is a symplectic embedding which conjugates the standard integrable system with the restriction of the system in question onto $\psi(A_D)$. This implies that

(v′) $\Omega|_U$ *is exact, i.e. there is a 1-form* θ *such that* $\Omega|_U = -d\theta$.

Since U admits a foliation into invariant tori, one can choose a diffeomorphism (not necessarily symplectic) $g: \mathbb{T}^n \times V \longrightarrow U$. Take the standard circles $C_k^{co} \subset \mathbb{T}^n$, $1 \leqq k \leqq n$, which are the images of the coordinate axes under the canonical projection $\mathbb{R}^n \longrightarrow \mathbb{T}^n = \mathbb{R}^n/\mathbb{Z}^n$, and let

(7.19) $$C_j(y) = g(C_j^{co} \times \{y\}), \quad y \in V.$$

An essential notion which we will now take into account is that, in view of Proposition 7.13, the angle–action coordinates are not uniquely defined, an $n \times n$-matrix A with integer coefficients and determinant ± 1, a constant vector b, and a smooth function S parametrising the diversity of different choices of the said co-ordinates. We can eliminate the first two of these arbitrary choices by introducing the condition

(vi′) *we look for angle–action coordinates* $(\varphi^1, \varphi^2, \ldots, \varphi^n, I_1, I_2, \ldots, I_n)$, *or equivalently for a symplectic embedding* $\psi: A_D \longrightarrow X$ *satisfying the appropriate conjugating conditions, such that*
(vi′a) *for each* $j \in \{1, 2, \ldots, n\}$ *the circle* $C_j(y) \subset \mathscr{T}_y$ *is of the same homotopy class as the circle defined by the equations* $\varphi^i = 0$, $i \neq j$ *(the image under* $\psi: A_D \longrightarrow X$ *of the circle* S_j *defined at the beginning of the proof of Proposition* 2.5a),
(vi′b) $\psi^*\theta = \sum_{i=1}^n I_i d\varphi^i$ *where* θ *is the* 1*-form of condition* (v′).

The condition (vi′a) fixes the matrix A of 7.13 and (vi′b) fixes the vector b of 7.13, if one makes the transformation from arbitrary angle action coordinates to those satisfying (vi′). As an immediate consequence of (vi′) we obtain the following formula for the action variables:

$$I_j(y) = \int_{C_j(y)} \theta, \quad 1 \leqq j \leqq n,\ y \in V, \tag{7.20}$$

the $C_j(y)$ being supplied with the orientation induced by the positive direction of the corresponding axis in $\mathbb{R}^n$ (see (7.19)). Note that the path $C_j(y)$ may be changed for another path lying on the same torus and homotopic to $C_j(y)$. We suppose that

(vii′) *the map* $y \longmapsto I(y) = (I_1(y), I_2(y), \ldots, I_n(y))$ *is a local diffeomorphism.*

The inverse map to that of (vii′) will be denoted $I \longmapsto h(I)$.

To find the angle coordinates we fix the third variable element of the angle–action coordinates, the zero surface of φ-variables (the varying of the latter is governed by the function S in 7.13). To do this we suppose that

(viii′) *there is a symplectic embedding* $\chi: A_B \longrightarrow X$, *$B$ being open and simply connected, such that* $U \subset \chi(A_B)$ *and the preimage of each torus,* $\chi^{-1}(\mathcal{T}_y)$, $y \in B$, *has a nondegenerate one-to-one projection onto a torus* $\mathbb{T}^n \times \{b\} \subset A_B = \mathbb{T}^n \times B$ *in that product, and*

$$\chi^*\theta = \sum_{i=1}^{n} p_i\, dq^i. \tag{viii′a}$$

Here θ is the form of (v′) *and* (vi′b), *and we use the notation* $(q^1, q^2, \ldots, q^n, p_1, p_2, \ldots, p_n)$ *for the angle–action coordinates on the auxiliary annulus* A_B.

We are now going to obtain the expression for $(\varphi^1, \varphi^2, \ldots, \varphi^n)$ which we look for in terms of $(q^1, q^2, \ldots, q^n, p_1, p_2, \ldots, p_n)$, or, in other words, to find a symplectic invertible map $\tilde{\upsilon}: \mathbb{R}^n \times D \longrightarrow \mathbb{R}^n \times B$ commuting with the actions of $\mathbb{Z}^n \times 1_D$ and $\mathbb{Z}^n \times 1_B$ on the domain and the image such that ψ, which we are looking for, equals $\psi = \chi \circ \upsilon$ where $\upsilon: A_D \longrightarrow A_B$ is the quotient of $\tilde{\upsilon}$.

Consider the function

$$G(q^1, q^2, \ldots, q^n, I_1, I_2, \ldots, I_n) = \int_0^{(q^1, q^2, \ldots, q^n)} \sum_{i=1}^{n} p_i dq^i. \tag{7.21}$$

where the integration is carried out the path in the covering space $\mathbb{R}^n \times B$ of A_B belonging to the preimage of the torus $\mathcal{T}_{h(I_1, I_2, \ldots, I_n)}$ under the map $\chi \circ p: \mathbb{R}^n \times B \longrightarrow X$, $p: \mathbb{R}^n \times B \longrightarrow A_B$ being the projection defined in 2.5, the path joining 0 with the point with first coordinates $(q^1, q^2, \ldots, q^n)$. If we demand that the angle coordinates which we seek satisfy

(ix′) $\tilde{\upsilon}$ *maps the Lagrangian slice* $(\varphi^1, \varphi^2, \ldots, \varphi^n) = 0$ *onto the Lagrangian slice* $(q^1, q^2, \ldots, q^n) = 0$

then, due to (vi′b), (viii′a), and definition (7.21), the angle–action coordinates (φ, I)

are linked to the auxiliary coordinates (q,p) by the formulas

$$p_i = \frac{\partial G}{\partial q^i}(q,I), \qquad \varphi^i = \frac{\partial G}{\partial I_i}(q,I), \quad 1 \leqq i \leqq n, \tag{7.22}$$

i.e. G is the generating function of the desired transformation $\tilde{v}$. The nondegeneracy condition of (viii′) yields the unique solvability of (7.22), so $\tilde{v}$ is correctly defined. We have the formulae (7.19)–(7.22) to solve the stated problem under the posed conditions (iv′)–(ix′), the conditions (vi′) and (ix′) being introduced only for eliminating the arbitrary nature of in the choice of the angle–action coordinates.

Remark 7.23. The choice of the form θ plays a crucial role in the construction of action coordinates by formula (7.19). In many cases such a form is given or naturally arises. For instance, if we deal with systems defined in the cotangent bundle, there is the Liouville form Θ_Q. In the case of generalized geodesic flow described in §6 the form of most interest is the post-Liouville form defined by (6.13). Unifortunately, this form Θ is discontinuous and multivalued if $\partial Q \neq \varnothing$. Nevertheless, it may be used for θ in (7.19) provided the path of integration is chosen to be transversal to Π (in order to eliminate multivalued images). Indeed, the form Θ differs from a smooth form θ (which is known to exist) by a closed piece-wise differentiable 1-form defined in U. The integration over closed paths crossing Π transversally gives the constants independent of y. To prove this notice that the integrals of $\Theta - \theta$ over two isotopic closed paths $C_j(y_1) \subset \mathscr{T}_{y_1}$ and $C_j(y_2) \subset \mathscr{T}_{y_2}$ differ by the integrals over two pieces 1 and -1 shown in Fig. 23 taken over the limit values of the integrand at points of Π lying on the corresponding sides of Π. So the usage of a discontinuous Θ in comparison to a smooth θ results in the adding of a constant vector of I, i.e. in another choice of action variables.

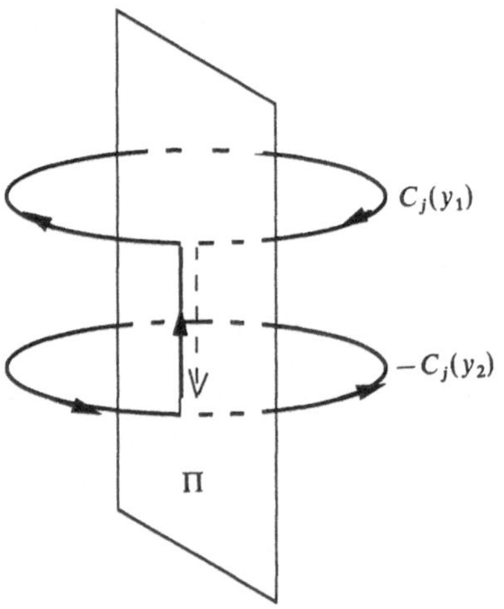

Fig. 23.

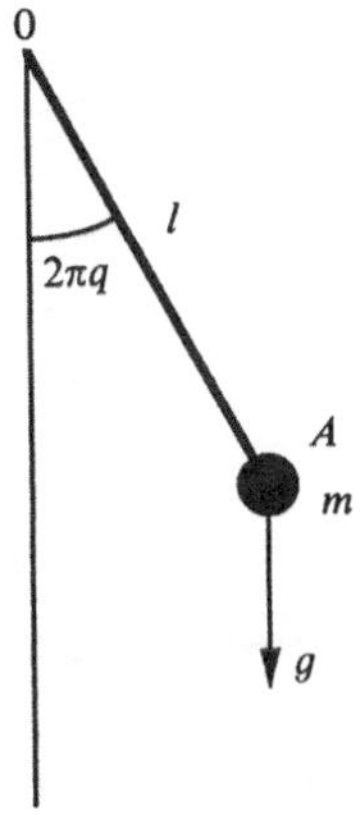

Fig. 24. A pendulum in a homogeneous gravitational field

Some Examples

7.24. Any two-dimensional Hamiltonian system with continuous time is integrable provided the set of its fixed points is no where dense. Indeed, the Hamiltonian function H itself forms the complete system of integrals in involution.

Example 7.25. Pendulum. Consider the mathematical pendulum in the homogeneous earth gravitational field g. It consists of a weightless rigid straight arm of length l with one fixed end at 0 and a pointwise mass m attached to the other end A (see Fig. 24). Take the angle variable $2\pi q$ which describes the deviation of the arm from the vertical stable position of the equilibrium. Then the Lagrangian of the system is given by (see Landau and Lifschitz (1963))

$$L(q, \dot{q}) = \frac{4\pi^2 m l^2 \dot{q}^2}{2} + mgl\cos(2\pi q).$$

Here the dot denotes the derivative with respect to time. Making the Legendre transform $\dot{q} \longmapsto p = \partial L/\partial \dot{q}$, we obtain the Hamiltonian function

$$H(q, p) = \frac{1}{8\pi^2} \frac{p^2}{ml^2} - mgl\cos(2\pi q). \tag{7.25a}$$

We may regard (7.25a) as being defined on the two-dimensional standard annulus $A_{\mathbb{R}}$, taking this to be the phase space of the system. The system is integrable due to 7.24. Figure 25 contains the picture of the tori $H = y = \text{const}$. There are three regions U_0, $U_\pm$ filled by the tori $\mathscr{T}_y$ satisfying (iv′). They are separated by the separatrices C_+ and C_-, which intersect at the fixed point F with coordinates $(0, 0)$. Another fixed point $G = (\frac{1}{2} \bmod \mathbb{Z}, 0)$ belongs to U_0. The regular set here is $X_r = A_{\mathbb{R}} \setminus \{F, G\}$. For each U_α, $\alpha \in \{0, +, -\}$, the above construction for finding the angle–action variable may be applied. In particular, formula (7.19) yields for the action variable to be equal to the area of the oval bounded by the corresponding tori in U_0 and the area between the tori and the q-axis in $U_\pm$.

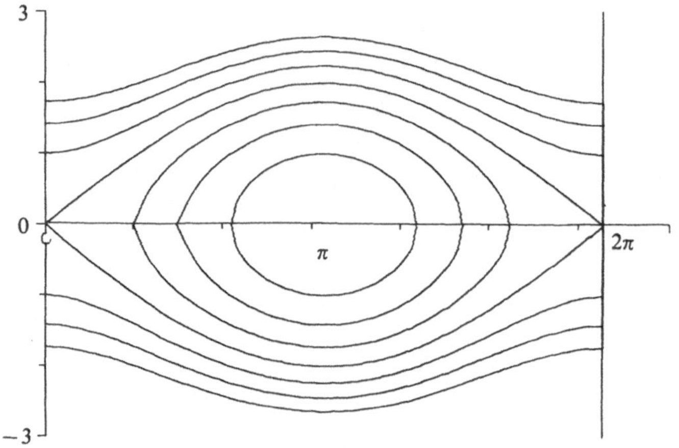

Fig. 25. The foliation on invariant tori of the phase space of a pendulum

7.26. Any four-dimensional Hamiltonian system with continuous time is integrable if it has one integral J such that $dJ(x)$ and $dH(x)$ are almost everywhere independent. Indeed, the system of integrals H, J is in involution due to Proposition 7.16.

Example 7.27. Billiard in the circle. Consider the billiard in the domain $Q = \{(x^1, x^2): (x^1)^2 + (x^2)^2 \leqq R^2\}$ supplied with the Euclidean metric induced from $\mathbb{R}^2$. The projections of billiard trajectories onto the coordinate space Q are straight lines (rays) reflecting from the circle boundary by the law: "the reflected angle equals the incident one" (see the general formulae (6.8), (6.9a)). One can easily verify that given a circle C_J with the same centre and of radius $\sqrt{J} < R$, the rays tangent to C_J remain tangent to C_J after reflections (see Fig. 26). So, the number J, as a function of a point on a trajectory, is the integral. The explicit expressions

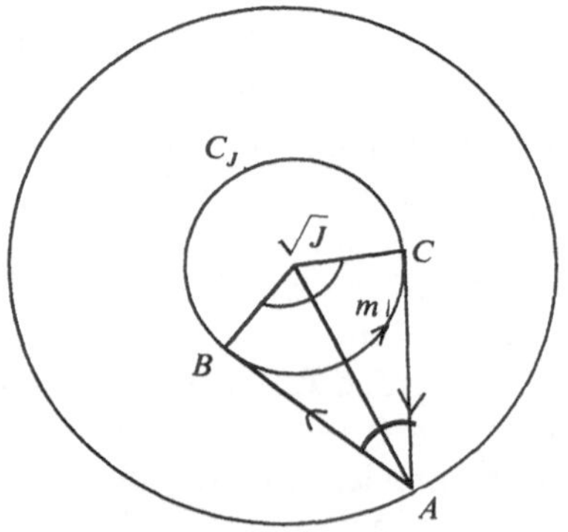

Fig. 26. Circular billiards

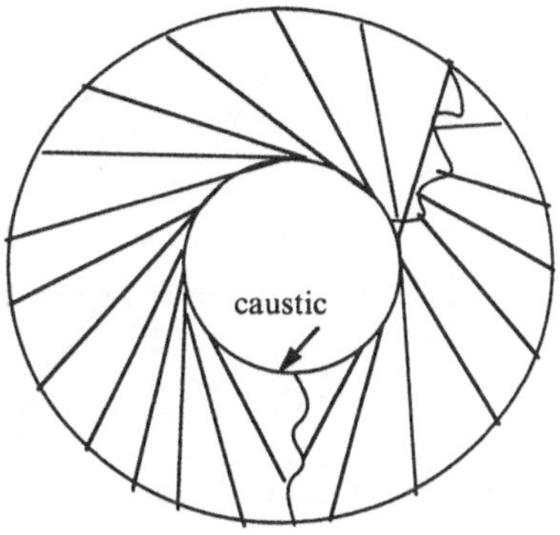

Fig. 27. Caustic and the field of rays in a circular domain

for H and J:

$$H = \frac{p_1^2 + p_2^2}{2}, \qquad J = \frac{(q^1 p_2 - q^2 p_1)^2}{p_1^2 + p_2^2}, \tag{7.27a}$$

enable us to directly verify the mutual independence of dH and dJ. The regularity condition (iii′) fails only at points satisfying $p_2 q^1 - q^2 p_1 = 0$ which correspond to the rays passing through the centre of the circle ($J = 0$). The remaining part of the phase space Z (X_r of this example) may be taken as the domain U of the construction of the preceding subsection. The collection of projections onto Q of trajectories filling one torus is shown in Fig. 27. The torus covers the annulus between two circles twice. The said circles being Q and the concentric circle of radius $\sqrt{J}$, the so-called *caustic*. The rays of this torus are all tangent to the caustic. In keeping with Remark 7.23 we use the post-Liouville form Θ for calculating the action coordinates. For closed paths of integration we take the paths C_1 and C_2 whose projections on Q are correspondingly the caustic and the triangle consisting of the two straight lines AB and AC, and the piece of the caustic BmC (see Fig. 26). We have the following expressions for the corresponding action variables:

$$I_1 = 2\pi\sqrt{2H}\sqrt{J}, \qquad I_2 = \left(2\sqrt{R^2 - J} - 2\sqrt{J}\arccos\frac{J}{R}\right)\sqrt{2H}. \tag{7.27b}$$

Example 7.28. A billiard in the ellipse $Q = \{(x^1, x^2)\colon (x^1)^2/a^2 + (x^2)^2/b^2 \leqq 1\} \subset \mathbb{R}^2$, $a^2 > b^2$, with induced Euclidean metric is also integrable. It is well known that the curves C_J:

$$\frac{(x^1)^2}{J} + \frac{(x^2)^2}{J - c^2} = 1, \quad c^2 = a^2 - b^2, \tag{7.28a}$$

which are confocal ellipses if $c^2 < J < a^2$ and confocal hyperbolas if $0 < J < c^2$, possess the optical property: each ray tangent to C_J remains tangent to C_J after reflection from the boundary (Fig. 28). As in the previous case of the circle, we

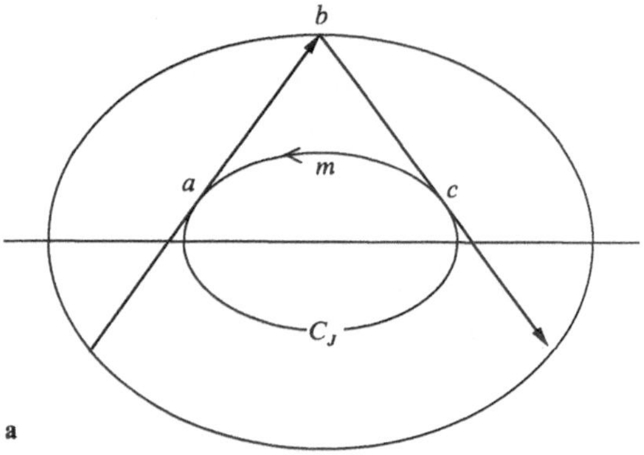

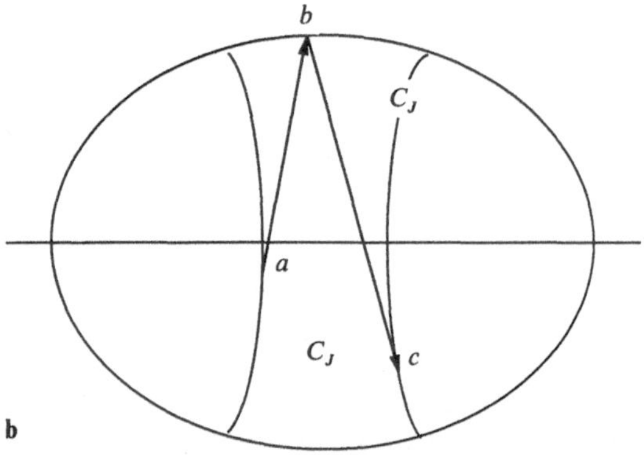

Fig. 28. Elliptic billiards. Two kinds of motion; (a) whispering gallery, (b) bouncing ball

may consider the parameter J as the integral of the billiard. To obtain the value of J at a point $z \in Z$ one must take the projection of the trajectory of z onto Q and find the uniquely defined curve C_J to which the projection is tangent. Here the regular set Z_r is divided into two regions U_{WG} where $c^2 < J < a^2$, and U_{BB} where $0 < J < c^2$ ("whispering gallery" and "bouncing ball" regions). The complement to $U_{WG} \cup U_{BB}$ consists of trajectories whose projections are rays passing through the foci $(\pm c, 0)$. The linear independence of dH and dJ fails only on two periodic trajectories which project onto the diameters of the ellipse. The tori in U_{WG} have the ellipses C_J, $c^2 < J < a^2$, as caustics. Each such torus projects twice onto the annulus between ∂Q and C_J (Fig. 29a). The torus of U_{BB} is drawn in Fig. 29b. Its caustic, C_J, is a hyperbola. The torus itself projects four times onto the domain between two branches of C_J. The preimages of a point $q \in Q$ are shown on the diagram as arrows. They correspond to the four directions of the straight lines tangent to C_J and passing through q. Starting with two integrals H and J, one can construct the action variables as in Example 7.27. In the case of the "whispering gallery" region, I_1 is equal to the length of the arc of the caustic C_J multiplied by $\sqrt{2H}$ and $I_2 = \sqrt{2H}\ (|ab| + |bc| - |\widehat{amb}|)$ (see Fig. 28a). Here $|\cdot|$ denotes the length of the straight segment, $|\hat{\cdot}|$ denotes the length of the arc. In the

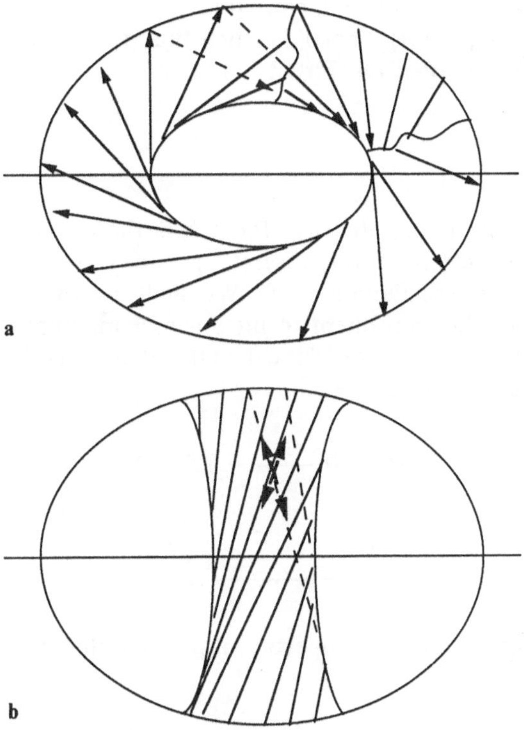

Fig. 29. Caustics and corresponding rays in an elliptic domain of (a) whispering gallery and (b) bouncing ball types

case of the "bouncing ball" torus, I_1 is equal the length of the arc of $C_J \cap Q$ multiplied by $\sqrt{2H}$, and I_2 is the common length of the arc of ∂Q which is covered by the torus also multiplied by $\sqrt{2H}$. This answer corresponds to the special choice of basic cycles C_1 and C_2 on tori. Another choice will give us other values of I_1, I_2, the new values being linked to the old ones by a linear transformation whose 2×2-matrix has integer coefficients and determinant ± 1.

§8. Systems in an Annulus

The objects of interest here are symplectic dynamical systems whose phase space is a symplectic annulus $A_D = \mathbb{T}^n \times D$ where $D \subset \mathbb{R}^n$ is a non-empty open set. We assume in this section that D is simply connected and bounded. In the case of continuous time the generator of a system is a locally Hamiltonian vector field ξ defined on A_D, in the case of discrete time a local symplectic diffeomorphism $f: A_D \longrightarrow A_{D'}$, $D' \supset D$ being another simply connected open set, plays the role of a generator. We restrict ourselves to symplectic embeddings f which are isotopic to the inclusion $i: A_D \hookrightarrow A_{D'}$. This means that there is a smooth family of local diffeomorphisms $f_\tau: A_D \longrightarrow A_{D'}$, $\tau \in [0,1]$, such that $f_0 = i$ and $f_1 = f$. The set of all such maps will be denoted $\mathrm{Diff}^{\mathrm{symp}}_{\mathrm{loc,o}}(A_D, A_{D'})$.

Lifts

Recall that the symplectic annulus A_D, $D \subset \mathbb{R}^n$, is the quotient space of $\mathbb{R}^n \times D$ with respect to the action of the discrete group Γ generated by the shifts $x \longmapsto x + \mathrm{ort}_i$, $1 \leqq i \leqq n$ (see Example 2.5). We shall use the notations of 2.5′ for the projections. It will be convenient to lift our systems onto the covering space $\mathbb{R}^n \times D$. A map $\tilde{f}: \mathbb{R}^n \times D \longrightarrow \mathbb{R}^n \times D'$ is called the *lift* of a map $f \in \mathrm{Diff}^{\mathrm{symp}}_{\mathrm{loc,o}}(A_D, A_{D'})$ if the following diagram commutes

$$\begin{array}{ccc} \mathbb{R}^n \times D & \xrightarrow{\ \tilde{f}\ } & \mathbb{R}^n \times D' \\ \Big\downarrow{\scriptstyle p} & & \Big\downarrow{\scriptstyle p} \\ A_D & \xrightarrow{\ f\ } & A_{D'} \end{array}$$

It immediately follows from the properties of p and f that $\tilde{f}$ is a local symplectic diffeomorphism.

Proposition 8.1. (i) *Each $f \in \mathrm{Diff}^{\mathrm{symp}}_{\mathrm{loc,o}}(A_D, A_{D'})$ has a lift.* (ii) *Any two lifts of the same f differ by a constant vector-valued function whose value is an element of Γ (we identify a shift $x \longmapsto x + v$ with a vector v).*

(iii) *Necessary and sufficient condition for a symplectic local diffeomorphism $\tilde{f}: \mathbb{R}^n \times D \longrightarrow \mathbb{R}^n \times D'$ to be a lift of some map $f \in \mathrm{Diff}^{\mathrm{symp}}_{\mathrm{loc,o}}(A_D, A_{D'})$ is its commutativity with the action of Γ:*

$$\tilde{f} \circ \gamma = \gamma \circ \tilde{f} \quad \forall \gamma \in \Gamma, \tag{8.1a}$$

or, equivalently, that $\tilde{f}$ may be represented as

$$\tilde{f}(x) = x + g(x), \quad x \in \mathbb{R}^n \times D, \tag{8.1b}$$

where $g(x)$ is a smooth vector-valued periodic function.

We say a function g defined on $\mathbb{R}^n \times D$ is *periodic* if

$$g(x + \gamma) = g(x) \quad \forall x \in \mathbb{R}^n \times D,\ \forall \gamma \in \Gamma,$$

or, equivalently, in (φ, I) coordinates: $g(\varphi + \mathrm{ort}_i, I) = g(\varphi, I)$, $1 \leqq i \leqq n$, $(\varphi, I) \in \mathbb{R}^n \times D$. The space of all smooth periodic functions defined on $\mathbb{R}^n \times D$ will be denoted $C^\infty_{\mathrm{per}}(\mathbb{R}^n \times D)$. Those functions in $C^\infty_{\mathrm{per}}(\mathbb{R}^n \times D)$ whose supports are contained in $\mathbb{R}^n \times [D]_a$, $a > 0$, (see List of Notations), form a subspace $C^\infty_{\mathrm{per},a}(\mathbb{R}^n \times D) \subset C^\infty_{\mathrm{per}}(\mathbb{R}^n \times D)$. All diffeomorphisms $f: \mathbb{R}^n \times D \longrightarrow \mathbb{R}^n \times D$ whose components belong to $C^\infty_{\mathrm{per}}(\mathbb{R}^n \times D)$ (or $C^\infty_{\mathrm{per},a}(\mathbb{R}^n \times D)$) form a set $\mathrm{Diff}_{\mathrm{per}}(\mathbb{R}^n \times D)$ (or a set $\mathrm{Diff}_{\mathrm{per},a}(\mathbb{R}^n \times D)$ respectively).

Proof of Proposition 8.1. The assertion (i) is an immediate consequence of Theorem 5.6.19 of Fuks and Rohklin (1984). $\tilde{f}$ is smooth and symplectic due to the equation

$$p \circ \tilde{f} \circ (p|_U)^{-1} = f|_{p(U)},$$

U being any sufficiently small open set, such that $p|_U$ is injective. This equation also shows that any two lifts of f at each point $x \in \mathbb{R}^n \times D$ differ by an element of Γ, which depends on x continuously and is therefore locally constant. This follows from the property $\mathbb{R}^n \times D$ is corrected. This proves (ii).

The equivalence between (8.1a) and (8.1b) with periodical g is obvious. One can easily see that a symplectic embedding with these properties is a lift of some $f \in \mathrm{Diff}^{\mathrm{symp}}_{\mathrm{loc,o}}(A_D, A_{D'})$. Conversely, let $\tilde{f}$ be a lift of $f \in \mathrm{Diff}^{\mathrm{symp}}_{\mathrm{loc,o}}(A_D, A_{D'})$. Then for each $\gamma \in \Gamma$ there is $\gamma' \in \Gamma$ such that

$$\tilde{f} \circ \gamma = \gamma' \circ \tilde{f}. \tag{8.1c}$$

This is an immediate consequence of the definition of a lift and from $\mathbb{R}^n \times D$ being connected. One can smoothly deform f into the inclusion i, the lift $\tilde{f}$ of f being deformed into the lift of i (to prove this latter assertion about $\tilde{f}$ it is sufficient to apply the cited theorem by Fuks and Rokhlin to the products of our spaces with the segment $[0, 1]$). The lift of i coincides with some element $\gamma'' \in \Gamma$. The elements γ and γ' in (8.1c) remain constant during the deformation. Due to Γ being commutative one can cancel $\tilde{f} = \gamma''$ in (8.1c). Therefore, we have $\gamma = \gamma'$, i.e. (8.1a). □

We conclude this subsection by the simple note that any vector field ξ defined on A_D can be uniquely lifted onto the covering space $\mathbb{R}^n \times D$ to a vector field $\tilde{\xi}$, the *lift* of ξ, so that the following diagram commutes

$$\begin{array}{ccc} \mathbb{R}^n \times D & \xrightarrow{\tilde{\xi}} & T(\mathbb{R}^n \times D) \\ \downarrow p & & \downarrow Tp \\ A_D & \xrightarrow{\xi} & TA_D \end{array}$$

The lift of a locally Hamiltonian vector field is locally Hamiltonian.

The Nullity of the Calabi Invariant

The goal of KAM theory is to find invariant tori. In the case of systems on an annulus we are interested in tori which are Lagrangian and isotopic to the torus of the form $\mathbb{T}^n \times \{I^0\}, I^0 \in D$. As a matter of fact an arbitrary symplectic dynamical system on A_D may have no invariant tori at all. The obstruction to the existence of invariant tori of the said type is the Calabi invariant. (See §3 for the definition.)

Proposition 8.2. *Let $f \in \mathrm{Diff}^{\mathrm{symp}}_{\mathrm{loc,o}}(A_D, A_{D'})$, or a locally Hamiltonian vector field ξ on A_D, have at least one invariant Lagrangian torus isotopic to the torus $\mathbb{T}^n \times \{I^0\}, I^0 \in D$. Then the Calabi invariant of the system in question is zero.*

Proof. Firstly consider a local symplectic diffeomorphism $f \in \mathrm{Diff}^{\mathrm{symp}}_{\mathrm{loc,o}}(A_D, A_{D'})$. Let $\mathscr{T} \subset A_D$ be a Lagrangian torus which is invariant under f and isotopic to $\mathbb{T}^n \times \{I^0\}$. Then there are circles S_i', $1 \leq i \leq n$, on $\mathscr{T}$ which are mapped to the circles S_i on $\mathbb{T}^n \times \{I^0\}$ defined in the proof of Proposition 2.5a under the isotopy which carries $\mathscr{T}$ to $\mathbb{T}^n \times \{I^0\}$. We have by (2.5g):

(8.2a) $$\mathrm{Ca}(f_i) = \int_{S_i'} f^*\Theta - \int_{S_i'} \Theta = \int_{f(S_i')} \Theta - \int_{S_i'} \Theta.$$

Since f is smoothly isotopic to the inclusion map, it induces the identity transformation in $\pi_1(A_D) \cong \pi_1(A_{D'})$. The inclusion $\mathscr{T} \hookrightarrow A_D$ is a homotopical equivalence (recall that D is simply connected). Therefore, $f|_{\mathscr{T}}: \mathscr{T} \longrightarrow \mathscr{T}$ induces the identity transformation in $\pi_1(\mathscr{T})$. So, there is a deformation inside $\mathscr{T}$ which carries S_i' into $f(S_i')$. This deformation can be smoothened, and we have a smooth map $\alpha: S^1 \times [0,1] \longrightarrow \mathscr{T}$ with $\alpha(S^1 \times \{0\}) = S_i'$ and $\alpha(S^1 \times \{1\}) = f(S_i')$. The right-hand side of (8.2a) may be rewritten by Stokes formula as

$$\int_{f(S_i')} \Theta - \int_{D_i'} \Theta = \int_{(S^1 \times [0,1])} d\Theta = 0$$

for $d\Theta|_{\mathscr{T}} = 0$ since $\mathscr{T}$ is Lagrangian.

In the case of continuous time, we must prove that a locally Hamiltonian vector field ξ is a globally Hamiltonian one if it leaves some Lagrangian torus $\mathscr{T}$, isotopic to $\mathbb{T}^n \times \{I^0\}$, invariant. Take $x_0 \in \mathscr{T}$ and define

(8.2b) $$H(x) = \int_{x_0}^{x} \xi^b.$$

The integration is carried out over an arbitrary path joining x_0 and x. The right-hand side of (8.2b) does not change if one deforms the path leaving x and x_0 fixed since ξ is locally Hamiltonian. It does not depend on the choice of the homotopical class of the path since any closed path may be deformed into that lying on $\mathscr{T}$, and $\xi^b|_{T\mathscr{T}} = 0$. Let us now prove the latter equality. Take $y \in \mathscr{T}, v \in T_y\mathscr{T}$. We have $\xi_y \in T_y\mathscr{T}$ due to $\mathscr{T}$ being invariant under, and $\xi_y^b(v) = \Omega(\xi_y, v) = 0$ due to $\mathscr{T}$ being Lagrangian. The single valued function (8.2b) is the Hamiltonian of ξ. □

We denote by $\mathrm{Diff}^{\mathrm{symp}}_{\mathrm{loc,o,o}}(A_D, A_{D'})$ the set of all maps in $\mathrm{Diff}^{\mathrm{symp}}_{\mathrm{loc,o}}(A_D, A_{D'})$ with zero Calabi invariant.

Factorizable Maps

Let $f \in \mathrm{Diff}^{\mathrm{symp}}_{\mathrm{loc,o}}(A_D, A_{D'})$ and $\tilde{f}: \mathbb{R}^n \times D \longrightarrow \mathbb{R}^n \times D'$ be its lift. The map f is called *factorizable* if graph $\tilde{f} \subset \mathbb{R}^n \times D \times \mathbb{R}^n \times D'$ has a nondegenerate projection onto the Lagrangian subspace $\mathbb{R}^n \times \{0\} \times \{0\} \times \mathbb{R}^n$ of $(\mathbb{R}^{2n} \times \mathbb{R}^{2n}, \Omega_{st} \oplus (-\Omega_{st}))$ (see 2.34). The set of all factorizable maps in $\mathrm{Diff}^{\mathrm{symp}}_{\mathrm{loc.o}}(A_D, A_{D'})$ will be denoted $\mathrm{Diff}^{\mathrm{symp}}_{\mathrm{loc.fact}}(A_D, A_{D'})$.

The condition $f \in \mathrm{Diff}^{\mathrm{symp}}_{\mathrm{loc,fact}}(A_D, A_{D'})$ can be equivalently expressed as follows:

The map $\tilde{f}_{1/2}: \mathbb{R}^n \times D \longrightarrow \mathbb{R}^n \times D'$ defined by the formula

$$\tilde{f}_{1/2}(\varphi, I) = (\varphi^1, \varphi^2, \ldots, \varphi^n, f_{n+1}(\varphi, I), f_{n+2}(\varphi, I), \ldots, f_{2n}(\varphi, I)),$$

where $f_i = \mathrm{pr}_i \circ f$, is a local diffeomorphism.

For $f \in \mathrm{Diff}^{\mathrm{symp}}_{\mathrm{loc,fact}}(A_D, A_{D'})$, denote $G_f = \tilde{f}_{1/2}(\mathbb{R}^n \times D) \subset \mathbb{R}^n \times D'$ and define $l = (\tilde{f}_{1/2})^{-1}: G_f \longrightarrow \mathbb{R}^n \times D$ and $r = \tilde{f} \circ l: G_f \longrightarrow \mathbb{R}^n \times D'$, the *left* and the *right* maps of f. We have the following commutative diagram

$$
\begin{array}{c}
\mathbb{R}^n \times D' \supset G_f \\
{}^{l} \swarrow \qquad \searrow {}^{r} \\
\mathbb{R}^n \times D \xrightarrow{\ \ \tilde{f}\ \ } \mathbb{R}^n \times D'
\end{array}
$$

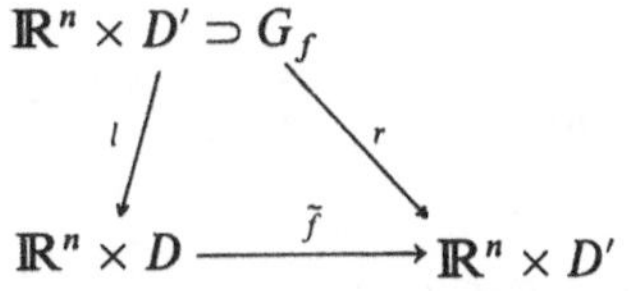

This justifies the word "factorizable" in the definition. It follows from 8.1 that

8.3 *the maps l and r commute with Γ.*

Applying the statement of 2.34 to $\tilde{f}$, we obtain the existence of the generating function $S: G_f \longrightarrow \mathbb{R}$ for $\tilde{f}$ depending on the old angles φ and new actions $\bar{I}$. It will be more convenient to introduce the *Hamiltonian function H* which is connected to S by the formula

$$S(\varphi, \bar{I}) = \sum_{i=1}^{n} \varphi^i \cdot \bar{I}_i + H(\varphi, \bar{I}).$$

In terms of H the equations for the left and right maps read:

$$l(\varphi, \bar{I}) = (\varphi, I), \qquad r(\varphi, \bar{I}) = (\bar{\varphi}, \bar{I}),$$

where

(8.4)
$$I_i = \bar{I}_i + \frac{\partial H(\varphi, \bar{I})}{\partial \varphi^i},$$

$$\bar{\varphi}^i = \varphi^i + \frac{\partial H(\varphi, \bar{I})}{\partial \bar{I}_i}, \qquad 1 \leq i \leq n.$$

The equations (8.4) are discrete analogue of the Hamiltonian equations (3.3). They may be expressed in a shorter form as

(8.5)
$$r^* \theta_1 + l^* \theta = d(\kappa + H),$$

where

$$\theta = \sum_{i=1}^{n} I_i d\varphi^i, \qquad \theta_1 = \sum_{i=1}^{n} \varphi^i dI_i,$$

and

$$\kappa = \sum_{i=1}^{n} \varphi^i \bar{I}_i.$$

The property 8.3 and the equations (8.4) imply that dH is periodic. The Calabi invariant of f may be expressed in terms of the Hamiltonian function:

$$\mathrm{Ca}(f_i) = -H(x + \mathrm{ort}_i) + H(x), \quad 1 \leqq i \leqq n. \tag{8.6}$$

The right-hand side of (8.6) is locally independent of x because of the said periodicity of dH. The word "locally" here may be omitted in virtue of $G_f = f_{1/2}(\mathbb{R}^n \times D)$ being simply connected.

Proof of (8.6). Fix $x_0 \in D$ and denote by l_i the segment joining x_0 to $x_0 + \mathrm{ort}_i$, $1 \leqq i \leqq n$. Then (2.5g) and (3.16) give us

$$\begin{aligned}
\mathrm{Ca}(f)_i &= \int_{l_i} (\tilde{f}^*\theta - \theta) \\
&= \int_{l^{-1}(l_i)} r^*\theta - l^*\theta \\
&= \int_{l_i} r^*\theta - l^*\theta \quad \text{(due to } l^{-1}(l_i) \text{ and } l_i \text{ being homotopies)} \\
&= \int_{l_i} r^*d(\kappa) - d(\kappa + H) \quad \text{((8.5) is used)} \\
&= -H(x_0 + \mathrm{ort}_i) + H(x_0) - (\kappa - r^*\kappa)|_{x_0}^{x_0 + \mathrm{ort}_i}.
\end{aligned}$$

Since r commutes with Γ and does not change the action variables I_i, both $\kappa|_{x_0}^{x_0+\mathrm{ort}_i}$ and $r^*\kappa|_{x_0}^{x_0+\mathrm{ort}_i}$ are equal to I_i^0, the ith action component of x_0. So, the last term in the above formula cancels. □

Note that formula (8.6) is also true in the case of continuous time.

A factorizable symplectic map $f: A_D \longrightarrow A_{D'}$ (a symplectic dynamical system with f as the generator) is called *Hamiltonian* if $\mathrm{Ca}(f) = 0$. Its Hamiltonian function is periodic due to (8.6). Its image under the canonical map $p: \mathbb{R}^n \times D \longrightarrow A_D$, a function $\hat{H} \in C^\infty(p(G))$ such that $p^*\hat{H} = H$, is also called the *Hamiltonian* of f (of a system with generator f). The Hamiltonian is defined uniquely up to an additive constant.

Remark 8.7. A curious feature of the special symplectic diffeomorphism $g_M: A_D \longrightarrow A_{D_1}$, $M \in \mathrm{GL}(n, \mathbb{Z})$, $D_1 = M(D)$, defined in 2.5*i*, is that the law of transformation of Hamiltonian functions also holds for this map in the case of discrete time: if $f: A_D \longrightarrow A_{D'}$, $D \subset D'$ is a factorizable symplectic map with Hamiltonian function H then the map $f_1 = g_M \circ f \circ g_M^{-1}: A_{D_1} \longrightarrow A_{D_1'}$, $D_1 = M(D)$, $D_1' = M(D')$ is also factorizable, and its Hamiltonian function is $H_1 = \tilde{g}_M^* H$, the Calabi invariant also being transformed: $\mathrm{Ca}(f_1) = M\,\mathrm{Ca}(f)$.

Diffeomorphisms

Up to this point we have considered local diffeomorphisms $f: A_D \longrightarrow A_{D'}$ which are not necessarily onto. Here we shall discuss the case when $D = D'$ and the functions f are diffeomorphisms of a symplectic annulus. The set of all symplectic diffeomorphisms of a symplectic annulus A_D onto itself will be denoted $\mathrm{Diff}^{\mathrm{symp}}(A_D)$. We single out the following subsets

$$\mathrm{Diff}^{\mathrm{Ham}}(A_D) \subset \begin{matrix} \mathrm{Diff}^{\mathrm{symp}}_{0,0}(A_D) \\ \mathrm{Diff}^{\mathrm{symp}}_{\mathrm{fact}}(A_D) \end{matrix} \subset \mathrm{Diff}^{\mathrm{symp}}_{0}(A_D) \subset \mathrm{Diff}^{\mathrm{symp}}(A_D),$$

$\mathrm{Diff}^{\mathrm{symp}}_{0}(A_D)$ consisting of all $f \in \mathrm{Diff}^{\mathrm{symp}}(A_D)$ isotope to 1_{A_D}, $\mathrm{Diff}^{\mathrm{symp}}_{0,0}(A_D)$ consisting of those with zero Calabi invariant, $\mathrm{Diff}^{\mathrm{symp}}_{\mathrm{fact}}(A_D)$ contains all factorizable diffeomorphisms, and, finally, $\mathrm{Diff}^{\mathrm{Ham}}(A_D) = \mathrm{Diff}^{\mathrm{symp}}_{0,0}(A_D) \cap \mathrm{Diff}^{\mathrm{symp}}_{\mathrm{fact}}(A_D)$ is the set of *Hamiltonian maps*. The these latters may be characterized, in view of the considerations of the preceding subsection, as those possessing the Hamiltonian function $H(\varphi, I)$ periodic in the φ-variables with period 1. We denote the space of all such functions $C^{\infty}_{\mathrm{per}}(\mathbb{R}^n \times D)$. This space may be canonically identified with $C^{\infty}(A_D)$, but we mainly prefer to deal with functions defined on $\mathbb{R}^n \times D$.

Example 8.8 (Standard Map). Let V be a real valued smooth function defined on the real axis, periodic with period 1. Consider the map $\tilde{f}: \mathbb{R}^2 \longrightarrow \mathbb{R}^2$ defined by the formula

$$\tilde{f}(\varphi, I) = (\bar{\varphi}, \bar{I}),$$

where

$$\bar{\varphi} = \varphi + \bar{I}$$
$$\bar{I} = I - \frac{d}{d\varphi} V(\varphi).$$

This map preserves Ω_{st} and commutes with the shift $(\varphi, I) \longmapsto (\varphi + 1, I)$, so it defines a symplectic map $f: A_{\mathbb{R}} \longrightarrow A_{\mathbb{R}}$, the *standard map*. f is Hamiltonian with Hamiltonian function

$$H(\varphi, \bar{I}) = \frac{\bar{I}^2}{2} + V(\varphi).$$

This system is reversible (see §3). One of the possible reversors is given by the formula

$$\sigma(\varphi, I) = (\varphi - I, -I).$$

If $V(\varphi)$ is symmetric: $V(\varphi) = V(-\varphi)$, then there is another useful reversor given by

$$\sigma_1: (\varphi, I) \longmapsto (-\varphi, I - V'(\varphi)).$$

Note that f commutes with the shift $(\varphi, I) \longmapsto (\varphi, I + 1)$ and therefore defines the map $\hat{f}: \mathbb{T}^2 \longrightarrow \mathbb{T}^2$ and the standard 2-dimensional symplectic torus.

One obtains an interesting one parameter family of examples by setting $V(\varphi) = K/4\pi^2\,(1 - \cos 2\pi\varphi)$.

$C^r - \gamma$ norms

To measure the distances between maps we shall use the r-norms defined in Appendix II. Let $f \in \mathrm{Diff}_0^{\mathrm{symp}}(A_D)$. We define the *$C^r$-norm of the deviation of f from the identity map* as follows:

$$\|f - \mathrm{id}\|_r = \inf_{\tilde{f}} \max_{1 \leqq i \leqq 2n} \|\mathrm{pr}_i(\tilde{f} - \mathrm{id})\|_r. \tag{8.9}$$

Here infimum is taken over all lifts $\tilde{f}$ of f, id in both sides of the formula standing for the identity maps in the corresponding spaces.

For the sake of future use we include an additional parameter $\gamma \in]0, 1]$ in the definition of the norm when we deal with an annulus or with its covering space. Consider the *stretching map* $\tilde{S}_\gamma: \mathbb{R}^n \times \mathbb{R}^n \longrightarrow \mathbb{R}^n \times \mathbb{R}^n$ which only acts on the action variables by the formula

$$\tilde{S}_\gamma(\varphi, I) = (\varphi, \varphi I). \tag{8.10}$$

Since $\tilde{S}_\gamma$ commutes with the action of Γ, it defines the *stretching map of the annulus* $S_\gamma: A_{\mathbb{R}^n} \longrightarrow A_{\mathbb{R}^n}$. We define the $C^r - \gamma$ norms as follows: if $u \in C^\infty(\mathbb{R}^n \times D)$ then

$$\|u\|_{r,\gamma} = \|u \circ \tilde{S}_\gamma\|_r, \tag{8.11}$$

and if $f \in \mathrm{Diff}_0^{\mathrm{symp}}(A_D)$ then

$$\|f - \mathrm{id}\|_{r,\gamma} = \|S_\gamma^{-1} \circ f \circ S_\gamma - \mathrm{id}\|_r. \tag{8.12}$$

Small Perturbations of an Integrable Hamiltonian

Here we consider the question of when a function $H \in C^\infty_{\mathrm{per}}(\mathbb{R}^n \times D)$ is the Hamiltonian function of some $f \in \mathrm{Diff}^{\mathrm{Ham}}(A_D)$. A satisfactory answer will be obtained provided H is sufficiently close to the Hamiltonian function of an integrable system (see Example 8.7). This function being independent of the angle variable φ. We represent H in the form

$$H(\varphi, I) = H_0(I) + H_1(\varphi; I). \tag{8.13}$$

Proposition 8.14. *Let H_1 in* (8.13) *belong to $C^\infty_{\mathrm{per},\gamma}(\mathbb{R}^n \times D)$ and let $\|H_1\|_{2,\gamma} < n^{-1}\gamma^2$, γ being a positive number. Then there is a unique $f \in \mathrm{Diff}^{\mathrm{Ham}}(A_D)$ with Hamiltonian function* (8.15).

Proof. Define the map $l, r: \mathbb{R}^n \times D \longrightarrow \mathbb{R}^n \times D$ by the formulae (8.4) with H of (8.13). One can easily check that the conditions of Lemma AII.24.25 are satisfied. It follows that l, r belong to $\mathrm{Diff}_{\mathrm{per},\gamma}(\mathbb{R}^n \times D)$, and hence $f = r \circ l^{-1}$ is the lift of a Hamiltonian map f. The validity of (8.4) with $H \in C^\infty_{\mathrm{per}}(\mathbb{R}^n \times D)$ then implies that f is Hamiltonian with H as its Hamiltonian function. □

Notes to Chapter I

The aim of Chapter I is to give necessary mathematical and mechanical preliminaries to KAM theory. These are Symplectic Geometry and the Theory of Hamiltonian Systems. There is no room here to expose the history of the subject, but we cannot avoid to mention the famous Poincaré' book (1892). The books by Abraham and Marsden (1972), and by Arnol'd (1978) have already become classical manuals on the modern treatment of classical mechanics. For some mathematical aspects of the theory the reader can also refer to Godbillon (1969). The largest part of linear symplectic theory is adopted from Guillemin and Sternberg (1977). Our treatment of nonlinear symplectic geometry (§2) follows mainly from Weinstein (1971) (see also Weinstein (1977)). We refer the reader for curious facts and fine points of this theory to Gromov (1985) and Sikorav (1989). The statement 2.15 was communicated to me by Ya. M. Eliashberg. The compact case is contained in Moser (1965) (see also Krygin (1971) and Gromov and Eliashberg (1973)).

The notion of a Lagrangian manifold was first introduced, in my opinion, in Maslov (1965).

Our terminology, given in §3, is not common. In order to get a natural generalisation, we prefer to talk about symplectic dynamical systems instead of Hamiltonian ones, while usually one only investigates Hamiltonian ones. However, to develop KAM theory it is not necessary to restrict ourselves with globally Hamiltonian systems, it is sufficient only to impose "Hamiltonian" condition locally, as explained in §§12,15.

The main sources for the theory of reversible systems are Arnol'd and Sevryuk (1986), and Sevryuk (1986).

The contents of §4 devoted to the construction of the "symplectic gluing", is original and never previously published.

Some details of the constructions and the proofs of §5 are taken from the thesis of R. Douady (1982).

There is a vast literature on planar billiards. The "stadium" was first considered by Bunimovich (1974, 1979). The generalisation of the stadium are the billiards in the domains whose boundaries are C^1 and consist of four arcs and circles (see Benettin and Strelcyn (1978), Hayli and Dumon (1986)). More general families of billiards with positive Lyapunov exponents were considered by Wojtkovski (1985, 1986) and recently by V. Donnay.

For billiards in polygons see Zemlyakov and Katok (1975), Boldrigini, Kean and Marchetti (1978), and Gutkin (1986).

There are several papers on multidimensional billiards: Bunimovich (1988, Stoyanov (1985), Gruber (1990).

KAM theory is usually treated as a perturbation theory, the unperturbed system being an integrable one. The theory of completely integrable Hamiltonian systems goes back to Liouvilles work (Liouville (1855)). The proof of the main theorem is in Arnol'd (1978) (see also Markus and Meyer (1974), and Duistermaat (1980)). A generalization of the theorem was given by Nekhoroshev (1972). We refer the reader to Dubrovin, Krichever, Novikov (1988) for an account of modern development in integrable systems.

Moser (1956) proved that in the analytical case of Example 7.17 there exist analytical symplectic coordinates (X_1, X_2) in a neighbourhood of the origin such that the map has, in these coordinates, the form (7.17a) with Λ an analytical function of the product $X_1 \cdot X_2$.

Recently Amiran (1991) proved that a smooth convex planar integrable billiard must be an ellipse. Hence Examples 7.27 and 7.28 are exceptional.

Chapter II. KAM Theorems

This chapter contains the formulations, simple consequences, and some applications of the main theorem of KAM theory. It asserts that there is a massive set of invariant tori, bearing quasiperiodic motions with suitable frequencies, which persists under small perturbations. The typical situation is a small Hamiltonian perturbation of a completely integrable system which has a foliation on invariant Lagrangian tori, as we have seen in §7. Roughly speaking, some of these tori persist and some are destroyed as one switches on a small perturbation. In the more general situation, the notion of a KAM set, which we shall derive in §10, appears to be an adequate one in describing the phenomenon in question. KAM sets exist in many familiar dynamical systems (see §§11–14) and they also persist under small perturbations (§15).

Pedagogical reasons force us to start with more simple notion of a KAM torus. A KAM set is a union of KAM tori satisfying some additional conditions. As we shall see in §12, any KAM torus can be included as a member in some KAM set.

We consider both the cases of continuous and discrete time: the variable t ranges over $\mathbb{R}$ or $\mathbb{Z}$.

§9. The KAM Torus

A KAM torus in a symplectic dynamical system is a particular case of an invariant set with complete dynamics. It is diffeomorphic to an n-torus ($n = (1/2) \times$ the dimension of the phase space), and bears a quasiperiodic motion. At first we consider the latter notion in the general contexts of topological and differential dynamics.

Quasiperiodic Motion on a k-Torus

Let $\mathbb{T}^k = \mathbb{R}^k/\mathbb{Z}^k$ be the standard k-torus. Fix $\omega \in \mathbb{R}^k$. A *quasiperiodic motion* $g^t_\omega : \mathbb{T}^k \longrightarrow \mathbb{T}^k$ is a smooth dynamical system (cascade or flow) defined globally by the following explicit expression for its lift $\tilde{g}^t_\omega : \mathbb{R}^k \longrightarrow \mathbb{R}^k$:

$$\tilde{g}^k_\omega(x) = x + \omega t, \quad t \in \mathbb{Z} \text{ or } \mathbb{R}.$$

The vector $\omega = (\omega^1, \omega^2, \ldots, \omega^k) \in \mathbb{R}^k$ is called the *frequency vector*. In the case of continuous time we also refer to it as the *frequency*. In the case of discrete time the adding of an integer vector to ω does not change g^t_ω. Hence it is better to consider the class $\{\omega\}\in\mathbb{R}^k/\mathbb{Z}^k$ of ω as a characteristic of a quassiperiodic motion rather than ω itself. So in the case of a cascade we call the *frequency* this element $\{\omega\}$ of $\mathbb{R}^k/\mathbb{Z}^k$.

Let us now investigate topological properties of a quasiperiodic motion. To do this we need some general definitions and results from topological dynamics. Consider a continuous flow/cascade $g^t: X \longrightarrow X$ where X is a compact topological space. It is called *topologically transitive* if there exists a dense trajectory, i.e. there is $x_0\in X$ such that $\overline{\{g^t(x_0)\}} = X$. A subset $A\subset X$ is called *completely invariant* if $g^t(A)=A$ for all t. A closed non-empty completely invariant set $A\subset X$ is called *minimal* if there are no other closed non-empty invariant subsets $A'\subset A$. A system $g^t: X \longrightarrow X$ is said to be *minimal* if its phase space X is itself minimal. This property is equivalent to saying that each trajectory is dense. Hence a minimality implies the topological transitivity. Denote by $C(X)$ the space of continuous real valued functions on X supplied with the norm $\|\varphi\| = \sup_{x\in X}|\varphi(x)|$. A linear continuous map $\mu: C(X)\longrightarrow\mathbb{R}$ is called a *measure* on X if (a) $\mu(\varphi)\geqq 0$ if $\varphi\geqq 0$, (b) $\mu(1)=1$ (the 1 in the argument of μ is meant to be the constant function equal to 1). Denote by $\mathfrak{M}(X)$ the space of all measures on X. Any continuous map $f: X\longrightarrow Y$, Y being another compact space, induces the map $f_*: \mathfrak{M}(X)\longrightarrow\mathfrak{M}(Y)$ defined by the formula

$$(f_*\mu)(\varphi) = \mu(\varphi\circ f), \qquad \mu\in\mathfrak{M}(X),\ \varphi\in C(Y).$$

A measure $\mu\in\mathfrak{M}(X)$ is called *invariant* under $g^t: X\longrightarrow X$ if $g^t_*\mu=\mu$ for all t. All the measures that are invariant under $\{g^t\}$ form a subset $\mathfrak{M}_{\text{inv}}(\{g^t\})\subset\mathfrak{M}(X)$. Bogoluboff–Krilov's theorem [Bogolubov, Krilov [1937] (see also Bourbaki [1966] Appendix to Chapter II Fixed points of compact convex sets)] asserts that $\mathfrak{M}_{\text{inv}}(g^t)\neq\varnothing$. A system $\{g^t\}$ is called *uniquely ergodic* if $\mathfrak{M}_{\text{inv}}(\{g^t\})$ consists of a single element: $\mathfrak{M}_{\text{inv}}(g^t) = \{\mu_{\text{inv}}(\{g^t\})\}$.

The *support* of a measure μ on X, $\operatorname{supp}\mu$, is a closed subset defined as follows: $x\notin\operatorname{supp}\mu \Leftrightarrow$ there is an open neighbourhood V of x such that for any $\varphi\in C(X)$, if $\operatorname{supp}\varphi\subset V$ then $\mu(\varphi)=0$.

Lemma 9.1. *Let $A\subset X$ be a non-empty closed subset that is invariant under $\{g^t\}$. Then there exists a measure $\mu\in\mathfrak{M}_{\text{inv}}(\{g^t\})$ such that* $\operatorname{supp}\mu\subset A$.

Proof. $\mu = i_*\mu'$ where $i: A\hookrightarrow X$ is the inclusion map, $\mu'\in\mathfrak{M}_{\text{inv}}(\{g^t|_A\})$. □

Let the phase space X of our system $\{g^t\}$ be supplied with a metric $d: X\times X\longrightarrow\mathbb{R}_+$, the usual axioms of a distance being assumed to be satisfied. We say that $\{g^t\}$ *preserves* d if $d(g^tx, g^ty) = d(x,y)$ for all $x,y\in X$ and all t.

Proposition 9.2. *If $\{g^t\}$ is uniquely ergodic and preserves a metric then it is minimal.*

Proof. Let $A_1\subset X$, $A_1\neq X$ be a non-empty closed subset invariant under $\{g^t\}$. Take $x_0\in X\backslash A$ and denote the closure of the trajectory of x_0 by A_2. Since $\operatorname{dist}(x_0, A_1) = \min_{x\in A_1} d(x_0, x) = \delta > 0$, we have $\operatorname{dist}(A_1, A_2) = \min_{x\in A_1, x'\in A_2} d(x, x') =$

$\delta > 0$. Here we have used the fact that d is invariant. Take invariant measures μ_i with supp $\mu_i \subset A_i$, $i = 1, 2$ (use Lemma 9.1). These are distinct members of $\mathfrak{M}_{\text{inv}}(\{g^t\})$. $\square$

Let us return to the study of quasiperiodic motion $g^t_\omega: \mathbb{T}^k \longrightarrow \mathbb{T}^k$. Obviously g^t_ω preserves the standard distance $d(x, y) = \min \|\tilde{x} - \tilde{y}\|$, $\tilde{x} \in x$, $\tilde{y} \in y$, on the torus. It also preserves Lebesgue measure $\mu_0(\varphi) = \int \varphi(x)\,dx$. We are going to discover when the system in question is topologically transitive, minimal, and uniquely ergodic. It turns out that the presence of these properties is completely determined by the frequency. A frequency $\omega \in \mathbb{R}^k$, or $\{\omega\} \in \mathbb{T}^k$, is said to be *resonant* if there is an $m \in \mathbb{Z}^k$, $m \neq 0$, such that $\langle m, \omega \rangle = 0$, or correspondingly $\{\langle m, \omega \rangle\} = 0$. Otherwise, a frequency is called *nonresonant.*

Proposition 9.3. *The following properties of a quasiperiodic motion are equivalent:*
(a) *its frequency is nonresonant,*
(b) *it is uniquely ergodic,*
(c) *it is minimal,*
(d) *it is topologically transitive.*

A quasiperiodic motion which satisfies these conditions will be called *transitive.*

Proof of Proposition 9.3. (a)$\Rightarrow$(b) Let μ be an invariant measure. Consider the values of μ on the *eigenfunctions*

(e) $$\psi_m(x) = \exp(2\pi i \langle m, x \rangle), \quad m \in \mathbb{Z}^k.$$

One may consider (e) as functions defined on $\mathbb{T}^k$. We have

(f) $$(g^t_\omega)^* \psi_m(x) = \exp(2\pi i \langle m, \omega \rangle t) \cdot \psi_m(x),$$

and

(g) $$\mu(\psi_m) = [(g^t_\omega)_* \mu](\psi_m) = \mu((g^t_\omega)^* \psi_m) = \exp(2\pi i \langle m, \omega \rangle t) \cdot \mu(\psi_m).$$

Let ω be nonresonant. Given $0 \neq m \in \mathbb{Z}^k$, there exists t such that $\exp(2\pi i \langle m, \omega \rangle t) \neq 1$. It follows that $\mu(\psi_m) = 0$ for all $m \neq 0$. Besides, we know that $\mu(\psi_0) = \mu(1) = 1$. The well known Fejér's theorem, which asserts that every continuous function can be approximated in $C(\mathbb{T}^k)$ by linear combinations of the eigenfunctions ψ_m, implies that a measure is completely defined by its values on ψ_m, $m \in \mathbb{Z}^k$. Hence $\mu = \mu_0$.

(b)$\Rightarrow$(c) follows from Proposition 9.2.

(c)$\Rightarrow$(d) is obvious.

(d)$\Rightarrow$(a). Let the frequency be resonant. Then there exists $m \in \mathbb{Z}^k$, $m \neq 0$, such that $\langle m, \omega \rangle = 0$ in the case of flow, and $\{\langle m, \omega \rangle\} = 0$ in the case of cascade. Given $C \in \mathbb{R}$, the family of hyperplanes $\{x \in \mathbb{R}^k$: $\langle m, x \rangle = C + n\}$, $n \in \mathbb{Z}$ is invariant under the action of $\tilde{g}^t_\omega$ and projects onto a compact submanifold of codimension 1 in $\mathbb{T}^k = \mathbb{R}^k/\mathbb{Z}^k$ which is invariant under g^t_ω. Each point of $\mathbb{T}^k$ belongs to such a submanifold with an appropriate value of C. Hence there is no dense trajectory. $\square$

In the preceding proof we used the eigenfunctions (e) satisfy the equations (f). In the transitive case the latter property determines them completely.

Proposition 9.4. *Let g^t_ω be a transitive quasiperiodic motion and ψ be a continuous complex valued function on $\mathbb{T}^k$, which is not identically zero. Let ψ satisfy the equation*

$$(9.4a) \qquad (g^t_\omega)^*\psi = \lambda(t)\psi, \quad t \in \mathbb{R},$$

where λ is a complex valued function of the variable t. Then there are $m \in \mathbb{Z}^k$ and all $C \in \mathbb{C} \setminus \{0\}$ such that *$\psi = C\psi_m$ and $\lambda(t) = \exp(2\pi i \langle m, \omega \rangle t)$.*

Proof. Consider the expansion $\psi = \sum_{s \in \mathbb{Z}^k} c_s \psi_s$. It can be regarded as a series in $L_2(\mathbb{T}^k, \mu_0)$. Substituting this into (9.4a) yields

$$\sum_s (\exp(2\pi i \langle \bar{s}, \omega \rangle t) - \lambda(t)) c_s \psi_s = 0.$$

Since ψ is not identically zero, there is an $m \in \mathbb{Z}^k$ such that $c_m \neq 0$. It follows that (9.4b) holds for this value of m. Then the nonresonant property of ω implies that $\exp(2\pi i \langle s, \omega \rangle t) - \lambda(t) = 0$ if and only if $s = m$. Hence $c_s = 0$ if $s \neq m$. □

The natural question that arises is when are two quasiperiodic motions $\{g^t_\omega\}$ and $\{g^t_{\omega'}\}$ on $\mathbb{T}^k$ with the same kind of time isomorphic as topological dynamical systems, i.e. there is a homeomorphism $h: \mathbb{T}^k \longrightarrow \mathbb{T}^k$ such that $g^t_{\omega'} = h^{-1} \circ g^t_\omega \circ h$ for all t. Transitivity is preserved under isomorphisms, so g^t_ω and $g^t_{\omega'}$ are within both transitive or both not. We restrict ourselves to the transitive case only.

Proposition 9.5. *Let two transitive quasiperiodic motions g^t_ω and $g^t_{\omega'}$ on $\mathbb{T}^k$ be isomorphic via a homeomorphism h, i.e. $g^t_{\omega'} = h^{-1} \circ g^t_\omega \circ h$ satisfying the norming condition $h(0) = 0$. Let $\tilde{h}: \mathbb{R}^k \longrightarrow \mathbb{R}^k$ be the lift of h also satisfying the norming condition, $\tilde{h}(0) = 0$. Then* (a) *this $\tilde{h}$ is a linear map, moreover $\tilde{h} \in \mathrm{GL}(k, \mathbb{Z})$, i.e. the matrix of $\tilde{h}$ has integer coefficients and* $\det \hat{h} = \pm 1$; (b) *$\omega = \tilde{h}\omega'$ for a flow and $\{\omega\} = \{\tilde{h}\omega'\}$ for a cascade.*

Proof. Applying h^* to equation 9.3(f), we obtain $g^{t*}_{\omega'} h^* \psi_m = \exp(2\pi i \langle m, \omega \rangle t) h^* \psi_m$. Proposition 9.4 gives us

$$(+) \qquad \psi_m = \text{const}\ \psi_{m'}$$

for some $m' \in \mathbb{Z}^k$ and

$$(++) \qquad \exp(2\pi i \langle m, \omega \rangle t) = \exp(2\pi i \langle m', \omega' \rangle t) \quad \text{for all } t.$$

The explicit formula 9.3 (e) for $\psi_{m'}(+)$, and the norming condition for $\tilde{h}$ yield $\langle m, \tilde{h}(x) \rangle = \langle m', x \rangle$. Writing $m' = A(m)$, we obtain $\tilde{h}^i(x) = \langle A(\mathrm{ort}_i), x \rangle$, i.e. $\tilde{h}$ is linear and its matrix has integer coefficients. The same is true for $\tilde{h}^{-1}$. Hence $\tilde{h} \in \mathrm{GL}(k, \mathbb{Z})$, and $m' = A(m) = \tilde{h}^t(m)$ (we identify a linear map with its matrix). Substituting this equality into $(++)$, one obtains assertion (b). □

Define the right action A of $\mathrm{GL}(k, \mathbb{Z})$ on the set of all frequencies (on $\mathbb{R}^k$ in the case of continuous time and on $\mathbb{T}^k$ in the case of discrete time) by the formula

$$A(g)(\omega) = g^{-1}\cdot\omega, \quad g\in \mathrm{GL}(k, \mathbb{Z}).$$

Here the dot denotes, in the continuous time case, the application of a matrix to a vector. In discrete time we use the property that a matrix with integer entries maps a class with respect to the $\mathbb{Z}^k$-section onto a class, so $g^{-1}\cdot\omega$ is well defined in this case also. The action A leaves the set of nonresonant frequencies invariant. The *frequency class* Ω of a quasiperiodic motion $\{g^t_\omega\}$ is the orbit of its frequency ω under A. Note that all the members of Ω for a transitive $\{g^t_\omega\}$ are nonresonant, and hence the action A is free when restricted to Ω. The above Proposition implies that, in the transitive case, the frequency class Ω is the only topological invariant of a quasiperiodic motion.

Remark 9.6. In the case of discrete time the following more general notion will be useful. A quasiperiodic motion on an N-chain of k-tori is a cascade $g^t_{\boldsymbol{\omega},N}\colon \bigcup_{j=0}^{N-1}\mathbb{T}_j \longrightarrow \bigcup_{j=0}^{N-1}\mathbb{T}_j$ where each $\mathbb{T}_j$, $0\leqq j\leqq N-1$, is an identical copy of the standard k-tori: $\mathbb{T}_j = \mathbb{T}^k = \mathbb{R}^k/\mathbb{Z}^k$, $\bigcup$ stands for a topological sum (disjoint union), and $g^t_{\boldsymbol{\omega},N}$ acts as follows. It is sufficient to describe the action of $g^t_{\boldsymbol{\omega},N}$ as $g^t_{\boldsymbol{\omega},N}(\mathbb{T}_j) = \mathbb{T}_{j+1(\mathrm{mod}\,N)}$, $g^t_{\boldsymbol{\omega},N}$ carries a point $x\in\mathbb{T}_j$ onto a point $x'\in\mathbb{T}_{j+1}$ with the same coordinates as x if $j<N-1$, and $g^t_{\boldsymbol{\omega},N}(x) = x+\boldsymbol{\omega}\in\mathbb{T}_1$ if $x\in\mathbb{T}_{N-1}$. Here $\boldsymbol{\omega}\in\mathbb{T}^k$ is the *frequency* of the motion, N is its *period*. One may consider a periodic motion with period N as a particular case of the motion in question with $k=0$. On the other hand the above definition of a quasiperiodic motion on $\mathbb{T}^k$ is again a particular case with $N=1$. All the definitions and classifications given above may be extended to this more general definition when applied to the cascade $\hat{g}^t = g^{Nt}_{\boldsymbol{\omega},NT_1}$ which obviously coincides with g^t_ω, $\omega\in\boldsymbol{\omega}$.

The Diophantum Set $\mathscr{E}_{\sigma\gamma}$

To divide quasiperiodic motions into two classes: resonant and transitive, is insufficient in KAM theory. More subtle classification is necessary which takes into account how closely the transitive motion can be approximated by resonant ones. We will impose the condition upon the frequency vector ω of a quasiperiodic motion that it must belong to $\mathscr{E}_{\sigma\gamma}$ in order that the motion might suit us.

Let us first define this subset $\mathscr{E}_{\sigma\gamma}$ in $\mathbb{R}^k$ which depends on the two positive parameters σ and γ. The definition is different for different kinds of time. In the case of continuous time a point $y=(y^1, y^2, \ldots, y^k)\in\mathbb{R}^k$ belongs to $\mathscr{E}_{\sigma\gamma}$ if and only if the inequality

$$|\langle y, m\rangle| \geqq \gamma\|m\|^{-\sigma} \tag{9.7}$$

holds for all integer nonzero vectors $m=(m_1, m_2, \ldots, m_k)\in\mathbb{Z}^k$. In the case of discrete time the inequality (9.7) must be changed to

$$|\langle y, m\rangle - m_0| \geqq \gamma\|m\|^{-\sigma} \tag{9.8}$$

which must be satisfied for all $(m_0, m_1, m_2, \ldots, m_k) \in \mathbb{Z}^{k+1}$ such that $m = (m_1, m_2, \ldots, m_k) \neq 0$.

One can prove the following proposition by estimating the volumes of consecutively removed slices.

Proposition 9.9. *Let Y be a measurable bounded set in* $\mathbb{R}^k$, *and let* $\gamma \in]0, 1[$, $\sigma > d - 1$, *where* $d = k$ *in the case of continuous time, and* $d = k + 1$ *in the case of discrete time. Then*

(9.9a) $$|\mathcal{M}es\, Y - (Y \cap \mathcal{E}_{\sigma\gamma})| \leqq \text{const}\, \gamma \max\{(\text{diam } Y)^{d-1}, (\text{diam } Y)^{k-1}\},$$

where const *depends only on k and* σ.

Invariant Tori in a Symplectic Dynamical System. Definition of a KAM Torus

Consider a symplectic dynamical system defined in a symplectic manifold X of dimension $2n$ (for definition see §3). An *invariant torus with quasiperiodic motion* is a subset $\mathcal{T} \subset X$ satisfying the conditions:

(IT1) $\mathcal{T}$ is an invariant set with complete dynamics. We denote by f^t, $t \in \mathbb{R}$ (or $\mathbb{Z}$) the flow (or cascade) generated on $\mathcal{T}$ by the system.

(IT2) $\mathcal{T}$ is an isotropic submanifold.

(IT3) There exists a transitive quasiperiodic motion $g^t_\omega : \mathbb{T}^k \longrightarrow \mathbb{T}^k$, $k \leqq n$, and an embedding $\chi : \mathbb{T}^k \longrightarrow X$ such that (1) $\mathcal{T} = \chi(\mathbb{T}^k)$, (2) χ conjugates $\{g^t_\omega\}$ and $\{f^t\}$, i.e. the following diagram commutes for all t:

(9.10) $$\begin{array}{ccc} \mathcal{T} & \xrightarrow{\ f^t\ } & \mathcal{T} \subset X \\ \uparrow \chi & & \uparrow \chi \\ \mathbb{T}^k & \xrightarrow{\ g^t_\omega\ } & \mathbb{T}^k \end{array}$$

We also say that the torus $\mathcal{T}$ *bears a quasiperiodic motion.*

Remark 9.11. In the case of discrete time one obtains the more general definition of a chain of invariant tori with quasiperiodic motion by substituting $\sum_{j=0}^{N-1} \mathbb{T}_j$ for $\mathbb{T}^k$ and $g^t_{\omega,N}$ for g^t_ω in (IT3) (see the definitions in Remark 9.6). Again we say that an N-chain of tori $\mathcal{T}$ *bears a quasiperiodic motion* and refer to ω as the *frequency* and to N as the *period* of that motion. All subsequent theory can be developed for such chains as well as for invariant tori.

We are going to study the behaviour of the system near $\mathcal{T}$. Here we shall only discuss the case $k = n$, a full-dimensional tori. $\mathcal{T}$ turns out to be a Lagrangian submanifold. Theorem 2.27 enables us to restrict our considerations to a canonical neighbourhood of $\mathcal{T}$ in X. This means that there are: an open neighbourhood U of $\mathcal{T}$ in X, an open ball $D \subset \mathbb{R}^n$ with its center at 0, and a symplectic diffeomorphism $\Psi : A_D \longrightarrow U$ such that $\Psi^{-1} \circ \chi$ coincides with the standard inclusion $x \longmapsto (x, 0)$ of

$\mathbb{T}^n$ into $A_D = \mathbb{T}^n \times D$ (see Example 2.29). Proposition 8.2 is applicable to the preimage of the system under Ψ, its Calabi invariant being zero. So the said preimage possesses a Hamiltonian H. In the case of discrete time we need to take a smaller ball D, if necessary, to be sure that the generator of the preimage is factorizable. The Hamiltonian H is constant on $\mathbb{T}^n \times \{0\}$ (use transitivity), and we normalize it by demanding to be zero on $\mathbb{T}^n \times \{0\}$. We call the collection of these objects: $\Psi : A_D \longrightarrow U$, H, a *clothing* of the torus $\mathscr{T}$. The Hamiltonian H of a clothing has the following asymptotic Taylor expansion near $\mathbb{T}^n \times \{0\}$:

$$H(\varphi, I) \sim \langle \omega, I \rangle + \sum_{s=2}^{\infty} h_s(\varphi, I), \tag{9.12}$$

where $h_s(\varphi, I)$ is a homogeneous polynomial of degree s in the variables I with coefficients depending periodically on φ. The explicit form of the first term in (9.12) is justified as follows: the restriction to $\mathbb{T}^n \times \{0\}$ of the image of our system under Ψ^{-1} coincides with g^t_ω, and the Hamiltonian equations (3.3), or (8.4), then give the values of the first derivatives of H on $\mathbb{T}^n \times \{0\}$. Note that a clothing is not defined uniquely. We say a clothing is of a normal form if every h_s in (9.12) is independent of φ. The existence of a normal form clothing essentially depends on arithmetical properties of the frequency.

Proposition 9.13. *Let a Lagrangian invariant torus $\mathscr{T}$ bear a quasiperiodic motion g^t_ω and let ω belong to the Diophantum set $\mathscr{E}_{\sigma\gamma}$ for some positive σ and γ. Then $\mathscr{T}$ possesses a normal form clothing. The coefficients $h_s(I)$ of the Hamiltonian in the normal form are uniquely determined by an embedding $\chi : \mathbb{T}^n \longrightarrow X$ of* (IT3).

Of course, the normal form clothing is not determined uniquely by χ. The proof of 9.13 will be derived later. First we discuss some consequences. Let $\chi' : \mathbb{T}^n \longrightarrow X$ be another embedding with the same image as χ, and let χ' conjugate the motion on $\mathscr{T}$ with another quasiperiodic motion $g^t_{\omega'} : \mathbb{T}^n \longrightarrow \mathbb{T}^n$. Without loss of generality we may set $\chi'(0) = \chi(0)$. Then, due to 9.5, $\chi^{-1} \circ \chi'$ has the lift $\tilde{g} \in GL(n, \mathbb{Z})$ and $\omega' = \tilde{g}^{-1} \cdot \omega = A(\tilde{g})(\omega)$. We can construct a normal form clothing $\Psi' : A_{D'} \longrightarrow U$, H', to a new embedding χ' by setting $D' = \tilde{g}^T(D)$, $\Psi' = \Psi \circ \theta$, where $\theta : A_{D'} \longrightarrow A_D$ has the lift $\tilde{\theta} : (\varphi', I') \longmapsto (\tilde{g}\varphi, (\tilde{g}^{-1})^T I')$, and $H' = \tilde{\theta}^* H$. Note that this latter transformation formula fits both continuous and discrete time because of the linearity of $\tilde{\theta}$. This implies that the corresponding quadratic form h_2 is transformed to $h'_2 = B(g)h_2$ where $(B(g)h_2)(I') = h_2((g^{-1})^T I')$. This formula defines a right action B of $GL(n, \mathbb{Z})$ on the space $S^2(\mathbb{R}^n)$ of real quadratic forms on $\mathbb{R}^n$. Let Ω be the frequency class of $\mathscr{T}$. Inasmuch as $GL(n, \mathbb{Z})$ acts freely on Ω, a frequency $\omega \in \Omega$ determines an inclusion χ (see 9.5), and hence the normal form term h_2, uniquely. Consider the map $s : \Omega \longrightarrow S^2(\mathbb{R}^n)$ which assignes to $\omega \in \Omega$ the quadratic form $2h_2$ of a corresponding normal form Hamiltonian. The above considerations yield that s is equivariant with respect to the actions $A(g)|_\Omega : \Omega \longrightarrow \Omega$ and $B(g) : S^2(\mathbb{R}^2) \longrightarrow S^2(\mathbb{R}^n)$, $g \in GL(n, \mathbb{Z})$, i.e. $s \circ A(g) = B(g) \circ s$ for all g. The last assertion of Proposition 9.13 implies that this map is a symplectic invariant of our system near the torus. We call s the *shear of* $\mathscr{T}$.

A Lagrangian invariant torus T, bearing a quasiperiodic motion with Diophantum frequency, is called a *KAM torus* if the values of its shear are nonsingular quadratic forms. This is equivalent to $\mathrm{grad}_I h_2 : \mathbb{R}^n \longrightarrow \mathbb{R}^n$ being an invertible linear map.

Proof of Proposition 9.13. We start by describing an inductive step which enables us to normalize the mth term of the expansion (9.12) provided all preceding terms are already in normal form. Let

$$H(\varphi, I) \sim \langle \omega, I \rangle + \sum_{s=2}^{m-1} \hat{h}_s(I) + \sum_{s=m}^{\infty} h_s(\varphi, I), \tag{9.14}$$

where $m \geqq 2$. Pass to the new angle–action variables (ψ, T):

$$T_i = I_i + \frac{\partial g_m(\psi, I)}{\partial \psi^i}, \quad \varphi^i = \psi^i + \frac{\partial g_m(\psi, I)}{\partial I_i}, \tag{9.15}$$

$1 \leqq i \leqq n$, where the Hamiltonian $g_m(\psi, I)$ is a homogeneous polynomial of degree m with respect to the old action variables I_i with coefficients periodically dependent on the new angle variables ψ^i. The transformation $\theta : (\psi, T) \longmapsto (\varphi, I)$ is symplectic. In the case of continuous time the Hamiltonian (9.14) is transformed into $H \circ \theta$, which is the Hamiltonian of the new clothing $\Psi \circ \theta$. Using (9.15), it is easy to obtain the following expression for $H \circ \theta$ up to terms of order $\| I \|^{m+1}$ which we denote by $\mathcal{O}(m+1)$:

$$H \circ \theta(\psi, T) = \langle \omega, T \rangle + \sum_{s=2}^{m-1} \hat{h}_s(T) + h_m(\psi, T) - \left\langle \omega, \frac{\partial g_m(\psi, T)}{\partial \psi} \right\rangle + \mathcal{O}(m+1).$$

We set $\hat{h}_m = [\![h_m]\!]$, where $[\![f]\!](T) = \int_0^1 f(\psi, T)\, d\psi$ is the mean value of f with respect to the angle variables, and define g_m as being a solution to

$$\left\langle \omega, \frac{\partial g_m}{\partial \psi} \right\rangle = h_m - \hat{h}_m. \tag{9.16}$$

This solution is given by the formula

$$g_m(\psi, T) = \sum_{k \in \mathbb{Z}^n \setminus \{0\}} \frac{h_{mk}(T)}{\langle \omega, k \rangle} e^{2\pi i \langle k, \psi \rangle}, \tag{9.17}$$

$$h_{mk}(T) = \int_{0 \leqq \psi \leqq 1} \exp(-i\langle k, \psi \rangle) h_m(\psi, T)\, d\psi.$$

Since ω is Diophantum and h_m is C^∞, the series (9.17) converges uniformly as well as the series of all its derivatives. Note that all the terms involved are homogeneous polynomials of degree m with respect to the variable T. The transformation thus defined makes the Hamiltonian of the clothing normalized up to terms of order m. The same procedure also works in the case of discrete time. We omit several tedious calculations where one obtains a new Hamiltonian of discrete time transformation after the changing the variables of form (9.15). The result is the same, but equations (9.16), (9.17) have to be changed for

$$g_m(\psi + \omega, T) - g_m(\psi, T) = h_m(\psi, T) - \hat{h}_m(T), \tag{9.18}$$

$$g_m(\psi, T) = \sum_{k \in \mathbb{Z}^n \setminus \{0\}} \frac{h_{mk}(T) \exp(2\pi i \langle k, \psi \rangle)}{\exp(2\pi i \langle k, \omega \rangle) - 1}. \tag{9.19}$$

Thus, we can normalize, step by step, all the members of the Hamiltonian (9.12). The sequence of superpositions of these transformations do not necessarily have a limit in the usual sense, but, inasmuch as the mth transformation involves only the terms of order m and higher and does not affect the terms of lower order, there does exist a formal series for the transformating Hamiltonian which would transform our Hamiltonian to the normal form. Applying Whitney's extension theorem to this formal series (see, for instance, Malgrange (1966), Chapter 1), one obtains a smooth Hamiltonian with the said series as its formal Taylor expansion. The transformation defined by this Hamiltonian normalizes our system in the desired sense.

Now there be two normal form Hamiltonian corresponding to the same embedding χ. Then the first terms are the same: $\langle \omega, T \rangle$. There exists a symplectic transformation Σ which reduces one of these Hamiltonians to the another and which is the identity on $\mathbb{R}^n \times \{0\}$. This Σ can be represented as a superposition of elementary steps, each being of the form (9.15) and acting on the terms of order m and higher, m being the number of steps. Let all the terms of our two normal form Hamiltonian coincide up to order $m-1$, and let the elementary transformation of ith step, $1 \leqq i \leqq m-1$, do not change action variables. Here $m \geqq 2$. Then the equations (9.16) or (9.18) with h_m and $\hat{h}_m$ being the mth terms of normal form Hamiltonians which we are comparing show that the mth terms of the normal form Hamiltonians as a matter of that coincide, and g_m is independent of angle variables, so an m-step elementary transformation also does not change angle–action variables. This proves by induction the claim concening uniqueness of the normal form expansion.

We shall continue the study of the motion of a symplectic dynamical system near a KAM torus in §12.

§10. KAM Set

It is preferable in KAM theory to deal with a more complicated object which consists of many KAM tori collected together in a special kind of family. A KAM set is the image under a symplectic embedding of tori bearing quasiperiodic motions with Diophantum frequencies of a standard integrable system. In this section we give the definition to this central notion.

Three constituents are necessary to define a KAM set: a *model system*, a *frequency set*, and a *conjugating map*. A *model system* to a KAM set is a standard integrable system in a symplectic annulus $g^t: A_D \longrightarrow A_D$ (see Example 7.14), which satisfies the twist condition (TC) formulated below. The domain D, the *action domain*, must be non-empty and open. Recall that the covering $\tilde{g}^t : \mathbb{R}^n \times D \longrightarrow \mathbb{R}^n \times D$ has the form $\tilde{g}^t(\varphi, I) = (\varphi + t\omega(I), I)$ where $\omega(I) = \operatorname{grad} K(I)$. The function $K : D \longrightarrow \mathbb{R}$ is called the *Hamiltonian* of $\{g^t\}$, the map $\omega : D \longrightarrow \mathbb{R}^n$, defined as $D \ni I \longmapsto \omega(I) = \operatorname{grad} K(I)$ is called the *frequency map*, and its image $B = \omega(D) \subset \mathbb{R}^n$, the *frequency domain* of $\{g^t\}$.

(TC) We say $\{g^t\}$ satisfies the *twist condition* if its frequency map is a diffeomorphism of the action domain onto the frequency domain.

A *frequency set* $\mathscr{E}$ is an arbitrary compact subset of $B \cap \mathscr{E}_{\sigma\gamma}$, where $\mathscr{E}_{\sigma\gamma}$ is the Diophantum set defined in §9, $\sigma > d - 1$ and γ being some positive constants. The tori $\mathbb{T}^n \times \{I\}, \omega(I) \in \mathscr{E}$ are called *$\mathscr{E}$-tori*. The set of all $\mathscr{E}$-tori, $\mathbb{T}^n \times \omega^{-1}(\mathscr{E})$ will be denoted Ξ. Also we adopt the notation $\tilde{\Xi}$ for the preimage $p^{-1}(\Xi)$ of the set of $\mathscr{E}$-tori under the canonical projection $p : \mathbb{R}^n \times D \longrightarrow A_D$ (see (2.5′)).

Now consider a symplectic dynamical system defined in a $2n$-dimensional symplectic manifold X. A *conjugating map* is a symplectic embedding $\psi : A_D \longrightarrow X$ such that

(CM1) the image $A = \Psi(A_D)$ is contained within the domain of the system (in the domain of the generator in the case of discrete time, in the case of continuous time this condition says nothing);

(CM2) the image $\mathscr{K} = \Psi(\Xi)$ of $\mathscr{E}$-tori (see definitions above) is an invariant set with complete dynamics (§3). Denote by $f^t : \mathscr{K} \longrightarrow \mathscr{K}$ the flow generated by the system on the set $\mathscr{K}$;

(CM3) the map Ψ conjugates $g^t|_\Xi$ and f^t, i.e. the following diagram commutes for all t

$$\begin{array}{ccccc} X & \supset & \mathscr{K} & \xrightarrow{\quad f^t \quad} & \mathscr{K} \\ \uparrow \Psi & & \uparrow \Psi|_\Xi & & \uparrow \Psi|_\Xi \\ A_D & \supset & \Xi & \xrightarrow{\quad g^t|_\Xi \quad} & \Xi \end{array} \tag{10.1}$$

Here one tacitly supposes that both the systems in question have the same kind of time: discrete or continuous.

Note that the image $\mathscr{T}_I$ under Ψ of a torus $\mathbb{T}^n \times \{I\}$, $I \in \omega^{-1}(\mathscr{E})$ is a KAM torus in the sense of the previous section, and Ψ conjugates quasiperiodic motion on $\mathbb{T}^n \times \{I\}$ with the restriction of f^t to $\mathscr{T}_I$. The tori $\mathscr{T}_I, I \in \omega^{-1}(\mathscr{E})$, form a Cantor like foliation of $\mathscr{K}$.

To complete the formulation of the definition of a KAM set, we need an additional condition to be posed on the conjugating map. As we saw in the previous section, a KAM torus $\mathscr{T}_I$ has a tubular neighbourhood where the motion is governed by a Hamiltonian. The conjugating map Ψ being restricted to a set of the form $\mathbb{T}^n \times U, U \ni I$ homeomorphic to an open n-ball, may serve as the first constituent of a clothing of $\mathscr{T}_I$ (see §9). There also exists the second one, i.e. the Hamiltonian of the clothing. The normed Hamiltonian has the Taylor expansion of form (9.12), but we do not here fix its value on the preimage of $\mathscr{T}_I$. We call it the *induced Hamiltonian* and denote it by H_{ind}. It is defined uniquely up to an additive constant. Such a construction may be performed for each KAM torus

$\mathscr{T}_I, I \in \omega^{-1}(\mathscr{E})$. Note that each H_{ind} and the function $K \circ p_2$ are constants on $\mathscr{E}$-tori. We also impose the following Hamiltonian Agreement Condition:

(HAC) The induced Hamiltonian H_{ind} can be chosen so that the jets of infinite order of H_{ind} and $K \circ p_2$ coincide at points of Ξ, that is, for each multiindex $k=(k^1, k^2, \ldots, k^n) \in \mathbb{Z}_+^n$, $D^k[(K \circ p_2) - H_{\text{ind}}] = 0$ at points of $\Xi \cap G$, G being the domain of definition of H_{ind}.

A *KAM set* in a symplectic dynamical system defined in a symplectic manifold X is a completely invariant set $\mathscr{K} \subset X$ such that there exists a model system $g^t: A_D \longrightarrow A_D$, obeying (TC), a frequency set $\mathscr{E}$, and a conjugating map Ψ, such that $\psi(\Xi) = \mathscr{K}$, Ξ being the set of $\mathscr{E}$-tori, the conditions (CM1), (CM2), (CM3), and (HAC) being satisfied.

Remark 10.2. I do not know whether the condition (HAC) is a consequence of all the other conditions. Following from the results of §12, to answer this question, it is sufficient to prove the uniqueness of a KAM torus with the prescribed frequency vector in a little perturbated standard integrable system, provided the said torus is sufficiently close to its nonperturbed counterpart. This uniqueness hypothesis is valid in the case of two degrees of freedom (see Katok (1989)). Hence (HAC) in superfluous for two-degrees-of-freedom systems.

In the next section we shall establish the existence of a KAM set in the special situation when one perturbs the Hamiltonian of a standard integrable system is a symplectic annulus by a small amount.

§11. The KAM Theorem in an Annulus

The formulation below, although appearing to be particular, is nevertheless very useful in applications. All the systems in question here are defined globally on a $2n$-dimensional symplectic annulus A_D where $D \subset \mathbb{R}^n$ is a non-empty, open, bounded, and simply connected domain. We fix a number $\sigma > d-1$ where d is the number of degrees of freedom (see §3). The other parameter γ entering the definition of $\mathscr{E}_{\sigma\gamma}$ takes its values in the interval $]0, 1]$, and our estimates will include it explicitly. The number r_0 of derivatives of a perturbation which we are to control equals to

$$r_0 = \begin{cases} \left[\dfrac{80}{3}\sigma + \dfrac{64}{3}n\right] + \dfrac{2}{3} + 27 & \text{for flows,} \\ \left[\dfrac{80}{3}\sigma + \dfrac{64}{3}n\right] + 49 & \text{for cascades.} \end{cases} \tag{11.1}$$

Here $[x]$ denotes the integral part of a number x.

We premise the theorem with dramatis personae:

(I) The *unperturbed integrable system* (coincides with its model) $g_0^t: A_D \longrightarrow A_D$ is defined by the Hamiltonian function $K_0: D \longrightarrow \mathbb{R}$, it satisfies the twist condition

(TC) (see §10) and there is a constant $R>0$ such that

$$\max_{i,k}\left\|\frac{\partial^2 K_0}{\partial I_i \partial I_k}\right\|_{r_0,\gamma} \leqq R, \qquad \left|\det\left\{\frac{\partial^2 K_0}{\partial I_i \partial I_k}\right\}\right| \geqq R^{-1} \tag{11.2}$$

(see §8 for the definitions of $C^r-\gamma$ norms $\|\cdot\|_{r,\gamma}$). The frequency domain $B=\omega_0(D)$ is to be convex and bounded where $\omega_0=\operatorname{grad} K_0$ is the frequency map of the unperturbed system. We single out four subdomains of B, namely

$$[B]_{i\gamma}=\{y\in B\colon \operatorname{dist}(y,\partial B)>i\gamma\}, \quad 1\leqq i\leqq 4.$$

Their preimages in D under the frequency map will be denoted by $D_{i\gamma}=\omega_0^{-1}([B]_{i\gamma})$.

($\mathscr{E}$) *The Diophantum frequency set* is given by

$$\mathscr{E}=\mathscr{E}_{\sigma,\gamma}\cap[B]_{4\sigma}.$$

(II) The *perturbed system* $f^t\colon A_D\longrightarrow A_D$ is defined by a Hamiltonian function of the form $H=K_0\circ\tilde{p}_2+H_1$ where H_1 satisfies the conditions

$$\operatorname{supp} H_1\subset\mathbb{R}^n\times D_{2\gamma}, \tag{11.3}$$

$$\|H_1\|_{r_0,\gamma}\leqq\gamma^2\delta. \tag{11.4}$$

We suppose, in view of 8.17, that $0<\delta<1/d$.

Both the unperturbed and the perturbed systems are given and we wish to find seek for a KAM set to (II) with frequency set $\mathscr{E}$, model system g^t, and conjugating map Σ.

(III) The *model system* $g^t\colon A_D\longrightarrow A_D$ is defined by the Hamiltonian function $K\colon D\longrightarrow\mathbb{R}$ satisfying the twist condition. Its frequency map $\omega=\operatorname{grad} K$ has the same domain D and the same image B as those of the unperturbed system. The set of $\mathscr{E}$-tori and its lieft will be denoted

$$\Xi=\mathbb{T}^n\times\omega^{-1}(\mathscr{E}) \quad\text{and}\quad \tilde{\Xi}=\mathbb{R}^n\times\omega^{-1}(\mathscr{E}).$$

(IV) The *conjugating map* $\Sigma\in\operatorname{Diff}_0^{\text{symp}}(A_D)$ conjugates $g^t|_\Xi$ and $f^t|_{\Sigma(\Xi)}$, i.e. the diagram

$$\begin{array}{ccc} \Xi & \xrightarrow{\;g^t|_\Xi\;} & \Xi \\ \Big\downarrow{\scriptstyle \Sigma|_\Xi} & & \Big\downarrow{\scriptstyle \Sigma|_\Xi} \\ \Sigma(\Xi) & \xrightarrow{\;f^t|_{\Sigma(\Xi)}\;} & \Sigma(\Xi). \end{array} \tag{11.5}$$

commutes.

Theorem 11.6. *Given n, σ, and R, there are positive numbers δ and c such that the following is true.*

For each pair of an unperturbed system (I) *and a perturbed one* (II), *the* (10.2)–(10.4) *being satisfied with the said constants and with some $\gamma\in]0,1]$, there exists a model system* (III) *and a conjugating symplectic diffeomorphism* (IV) *such that*

(i) *the diagram* (1.5) *commutes,*

(ii) *the system $\Sigma^{-1}\circ f^t\circ\Sigma$ possesses the Hamiltonian H_{ind} such that the jets of* *infinite order of functions* $K \circ p_2$ *and* H_{ind} *coincide at points of* $\tilde{\Xi}$. *In the case of continuous time one may take* $H_{\text{ind}} = H \circ \Sigma$.

(iii) *the estimates*

$$\| K \circ \tilde{p}_2 - K_0 \circ \tilde{p}_2 \|_{2,\gamma} \leqq c^{16/15} \| H_1 \|_{r_0,\gamma}^{7/15}, \tag{11.7}$$

$$\| \Sigma - \text{id} \|_{1,\gamma} \leqq c\gamma^{-8/15} \| H_1 \|_{r_0,\gamma}^{4/15}, \tag{11.8}$$

and inclusions

$$\text{supp}(K - K_0) \subset D_\gamma, \tag{11.9}$$

$$\text{supp}(\tilde{\Sigma} - id) \subset \mathbb{R}^n \times D_\gamma, \tag{11.10}$$

$$\Sigma(\Xi) \subset A_{D_3} \quad \textit{hold.} \tag{11.11}$$

Thus we have that $\Sigma(\Xi)$ appears to be a KAM set to the perturbed system. It has the same action and frequency domains and the same Diophantum frequency set as those of the unperturbed system. The theorem asserts, therefore that the Kolmogorov set to an integrable system in the annulus persists under pertinent small perturbation, the model system varying only slightly.

The whole of Chapter IV is devoted to the proof of Theorem 11.6. There are other formulations of the KAM theorem which are reviewed in Bost (1985). The essential feature of this one is the weak condition (11.2) on the unperturbed Hamiltonian, it need not be analytical. Of course, the r_0 (see (11.1)) and the exponents in the right-hand sides of (11.7) and (11.8) do not seem to be the best ones, the values of which are due to our method of proof. Somewhat better estimates may be obtained under the additional restrictions on σ. This is also discussed in Bost (1985).

§12. Near a Torus

We now return to the study of the motion of a general symplectic dynamical system near a single KAM torus $\mathcal{T}$. We will show that $\mathcal{T}$ can always be included, as a member, in some KAM set.

Let Ω be the frequency class of $\mathcal{T}$. Choose some frequency $\omega \in \Omega$ and let the corresponding frequency vector belong to $\mathscr{E}_{\sigma\gamma}$. We look for a KAM set in the neighbourhood of $\mathcal{T}$ whose tori have the frequency vectors ranging over the set $\mathscr{E}_{\sigma'\gamma'}$ with $\sigma' \geqq \sigma, \gamma' \geqq \gamma$. We fix σ' and will choose γ' later. Fix some normal clothing to $\mathcal{T}$ which corresponds to the chosen ω (see §9), and represent the Hamiltonian as follows

$$H(\varphi, I) = K_0(I) + H_1(\varphi, I)$$

where

$$K_0(I) = \langle \omega, I \rangle + \sum_{s=2}^{N} \hat{h}_s(I) \tag{12.1}$$

and the Taylor expansion for H_1 in action variables begins with terms of order $N+1$. The number N will be chosen sufficiently large so that $N \geqq r_0 + 2$ where r_0 is given by (11.1) with σ exchanged for σ'. Obviously H_1 satisfies the estimates:

$$(12.2) \qquad |D_\varphi^{\rho_1} D_I^{\rho_2} H_1| \leqq \text{const}\,|I|^{N+1-r_0}, \quad 0 \leqq \rho_1 + \rho_2 \leqq r_0,$$

where const depends only on the choice of the clothing and the number N.

The frequency domain of the type of a KAM set which we are looking for will be the open ball B_ε of radius ε with the center at ω, and ε will be chosen later. Fix $p \in]1, \frac{1}{2}(N+1-r_0)[$ and set $\gamma' = \varepsilon^p$. Then the estimate (9.9a) gives us

$$|\mathcal{M}e\mathfrak{s} B_\varepsilon - \mathcal{M}e\mathfrak{s}(B_\varepsilon \cap \mathscr{E}_{\sigma'\gamma'})| \leqq \text{const}\, \varepsilon^{d-1+p},$$

const only depending on n and σ'.

Since $\hat{h}_2(I)$ is a nonsingular quadratic form, the frequency map $\nu_0 : \longrightarrow \operatorname{grad} K_0(I)$ is a local diffeomorpism, being restricted onto some neighbourhood of zero, and it maps zero to ω. Denote $D_\varepsilon = \nu_0^{-1}(B_\varepsilon)$ and let ε be so small such that ν_0 maps D_ε diffeomorphically onto B_ε and there is a positive constant R such that K_0 given by (12.1) satisfies (11.2) with γ' standing for γ. There is an $\varepsilon_0 > 0$ such that for $\varepsilon \in]0, \varepsilon_0]$ such a constant exists and is independent of ε. To now force the perturbative Hamiltonian to satisfy (11.3) let us introduce a switching multiplier of the form

$$v(I) = v_0((2\gamma')^{-1}(|\nu_0(I) - \omega| - \varepsilon))$$

where $v_0(x)$ is a C^∞-function which is zero if $x \geqq 0$, and which is equal to 1 if $x < -1$. Then the estimates (12.2) are transformed to

$$\| v \cdot H_1 \|_{r_0, \gamma'} \leqq \text{const} \cdot \varepsilon^{N+1-r_0-2p} \cdot (\gamma')^2$$

and the explicit form of the multiplier v yields

$$\operatorname{supp}(v \cdot H_1) \subset \mathbb{R}^n \times [D_\varepsilon]_{2\gamma'}$$

These results show that, if ε is sufficiently small, our Hamiltonian satisfies the conditions of Theorem 11.6. We formulate the consequence as follows.

Theorem 12.3. *Let $\mathcal{T}$ be a KAM torus with a frequency vector $\omega \in \mathscr{E}_{\sigma\gamma}$. Choose a clothing of $\mathcal{T}$ corresponding to ω. Let Ψ be the map of the clothing. Given $p > 1$, $M > 0$, and $\sigma' \geqq \sigma$, there exists $\varepsilon_0 > 0$ such that for any positive $\varepsilon < \varepsilon_0$ there exists a KAM set $\mathcal{K}$ to the system in question with frequency domain $B_\varepsilon = \{\omega' \in \mathbb{R}^n : \|\omega' - \omega\| < \varepsilon\}$ and with frequency set $\mathscr{E} = \mathscr{E}_{\sigma'\gamma'} \cap [B_\varepsilon]_{4\gamma'}$ where $\gamma' = \varepsilon^p$, such that $\mathcal{T}$ is a member of $\mathcal{K}$ corresponding to the frequency vector ω. Let Σ and ν be respectively the conjugating and the frequency maps of $\mathcal{K}$. They can be chosen so that the Liouville measure of $\Sigma(A_{\nu^{-1}(B_\varepsilon)}) \setminus \Sigma(\mathbb{T}^n \times \nu^{-1}(\mathscr{E}))$ is less than $\text{const} \cdot \varepsilon^{d-1+p}$, and*

$$\| \Psi^{-1} \circ \Sigma - \text{id} \|_{1,\gamma'} \leqq \text{const} \cdot \varepsilon^M,$$

the constants in these estimates depending only on $\mathcal{T}, \Psi, p, M$, and σ'.

§13. Near a Periodic Motion

As was stated in §5, the investigation of a symplectic dynamical system near a periodic trajectory can always be reduced to the case of a local symplectic diffeomorphism $f: U \longrightarrow X$, near a fixed point $x_0 \in U \subset X$. Inasmuch as the behaviour of trajectories near a hyperbolic fixed point resembles that of its tangent map – a linear hyperbolic map – there are no invariant tori in a small neighbourhood of such a point. Therefore we here restrict ourselves to the case of a simple elliptic fixed point.

The first step is to study the problem in question on the level of formal power series.

Formal Reduction to a Normal Form

Let $\mathscr{G}$ denote the group of formal symplectic transformations of $\mathbb{R}^{2n}$ of the form $g: x \longmapsto \sum_{k \geqq 1} a_k(x)$ where a_k are vector valued polynomials of degree k of variables $x = (q^1, q^2, \ldots, q^n, p_1, p_2, \ldots, p_n)$. The group operation in $\mathscr{G}$ is the formal superposition. Since g is symplectic it preserves $\Omega_{st} = \sum_{i=1}^n dq^i \wedge dp_i$ formally. In particular $a_1 \in \mathrm{Sp}(R^{2n})$. The complex variant ${}^{\mathbb{C}}\mathscr{G}$ of the group of formal symplectic transformations differs from the real one in that the coefficients of a_k are allowed to be complex. The real group $\mathscr{G}$ can be regarded as a subgroup of ${}^{\mathbb{C}}\mathscr{G}$. Given an integer $m \geqq 2$ we also consider the subgroups $\mathscr{G}_m$ and ${}^{\mathbb{C}}\mathscr{G}_m$, consisting of all formal transformations which differ from the identity by terms of degree m and higher: $x \longmapsto x + \sum_{k \geqq m} a_k(x)$. The change of variables $\rho: (q, p) \longmapsto (z, w) = (q + ip, q - ip)$ transforms $\mathscr{G}$ and $\mathscr{G}_m$ into the other subgroups $\hat{\mathscr{G}} = \rho \mathscr{G} \rho^{-1}$, $\hat{\mathscr{G}}_m = \rho \mathscr{G}_m \rho^{-1}$. The subgroup ${}^{\mathbb{C}}\mathscr{G}_m$ is invariant under that transformation. Any element $g \in {}^{\mathbb{C}}\mathscr{G}_m$ can be written with the help of a Hamiltonian function $h(q, p') = \sum_{k \geqq m+1} h_k(q, p')$, the h_k being homogeneous components of degree k, so that the following relations are true:

$$(13.1) \qquad g(q, p) = (q', p'), \quad q' = q + \frac{\partial h(q, p')}{\partial p'}, \quad p = p' + \frac{\partial h(q, p')}{\partial q}.$$

Elements of $\mathscr{G}_m$ are precisely those polynomials h_k with real coefficients.

Proposition 13.2. (1) *Let* $\Phi(\gamma) = \sum_{s=1}^{\infty} \varphi_s(\gamma)$ *be a formal real series in the variable* $\gamma = (\gamma_1, \gamma_2, \ldots, \gamma_n)$. *Then the formal transformation* $r_\Phi: (z, w) \longmapsto (z', w')$ *where* $z'_j = \exp(i(\partial \Phi(\gamma))/\partial \gamma_j) z_j$, $w'_j = \exp(-i(\partial \Phi(\gamma))/\partial \gamma_j) w_j$, $1 \leqq j \leqq n$, $\gamma = (z_1 w_1, z_2 w_2, \ldots, z_n w_n)$, *belongs to* $\hat{\mathscr{G}}$.

(2) *The homogeneous polynomial*

$$h_{m+1}(z, w) = \sum_{k+l=m+1} h_{m+1,k,l} z^k w^l, \quad k, l \in \mathbb{Z}^n_+,$$

is the leading term of the Hamiltonian of some transformation $g \in \hat{\mathscr{G}}_m$ *if and only if*

$$(13.2a) \qquad h_{m+1,k,l} = -\bar{h}_{m+1,l,k}.$$

Proof. (1) We can check that r_Φ is symplectic immediately by a long but simple calculation. The reverse change of variable $\rho^{-1}:(z,w)\longmapsto(q,p)=((1/2)(z+w),(1/2i)(z-w))$ transforms r_Φ into rotation through angle $(\partial/\partial\gamma)\Phi$ depending on the square of the radius $\gamma=p^2+q^2$.

(2) The condition $g\in\hat{\mathscr{G}}$ may also be described as the preservation of the plane $z=\bar{w}$. Writing out the equations $g(z,w)=(z',w')$, $z'=z+(\partial/\partial w')h_m(z,w')+\ldots$, $w=w'+(\partial/\partial z)h_m(z,w')+\ldots$, and setting $z=\bar{w}$, $z'=\bar{w}'$, one obtains

$$\frac{\partial h_{m+1}(z,w')}{\partial z}+\frac{\partial h_{m+1}(z,w')}{\partial w'}+\text{higher terms}=0.$$

Substituting $z=\bar{w}$ and taking into account that $\bar{w}'=z'=z+$ *higher order terms*, yields the desired relations (13.2a). One obtains the proof that (13.2a) is sufficient by noticing that the transformation $g\longmapsto\rho^{-1}\circ g\circ\rho$, $g\in\hat{\mathscr{G}}_m$, transforms the leading terms of Hamiltonians independently of the others. These may be corrected in ${}^{\mathbb{C}}\mathscr{G}_m$ to be real for the image of g in ${}^{\mathbb{C}}\mathscr{G}_m$, whilst the leading term is already real here due to (13.2a). Returning to the variables z, w, one obtains a corrected $g\in{}^{\mathbb{C}}\mathscr{G}_m$ with the given leading term h_m of the Hamiltonian, satisfying (13.2a). □

Now consider a local symplectic diffeomorphism $f:U\longrightarrow X$ with a simple elliptic fixed point x_0 (see Example 3.12). Take a symplectic chart $\beta:V\longrightarrow X$, $V\subset\mathbb{R}^{2n}$, such that $0\in V$ and $\beta(0)=x_0$. We may assume that the matrix of $T_0(\beta^{-1}\circ f\circ\beta)$ is already in the quasidiagonal form described in Proposition 1.30. Then the infinite jet (formal Taylor expansion) $J_0^\infty(\beta^{-1}\circ f\circ\beta)\in\mathscr{G}$ of $\beta^{-1}\circ f\circ\beta$ at 0 may be represented as the superposition $\rho^{-1}\circ r_\Phi\circ k\circ\rho$ where $k\in\hat{\mathscr{G}}_2$ and r_Φ is defined as in Proposition 13.2 (1), with $\Phi(\gamma)=\langle\alpha,\gamma\rangle$, $\alpha=(\alpha_1,\alpha_2,\ldots,\alpha_n)\in\mathbb{R}^n$, $\alpha_k\in]0,\pi[\cup]-\pi,0[$, $1\leqq k\leqq n$, being the invariants of $T_0(\beta^{-1}\circ f\circ\beta)$ given by Proposition 1.30. We call $(2\pi)^{-1}\alpha$ the *frequency vector* of *f at the point* x_0. This definition also holds for the more general case of a simple elliptic periodic orbit M.

Induction Lemma 13.3. *Consider a transformation of the form* $r_\Phi\circ k\in\hat{\mathscr{G}}$, *where* $k\in\hat{\mathscr{G}}_{m-1}$ *and* r_Φ *is defined in Proposition 13.2 with*

$$\Phi(\gamma)=\sum_{1\leqq s\leqq(m-1)/2}\gamma_s(\gamma),\quad \varphi_1(\gamma)=\langle\alpha,\gamma\rangle,$$

φ_s *is a homogeneous polynomial of degree s of variables* $\gamma_i=z_iw_i$, $m\geqq 3$. *Let the following condition for the absence of resonances be satisfied:*

(AR*m*) $$(2\pi)^{-1}\langle s,\alpha\rangle\notin\mathbb{Z}\ \textit{for each integer vector}\ s=(s_1,s_2,\ldots,s_n)\ \textit{with}\ 1\leqq\textstyle\sum_{i=1}^n|s_i|\leqq m.$$

Then there exists a map $g\in\hat{\mathscr{G}}_{m-1}$ *such that*

(13.3a) $$g\circ r_\Phi\circ k\circ g^{-1}=r_{\Phi'}\circ k'$$

where $k'\in\hat{\mathscr{G}}_m$, *and* $\Phi'=\Phi$ *if m is odd,* $\Phi'=\Phi+\varphi_{m/2}$ *if m is even,* $\varphi_{m/2}$ *being a homogeneous polynomial of degree m/2 uniquely determined by the leading term of the Hamiltonian of k.*

Proof. Let h_m be the leading term of the Hamiltonian of k. Define a map $g\in\hat{\mathscr{G}}_{m-1}$ as the Hamiltonian with the leading term $g_m(z,w)$ which is a solution to the equation

$$\text{(b)}\qquad g_m(z,w)-g_m(e^{i\alpha}z,e^{-i\alpha}w)=h_m(z,w)-i\varphi_{m/2}(\gamma).$$

In (b) the last term in the right-hand side must be dropped in the case of m being odd, $\gamma=(z_1w_1,z_2w_2,\ldots,z_nw_n)$, $e^{i\alpha}\cdot z$ stands for $(e^{i\alpha_1}z_1,e^{i\alpha_2}z_2,\ldots,e^{i\alpha_n}z_n)$, and similarly for $e^{-i\alpha}\cdot w$. Let us now prove that, under the above conditions on α, (b) has a solution, the polynomial $\varphi_{m/2}$ also being defined from (b). Writing out (b) in coefficients looks like

$$\text{(c)}\qquad g_{mkl}-g_{mkl}\exp(i\langle k-1,\alpha\rangle)=h_{mkl}-i\varphi_{m/2,k}\delta_{kl}.$$

Here $k,l\in\mathbb{Z}^n_+$ and range over all values satisfying $|k|+|l|=m$. The solution is

$$g_{mkl}=(\exp(i\langle k-l,\alpha\rangle)-1)^{-1}h_{mkl},\quad \text{if } k\neq 1,$$

$\varphi_{m/2,k}=-iH_{mkk}$, g_{mkk} remains arbitrary. The property (13.2a) implies that h_{mkk} are purely imaginary, so $\varphi_{m/2,k}$ are real. Also the coefficients g_{mkl} satisfy (13.2a) provided that the h_{mkl} do as well. For definitness we take $g_{mkk}=0$. Extending the series for the Hamiltonian with the leading term g_m, as it was done in the proof of Proposition 13.2, we obtain a map $g\in\hat{\mathscr{G}}_{m-1}$. It remains for us to check (13.3a). Let us first calculate $r_{\Phi'}^{-1}\circ g\circ r_\Phi\circ k\circ g^{-1}$. We use the following notations for the variables:

$$(\zeta',\chi')\xleftarrow{r_{\Phi'}^{-1}}(\zeta'',\chi'')\xleftarrow{g}(z'',w'')\xleftarrow{r_\Phi}(z',w')\xleftarrow{k}(z,w)\xleftarrow{g^{-1}}(\zeta,\chi),$$

and

$$\gamma=z'\cdot w'=z''\cdot w'',\ \gamma'=\zeta'\cdot\chi'=\zeta''\cdot\chi''.$$

Here $z\cdot w$, $z,w\in\mathbb{C}^n$, stands for $(z_1w_1,z_2w_2,\ldots,z_nw_n)\in\mathbb{C}^n$. Subsequently using Hamiltonian equations (13.1), one obtains the following expressions which link (ζ,χ) with (ζ',χ'):

$$\text{(d)}\qquad \begin{aligned}\zeta&=\zeta'\cdot\exp\left(i\frac{\partial\Phi'(\gamma')}{\partial\gamma'}-i\frac{\partial\Phi(\gamma)}{\partial\gamma}\right)-\frac{\partial g_m(z'',\chi'')}{\partial\chi''}\cdot\exp\left(-i\frac{\partial\Phi(\gamma)}{\partial\gamma}\right)\\&\quad-\frac{\partial}{\partial w'}h_m(z,w')+\frac{\partial}{\partial\chi}g_m(z,\chi)+\mathcal{O}(m),\\ \chi'&=\exp\left(i\frac{\partial\Phi'(\gamma')}{\partial\gamma'}\right)\cdot\left\{\exp\left(-i\frac{\partial\Phi(\gamma)}{\partial\gamma}\right)\cdot\left[\chi+\frac{\partial g_m(z,\chi)}{\partial z}-\frac{\partial h_m(z,w')}{\partial z}\right]\right.\\&\quad\left.-\frac{\partial g_m(z'',\chi'')}{\partial z''}\right\}+\mathcal{O}(m),\end{aligned}$$

where $\mathcal{O}(m)$ denotes the terms of order m and higher. It is easy to see that

$$\begin{aligned}z'&=\zeta'+\mathcal{O}(m-1)=\zeta+\mathcal{O}(m-1)=z+\mathcal{O}(m-1),\\ w'&=\chi'+\mathcal{O}(m-1)=\chi+\mathcal{O}(m-1)=w+\mathcal{O}(m-1),\\ \zeta''&=z''+\mathcal{O}(m-1)=\exp(i\alpha)z+\mathcal{O}(2),\\ \chi''&=w''+\mathcal{O}(m-1)=\exp(-i\alpha)w+\mathcal{O}(2).\end{aligned}$$

It follows that $\gamma' = \gamma + \mathcal{O}(m)$. Substituting these into (d) yields

$$\text{(e)}\quad \begin{aligned} \zeta &= \zeta' + i\frac{\partial \varphi_{m/2}(\gamma)}{\partial \gamma}\cdot z - e^{-i\alpha}\frac{\partial g_m(a,b)}{\partial b}\bigg|_{a=e^{i\gamma}\cdot z,\, b=e^{-i\alpha}\cdot w} - \frac{\partial h_m(z,w)}{\partial w} + \frac{\partial g_m(z,w)}{\partial w} + \mathcal{O}(m),\\ \chi' &= \chi + i\frac{\partial \varphi_{m/2}(\gamma)}{\partial \gamma} w + \frac{\partial g_m(z,w)}{\partial z} - \frac{\partial h_m(z,w)}{\partial z} - e^{i\alpha}\cdot\frac{\partial g_m(a,b)}{\partial a}\bigg|_{a=e^{i\alpha}\cdot z,\, b=e^{-i\alpha}\cdot w} + \mathcal{O}(m). \end{aligned}$$

Our choice of g_m and $\varphi_{m/2}$ (equations (c)) results in the main terms in the right-hand sides of (e) vanishing. We have $\zeta = \zeta' + \mathcal{O}(m)$ and $\chi' = \chi + \mathcal{O}(m)$, i.e. $r_{\Phi'}^{-1} \circ g \circ r_{\Phi} \circ k \circ g^{-1} \in \hat{\mathcal{G}}_m$. □

On being applied several times this Lemma gives the following

Proposition 13.4. *Let f be a symplectic local diffeomorphism defined in an open subset U of a $2n$-dimensional symplectic manifold X, and let x_0 be its simple elliptic fixed point. Let the frequency vector $(2\pi)^{-1}\alpha$ of f at x_0 satisfy the condition of absence of resonances (ARm), $m \geqq 4$, formulated in 13.3. Then there is a symplectic chart $\Psi: V \longrightarrow X$, $0 \in V \subset \mathbb{R}^{2n}$, which carries the origin to x_0, such that the image of f in that chart has the form*

$$\Psi^{-1}\circ f \circ \Psi(q,p) = (q', p'),$$

$$\text{(13.4a)}\quad \begin{aligned} (q')^i &= q^i \cos\left(2\pi\frac{\partial A(I)}{\partial I_i}\right) - p_i \sin\left(2\pi\frac{\partial A(I)}{\partial I_i}\right) + \mathcal{O}(m),\\ p_i' &= q^i \sin\left(2\pi\frac{\partial A(I)}{\partial I_i}\right) + p_i \cos\left(2\pi\frac{\partial A(I)}{\partial I_i}\right) + \mathcal{O}(m), \end{aligned}$$

$$1 \leqq i \leqq n,$$

where $I = (I_1, I_2, \ldots, I_n)$, $I_i = q^i p_i$,

$$\text{(13.4b)}\quad A(I) = (2\pi)^{-1}\langle \alpha, I\rangle + \sum_{s=2}^{m/2} a_s(I),$$

The components $a_s(I)$ are homogeneous polynomials of variable I of degree s. They do not depend on the choice of the chart Ψ or the number m. Here they are invariants of f at x_0 under a symplectic isomorphism.

Remark. The proof of the independence of $a_s(I)$ to the choice of a chart etc. follows in the same way as in the proof of the analogous statement in Proposition 9.13.

A fixed point of a local symplectic diffeomorphism is said to be *general elliptic* if (1) it is simple elliptic, (2) its frequency vector satisfies the condition of absence of resonances up to fourth order (AR4), (3) the quadratic form $a_2(I)$, given by Proposition 13.4, is nonsingular. This definition can naturally generalized to the case of an arbitrary periodic motion.

Application of the KAM Theorem

To investigate the motion near the fixed point x_0 it is convenient to change to the angle-action coordinates (φ, I) which are linked with (q,p) by the relations $q^i = \sqrt{I_i/\pi}\cos 2\pi\varphi^i$, $p_i = \sqrt{I_i/\pi}\sin 2\pi\varphi^i$, $1 \leqq i \leqq n$. These formulae define a symplectic map $\Psi_1 : A_D \longrightarrow \mathbb{R}^{2n}$, where $D = \{I \in \mathbb{R}^n\colon I_i > 0, \sum_{i=1}^n I_i < \gamma_0\}$, γ_0 being a sufficiently small positive number. The image of A_D under the map $\Psi \circ \Psi_1$ is an open neighbourhood of x_0 with x_0 omitted. Here we used the chart introduced in Proposition 13.4. Being regarded in the angle–action coordinates thus defined, the transformation $\hat{f} = \Psi_1^{-1} \circ \Psi^{-1} \circ f \circ \Psi \circ \Psi_1 : A_D \longrightarrow A_{\mathbb{R}^n}$ has a Hamiltonian function

$$H(\varphi, I) = A(I) + \sum_{|k| = m+1} (I^{1/2})^k h_k(\varphi, I^{1/2}) \tag{13.5}$$

where the h_k are smooth functions with bounded derivatives of all orders, $I^{1/2} = (I_1^{1/2}, I_2^{1/2}, \ldots, I_n^{1/2})$, $k = (k_1, k_2, \ldots, k_n)$ are multiindices. The function $A(I)$ is given by (13.4b).

Let us consider the second term in (13.5) as a perturbation. The unperturbed system here is integrable one with A as the Hamiltonian. It has invariant tori $\mathscr{T}_c$, $c \in D$, defined by the equations $I_i = c_i$, $1 \leqq i \leqq n$. Some of these tori survive under the perturbation and form KAM sets near x_0, for brevity we call them the nearby KAM tori if one can carry them back to X via $\Psi \circ \Psi_1$. The nearby KAM tori, whose existence we are going to establish, will form a countable collection of mutually nonintersecting KAM sets. Let $\mathscr{K}$ be the union of such a collection. We say the nearby KAM tori *accumulate densely* to x_0 if

$$\lim_{\varepsilon \to 0} \mathcal{M}es(B(\varepsilon) \setminus \mathscr{K}) / \mathcal{M}es\, B(\varepsilon) = 0.$$

Here we denote the ball of radius ε with centre at x_0 by $B(\varepsilon)$, the distance being calculated in some smooth chart.

Theorem 13.6. *Let x_0 be a general elliptic fixed point of a symplectic local diffeomorphism f. Then there exists a countable number of mutually nonintersecting KAM sets of f, consisting of nearby KAM tori, which accumulate densely to x_0. The difference between the tori in question and the tori $I = \text{const}$ being estimated by* (13.6cj). *Any open neighbourhood of x_0 contains infinitely many KAM sets with frequencies in $\mathscr{E}_{\sigma\gamma}$ with arbitrarily large σ and arbitrarily small γ.*

Proof. Fix $\sigma > n$. We shall use formula (13.5) with $m = 4$. The frequency map ν: $I \longmapsto \operatorname{grad} A(I)$ is linear and carries $D = \{I \in \mathbb{R}^n\colon I_i > 0, \sum I_i < \gamma_0\}$ to some open convex polyhedron B, the closure of B containing the frequency vector $(1/2\pi)\alpha$ of f at x_0. Let us divide D into the following countable collection of subsets:

$$D_j = \left\{ I \in D\colon \gamma_j < \sum_{k=1}^n I_k < \gamma_{j-1} \right\}, j \geqq 1, \tag{a}$$

where $\gamma_j = 2^{-(1/2)j(j+1)}\gamma_0$, the boundaries between these D_j's being neglected. The image B_j of D_j under v is also a convex polyhedron. There is j_0 such that

(b) $$\mathscr{E}_j = \mathscr{E}\sigma\gamma_j \cap [B_j]_{4\gamma_j}$$

is nonempty if $j \geqq j_0$. Consider the system with Hamiltonian (13.5) on A_{D_j}. To force the perturbation to satisfy the conditions of Theorem 11.6 we multiply the sum in the right-hand side of (13.5) by an appropriate switching function of the form $\chi(I/\gamma'_j)$ which vanishes in the strip of width $\gamma'_j = \text{const}\,\gamma_j$ near the boundary of D_j and is 1 outside the analogous strip of width $(3/2)\gamma'_j$, the const here being chosen in an appropriate manner for the images under v of the mentioned strips to have widths $2\gamma_j$ and $3\gamma_j$ respectively. The Hamiltonian obtained,

$$H_1 = \chi \sum_{|k|=5} (I^{1/2})^k h_k,$$

satisfies (11.3) with $\gamma = \gamma_j$. Let us calculate $C^{r_0} - \gamma_j$ norm of H_1. We have

$$H_1(\varphi, \gamma_j I) = \gamma_j^{5/2} \chi \sum_{|k|=5} (I^{1/2})^k h_k(\varphi, \gamma_j I),$$

and

$$\| H_1 \|_{r_0, \gamma_j} \leqq \text{const}\,\gamma_j^{5/2}(\gamma_{j-1}/\gamma_j)^{5/2} = \gamma_j^2 \delta_j$$

where

$$\delta_j = \text{const}\,\gamma_j^{1/2}(\gamma_{j-1}/\gamma_j)^{5/2} = \text{const}\,2^{-(1/4)j^2 + (9/4)j}.$$

The estimate (11.4) will be satisfied for $j \geqq j_0$ if j_0 is chosen to be sufficiently large.

Applying Theorem 11.6 yields the existence of a KAM set whose tori have frequency vectors in $\mathscr{E}_j$ given by (b) and with conjugating diffeomorphism $\Sigma_j : A_{D_j} \longrightarrow A_{D_j}$ satisfying the estimate

(13.6cj) $$\| \Sigma_j - \text{id} \|_{1, \gamma_j} \leqq \text{const}\,2^{-(1/15)j^2 + (1/5)j},$$

where const depends only on σ and f. The spoiling of the Hamiltonian of perturbation by multiplying by χ does not affect the tori $\mathscr{T}_I = \Sigma_j(\mathbb{T}^n \times \{I\})$, $v(I) \in \mathscr{E}_j$, as follows from (11.11). So the tori $\Psi \circ \Psi_1(\mathscr{T}_I)$ appear to be invariant KAM tori to f which are near x_0.

Let us now estimate the Liouville measure of the union of KAM tori. Due to $\Psi \circ \Psi_1$ being symplectic it is sufficient to consider the situation in A_D. Denote $\mathscr{K}_j = \Sigma_j(\mathbb{T}^n \times \mathscr{E}_j)$. It follows from (a) that

$$\mathscr{M}es\, D_j = \frac{1}{n!}(2^{-nj(j-1)/2} - 2^{-nj(j+1)/2})\gamma_0^n \sim \frac{1}{n!}\gamma_0^n 2^{-nj(j-1)/2}.$$

Taking into account the estimate (9.9a) and adding the volume of the boundary strips of width $2\gamma_j$ gives us

$$\text{Vol}(A_{D_j} \backslash \mathscr{K}_j) \leqq \text{const}\,\gamma_0^{n-1} 2^{-(n-1)j(j-1)} \gamma_j.$$

Given $\alpha > 0$ and real β such that $\alpha j^2 - \beta j > 0$ for $j \geqq j_0$, the inequalities

$$2^{-\alpha j_0^2 + \beta j_0} \leqq \sum_{j=j_0}^{\infty} 2^{-\alpha j^2 + \beta j} \leqq 2^{-\alpha j_0^2 + \beta j_0 + 1}$$

hold. Taking this into account and summing over $j \geqq j_0$ yields $1 \leqq (\mathcal{M}es \bigcup_{j \geqq j_0} D_j) n! \gamma_0^{-n} 2^{nj_0(j_0-1)/2} \leqq 2$,

$$\mathrm{Vol}\Big(\bigcup_{j \geqq j_0} (A_{D_j} \backslash \mathcal{K}_j)\Big) \leqq \mathrm{const}\, \gamma_0^{n-1} 2^{-(n-1)j_0(j_0-1)/2} \gamma_{j_0}$$

provided j_0 is sufficiently large, and we have for the ratio:

$$\mathrm{Vol}\Big(\bigcup_{j \geqq j_0} (A_{D_j} \backslash \mathcal{K}_j)\Big) \Big/ \mathrm{Vol}\Big(\bigcup_{j \geqq j_0} A_{D_j}\Big) \leqq \mathrm{const}\, 2^{(1/2)j_0(j_0-1)-(1/2)j_0(j_g+1)} = \mathrm{const}\, 2^{-j_0}.$$

The latter estimate means that the tori of $\bigcup_{j \geqq j_0} \mathcal{K}_j$ accumulate densely to x_0.

The Two-Dimensional Case

Let us now consider in detail the formulae which arise in the case $n = 2$. Below we follow the paper of Churchill et al. (1983). Choose a symplectic chart with coordinates (q, p) near a fixed point x_0 so that x_0 becomes the origin $(0, 0)$. It is convenient to use one complex variable $z = q + ip$. Then the transformation $f: z \longmapsto z'$ has the Taylor expansion

$$\begin{aligned} z' = {} & A_1 z + A_2 \bar{z} + A_3 z^2 + A_4 z\bar{z} + A_5 \bar{z}^2 \\ & + A_6 z^3 + A_7 z^2 \bar{z} + A_8 z\bar{z}^2 + A_9 \bar{z}^3 + \mathcal{O}(4). \end{aligned} \tag{13.7}$$

The symplectic property implies the following constraints on the coefficients:

(a) $|A_1|^2 - |A_2|^2 = 1$,
(b) $2\bar{A}_1 A_3 + A_1 \bar{A}_4 - \bar{A}_2 A_4 - 2A_2 \bar{A}_5 = 0$,
(c) $3\bar{A}_1 A_6 + 2A_3 \bar{A}_4 + A_1 \bar{A}_8 - \bar{A}_2 A_7 - 2A_4 \bar{A}_5 - 3A_2 \bar{A}_9 = 0$,
(d) $\mathrm{Re}(A_1 \bar{A}_7) - \mathrm{Re}(\bar{A}_2 A_8) + |A_3|^2 - |A_5|^2 = 0$.

The eigenvalue equation for $T_{x_0} f$ reads $\lambda^2 - 2\,\mathrm{Re}\, A_1 \lambda + 1 = 0$. The transformation is simple elliptic if and only if $|\mathrm{Re}\, A_1\| < 1$. For the remainder of subsection we assume this condition is satisfied.

Let ν denote the solution of the eigenvalue equation which has the same sign for the imaginary part as A_1. If $A_2 = 0$ then $\nu = A_1$. If $A_2 \neq 0$ then the change of variables

$$z = i(2(1 - \mathrm{Re}(\nu A_1)))^{-1/2}((\bar{A}_1 - \nu)u + A_2 \bar{u})$$

which is symplectic transforms (13.7) to

$$u_1 = \nu u + \tilde{A}_3 u^2 + \tilde{A}_4 u\bar{u} + \tilde{A}_5 \bar{u}^2 + \tilde{A}_6 u^3 + \tilde{A}_7 u^2 \bar{u} + \tilde{A}_8 u\bar{u}^2 + \tilde{A}_9 \bar{u}^3 + \mathcal{O}(4). \tag{13.8}$$

Let ν satisfy the condition of absence of resonances up to order 4 which now reads $\nu^j \neq 1, j = 1, 2, 3, 4$. Then the normalizing procedure described above gives a symplectic change of variables $u = w + P_2(w, \bar{w}) + P_3(w, \bar{w}) + \mathcal{O}(4)$ which converts (13.8) to

$$w_1 = \nu \exp(i\gamma w\bar{w}) w + \mathcal{O}(4), \tag{13.9}$$

where the coefficient γ is given by:

$$\gamma = -i\left\{i\,\mathrm{Im}(\bar{\nu}\tilde{A}_7) + 3|\tilde{A}_3|^2\left(\frac{\nu+1}{\nu-1}\right) + |\tilde{A}_5|^2\left(\frac{\nu^3+1}{\nu^3-1}\right)\right\} \tag{13.10}$$

Stability Problem

A periodic motion M of a dynamical system f^t is said to be *stable* if, given a neighbourhood U of M, there is a neighbourhood V of M such that there are local dynamics $f^t: V \longrightarrow X$ defined for all t and such that $f^t(V) \subset U$ for all t. Consider the case where $M = \{x_0\}$ is a fixed point of a two-dimensional symplectic diffeomorphism f. If x_0 is general elliptic then the application of the KAM theorem provides us with a large collection of invariant closed curves surrounding x_0. Since a point taken inside a circle remains there after the application of f, it follows that x_0 is stable. More generally, one may assert that:

13.11. *Any general elliptic periodic motion in a symplectic dynamical system with two degrees of freedom is stable.*

The situation is quite different in systems with more than two degrees of freedom. This is because of the fact that KAM tori do not divide the phase space (or the energy surface in the case of continuous time). So called Arnold diffusion generically yields a slow transport which densely penetrates the whole phase space in the discrete time case and an energy surface in the case of Hamiltonian flow. This phenomenon will be discussed in more detail in §16.

§14. Near the Boundary of Planar Convex Billiards

The billiards in a convex domain $Q \subset \mathbb{R}^2$ (see 6.21) possesses a family of invariant tori which correspond to the motion close to boundary. The projections onto Q of trajectories belonging to one such torus, the *rays*, form the family of straight lines tangent to a curve called the caustic; Fig. 32. The caustics form, in turn, a disjoint family accumulating to the boundary of Q. We shall derive this as being a consequence of the KAM theorem.

It is convenient here to consider the Poincaré map $f: \Pi \longrightarrow \Pi$ induced by the billiard flow on the cross-section Π described in 6.21. Recall that Π is the image of $\partial S_1^* Q$, the bundle of unit cotangent vectors attached to the boundary of Q, under the gluing projection. We use the coordinates (s, ϑ) in Π introduced in 6.21.

Proposition 14.1. *In coordinates* (s, ϑ) *the map* f *reads* $(s, \vartheta) \longmapsto (s_1, \vartheta_1)$ *where* s_1, ϑ_1 *are defined by the equations*

$$\vartheta + \vartheta_1 = \int_s^{s_1} \frac{ds'}{\rho(s')}, \int_s^{s_1} \sin\left(\vartheta - \int_s^{s'} \frac{ds''}{\rho(s'')}\right) ds' = 0, \tag{14.1a}$$

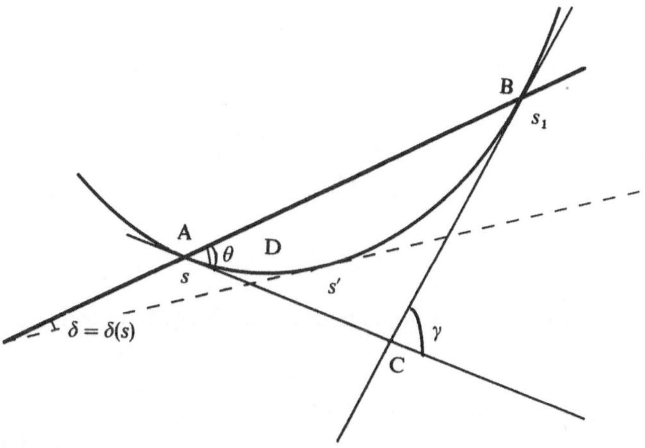

Fig. 30

and satisfy the estimates

(14.1b)
$$2\rho_{\min}\vartheta \leqq s_1 - s \leqq 2\rho_{\max}\vartheta,$$

(14.1c)
$$0 \leqq \vartheta_1 \leqq \left(2\frac{\rho_{\max}}{\rho_{\min}} - 1\right)\vartheta.$$

Here $\rho(s)$ is the curvature radius of ∂Q as a function of the arc length, $\rho_{\max}$ and $\rho_{\min}$ are its maximum and minimum values.

Proof. Consider Fig. 30. The angle γ, formed by the tangent line to ∂Q at points A with coordinate s and B with coordinate s_1, is equal to $\int_s^{s_1} ds'/\rho(s')$. One obtains the first equation, (14.1a), by examining ΔABC. The angle $\delta = \delta(s')$ formed between AB and the line tangent to ∂Q at a point D with coordinate $s' \in [s, s_1]$ being calculated as

$$\delta = \vartheta - \int_s^{s'} \frac{ds''}{\rho(s'')}.$$

Projecting the arc element onto the axis orthogonal to AB (which is equivalent to multiplying by $\sin\delta$) and integrating with respect to the variable s' between the limits s and s_1 yields the second equation (14.1a) as a consequence of the fact that both the points A and B are projected to the same point of the mentioned axis.

To prove the inequalities (14.1b) and (14.1c) let us construct the circles $K_{\min}$ and $K_{\max}$ with radii $\rho_{\min}$ and $\rho_{\max}$ respectively which are tangent to ∂Q at a point A with coordinate s and lie on the same side as Q with respect to the tangent line to ∂Q at A (Fig. 31). The circle $K_{\min}$ lies entirely within Q, $K_{\max}$ surrounds Q. Let C be the point on ∂Q with the coordinate s_1, B and D be the points of intersection with $K_{\min}$ and $K_{\max}$ respectively of the straight line passing through A and C and different from A. One can see from Fig. 31 that the length of the arc AC is between the length of the arc AB and the length of arc AD of the circles $K_{\min}$ and $K_{\max}$

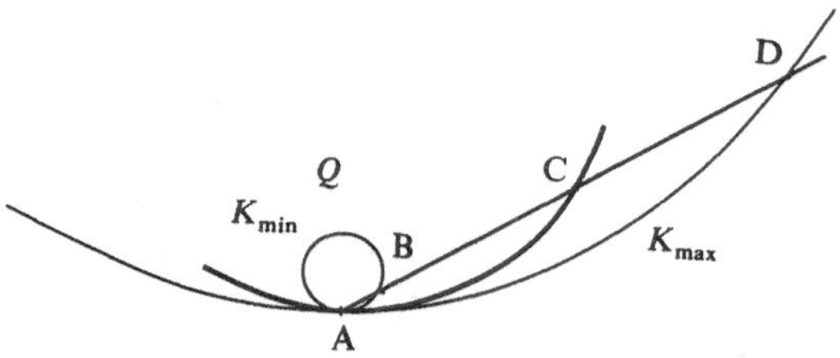

Fig. 31

respectively which results in (14.1b). The inequalities (14.1c) follows from the first equation (14.1a) and (14.1b). □

As was pointed out earlier in 6.21, the surface Π is topologically a two dimensional annulus. Let us attach two boundary circles into Π order to make it a closed annulus $\bar{\Pi}$. One can perform this operation by use of the coordinates (s, ϑ) and by allowing the coordinate ϑ to range over the closed interval $[0, \pi]$. Let us extend f to $\bar{\Pi}$ by setting $f|_{\partial\bar{\Pi}} = \mathrm{id}$ (we use the same letter for denoting the extended map). The above inequalities show us that the extended f is continuous at the points of the component $\vartheta = 0$ of $\partial\bar{\Pi}$. Symmetry arguments give the continuity on the other component of $\partial\bar{\Pi}$. So the extended f is continuous. We define a smooth structure on $\bar{\Pi}$ by use of the coordinates (s, ϑ).

Proposition 14.2. *The map* $f: \bar{\Pi} \longrightarrow \bar{\Pi}$ *is smooth. Its derivatives of order k in the coordinates* (s, ϑ) *are bounded by constants depending only on k and* $\max\{\rho_{\max}, \rho_{\min}^{-1} \max_s |\rho^{(i)}(s)|, 1 \leqq i \leqq k-1\}$.

Proof. We shall use the notations of Proposition 14.1 and Fig. 30. Set $s_1 = s + a(s, \vartheta)$. It is sufficient to prove the smoothness and the estimates for $a(s, \vartheta)$, since those for $\vartheta_1(s, \vartheta)$ follow from (14.1a). Let us firstly prove the statement for the domain $0 \leqq \vartheta \leqq (\pi/3)(2\rho_{\max}\rho_{\min}^{-1} - 1)^{-1}$. Then also $0 \leqq \vartheta_1 \leqq \pi/3$ as follows from (14.1a). The second equation (14.1a) may be written in the form

$$F(a, \vartheta, s) = \int_0^1 \sin\left(\vartheta - \int_s^{s+ta} \frac{ds''}{\rho(s')}\right) dt = 0. \tag{14.2a}$$

The function F is smooth and its derivative with respect to a satisfies the estimate

$$-\frac{\partial F}{\partial a} = \int_0^1 \cos\left(\vartheta - \int_s^{s+ta} \frac{ds''}{\rho(s'')}\right) \frac{tdt}{\rho(s+ta)} \geqq \frac{1}{4\rho_{\max}} \tag{14.2b}$$

since

$$\vartheta' = \vartheta - \int_s^{s+ta} \frac{ds''}{\rho(s'')}$$

varies within the interval $[-\pi/3, \pi/3]$. Applying the implicit function theorem to (14.2a) and using the estimate (14.2b) yields that $a = a(s, \vartheta)$ is smooth at $\vartheta = 0$ and the desired estimates for its derivatives in the domain considered. Symmetry arguments show that a similar result is true in the domain obtained by replacing ϑ by $\pi - \vartheta$. To obtain the estimates in the remaining domain it is convenient to use the second equation (14.1a) immediately. Differentiating, one obtains

$$-\frac{ds_1}{ds}\sin\vartheta_1 + \frac{1}{\rho(s)}\int_s^{s_1}\cos\left(\vartheta - \int_s^{s'}\frac{ds''}{\rho(s'')}\right)ds' - \sin\vartheta = 0. \tag{14.2c}$$

It remains to note that, by symmetry, $\vartheta_1 \geqq (2\rho_{\max}\rho_{\min}^{-1} - 1)^{-1}$ and therefore ϑ_1 lies in the interval $[\pi - \vartheta^*, \vartheta^*]$ where $\vartheta^* = (\pi/3)(2\rho_{\max}\rho_{\min}^{-1} - 1)^{-2}$. The $\sin\vartheta_1$ turn is bounded from below in that interval by a constant only depending on $\rho_{\max}$ and $\rho_{\min}$. The result follows from (14.2c). □

Later we will need explicit formulae describing the transformation f near the boundary of $\bar{\Pi}$. We choose the component of $\partial\bar{\Pi}$ where $\vartheta = 0$. Denote it $\partial\Pi_+$. Then, as follows from the above Proposition, the transformation can be formally expanded in Taylor series

$$(s,\vartheta) \longmapsto \left(s + \sum_{k\geqq 1}\alpha_k(s)\vartheta^k (\bmod |\partial Q|),\ \sum_{k\geqq 1}\beta_k(s)\vartheta^k\right). \tag{14.3}$$

The results of immediate calculation of the coefficients in (14.3) with $k \leqq 4$ are the following:

$\alpha_1 = 2\rho, \qquad \beta_1 = 1,$
$\alpha_2 = \frac{4}{3}\rho'\rho, \qquad \beta_2 = -\frac{2}{3}\rho',$
$\alpha_3 = \frac{2}{3}\rho''\rho^2 + \frac{4}{3}\rho'^2\rho, \qquad \beta_3 = -\frac{2}{3}\rho''\rho + \frac{4}{9}\rho'^2,$
$\alpha_4 = \frac{4}{15}\rho'''\rho^3 + \frac{76}{45}\rho''\rho'\rho^2 - \frac{2}{45}\rho'\rho + \frac{16}{135}\rho'^3\rho, \quad \beta_4 = -\frac{2}{5}\rho'''\rho^2 - \frac{44}{45}\rho''\rho'\rho - \frac{2}{45}\rho' - \frac{44}{135}\rho'^3.$

Let us pass to the new variables x, y depending on s, by the formulae

$$x = C_1\int_0^s \rho^{-2/3}(s')\,ds', \qquad y = C_2\rho^{1/3}(s)\sin(\vartheta/2), \tag{14.4}$$

where $C_1 = (\int_0^{|\partial Q|}\rho^{-2/3}(s)\,ds)^{-1}$ and $C_2 = 4C_1$. The choice of such obscure expressions is motivated by some formulae in diffraction theory which describe exponential decay in the shadow (see Babitch and Buldirev (1972)). In these new coordinates the Taylor series for f looks like

$$(x, y) \longmapsto \left(x + y + \sum_{k\geqq 3}\tilde{\alpha}_k(x)y^k (\bmod 1),\ y + \sum_{k\geqq 4}\tilde{\beta}_k(x)y^k\right), \tag{14.5}$$

where $\tilde{\alpha}_k(x)$ and $\tilde{\beta}_k(x)$ are smooth periodic functions with period 1. A remarkable fact is the vanishing of the coefficients at y^2 in the first component and at y^2 and y^3 in the second one. To obtain (14.5) one must use the above explicit expressions for the coefficients of (14.3). Also, it is not difficult to calculate the symplectic structure in coordinates (x, y):

$$\Omega|_{T\Pi} = 4C_1^{-1}C_2^{-2}y\,dx \wedge dy.$$

To transform (14.5) into a more convenient form we use the following inductive Lemma.

Lemma 14.6. *Let the transformation*

$$g:(x, y)\longmapsto(x+y+y^{m+1}a(x, y), y+y^{m+2}b(x, y))$$

be defined in a strip $|y| \leqq \varepsilon, m \geqq 1$, *a and b being smooth, periodic functions with period 1 in the variable x, and let g preserve the form* $\rho(x, y)\,dx \wedge dy$ *where* $\rho(x, y)$ *has the Taylor expansion near* $y=0$ *of the form*

$$\rho(x, y)=y+\rho_3(x)y^2+\rho_3(x)y^3+\cdots$$

Then

(a) $$\rho_k'(x)=0 \text{ if } 2 \leqq k \leqq m.$$

(b) $$\int_0^1 b(x, 0)\,dx=0.$$

(c) *The change of variables*

$$x=\xi+\eta^m u(\xi), \qquad y=\eta+\eta^{m+1}v(\xi),$$

where u and v are periodic, with period 1, *solutions of the system*

(d) $$u'(\xi)=v(\xi)+a(\xi, 0), \quad v'(\xi)=b(\xi, 0)$$

converts g into the form

$$\tilde{g}:(\xi, \eta)\longmapsto(\xi+\eta+\eta^{m+2}a_1(\xi, \eta), \eta+\eta^{m+3}b_1(\xi, \eta)),$$

a_1, b_1 *being smooth and being periodic with period* 1 *in the variable* ξ.

Proof. The preservation of $\rho(x, y)\,dx \wedge dy$ may be equivalently expressed as

(e) $$\frac{\rho(g(x, y))}{\rho(x, y)}=\det\begin{vmatrix}1+y^{m+1}a_x' & 1+(m+1)y^m a+y^{m+1}a_y' \\ y^{m+2}b_x' & 1+(m+2)y^{m+1}b+y^{m+2}b_y'\end{vmatrix}$$
$$=1+y^{m+1}(a_x'(x, 0)+(m+1)b(x, 0)=\mathcal{O}(y^{m+2}).$$

Given $k, 2<k \leqq m+1$, we have proved $\rho_s'(x)=0$ for all $s \leqq k-1$. Then we have

$$\begin{aligned}\rho(g(x, y))/\rho(x, y)=&(\rho'(x, y)+\rho_k'(x)(y+y^{m+1}a)y^k+y^{m+2}b \\ &+\mathcal{O}(y^{m+3})+\mathcal{O}(y^{k+2}))/\rho(x, y) \\ =&1+\rho_k'(x)y^k+b(x, 0)y^{m+1}+\mathcal{O}(y^{m+2})+\mathcal{O}(y^{k+1}).\end{aligned}$$

Comparing this with the right-hand side of (e) yields $\rho_k'(x)=0$ if $k \leqq m$, and $\rho_{m+1}'(x)+b(x, 0)=a_x'(x, 0)+(m+2)b(x, 0)$ if $k=m+1$. The first equality proves (a) by induction, the second one proves (b). Note that (b) implies solvability of the system (d) in the class of periodic functions. Then one can check the assertion (c) immediately. □

Fix positive integers μ and $v \geqq 4$ and let $N=4(v+r_0+1)\mu+1$ where r_0 is given by (11.1) for cascades with $n=d-1=1$ and with some $\sigma>1$. Apply Lemma 14.6 $N-3$ times in order to convert f into the form

(14.7) $$(\xi, \eta)\longmapsto(\xi+\eta+\eta^N a(\xi, \eta), \eta+\eta^{N+1}b(\xi, \eta))$$

whilst the symplectic structure becomes

$$\Omega = \left[\sum_{i=1}^{N-1} \rho_i \eta^i + \eta^N \tau(\xi, \eta)\right] d\xi \wedge d\eta, \tag{14.8}$$

ρ_i being constants, $\rho_1 > 0$. The transformation which joins (x, y) of (14.4) and (ξ, η) reads

$$h:(\xi, \eta) \longmapsto (\xi + \eta^2 u(\xi, \eta), \eta + \eta^3 v(\xi, \eta)), \tag{14.9}$$

u and v being smooth near $\partial\Pi_+$. Consider the mean of the density of (14.8):

$$\hat{\rho}(\eta) = \sum_{i=1}^{N-1} \rho_i \eta^i + \eta^N [\tau](\eta), \; [\tau](\eta) = \int_0^1 \tau(\xi, \eta) d\xi$$

and define a smooth function $S(z)$ by the equality

$$z^2 = \int_0^{S(z)} \hat{\rho}(y)\, dy = \tfrac{1}{2}\rho_1 S^2 + \cdots \tag{14.10}$$

The change of variables

$$k:(\xi, \eta) \longmapsto \left(X = \xi, Y = \int_0^\eta \hat{\rho}(y)\, dy\right) \tag{14.11}$$

converts (14.7) and (14.8) into a more convenient form

$$\begin{gathered}(X, Y) \longmapsto (X + S(\sqrt{Y}) + Y^{N/2} A(X\sqrt{Y}), Y + Y^{(N+1)/2} B(X, \sqrt{T})),\\ \Omega = [1 + Y^{(N-1)/2} C(X, \sqrt{Y})]\, dX \wedge dY,\end{gathered} \tag{14.12}$$

the functions A, B, C belongs to $C^\infty_{\text{per}}(\mathbb{R} \times [0, \varepsilon])$, ε small and positive. Thus, the representation (14.12) for the billiard map f and invariant structure Ω is valid in a small neighbourhood of $\partial\Pi_+$. Let us divide this neighbourhood $0 \leqq Y \leqq \varepsilon$ into pieces $\Pi^{(j)}, j = 0, 1, 2, \ldots$, defined by the inequalities $\theta_{j+1} \leqq Y \leqq \theta_j$ where $\theta_j = \varepsilon 2^{2j}$. Consider one such piece $\Pi^{(k)}$. Denote $\theta = \theta_j$. It is useful to change the scale of the variable Y:

$$s:(\bar{X}, \bar{Y}) \longmapsto (X = \bar{X}, Y = \sqrt{\theta}\bar{Y}). \tag{14.13}$$

In the new variables the billiard map and the invariant symplectic form read:

$$\begin{gathered}(\bar{X}, \bar{Y}) \longmapsto (X + S(\theta^{1/4}\sqrt{\bar{Y}}) + \theta^{N-4}\bar{A}(\bar{X}, \bar{Y}), \bar{Y} + \theta^{(N-1)/4}\bar{B}(\bar{X}, \bar{Y})),\\ \bar{\Omega} = \theta^{1/2}\Omega = [1 + \theta^{(N-1)/4}\bar{C}(\bar{X}, \bar{Y})]\, d\bar{X} \wedge d\bar{Y},\end{gathered} \tag{14.14}$$

A, B, C being smooth periodic functions defined on the strip $\tilde{\Pi}_0 = \tilde{\Pi}^{(j)} = \{(\bar{X}, \bar{Y}): \sqrt{\theta}/2 \leqq \bar{Y} \leqq \sqrt{\theta}\}$ with moduli bounded there by constants independent of j. Further on in this section const denotes a constant independent of the number j of the strip $\tilde{\Pi}^{(j)}$.

Set $\gamma = \theta^\mu$ and define the strips

$$\tilde{\Pi}_s = \{(\bar{X}, \bar{Y}): \tfrac{1}{2}\theta^{1/2} + s\gamma \leqq \bar{Y} \leqq \theta^{1/2} - s\gamma\} \subset \tilde{\Pi}_0: \quad 0 \leqq s \leqq 8.$$

Our first aim is to straighten the 2-form (14.14) in $\tilde{\Pi}_3$ by use of Proposition 2.38. Multiplying $\bar{C}(\bar{X}, \bar{Y})$ by an appropriate smooth switching function $\chi(\bar{Y})$ which is zero on $\tilde{\Pi}_0 \backslash \tilde{\Pi}_1$ and 1 on $\tilde{\Pi}_2$, and whose derivatives satisfy the estimates $|\chi^{(k)}| \leqq C_k \gamma^{-k}$, one obtains a new 2-form $\hat{\Omega}$ satisfying the conditions of 2.38. Then we have

$$\| \chi \theta^{(N-1)/4} \bar{C} \|_r \leqq \mathrm{const}\, \theta^{(N-1)/4} \gamma^{-r} = \mathrm{const}\, \gamma^{v+r_0-r+1},$$

$1 \leqq r \leqq r_0 + 1$, and Proposition 2.38 yields, for sufficiently large j, the existence of a smooth map $g_2 \in \mathrm{Diff}^{\infty}_{\mathrm{per}}(\tilde{\Pi}_0)$ which satisfies

$$\text{(14.15)} \qquad \| g_2 - \mathrm{id} \|_r \leqq \mathrm{const}\, \gamma^{v+r_0-r+1}, \quad 1 \leqq r \leqq r_0, \qquad \mathrm{supp}\,(g_2 - \mathrm{id}) \subset \tilde{\Pi}_1,$$

and such that $g_2^* \bar{\Omega}|_{\tilde{\Pi}_3} = \Omega_{st}$. Thus, $\tilde{\Pi}_3/\mathbb{Z}$ supplied with an induced symplectic structure becomes a symplectic annulus. We shall use there the angle–action variables (φ, I). In these variables, after the substitution $(\bar{X}, \bar{Y}) = g_2(\varphi, I)$, the map (14.14) becomes Hamiltonian. To substantiate this, denote $\Pi_s = \tilde{\Pi}_s/\mathbb{Z}$ and consider the map $f_1 : \Pi_4 \longrightarrow \Pi_3$ induced by (14.14). It is obviously factorizable (§8) provided θ is sufficiently small. Also, the map f_1 has zero Calabi invariant. Indeed, the initial Poincaré map $f : \Pi \longrightarrow \Pi$ possesses this property, and f_1 is conjugated to f, the latter being restricted onto some subannulus of Π (see the interpretation in 3.15). We conclude that there is a Hamiltonian function H to the map f_1, i.e. $f_1 = r \circ l^{-1}$ where r and l are determined by H via the formulae (8.4). Calculation shows that H is of the form

$$H(\varphi, I_1) = \theta^{-1/2} K_0(I_1 \theta^{1/2}) + H_1(\varphi, I_1),$$

where

$$\text{(14.16)} \qquad K_0(I_1) = \int_0^{I_1} S(\sqrt{I})\, dI,$$

S being defined by (14.10), and H_1 is smooth, periodic in the variable φ, and satisfies the estimates

$$\text{(14.17)} \qquad \| H_1 \|_r \leqq \mathrm{const}\, \gamma^{v+r_0-r+2}, \quad 1 \leqq r \leqq r_0.$$

Again, multiplying H_1 by an appropriate smooth function $\chi_1(I_1)$ which is zero outside $\tilde{\Pi}_5$ and 1 inside $\tilde{\Pi}_6$, we obtain a Hamiltonian function which generates, by the formulae (8.4), a symplectic diffeomorphism $f_2 : \Pi_3 \longrightarrow \Pi_3$ such that $f_2|_{\Pi_7} = f_1|_{\Pi_7}$ and $\mathrm{supp}\,(f_2 - f_0) = 0$ outside Π_5. Here f_0 is an integrable map defined by $\bar{K}_0(I_1) = \theta^{-1/2} K_0(I_1 \theta^{1/2})$ as a Hamiltonian. It is not difficult to show that this satisfies the estimates (11.2) in Π_3 with some constant R that is independent of j. The switched perturbation Hamiltonian $\chi_1 H_1$ satisfies the estimate similar to (14.17). So, Theorem 11.6 is applicable to $f_2 : \Pi_3 \longrightarrow \Pi_3$ provided j is sufficiently large (one must take $\delta = \mathrm{const}\, \gamma^v$). As a consequence we obtain the existence of a symplectic conjugating map $\Sigma : \Pi_3 \longrightarrow \Pi_3$ and a smooth model Hamiltonian $\bar{K}(I) = \theta^{-1/2} K(\theta^{1/2} I)$, $K : \Delta_3 \longrightarrow \mathbb{R}$, which define a KAM set $\mathscr{K} \subset \Pi_7$ for f_2 whose frequency set is

$$\text{(14.18)} \qquad \mathscr{E}_j = \mathscr{E}_{\sigma,\gamma} \cap \mathrm{grad}\, K_0(\Delta_8).$$

Here we have used the notation

$$\Delta_s = \theta/2 + s\gamma\sqrt{\theta}, \theta = s\gamma\sqrt{\theta}], \quad s = 1, 2, \ldots \tag{14.19}$$

The following estimates and inclusions can be shown to be true

$$\begin{aligned} &\| \Sigma - \mathrm{id} \|_{1,\gamma} \leqq C\gamma^{4/15\nu}, &&\mathrm{supp}\,(\Sigma - \mathrm{id}) \subset \Pi_4, \\ &\| K - K_0 \|_{2,\tilde{\gamma}} C\gamma^{2+7/15\nu+1/2\mu}, &&\tilde{\gamma} = \sqrt{\theta}\gamma, \mathrm{supp}\,(K - K_0) \subset \Delta_4. \end{aligned} \tag{14.20}$$

To restore the property that superpositions of the above maps are symplectic which was removed by the stretching map s defined by (14.13), we also apply the inverse: $s^{-1}: \mathbb{T}^1 \times \Delta_3 \longrightarrow \Pi_3$. We summarize our construction in the following result:

Theorem 14.21. *Let $Q \subset \mathbb{R}^2$ be a bounded strictly convex domain with smooth boundary. Consider the corresponding billiard ball map $f: \bar{\Pi} \longrightarrow \bar{\Pi}$, and let $\partial\bar{\Pi}_+$ be one of the two components of $\partial\bar{\Pi}$. Fix $\delta > 1$, integers $\mu \geqq 1$, $\nu \geqq 4$. Then there exists a sequence of KAM sets $\mathscr{K}_j, j \geqq j_0$, for f, each consisting of invariant circles isotopic to $\partial\bar{\Pi}_+$, the sequence $\{\mathscr{K}_j\}$ accumulating to $\partial\bar{\Pi}_+$. The KAM set $\mathscr{K}_j$ is determined by the conjugating map $\Psi_j: A_{\Delta_3} \longrightarrow \Pi$ where Δ_3 is given by (14.19) with $\theta = \varepsilon 2^{-j}$, $\gamma = \gamma_j = \theta^\mu$, ε being some positive number. The map Ψ_j is factorized as follows:*

$$\Psi_j = g_1^{-1} \circ h \circ k \circ s \circ g_2 \circ \Sigma \circ s^{-1}.$$

Here g_1 is a coordinate map in Π near $\partial\Pi_+$ defined by (14.4), h, k, s are defined above by the formulae (14.9), (14.11) and (14.13) respectively. The maps g_2, Σ, and the model Hamiltonian K satisfy (14.15) and (14.20), where K_0 is given by (14.16). The frequency set is $\mathscr{E}_j$ given in (14.18).

The following estimate for the Liouville measure of the complement to $\mathscr{K}_j$ holds

$$\mathrm{Vol}\,(\Pi^{(j)} \backslash \mathscr{K}_j) \leqq \mathrm{const}\, \gamma_j \sqrt{\theta_j},$$

where $\Pi^{(j)}$ form the partition of a neighbourhood of $\partial\bar{\Pi}_+$ in Π described on page 147. The density of the said KAM tori tends to 1 when one approaches $\partial\bar{\Pi}_+$: the ratio

$$\mathrm{Vol}\left(\bigcup_{j=k}^{\infty} \mathscr{K}_j\right) \bigg/ \mathrm{Vol}\left(\bigcup_{j=k}^{\infty} \Pi^{(j)}\right)$$

differs from 1 by $\mathrm{const}\, 2^{-k(\mu - 1/2)}$.

Corollary 14.22 (Stability of the Border). *Since there are invariant circles in $\bar{\Pi}$ accumulating to $\partial\bar{\Pi}_+$, given a neighbourhood U of $\partial\bar{\Pi}_+$ in $\bar{\Pi}$ there is another neighbourhood V of $\partial\bar{\Pi}_+$ such that $f^t(z) \in U$ for all $t \in \mathbb{Z}$ provided $z \in V$. Here f^t denotes the billiard cascade induced on the Poincaré cross-section $\bar{\Pi}$.*

Corollary 14.23 (Completeness of the Dynamics of the Billiard Flow). *Consider the vector field ξ defined in §6 in the space phase space Z of the generalized geodesic flow for the case of a strictly convex domain $Q \subset \mathbb{R}^2$ with smooth boundary. Note that Z is noncompact. A priori ξ need not be complete. Nevertheless it is. Indeed, as follows from Corollary 14.23, given the initial condition $z \in \Pi \subset Z$, there exists $a = a(z) > 0$*

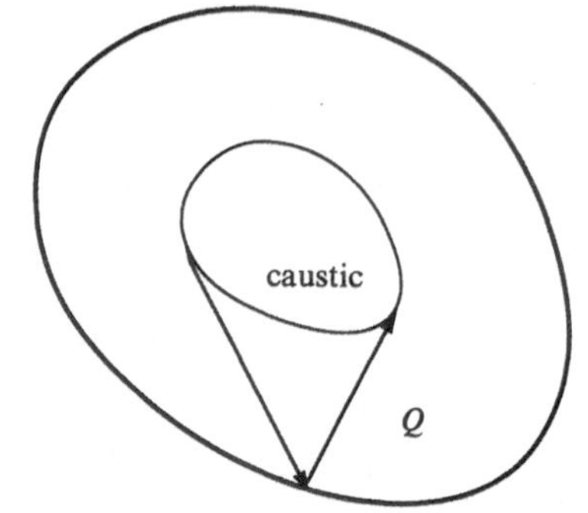

Fig. 32. A caustic in a convex domain Q. The rays tangent to a caustic remain tangent to it after the reflection at points of the boundary ∂Q of Q

such that $\operatorname{dist}(\pi' f^{t+1}(z), \pi' f^t(z)) \geqq a$ *for all* $t \in \mathbb{Z}$. *Here* $\pi' : Z \longrightarrow Q$ *is the coordinate projection (see* 6.12)). *Therefore, the trajectory of billiard flow starting at z may be extended infinitely in the both time directions. Since each trajectory has an intersection with* Π, *the latter assertion is true for each point* $z \in Z$.

Note 14.24. Invariant circles belonging to $\mathscr{K}_j$ are the traces on Π of KAM tori of the billiard flow in Q. The projection onto Q of trajectories of one such torus form a one parameter family of rectilinear segments in Q with common tangent curve called the caustic. If the corresponding invariant curve is sufficiently close to $\partial\bar{\Pi}_+$ the caustic is smooth, convex, and close to ∂Q (see Fig. 32).

§15. The Robustness of a KAM Set

In this section we establish the claimed persistence of KAM sets under small Hamiltonian perturbations. Let $\{f_0^t\}$ be a symplectic dynamical system, X its phase space, and let $\mathscr{K}_0$ be a KAM set of f_0^t with a model system $g_0^t : A_D \longrightarrow A_D$, frequency set $\mathscr{E}$, and conjugating map Ψ_0. Our goal is to find a suitable neighbourhood of $\{f_0^t\}$ in the space of symplectic dynamical systems such that any system $\{f^t\}$ in that neighbourhood possesses a KAM set $\mathscr{K}$ close to $\mathscr{K}_0$, the corresponding model system $g^t : A_D \longrightarrow A_D$ and conjugating map $\Psi : A_D \longrightarrow X$ being close to $\{g_0^t\}$ and Ψ_0 respectively. The additional restriction on $\{f^t\}$ is that it is to differ from $\{f_0^t\}$ by a "Hamiltonian" perturbation, i.e. the corresponding induced system $\Psi_0^{-1} \circ f^t \circ \Psi_0$ is to have zero Calabi invariant. Such a reservation is necessary since otherwise the drift phenomena described in 3.15 occurs and lays obstacles to the existence of KAM tori in the perturbed system. The simplest situation is that with $\{f_0^t\}$ and $\{f^t\}$ both being Hamiltonian flows whose Hamiltonian functions H and H_0 only differ slightly in the appropriate metric in $C^\infty(X)$.

The following objects will be common for the KAM sets of the unperturbed system and the perturbed one: the action domain D, the frequency set $\mathscr{E}$, the frequency domain B, and the image A of both the conjugating maps Ψ_0 and Ψ.

The notations K_0 and K will be used for denoting the Hamiltonians of the model systems, and Ξ_0 and Ξ for the sets of $\mathscr{E}$-tori, the index 0 corresponding to the unperturbed system, and the lack of an index to the perturbed one.

The key idea is that one can decompose $\mathscr{K}_0$ in an arbitrary finite number of fewer KAM sets due to the Cantor like nature of a frequency set. If the units of such a decomposition are sufficiently small, the Hamiltonian Accordance Condition (HAC) involves the proximity of the unperturbed induced Hamiltonian to K_0. So, within every mentioned unit, the perturbed induced Hamiltonian may be represented as in Theorem 11.6, as the sum of an integrable and a small summand. Multiplying by a suitable switching function reduces the situation in each unit to that considered in Theorem 11.6.

Construction

Since $\mathscr{E}$ is compact, one may assume without loss of generality that B is a bounded subset of $\mathbb{R}^n$. Let $\square$ be the closed cube in $\mathbb{R}^n$ with centre at the origin and with the length of each side being $2L$. Choose $L>1$ sufficiently large so that $\square \supset B$. We are going to divide $\square$ into small pieces which cover $\mathscr{E}$. For $0 \neq m \in \mathbb{Z}^n$ and $m_0 \in \mathbb{Z}$ define the *resonant strips*

(15.1a) $$S_m = \{y \in \mathbb{R}^n : |\langle m, y\rangle| < \gamma |m|^{-\sigma}\},$$

(15.1b) $$S_{m,m_0} = \{y \in \mathbb{R}^n : |\langle m, y\rangle - m_0| < \gamma |m|^{-\sigma}\},$$

where we use S_m in the case of continuous time, and S_{m,m_0} in the case of discrete time. Setting $\mu = m$ or $\mu = (m, m_0)$, let us introduce the general notation S_μ for the appropriate strip. Note that the strips S_μ are precisely those one must remove from $\mathbb{R}^n$ to obtain the Diophantum set $\mathscr{E}_{\sigma,\gamma}$.

Define the norm $\|\mu\| = \|m\| = (\sum_{i=1}^n (m_i)^2)^{1/2}$. Let $\mathfrak{M}^{(k)}$ be the set of all μ with $\|\mu\| \leqq k$ such that $S_\mu \cap \square \neq \varnothing$. Obviously $\mathfrak{M}^{(k)}$ is a finite non-empty set for each positive integer k. Consider the monotone non-increasing sequence of compact sets

$$F^{(k)} = \square \setminus \left(\bigcup_{\mu \in \mathfrak{M}^{(k)}} S_\mu \right).$$

Each $F^{(k)}$ may be represented as the union of a finite number of disjoint compact convex polyhedra $\Delta_j^{(k)}$. To measure the size of $\Delta_j^{(k)}$, we define the transversal diameter of such a body as follows. In the case of discrete time this is the usual diameter:

$$\operatorname{trdiam} \Delta_j^{(k)} = \operatorname{diam} \Delta_j^{(k)} = \max\{\operatorname{dist}(x, y) : x, y \in \Delta_j^{(k)}\}.$$

In the case of continuous time trdiam $\Delta_j^{(k)}$ is the diameter of the intersection of $\Delta_j^{(k)}$ with the border $\partial\square$ of our cube. This intersection is non-empty as follows from 15.2c below.

Lemma 15.2. *If* $k \geqq 2$, *then*

(15.2a) $$\operatorname{trdiam} \Delta_j^{(k)} \leqq \frac{c_1}{k},$$

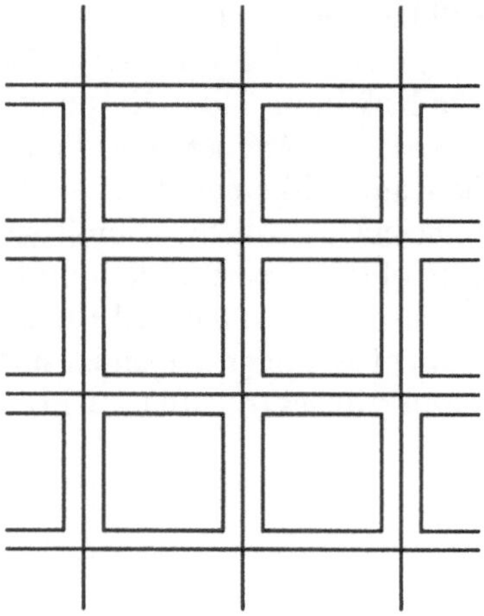

Fig. 33. Splintering of $\mathbb{R}^n$ by the cells of the complement of to S_{m,m_0}, $m = k\,\mathrm{ort}_i$

where $c_1 > 1$ is a constant depending only on n and (in the case of continuous time) on L.

Proof. In the case of discrete time the required estimate follows immediately from the fact that the strips S_{m,m_0}, where $m = k\,\mathrm{ort}_i$, $1 \leqq i \leqq n$, i and $m_0 \in \mathbb{Z}$ varying, k being kept fixed, form the lattice wise splintering of $\mathbb{R}^n$ as shown in Fig. 33, the diameter of each cell not exceding $\sqrt{n}/k$. Since each $\Delta_j^{(k)}$ is wholly contained in some such cell, the estimate (11.1a) is valid with $c_1 = \sqrt{n}$ for all $k \geqq 1$.

Consider the case of continuous time. We need to investigate the geometry of $\Delta_j^{(k)}$ in more detail. Let $\mathrm{cp}: \square \backslash \{0\} \longrightarrow \partial\square$ be the central projection which assignes to a point $y \in \square$ the point $\mathrm{cp}(y) \in \partial\square$ which lies on the line passing through 0 and y and such that the closed rectilinear segment $l(y)$ joining y and $\mathrm{cp}(y)$ does not contain 0 (see Fig. 34). We claim the following properties of $\Delta_j^{(k)}$:

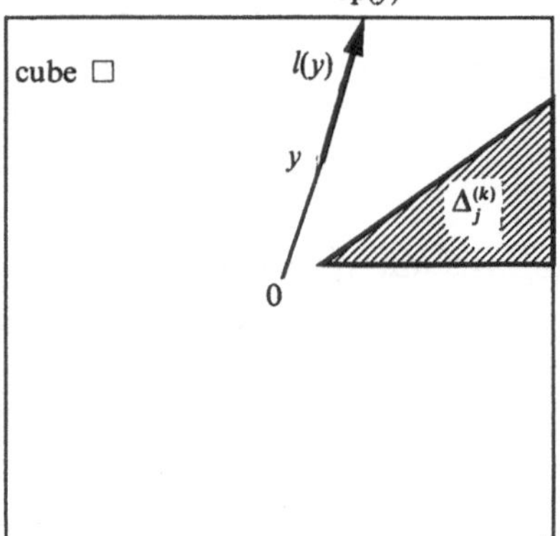

Fig. 34. Configuration in the case of continuous time

15.2b. *If* $y \in \Delta_j^{(k)}$ *then* $l(y) \subset \Delta_j^{(k)}$,

15.2c. $\Delta_j^{(k)} \cap \partial\Box = \mathrm{cp}(\Delta_j^{(k)})$ *and this set is wholly contained in some face* $\Box_i$ *of the cube* $\Box$.

The cube $\Box$ possesses the property 15.2b and the latter remains valid if one removes any strip S_m. This proves 15.2b for $\Delta_j^{(k)}$. The equality in 15.2c follows from 15.2b and the definition of $l(y)$. To prove the remaining part of 15.2c, since $\Delta_j^{(k)} \cap \partial\Box$ is connected, it is sufficient to show that $\Delta_j^{(k)} \cap \Box_{is} = \varnothing$ for all $\Box_{is} = \Box_i \cap \Box_s$, the intersection of faces. But $\Box_{is} \subset S_{m^*}$ where $m^* \in \mathbb{Z}^n$ has 1 at the ith and sth positions and 0 at all the remaining positions. Since k is at least 2, $m^* \in \mathfrak{M}^{(k)}$. So S_m is to be removed to obtain $\Delta_j^{(k)}$, and hence 15.2c is valid.

For simplicity suppose that $\mathrm{cp}(\Delta_j^{(k)}) \subset \Box_1 = \{y = (y^1, y^2, \ldots, y^n) \in \Box\colon y^1 = L\}$. Define the map $\tau\colon \Box_1 \longrightarrow \mathbb{R}^{n-1}$ by the formula $\tau(L, y^2, \ldots, y^n) = (y^2/L, \ldots, y^n/L)$. It is easy to see that τ maps $S_m \cap \Box_1$ onto $S_{m'-m_1} \cap \Box'$ where $\Box'$ is the cube in $\mathbb{R}^{n-1}$ with its centre at the origin and with the length of each edge being 2, $m' = (m_2, m_3, \ldots, m_n)$ if $m = (m_1, m_2, \ldots, m_n)$ and γ is replaced by another positive number. Take the splintering of $\mathbb{R}^n$ by S_m with $m = (m_1, m_2, \ldots, m_n)$ such that $m' = (m_2, \ldots, m_n)$ has the norm $\|m'\| = k' = [k/2]$ and $|m_1| \leqq k'$. We have $\|m\|^2 = \|m'\|^2 + m_1^2 \leqq k^2$, hence $m \in \mathfrak{M}^{(k)}$. Some of the mappings $\tau(S_m \cap \Box_1)$ (namely, those with $m' = k' \,\mathrm{ort}_i$) form the splintering of $\tau(\Box_1)$ like in the case of discrete time with k' standing for k, and with $n-1$ standing for n. The map τ reduces the distances by a factor of L^{-1}. Applying the result for discrete time, we establish (11.1a) with $c_1 = 2\sqrt{2}\sqrt{n-1}L$. $\Box$

In the case of continuous time we need to cut $\Delta_j^{(k)}$ into pieces once more. For simplicity, let $\mathrm{cp}(\Delta_j^{(k)}) \subset \Box_1$. Denote $\mathscr{E}_j^{(k)} = \mathscr{E} \cap \Delta_j^{(k)}$ and let $\mathscr{E}_j^{(k)} \neq \varnothing$. Consider the projection $\mathrm{pr}_1(\mathscr{E}_j^{(k)})$ of this set onto the y^1-axis. The complement $\mathbb{R}\backslash\mathrm{pr}_1(\mathscr{E}_j^{(k)})$ is the union of open disjoint intervals δ_1. Define the convex polyhedra $\Delta_{jr}^{(k)}$ as being the connected components of $\Delta_j^{(k)}\backslash(\bigcup_{|\delta_1| > c_1/k} \delta_1 \times \mathbb{R}^{n-1})$. Obviously each $\Delta_{jr}^{(k)}$ satisfies (15.2a). Denote $\mathscr{E}_{jr}^{(k)} = \mathscr{E} \cap \Delta_{jr}^{(k)}$. It follows by construction that

(15.3) *Given* $y \in \Delta_{jr}^{(k)}$, *there is a* $y^* \in \mathscr{E}_{jr}^{(k)}$ *such that* $|\mathrm{pr}_1(y) - \mathrm{pr}_1(y^*)| \leqq c_1/k$.

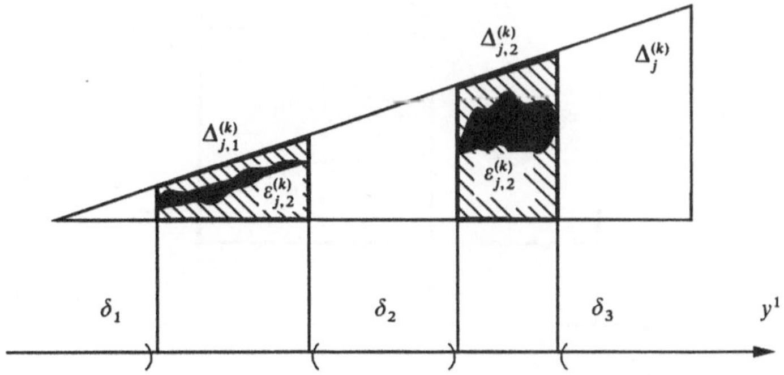

Fig. 35. Additional cutting of $\Delta_j^{(k)}$

Since cp(y), and cp(y) both belong to $\square_1$, the angles α and α^*, which $l(y)$ and $l(y^*)$ form with the y^1-axis, satisfy the inequalities

$$\cos\alpha \geqq 1/\sqrt{n}, \qquad \cos\alpha^* \geqq 1/\sqrt{n}. \tag{15.4}$$

It follows from (15.2a), (15.3), and (15.4) that

$$\operatorname{dist}(y^*, y) \leqq \frac{c_1}{k}(\sqrt{n}+1) \tag{15.5}$$

(use the triangle inequality for the triangle y^*y as in Fig. 36) Let $\imath$ denote the pair (j, r) in the case of continuous time and j in the case of discrete time, and let us adopt the common notations $\Delta_\imath^{(k)}$, $\mathscr{E}_\imath^{(k)}$. Consider those $\Delta_\imath^{(k)}$ where $\mathscr{E}_\imath^{(k)} \neq \varnothing$. The following assertion is then a consequence of (15.2), (15.3) and (15.5).

15.6. *Given positive* α, $\Delta_\imath^{(k)} \subset [\mathscr{E}_\imath^{(k)}]^\alpha$ *provided* $k > c_1(\sqrt{n}+1)/\alpha$.

Recall that $[\cdot]^\alpha$ denotes the α-dilation (see the List of Notations).
The constructed $\Delta_\imath^{(k)}$ have the property

15.7. $\operatorname{dist}(\Delta_\imath^{(k)}, \Delta_{\imath'}^{(k)}) \geqq \gamma k^{-\sigma}$ *if* $\imath \neq \imath'$.

This is because there is a strip S_μ with $\mu \in \mathfrak{M}^{(k)}$ which separates these two polyhedra. Take

$$\gamma_k = \frac{1}{11}\gamma k^{-\sigma} \tag{15.8}$$

and for each positive integer k define the collection of open sets in $\mathbb{R}^n$

$$\beta^{(k)} = \{[\Delta_\imath^{(k)}]^{5\gamma_k}\}, \tag{15.9}$$

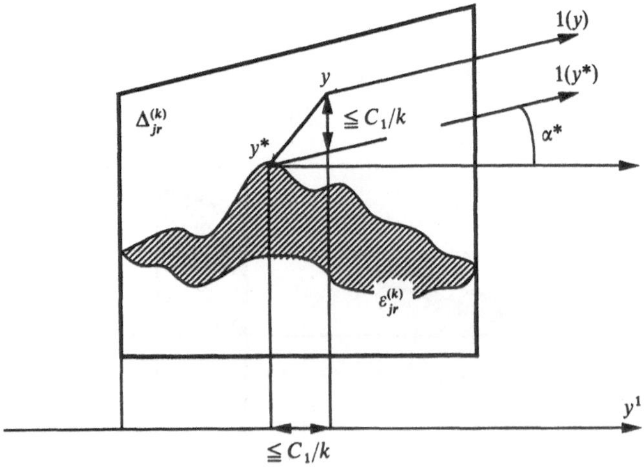

Fig. 36

only those $\Delta_\iota^{(k)}$ which contain non-empty $\mathscr{E}_\iota^{(k)}$ taking part in forming $\beta^{(k)}$. Given $C\in\beta^{(k)}$, we call the polyhedron Δ of our construction, such that $C=[\Delta]^{5\gamma_k}$, the core of C. Each core contains points of $\mathscr{E}$, and the cores cover $\mathscr{E}$. It is easy to see that the members of $\beta^{(k)}$ are disjoint open sets (follows from (15.7)–(15.9)).

Proposition 15.10. *Consider a symplectic dynamical system $\{f_0^t\}$ and a KAM set to $\{f_0^t\}$ with a model system Hamiltonian $K_0: D \longrightarrow \mathbb{R}$, frequency map ω_0, frequency domain B, frequency set $\mathscr{E}$, and the set of $\mathscr{E}$-tori $\mathscr{K}_0$. Let $\beta^{(k)}$ be the sequence of collections of sets constructed above for these data. Then there exists a positive integer k_0, such that the following is true for each $C\in\beta^{(k)}$ provided that $k \geqq k_0$:*

(i) *$C\subset B$;*

(ii) *system $\{f_C^t\}$ on $\mathbb{T}^n\times\omega_0^{-1}(C)$ the induced by $\{f_0^t\}$ via the conjugating map possesses a Hamiltonian H_C^0 defined on an open set $G_C\supset\mathbb{T}^n\times\omega_0^{-1}(C)$ ($G_C=\mathbb{T}^n\times\omega_0^{-1}(C)$ in the case of continuous time) such that condition* (HAC) *is fulfilled for $H_{\text{ind}}=H_C^0$ and $K=K_0$.*

Proof. If k is sufficiently large, the inclusion (i) is a consequence of 15.6 (take $\alpha=\operatorname{dist}(\mathscr{E},\partial B)$). To establish (ii), let us first consider the case of discrete time. It follows from the definition of a KAM set that, given $y\in\mathscr{E}$, there exist open sets $U'\supset U\ni\omega_0^{-1}(y)$ such that the generator of the induced system, $\omega_0^{-1*}f_0^1$ restricted to $\mathbb{T}^n\times U$ has a Hamiltonian function defined on an open set $G, \mathbb{T}^n\times U\subset G\subset\mathbb{T}^n\times U'$, and satisfying (HAC). The sets $\omega_0(U)$ form an open cover of $\mathscr{E}$. Let λ be the Lebesgue number of that cover. If k is sufficiently large then $\operatorname{diam} C<c_1k^{-1}+(5/11)\gamma k^{-\sigma}<\lambda$, $C\in\beta^{(k)}$, and C is contained wholly in some $\omega_0(U)$. Hence (ii) is true for this C.

In the case of continuous time fix k_1 so large that, for all $c\in\beta^{(k_1)}$, the inclusions (i) and $C\subset[\mathscr{E}]^\alpha$ hold (see 15.6), α being taken as above. Since the sets C are no longer small, the preceding arguments yield only the existence of a finite number of open sets U_s such that the induced system possesses a Hamiltonian H_s on U_s satisfying (HAC), and the union of $\omega_0(U_s)$ covers the union of $C\in\beta^{(k_1)}$. Let ξ be the induced vector field on A_D, the preimage under the conjugating map of the generator of $\{f_0^t\}$. The Calabi invariant of ξ is zero because of the existence of invariant tori (see 8.2), so ξ has a Hamiltonian H' which is uniquely defined up to an additive constant. It follows that $H'|_{T^n\times U_s}-H_s=\text{const}$, const depending on s. Since the H_s satisfy (HAC), there is a finite partition $\mathscr{E}=\sum_j\mathscr{E}_j$ into disjoint sets and a finite collection of constants c_j such that $H'-c_j$ satisfies (HAC) being restricted to some neighbourhood of $\mathbb{T}^n\times\omega_0^{-1}(\mathscr{E}_j)$. Let $2\alpha_0=\min\operatorname{dist}(\mathscr{E}_j,\mathscr{E}_{j'}), j\neq j'$ and choose $k_0\geqq k_1$ large enough so that $C\subset[\mathscr{E}]^{\alpha_0}$ if $C\in\beta^{(k)}, k\geqq k_0$. Then $C\subset[\mathscr{E}_j]^{\alpha_0}$ since C is connected, and we have that $H_C^0=H'-c_j$ satisfies (HAC). □

The following estimates will be used further, Compactness arguments show that there is a constant $R>1$ independent of $k\geqq k_0$, such that

$$\|K_0|_{\omega_0^{-1}(C)}\|_{r_1}\leqq R, \qquad \left|\det\left\{\frac{\partial^2 K_0}{I_iI_k}\right\}\right|\geqslant R^{-1}, \tag{15.11}$$

$$\|H_C^0|_{\mathbb{T}^n\times\omega_0^{-1}(C)}\|_{r_1}\leqq R, \qquad r_1=r_0+[2\sigma]2.$$

Switching Functions

To reduce the problem to Theorem 1.6 we need to supplement our construction with suitable switching functions.

Lemma 15.12. *There is a sequence $c_2^{(k)}$, $k = 1, 2, \ldots$, of positive constants depending only on n, such that given a non-empty set $A \subset \mathbb{R}^n$ and positive α there exists a function $\chi_{A,\alpha} \in C^\infty(\mathbb{R}^n)$ with properties*

(i) $0 \leqq \chi_{A,\alpha} \leqq 1$,
(ii) $\operatorname{supp} \chi_{A,\alpha} \subset [A]^{2\alpha}$,
(iii) $\chi_{A,\alpha}(x) = 1$ if $x \in [A]^\alpha$,
(iv) $|D^\rho \chi_{A,\alpha}| \leqq c_2^{(k)} \alpha^{-k}$, $|\rho| = k$.

Proof. Fix a smooth nonnegative function $\eta: \mathbb{R} \longrightarrow \mathbb{R}$ such that $\operatorname{supp} \eta \subset [1/2, 1]$ and $\int \eta(r) r^{n-1}\, dr = \Omega_n^{-1}$ where Ω_n is the $(n-1)$-volume of the unit sphere in $\mathbb{R}^n$. Set

$$\chi_{A,\alpha}(x) = \left(\frac{2}{\alpha}\right)^n \int_{[A]^{(3/2)\alpha}} \eta\left(\frac{2}{\alpha}|x - y|\right) dy.$$

The properties (i)–(iii) are obvious. As for (iv), one may take

$$c_2^{(k)} = \max_{|\rho| \leqq k} 2 \int_{\mathbb{R}^n} |D_x^\rho \eta(|x|)|\, dx.$$

□

We now return to our construction. Let $C \in \beta^{(k)}$ and Δ be its core. We supply C with the *switching function* $\chi_C: \mathbb{R}^n \times D \longrightarrow \mathbb{R}$ defined as follows

(15.13) $$\tilde{\chi}_C = \chi_{A,\alpha} \circ \omega_0 \circ \tilde{p}_2, \qquad A = [\Delta]^{\gamma_k}, \qquad \alpha = \gamma_k.$$

Proposition 15.14. *The switching function possesses the properties*

(15.14a) $$\operatorname{supp} \tilde{\chi}_C \subset \mathbb{R}^n \times \omega_0^{-1}([\Delta]^{3\gamma_k}),$$

(15.14b) $$\tilde{\chi}_C(x) = 1 \text{ if } x \in \mathbb{R}^n \times \omega_0^{-1}([\Delta]^{2\gamma_k}),$$

(15.14c) *for any smooth function $u: \mathbb{R}^n \times D' \longrightarrow \mathbb{R}$, $D' \subset D$,*

$$\|\tilde{\chi}_C u\|_{r_0,\gamma_k} \leqq c_3 \|u\|_{r_0,\gamma_k},$$

where $c_3 > 1$ depends only on n, r_0, and R_1.

Proof. Follows immediately from Lemma 15.12, definition (15.13), and the rule for differentiation. □

Switched Deviation

Suppose that $k > k_0$ as of Proposition 15.10. Take $C \in \beta^{(k)}$ and focus on the system induced on $\mathbb{T}^n \times \omega_0^{-1}(C)$. It possesses the induced Hamiltonian H_C^0 defined on some open $G_C \supset \mathbb{T}^n \times \omega_0^{-1}(C)$, such that the jets of infinite order of H_C^0 are the same as those of $K_0 \circ p_2$ at points of Ξ. Define the switched deviation to C as

(15.15) $$H_C^{\mathrm{swd}} = \tilde{\chi}_C \cdot (H_C^0 - K_0 \circ p_2)|_{\mathbb{T}^n \times \omega_0^{-1}(C)}.$$

The switched nonperturbed Hamiltonian is now

(15.16) $$H_C^{\mathrm{swn}} = K_0 \circ p_2 + H_C^{\mathrm{swd}}.$$

We are going to show that the proximity to Ξ and the (HAC) yield that the switched deviation is small. Further in this section const denotes a positive constant which only depends on n, R, and σ. Take $(\varphi, I) \in \mathbb{R}^n \times \omega_0^{-1}(C)$. It follows from (15.6) that there exists I^* such that $(\varphi, I^*) \in \tilde{\Xi}$ and $\mathrm{dist}(I, I^*) < \mathrm{const}/k$. We apply the Taylor formula at the point (φ, I^*) to estimate the derivatives $K_0 \circ p_2 - \tilde{H}_C^0$ at (φ, I), $\tilde{H}_C^0$ being the lift of H_C^0 onto $\mathbb{R}^n \times \omega_0^{-1}(C)$. Because of (HAC), they vanish at (φ, I^*), so only the reminder term must be taken into account. We take $r_0 + [2\sigma] + 1$ terms of the Taylor expansion. Let $\rho_1, \rho_2 \in \mathbb{Z}_+^n$ be multiindices with $|\rho_1| + |\rho_2| \leqq r_0$. Introducing the stretched variables $\bar{I} = \gamma_k^{-1} I$, we have

(15.17) $$\begin{aligned} &|D^{\rho_1} D_{\bar{I}}^{\rho_2}(K_0 \circ p_2 - H_C^0)(\varphi, \gamma_k \bar{I})| \\ &\leqq \mathrm{const} \max_{\substack{|m| = [2\sigma]+2 \\ \varphi, I'}} |D_\varphi^{\rho_1} D_{I'}^{\rho_2 + m}(K_0 \circ \tilde{p}_2 - \tilde{H}_C^0)(\varphi, \gamma_k I')(k\gamma_k)^{-[2\sigma]+2} \\ &\leqq \mathrm{const}\, k^{-[2\sigma]-2}. \end{aligned}$$

Here we used (15.11). The estimates (15.17), (15.14c), and formula (15.8) for γ_k give us

(15.18) $$\delta_d = \gamma_k^{-2} \| H_C^{\mathrm{swd}} \|_{r_0, \gamma_k} \leqq \mathrm{const}\, \gamma^{-2} k^{-1}.$$

Besides the estimate (15.18) the switched deviation satisfies the inclusion

(15.19) $$\operatorname{supp} H_C^{\mathrm{swd}} \subset \mathbb{T}^n \times \omega_0^{-1}([\Delta]^{3\gamma_k}),$$

and the switched unperturbed Hamiltonian coincides with the nonswitched one on the essential part of its domain:

(15.20) $$H_C^{\mathrm{swn}} = H_C^0 \quad \text{on } \mathbb{T}^n \times \omega_0^{-1}([\Delta]^{2\gamma_k}).$$

If the right-hand side of (15.18) is less than 1, and $\gamma_k < 1$ then the H_C^{swn} entirely governs the motion of the induced system on $\mathbb{T}^n \times \omega_0^{-1}([\Delta]^{\gamma_k})$.

Since we are going to apply Theorem 11.6 to a perturbation of (15.16) defined on $\mathbb{T}^n \times \omega_0^{-1}(C)$ let us fix $k = k^*$ so that the right-hand side of (15.18) is smaller than $\frac{1}{2}\delta$, where δ is as in Theorem 11.6, R being the same as in (15.11). Therefore k^* depends on γ. For brevity denote $\beta = \beta^{(k^*)}$ and $\gamma_* = \gamma_{k^*}$.

KAM Neighbourhood

We are now close to the main goal of our efforts. Let $\{f^t\}$ be a symplectic dynamical system of the same kind of time, and defined in the same phase space X, as $\{f_0^t\}$. We impose the following conditions on f^t:

15.21. *Domain condition.* The domain of the generator (the vector field ξ or 1-time map $f = f^1$) of f^t contains the image A of the conjugating map Ψ_0. In the case of discrete time we suppose in addition that $f \circ \Psi_0(A_{C'}) \subset A$ for each $C \in \beta^*$ $C' = \omega_0^{-1}(C)$.

Consider the preimage on $A_{C'}$ under Ψ_0 of the system $\{f^t\}$. Its generator is the vector field $\xi_C = (T\Psi_0)^{-1} \circ \xi \circ \Psi_0|_{A_{C'}}$ or the map $f_C = \Psi_0^{-1} \circ f \circ \Psi_0|_{A_{C'}} : A_{C'} \longrightarrow A_D$.

15.22. *Existence of a Hamiltonian.* For each $C \in \beta^*$ the generator ξ_C, or f_C, has zero Calabi invariant.

In the case of continuous time, it follows that there is a Hamiltonian function H_C to ξ_C, the partial Hamiltonian. In the case of discrete time, suppose in addition

15.23. *Existence of a partial Hamiltonian.* For each $C \in \beta^*$ the map f_C is factorizable, and the domain of its Hamiltonian H_C contains $A_{C'}$.

We define the switched perturbation $H_C^{\text{swp}} = \tilde{\chi}_C(H_C - H_C^0)$ where $\tilde{\chi}_C$ is the same as in (15.15). Our final condition is

15.24. *Small size of the perturbation.* For each $C \in \beta^*$

$$\delta_p = \gamma_*^{-2} \| H_C^{\text{swp}} \|_{r_0, \gamma_*} < \tfrac{1}{2}\delta$$

where δ is of Theorem 11.6 with R of (15.11).

The *KAM neighbourhood* of $\{f_0^t\}$ is the set of all symplectic dynamical system $\{f^t\}$ which satisfy 15.21–15.24. Obviously f_0^t belongs to this set. The conditions 15.21 and 15.24 are open, i.e. they do not fail if one perturbs f^t slightly. The condition 15.22 does, but its presence of 15.22 is inevitable if one desires the persistance of KAM tori (see Proposition 8.2).

Theorem 15.25. *Let $\{f_0^t\}$ be a symplectic dynamical system possessing a KAM set with model Hamiltonian $K_0 : D \longrightarrow \mathbb{R}$, frequency set $\mathscr{E}$, and conjugating map Ψ_0, and let $\{f^t\}$ belong to the KAM neighbourhood of $\{f_0^t\}$ constructed above. Then there exists a KAM set to $\{f^t\}$, the perturbed one, with model Hamiltonian K and conjugating map Ψ, the action and frequency domains, frequency set, domain and the image of the conjugating maps being common for the perturbed and the initial, unperturbed KAM sets, such that the following holds:*

(i) $$\| \Psi_0^{-1} \circ \Psi - \text{id} \|_{1,\gamma_*} \leqq c(\delta_d + \delta_p)^{4/15},$$

(ii) $$\| K - K_0 \|_{2,\gamma_*} \leqq c(\delta_d + \delta_p)^{7/15},$$

c being the constant of Theorem 11.6 with R of (15.11),

(iii) *K coincides with K_0 outside $D' = \bigcup_\Delta \omega_0^{-1}([\Delta]^{4\gamma_*})$ where Δ runs over all cores of the members of β^*, Ψ coincides with Ψ_0 outside $A_{D'}$.*

Proof. For each $C \in \beta^*$ apply Theorem 11.6 to the system on $A_{C'}$ taking $K_0|_{C'}$ for the Hamiltonian of the unperturbed system and $H_C^{\text{swd}} + H_C^{\text{swp}}$ for the perturbation (H_1 in 11.6). The inequality (15.18), the choice of k^*, and (15.24) show that the conditions of Theorem 11.6 are satisfied if one takes γ_* for γ. We obtain the collection of symplectic maps $\Sigma_C : A_{C'} \longrightarrow A_{C'}$ and integrable model systems

$g_C^t: A_{C'} \longrightarrow A_{C'}$ defined by the Hamiltonian functions $K_C: C' \longrightarrow \mathbb{R}$, $C \in \beta^*$, which are KAM sets to the systems spoiled by switching. The corresponding Diophantum frequency sets $\mathscr{E}_C^* \subset \mathscr{E}_{\sigma,\gamma_*}$ cover our $\mathscr{E}$ since $\gamma_* < \gamma$. The inclusions (11.9) and (11.10) enable us to join together the maps Σ_C, $C \in \beta^*$, into one map $\Sigma: A_D \longrightarrow A_D$, and the Hamiltonians K_C, $C \in \beta^*$, into one Hamiltonian K. In the latter case we use K_0 for patching the complement to the union of the sets C'. We set $\Psi = \Psi_0 \circ \Sigma$. The assertions (i)–(iii) are no more than paraphrases of the estimates (11.7), (11.8) and the inclusions (11.9) and (11.10) of Theorem 11.6. To conclude, one need only notice that spoilage by switching does not touch some open neighbourhood of the obtained KAM set. This is due to (11.11), the definition ($\mathscr{E}$) of the Diophantum frequency set in Theorem 11.6, and our choice of $\tilde{\chi}_C$. Hence, its tori are invariant under the original perturbed system, and the (HAC) is also satisfied for the original system. □

Notes to Chapter II

KAM theory has acquired its title from the initials of Kolmogorov, Arnol'd and Moser, who were the first to study in this area. The famous paper Kolmogorov (1954) contains the proof of the persistence of an invariant torus with Diophantine frequency for Hamiltonian flows in the nondegenerate case (The Hessian of an unperturbed Hamiltonian is detatched from zero). Arnol'd (1963) gave the detailed proof and enlarged this result to degenerate systems which occur in celestial mechanics. He obtained the existence of a set of positive measure of such tori in the phase space. Arnol'd investigated the case where both an unperturbed integrable system and a perturbed one were real analytical. Moser (1962) considered an area preserving map of finite class of smoothness (C^{333}) and proved the existence of a one-dimensional Diophantine torus. These were the first steps. We refer the reader to the review Bost (1986) for a more detailed historical account, and mention here only a few papers where this topic was developed.

Rüssmann (1970, 1983) diminished the number of derivatives in Moser's theorem (see also Herman (1983)).

The smooth dependence of KAM curves on a transversal parameter (say frequency) was established in Lazutkin (1973, 1974), generalized to higher dimensions by Svanidze (1980), and for the case of Hamiltonian flows in Pöschel (1982), Chierchia and Gallavotti (1982).

The reader can find a proof of Kolmogorov theorem using Lie method in Benettin, Galgani, Giorgilli, Strelcyn (1983).

R. Douady (1982) demonstrated the equivalence of KAM theorems for flows and cascades.

All the proof in the papers cited above used a sort of Newton method. An approach involving the implicit function theorem was suggested by E. Zehnder (1975, 1976).

Chapter III. Beyond The Tori

As was stated in the previous chapter, a Hamiltonian system frequently possesses KAM sets, the larger the part of the phase space occupied by these KAM sets, the closer the Hamiltonian system is to an integrable one. KAM sets bear a Cantor-like nature: they consist of KAM tori joined in families, the frequency playing the role of the parameter and ranging over a Cantor set $\mathscr{E}$. There are an infinite number of holes and gaps which are to be removed from the phase space to obtain the set occupied by KAM tori.

How does the system behave within these holes? There is no exhaustive answer to this question. The completely integrable case discussed in §7, where all trajectories behave regularly, is an exclusive one. In a system of general position, the motion between tori is quite complicated and manifests probabilistic, chaotic features as shown by numerical experiments.

The aim of this chapter is to give a plausible description of phenomena occurring beyond tori in contrast to, and as a complement to, the KAM tori theory developed above. Since there is no general theory, the exposition in this chapter will inevitably be incomplete and shallow. Also, I do not claim to comprehend the whole diversity of views and results available concerning this matter. As a rule, I do not present proofs, referring the reader to the literature cited in Notes.

§16. The General Picture of Stochasticity Near KAM Tori. The Case of More than Two Degrees of Freedom

The elementary bricks, which are collectively responsible for the creation of stochastic behaviour in the complement of KAM sets, are the so-called whiskered tori. We start with the definition of that object, and then we shall give a sketch of geometric and analytical ingredients of the phenomenon which takes place in systems with more than two degrees of freedom known as "Arnol'd diffusion".

Whiskered Tori

Let $\langle f^t \rangle$ be a symplectic dynamical system in a phase space X, $\dim X = 2n$, and let $\mathscr{T}$ be an invariant torus of $\langle f^t \rangle$ with a quasiperiodic motion. Recall that $\mathscr{T}$ is a completely invariant, isotropic submanifold, such that there exists an embedding χ of $\mathbb{T}^k$ with $\mathscr{T}$ as the image, which conjugates $f^t|_{\mathscr{T}}$ and a quasiperiodic motion g^t_ω on $\mathbb{T}^k$ (a more detailed definition was given in §9). Define the *stable*, $W^s(\mathscr{T})$, and *unstable*, $W^u(\mathscr{T})$, *manifolds* of $\mathscr{T}$ as follows:

$$W^s(\mathscr{T}) = \{x \in X : f^t(x) \longrightarrow \mathscr{T} \text{ as } t \longrightarrow \infty\},$$
$$W^u(\mathscr{T}) = \{x \in X : f^t(x) \longrightarrow \mathscr{T} \text{ as } t \longrightarrow -\infty\}.$$

Here $f^t(x) \longrightarrow \mathscr{T}$, $t \longrightarrow \infty$ (or $t \longrightarrow -\infty$) denotes that, for any neighbourhood U of $\mathscr{T}$ in X, there is a number t_0 such that $f^t(x) \in U$ for all $t > t_9$ (or $t < t_0$ correspondingly). We say $\mathscr{T}$ is a *whiskered torus* if there exist two smooth injective immersions $\Psi^{s,u}: \mathbb{T}^k \times V^{s,u} \longrightarrow X$, where $V^{u,s} \subset \mathbb{R}^{n-k}$ are open sets containing the origin, such that

(1) $\Psi^{s,u}(\mathbb{T}^k \times V^{s,u}) = W^{s,u}(\mathscr{T})$;

(2) $\Psi^{s,u}|_{\mathbb{T}^k \times \{0\}} \times \chi$, the embedding which determines $\mathscr{T}$;

(3) there exists $\varepsilon > 0$ such that $B_\varepsilon = \langle x \in \mathbb{R}^{n-k}: |x| < \varepsilon \rangle \subset V^{s,u}$, $\Psi^{s,u}(\mathbb{T}^k \times B_\varepsilon)$ are Lagrangian submanifolds, and

$$W^s(\mathscr{T}) = \bigcup_{t \geqq 0} f^{-t}\Psi^s(\mathbb{T}^k \times B_\varepsilon), \qquad W^u(\mathscr{T}) = \bigcup_{t \geqq 0} f^t\Psi^u(\mathbb{T}^k \times B_\varepsilon).$$

The stable and unstable manifolds (whiskers) of $\mathscr{T}$ may spread to different places of the phase space and, in such a manner, provide a link between them.

It is easy to deduce from the definition that $W^{s,u}(\mathscr{T})$ has no selfintersections. The stable manifolds of two distinct tori also cannot intersect, and the same is true with respect to their unstable manifolds. All other possibilities may occur. Let $\mathscr{T}_i$, $i = 1, 2$, be two (not necessarily different) whiskered tori. Let $x \in W^s(\mathscr{T}_1) \cap W^u(\mathscr{T}_2)$. Then, as follows from the definition of $W^{s,u}$, $f^t(x)$ is defined for all t, and the whole trajectory $\gamma = \bigcup_t f^t(x)$ lies in $W^s(\mathscr{T}_1) \cap W^u(\mathscr{T}_2)$. Such a γ is called a *homoclinic trajectory* if $\mathscr{T}_1 = \mathscr{T}_2$ (its points are called *homoclinical points*), and a *heteroclinic trajectory* if $\mathscr{T}_1 \neq \mathscr{T}_2$ (and its points are called *heteroclinical points*). In the case of discrete time a homo(hetero)clinic trajectory γ is called *transversal* if the corresponding stable and unstable manifolds intersect transversely at points of γ (see definition in AI.53). In the continuous time case to obtain the definition of a transversal homoclinical trajectory one must require the transversality of intersections of stable and unstable manifolds with respect to the manifold of constant energy, that is a Hamiltonian which always exists in a neighbourhood of the torus (see for example the subsection devoted to periodic trajectories), each homo(hetero)clinical trajectory partially entering into such a neighbourhood.

In the case of discrete time we define a *chain* of *whiskered tori* by substituting an invariant torus $\mathscr{T}$ with a chain of invariant tori $\bigcup_{k=0}^{N-1} \mathscr{T}_k$ (see Remarks 9.6 and 9.11). The definitions of stable and unstable manifolds, homo and heteroclinic trajectories can be extended to this more general object.

A special case of the preceding definition is that of a *hyperbolic periodic trajectory*. This corresponds to the value $k = 1$ in the case of continuous time and $k = 0$ in the case of discrete time.

Example 16.2. Hyperbolic fixed point of the standard Taylor–Chirikov–Greene map.

Consider the standard map of Example 8.8 with Frenkel–Kontorova potential

$$V(\phi) = \frac{k}{4\pi^2}(-1 - \cos 2\pi\phi).$$

The point with coordinates $(-1/2, 0) = (1/2, 0)$ *is* fixed. Since the trace of the tangent map at this point is equal to $2 + k$, it is a hyperbolic point. Denote by W^u and W^s its unstable and stable manifolds. Any reversor, which preserves a fixed point, maps its stable manifold onto the unstable one and vice versa. The reversor $\hat{\sigma}_1$ defined in 8.8 possesses this property with respect to our hyperbolic fixed point and leaves the symmetry line $\phi = 0$ invariant. It follows that any intersection of that line with W^u, if it exists, is also an intersection with W^s, i.e. is a homoclinic point.

Consider small values of $k > 0$. Changing the scale of the variable $I = \sqrt{k}J$ enables us to rewrite equations of motion as $(\phi, J) \longmapsto (\bar{\phi}, \bar{J})$,

$$\text{(16.2a)} \qquad \frac{\bar{\phi} - \phi}{\sqrt{k}} = \bar{J}, \qquad \frac{\bar{J} - J}{\sqrt{k}} = \frac{1}{2\pi} \sin 2\pi\phi.$$

If k tends to zero, the equations (16.2a) convert to the system of pendulum equations

$$\text{(16.2b)} \qquad \frac{d}{dt}\phi = J, \qquad \frac{d}{dt}J = -(2\pi)^{-1} \sin 2\pi\phi.$$

The latter has an integral $H_{FK} = \frac{1}{2}J^2 + (4\pi)^2(-1 - \cos 2\pi\phi)$, and the separatrices defined by the equation $H_{FK} = 0$. We conclude that W^s and W^u remain close to the said separatrices of the differential equation, provided k is small. At least, this is true for any finite initial segment of W^s and W^u. In particular, W^u intersects the line $\phi = 0$. Denote by z_Γ the first such intersection. As we have seen, z_Γ is a homoclinic point.

It has been proven that the point z_Γ is transversal provided k is sufficiently small.

Preparatory Reductions

Since we are going to investigate the motion near a KAM set, the devices described in §15 will be useful here. Thus, we may restrict ourselves to considering a Hamiltonian system defined in a symplectic annulus A_D, where $D = (\operatorname{grad} K_0)^{-1}(C)$, C being of the form $C = [\Delta]^{5\gamma}$, Δ being a convex polyhedron, one of those forming the complement to the union of resonant stripes S_μ, $\mu \in \mathfrak{M}^{(k)}$ (see the construction of §15). Using the cross-section construction of §5, the problem is reduced to those with discrete time. One may suppose that the Hamiltonian is of the form

$$\text{(16.3)} \qquad H(\phi, I) = K_0(I) + h(\phi, I),$$

where $K_0(I)$ is a smooth function, such that $\omega_0 = \operatorname{grad} K_0$ maps C diffeomorphically onto D. In addition we suppose that the matrix $\{\partial^2 K/\partial I_i \partial I_j\}$ is uniformly positively defined, since many questions look much simpler in that case. The perturbation term $h(\phi, I)$ is supposed to be small and vanish at a neighbourhood of the boundary of $\mathbb{R}^n \times D$. Such a representation may be formed for any symplectic dynamical system in a neighbourhood of its KAM set, the latter necessarily being divided into small pieces. The expounded model correctly reflects the movement of the actual system within a slightly smaller domain, say $\mathbb{T}^n \times \omega_0([\Delta]^{2\gamma})$, the portion of the KAM set in question having the form $\mathcal{K} = \mathbb{T}^n \times \omega_0(\mathscr{E})$, $\mathscr{E} = \Delta \cap \mathscr{E}_{\sigma,\gamma}$. The condition (HAC) implies that the function h of (16.3) vanishes at points of $\hat{\mathcal{K}} = \mathbb{R}^n \times \omega_0(\mathscr{E})$ as well as all of its derivatives.

Generic Properties

Denote by H the space of all $h \in C^{\infty}_{\text{per},a}(\mathbb{R}^n \times D)$, $a > 0$ being fixed, which satisfy two following conditions

(i) $\|h\|_{r_0} \leqq \varepsilon$, ε small and fixed,

(ii) $h(\phi, I)$ vanishes together with all of its derivatives at $(\phi, I) \in \mathbb{R}^n \times \omega_0(\mathscr{E})$.

Let us supply H with the metric

$$\rho(h_1, h_2) = \sum_{r=0}^{\infty} 2^{-r} \|h_1 - h_2\|_r (1 + \|h_1 - h_2|_r)^{-1},$$

where the norms $\|\cdot\|$ are defined in AII.3. Being supplied with this metric, H becomes a complete separable metric space.

This H contains "perturbations" h such that the system with Hamiltonian (16.3) has a quite degenerate behaviour of trajectories which cannot serve as a typical representative. For instance, if h does not depend on ϕ, we have an integrable system whose trajectories behave regularly, but this behaviour changes drastically if one perturbs h slightly in an appropriate manner. The common belief is that one must investigate the case of "general position", or the "generic" case.

Any complete separable metric space, and H is one of them, possesses the *Baire property*. This means that the intersection of any denumerable collection of open everywhere dense sets is again everywhere dense. Such an intersection is called a *residual set*. A property of elements in the space in question is called *generic* if it holds for all elements of some residual set. An important example of a generic property is the transversality of homo (hetero) clinic trajectories of whiskered tori. We say simply a "generic" element (map, system, etc.) if some generic property is meant implicity or explicitly. What we have to study is how a symplectic dynamical system with Hamiltonian (16.3) behaves provided h is generic.

Arnol'd Web

The most part of the phase space A_D of our model is occupied by the KAM set $\mathcal{K} = \mathbb{T}^n \times \omega_0(\mathscr{E})$. The complement of this set may be obtained by removing the *resonant gaps* $G_{m,m_0} = \mathbb{T}^n \times \omega_0^{-1}(S_{m,m_0})$, where S_{m,m_0} is the resonant strip defined by (15.1b), $m = (m_1, m_2, \ldots, m_n) \in \mathbb{Z}^n$, (m, m_0) ranges over a subset of $\mathbb{Z}^{n+1}$ consisting of all (m, m_0) with property $\omega_0^{-1}(S_{m,m_0}) \cap C \neq \emptyset$. The geometry of resonant gaps is quite different according to whether there are two or more degrees of freedom. In the former case the resonant gaps are combined in cluster gaps of the form $G_\alpha = \mathbb{T}^n \times \alpha$, $\alpha \subset \mathbb{R}$ being an open interval, and form a disjoint collection. In the latter case, with $n \geqq 2$, as one can observe without difficulty, the resonant gaps intersect and form a connected infinite system of canals which penetrate the phase space densely. The *resonant hyperplanes*

$$L_{m,m_0} = \langle y \in \mathbb{R}^n : \langle m, y \rangle = m_0 \rangle$$

constitute the *Arnol'd web*. The union of all such $L_{m,m}$ $(m, m_0) \in \mathbb{Z}^n$ is dense in $\mathbb{R}^n$. If $n \geqq 2$, one can travel in the system of resonant gaps and reach every point of the phase space starting near another arbitrary point. This fact is crucial for the formation of Arnol'd diffusion.

Diffusion along a Resonant Gap

To examine the mechanism of diffusion let us focus on one resonant gap which cleaves the phase space, say $G_{m,m_0} = \mathbb{T}^n \times \omega_0^{-1}(S_{m,m_0})$. Note that the width of this gap, $\gamma |m|^{-\sigma}$, may be very small, but the smaller the width, the smaller the perturbation term h in (16.3) when evaluated there, as follows from condition (ii) in the definition of the space H.

It is sufficient to consider a gap with $m = k_0 \operatorname{ort}_1$. A general gap G_{m,m_0} may be carried to this special one by applying a special annuli diffeomorphism g_M described in 2.5*i* while the Hamiltonian is transformed as in the case of continuous time: $H \longmapsto g_M^* H$ (see 8.7). The following Lemma provides us with the appropriate matrix M.

Lemma 16.5. *Let* $0 \neq m = (m_1, m_2, \ldots, m_n) \in \mathbb{Z}^n$, *and let* k_0 *be the greatest common divisor of* $m_1, m_2, \ldots, m_n$. *Then there exists a matrix* $M \in \mathrm{GL}(n, \mathbb{Z})$ *such that* $Mm = k_0 \operatorname{ort}_1$.

Proof. One may assume that $k_0 = 1$ (otherwise divide m by k_0). Consider the set $\mathfrak{M}_m$ of all matrices with integer entries and with first column equal to m. Their determinants form an ideal in $\mathbb{Z}$. Since every ideal in $\mathbb{Z}$ is principal, there is a positive integer p such that the set $\{\det A : A \in \mathfrak{M}_m\}$ coincides with $\{jp : j \in \mathbb{Z}\}$. For any $i \in \{1, 2, \ldots, n\}$ one can find an $A \in \mathfrak{M}_m$ such that $\det A = m_i$. Hence $p = 1$. Take $A \in \mathfrak{M}_m$ with $\det A = 1$ and put $M = A^{-1}$. □

As one can observe, the transformation induced by g_M on the frequency domain is linear, and its matrix is $(M^{-1})^T$. Hence the image of the strip S_{m,m_0} under the map g_M with M given by Lemma 16.5, will be the strip S defined by the inequality

$$|k_0 y^1 - m_0| < \gamma |m|^{-\sigma} \text{ (the old quantity).}$$

From this we assume that our gap is already reduced to the new variables. Denote it by G and the corresponding resonant hyperplane $y^1 = m_0/k_0$ by L. Choose a

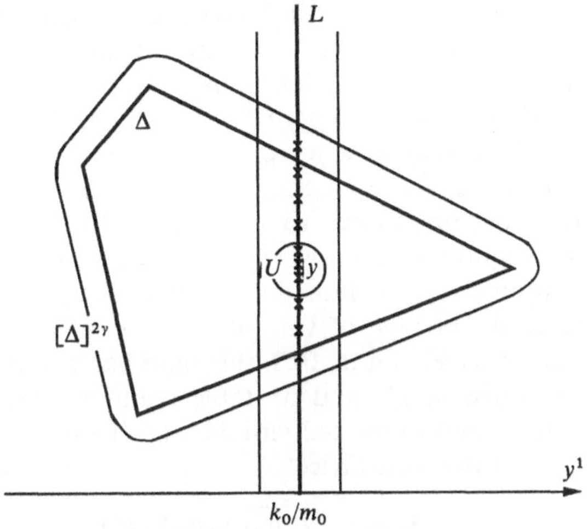

Fig. 37. A resonant strip cleaving $[\Delta]^{2\gamma}$

point $y \in L \cap [\Delta]^{2\gamma}$ focus on the part of G of the form $\mathbb{T}^n \times \omega_0^{-1}(U)$ where U is a neighbourhood of y in $[\Delta]^{2\gamma}$ (see Fig. 37). For convenience assume that $\omega_0^{-1}(y) = 0$ and $K(0) = 0_0$.

Expanding the Hamiltonian in a Taylor series in the action variables near $I = 0$, in a Fourier series in the angle variable, and applying the procedure analogous to that described in 9.13 to remove nonresonant harmonics in the term which does not contain I, one obtains the following principal terms of the Hamiltonian near $\mathbb{T}^n \times \{0\}$:

$$H(\phi, \bar{I}) \sim (m_0/k_0)\bar{I}_1 + \sum_{i=2}^{n} y^i \bar{I}_i + (1/2)\sum_{i,k} a_{ik}\bar{I}_i\bar{I}_k - b\cos 2\pi k_0\phi^1. \tag{16.6}$$

Here we have retained the members in the Taylor expansion of K_0 up to second order and the first resonant member in the Fourier expansion of $h(\phi, 0)$. The phase of the cosin in (16.6) is taken to be zero at $\phi^1 = 0$, since otherwise one can make it zero by an appropriate shift of ϕ^1. The matrix $\{a_{ik}\}$ is positively defined due to our assumption about K_0.

Denote the right-hand side of (16.6) by H_0 and let us first study the system defined by H_0 as a Hamiltonian function. The set F_y defined by the equations $I_i = 0$, $2 \leqq i \leqq n$, is completely invariant with respect to this system. Denote by $f_0: (\phi, I_1) \longmapsto (\bar{\phi}, \bar{I}_1)$ the restriction onto F_y of the generator. Then we have

$$\begin{cases} \bar{\phi}^1 = \phi^1 + m_0/k_0 + a_{11}\bar{I}_1, \\ I_1 = \bar{I}_1 + 2\pi b k_0 \sin 2\pi k_0 \phi^1, \\ \bar{\phi}^i = \phi^i + y^i, \quad 2 \leqq i \leqq n. \end{cases} \tag{16.7}$$

The system (16.7) breaks down into two independent systems: a quasiperiodic motion on the $(n-1)$-torus of variables ϕ^i, $2 \leqq i \leqq n$, with frequency vector $y' = (y^2, y^3, \ldots, y^n)$, and a system f_1 in the variables (ϕ^1, I_1). The latter is quite similar to the standard one (Example 8.8). To reduce the former to the latter, let us note that the map $f_1:(\phi^1, I_1)\longmapsto(\bar{\phi}^1, \bar{I}_1)$ commutes with the shift $\sigma_{k_0}:(\phi^1, I_1)\longmapsto(\phi^1+1/k_0, I_1)$. Hence the study of the iterations of f_1 is equivalent to that of the factor map $\hat{f}_1:(\mathbb{T}^1\times\mathbb{R})/\Gamma_{k_0}\longrightarrow(\mathbb{T}^1\times\mathbb{R})/\Gamma_{k_0}$ where Γ_{k_0} is the group generated by σ_{k_0}. The space $(\mathbb{T}^1\times\mathbb{R})/\Gamma_{k_0}$ is again a symplectic annulus. One obtain appropriate coordinates in this factor space by setting $\psi=k_0\phi^1$ and $J=2ak_0I_1$. Then the factor map $\hat{f}_1$ in this coordinates exactly coincides with the standard map with Frenkel–Kontorova potential $V(\phi)=k/(4\pi^2)^{-1}(-1-\cos 2\pi\psi)$, where $k=8\pi^2abk_0^2$. For generic perturbation h it is obvious that $b\neq 0$. Suppose, for definiteness, that $b>0$ (otherwise make the change of variable $\psi\longmapsto\psi+1/2$). Then k is a small positive quantity. As we have seen in 16.2, this map has a hyperbolic fixed point at $\psi=\pm 1/2$, $J=0$, whose stable and unstable manifolds are close to the line $H_{FK}=0$. Returning to f_1 and to the old variables (f, I), one obtains the following approximate equation of the separatrices:

$$I_1=\pm\sqrt{2b/a}\cos(\pi k_0\phi^1). \tag{16.8}$$

One may conclude that the system with Hamiltonian H_0 has the chain of whiskered $(n-1)$-dimensional tori $M_y^0=\bigcup_{j=0}^{n-1}\mathcal{T}_j$ with period k_0 and frequency vector $y'=(y^1, y^2, \ldots, y^n)$, the tori being defined by the equations $I_1=0$, $\phi^1=j/k_0$, $0\leqq j\leqq k_0-1$. Their stable and unstable manifolds $W^s(M_y^0)$ and $W^u(M_y^0)$ also lie in the invariant submanifold F_y. The motion of the system (16.7) in F_y near the resonant hyperplane L may be described as follows. In ϕ^i-variables, $2\leqq i\leqq n$, the motion is fast, quasiperiodic with frequency vector $y'=(y^1, y^2, \ldots, y^n)$. In (ϕ^1, I_1)-variables it is more convenient to consider the dynamics of a power $f_1^{k_0}$, the k_0 iteration. Its picture is portrayed schematically in Fig. 38. The mean velocity has order $\sqrt{k}=2\sqrt{2}\pi k_0\sqrt{ab}$. There is a diffusion layer in a small (of order $\exp(-\mathrm{const}/\sqrt{k})$ neighbourhood of the line which is approximately given by (16.8). The point travels chaotically there; whenever it meets a bifurcation, it chooses one of two ways, the infinite sequences of these choices may be arbitrary, and we can determine only a finite number of results of these choices if we know the approximate coordinates of the initial point. In contrast, the motion outside the drawn layer is simpler: points above and below run uniformly near the circles around the

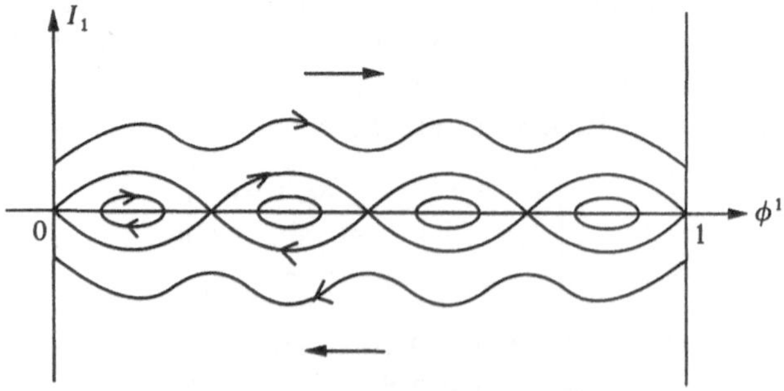

Fig. 38. Diffusion along φ^1-direction

annulus in the direction of the arrows. In the holes of the diffusion layer points also run near smaller circles around the elliptic fixed points at $\phi^1 = k_0^{-1}(j+1/2)$, $0 \leqq j \leqq n-1$. If one looks closer at the actual picture, not the schematic one, one discovers a lot of other very small holes in the diffusion layer, as well as other very small diffusion layers in the holes and between invariant curves. All these finer elements mimic the structure of the initial "large" diffusion layer and that of the "large" holes. A sort of hierarchy structure arises. The reader can get an idea of such a structure by looking at Fig. 1 or 41.

What we have described is also valid in the case of two degrees of freedom, when $n=1$ and F_y coincides with the whole phase space $\mathbb{T}^1 \times C$. We shall return to this case in the next section.

In the remaining part of this section we suppose $n>1$. Then the above described motion takes place in each invariant slice F_y, and no diffusion exists in the $(I_2, I_3, \ldots, I_n)$-direction provided the motion is governed by H_0. What happens if one switches on a small generic perturbation, i.e. considers the actual Hamiltonian H? To answer this question let us note that the stable and unstable manifolds of M_y^0 have intersections (see Figs. 38 and 43) but these are not transversal. Generic perturbation restores the transversality. Unfortunately it also destroys the chains M_y^0 themselves unless $y \in \mathscr{E}_{\gamma', \sigma'}$ for some pertinent γ' and σ'. Let us forget this unpleasant circumstance for now and consider what happens if all whiskered tori persist, but still undergo slight deformation (denote them M_y). Then we find that $W^u(M_y)$ has a transversal intersection with $W^s(M_y)$ provided y' is sufficiently close to y. The mechanism of diffusion in the $(I_2, I_3, \ldots, I_n)$-direction then works in the following way. *Transition sequences* $\langle M_{y_i}, 1 \leqq i \leqq n \rangle$ *of chains of whiskered tori* arise which have the property that $W^u(M_{y_i})$ intersects transversely with $W^s(M_{y_{i+1}})$. Not going into technical details (which are fairly complicated) we conclude that there is a slow diffusion along any such transition sequence. Since $W^u(M_{y_i}) \cap W^s(M_{y_{i+1}})$ is non-empty there exist points which run from M_{y_i} to $M_{y_{i+1}}$. The transversality of the intersections implies the existence of points which run along the whole transition sequence. The picture is similar to that described in Fig. 38 when we discussed the motion within the diffusion layer, but one must slightly move each torus in the $(I_2, I_3, \ldots, I_n)$-direction in an arbitrary manner. Since the direction is arbitrary, this diffusion bears a chaotic character similar to Brownian motion. Its rate is proportional to the maximal mutual distance between neighbouring tori such that transversal intersections of stable and unstable manifolds still occur. In analytical case one suspects this distance to be of order $\exp(-\text{const}/k^a)$, $a>0$. This is plausible estimate for the rate of diffusion in $(I_2, I_3, \ldots, I_n)$-direction along a resonant gap.

Returning to the delicate question about concerning the destrased M_y with bad frequencies $(y^2, y^3, \ldots, y^n) \notin \mathscr{E}_{\sigma', \gamma'}$, one may hope that the stable and unstable manifolds are more robust objects than the quasiperiodic motions they grow away from. Thus a k-torus bearing a resonant quasiperiodic motion may disintegrate into a finite number of k'-tori, $k'<k$, some of also having stable and unstable manifolds which are situated not far from those of the old torus. Again, the picture of Fig. 38 illustrates such a metamorphosis. One must imagine a quasiperiodic motion complete with hyperbolic whiskers in the complementary dimensions.

Starting with degenerate motion, namely with the rational m_0/k_0-shift on the ϕ^1-axis, the perturbation converts it to a finite number of elliptic and hyperbolic periodic trajectories, the 1-dimensional stable and unstable manifolds of the hyperbolic ones, together with the corresponding $(n-1)$-dimensional manifolds in the complementary space, constitute the full-dimensional perturbed stable and unstable manifolds which grow from the tori of smaller dimension. Of course, such discussion is quite speculative, but at present there is no rigorous general theory. Some references on this subject will be given in Notes to the this Chapter.

Arnol'd Diffusion

Since points of the action domain D are in one-to-one correspondence with points of the frequency domain C via the frequency map, one may use the frequency variables in place of the action ones. As we have already seen, in the case $n > 1$, the resonant strips S_{m,m_0} densely penetrate C. There are a lot of "chaotically" crossing gaps, or passages, with width tending to zero as the number m tends to infinity, and the motion in each such gap is as described in the previous subsection. In particular, it follows that the frequency coordinates of a point slowly evolve in a chaotic manner along each of the passages. In the crossings, a point may choose any possible direction for its further progression. So we have a sort of random walk in a labyrinth (see Fig. 39). Since the Arnol'd web is dense and connected, one can, starting near an arbitrary point $A \in [\Delta]^{2\gamma}$, reach, as close as one desires, another arbitrary point $B \in [\Delta]^{2\gamma}$. This plausible (but as yet not rigorously proved) assertion may also be expressed as follows. Place a small ball containing red ink in the phase space of our system and let the particles of ink travel according to the equations of motion. Then, after a sufficiently long time, the phase space will become pink. Applying a microscope, one can find pieces of the phase space which

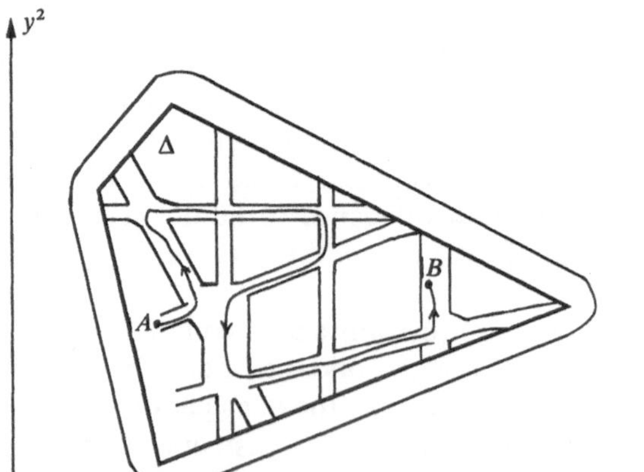

Fig. 39. Diffusion in the frequency domain along the Arnol'd web channels

are not touched by ink, but the larger the time of detection, the greater the magnification must be. I believe that this must be true for generic systems.

To complete the section, let us formulate two questions which sound a bit contradictory to the conjecture just outlined. The KAM theory provides us with KAM sets which occupy the part of phase space of positive Lebesgue measure. Let us collect together all KAM sets consisting of tori and of chains of tori and call the union the *total* KAM set.

Is the total KAM set dense provided the system is generic and close to an integrable one?

The last reservation about being close to an integrable system is necessary, since Anosov systems do not possess any KAM sets (see Examples 6.19, 6.20).

Is the Lebsegue measure of the complement to the total KAM set zero or positive, provided that the system is close to an integrable one?

If the answer to the first question is positive, then it means that KAM tori also pervade the phase space. A plausible conjecture is that it is generically true for all kinds of robust invariant sets with complete dynamics in a neighbourhood of a KAM torus.

§17. Picture of Stochastically Near KAM Tori in the Case of Two Degrees of Freedom

The entitled case drastically differs from that of more than two degrees of freedom. There is no Arnol'd diffusion since invariant tori divide the phase space (or the energy surface in the case of continuous time) and serve as natural obstacles for a trajectory to freely travel in the phase space. This and other specific features suggest that we discuss the case of two degrees of freedom separately. We restrict ourselves to the case of discrete time, KAM tori appearing to be closed curves, which we call KAM curves.

The quite different kinds of motion that coexist near a KAM curve are: *elliptic*, which includes

(a) other KAM curves of the same homotopical class as the initial KAM curve;

(b) elliptic periodic trajectories;

(c) KAM curves (or chains of KAM curves (see Remark 9.11)) which occur in "islands" surrounding stable elliptic periodic trajectories;

(d) "quasiperiodic motions" on nondifferentiable critical curves which bound, or divide, regions of stochastic motion; and *hyperbolic*, which includes

(e) hyperbolic periodic trajectories;

(f) cantori which bear a "generalized quasiperiodic motion";

(g) homoclinic and heteroclinic trajectories which are transversal intersections of stable and unstable manifolds of invariant sets of types (e) and (f);

(h) uniformly hyperbolic sets in neighbourhoods of sets of type (g) (Smale' horseshoes);

(i) stochastic layers.

All these types of invariant sets will be discussed in this section. Of course, other might exist, and the above list is not exhaustive.

Model

The phase space of a symplectic cascade with two degrees of freedom is a two dimensional manifold, KAM tori, which are the KAM curves, are smooth embedded circles which bear quasiperiodic motions, the frequency satisfying the Diophantum condition, and the shears being nonzero. Since KAM tori are not isolated (see §12), in order to study the motion near a KAM curve, it is sufficient to consider a domain situated between two nearby KAM curves. Such a domain is isomorphic to a two dimensional symplectic annulus, the symplectic structure becoming an area form. So the general problem is reduced to that of considering an area preserving map of an annulus which leaves the ends invariant, has zero Calabi invariant, and differs slightly from an integrable map with nonzero shear. Such a map f possesses a Hamiltonian of the form (cf. 16.3):

$$H(x, y_1) = K_0(y_1) + h(x, y_1), \tag{17.1}$$

the covering map $\tilde{f}: \mathbb{R}[a, b] \longrightarrow \mathbb{R} \times [a, b]$ of f being of the form $\tilde{f}(x, y) = (x_1, y_1)$, where

$$\begin{cases} x_1 = x + \omega_0(y_1) + \dfrac{\partial h(x, y_1)}{\partial y_1}, \\ y = y_1 + \dfrac{\partial h(x, y_1)}{\partial x}. \end{cases} \tag{17.2}$$

Here $\omega_0(y_1) = K_0'(y_1)$ is a strictly increasing function, i.e. $\omega_0'(y_1) > 0$, defined on $[a, b], h: \mathbb{R} \times [a, b] \longrightarrow \mathbb{R}$ is smooth, periodic in the variable x with period 1, constant on the boundary components $\mathbb{R} \times \{a\}$ and $\mathbb{R} \times \{b\}$. One may regard h as a small perturbation of the integrable part K_0. We shall assume h together with a necessary number of its derivatives to be as small as would be necessary. We shall illustrate the properties of such a map, the presence of the invariant sets in question, by use of the standard Taylor–Chirikov–Greene map (Example 8.8 and 16.2). It is not of the form discussed here, since the vertical variable ranges over an infinite interval. But one can reduce this map to the form (17.1), (17.2), making a smooth symplectic change of variables $(\phi, I) \longmapsto (x, y)$ in order to straighten two KAM curves, provided k is small enough, and restricting one's attention to the annulus between these two chosen curves. The pictures will be drawn in the old variables (ϕ, I), the Hamiltonian being

$$H(\phi, I_1) = \tfrac{1}{2} I_1^2 + \frac{k}{4\pi^2}(-1 - \cos 2\pi\phi), \tag{17.3}$$

and the law of transformation being

$$\begin{cases} \phi_1 = \phi + I_1 \\ I_1 = I - \dfrac{k}{2\pi} \sin 2\pi\phi. \end{cases} \tag{17.4}$$

This model contains one parameter k. I believe all the features of the general map (17.1, 2) may be illustrated by this example, by an appropriate choice of the parameter and by considering an appropriate piece of the phase space.

Rotational KAM Curves

Let h in (17.1) be sufficiently small. The KAM theorem 11.6, on being applied to the system (17.2), yields the existence of KAM curves which differ slightly from the curves $y = \text{const.}$ (To reduce (17.1) to the conditions of Theorem 11.6, it is sufficient to extend K_0 and h onto a larger strip and to multiply h by an appropriate switching multiplier, the latter being 1 on the old strip). It follows that these *rotational KAM curves* occupy the bulk of the phase space: the Lesbegue measure of the complement being estimated by the square root of the norm of the perturbation. The rotational KAM curves here are parameterized by their frequency ω, the motion on the rotational curve Γ_ω with frequency ω being conjugated with the shift $\psi \longmapsto \psi + \omega \pmod 1$. If the interval $[a, b]$ is not too large and h is sufficiently small, there is a unique rotational KAM curve Γ_ω for each $\omega = \mathscr{E}_{\sigma,\gamma}/\mathbb{Z}$. Their disposal in the annulus corresponds to the ordering of the frequencies in $\mathbb{T}^1$.

The rotational KAM curves do not fill all the phase space. Gaps exist near each "rational curve" given by the equation

$$\omega_0(y) = \frac{p}{q}, \quad p, q \in \mathbb{Z}, \quad q > 0. \tag{17.5}$$

Periodic Points

Each gap near the value (17.5) contains periodic trajectories. A typical picture is like that of Fig. 38 (see also 40). Generically, all the trajectories are of a definite kind: general elliptic (§13) or hyperbolic (Example 3.12), and there is a finite and equal number of them possessing the properties: (i) the period of a trajectory is equal to q, (ii) the map permutes their x-projections $(x_1, x_2, \ldots, x_q)$ by the shift $x_i \longmapsto x_{i+p \pmod q}$, x_i being situated in the resultant order on the x-axis.

Islands and the Hierarchy of KAM Sets

Each general elliptic periodic trajectory is surrounded by KAM curves, as follows from the considerations of §13. More precisely, the points of a general elliptic periodic trajectory of period q are surrounded by the components of the q-chain of KAM tori (see Remark 9.11). Such a component of a KAM set may be called in *island*. Islands always occur near general elliptic periodic points, whenever they appear: in a resonant gap (see Figs. 40, 41) or elsewhere. The reader can also see them in Fig. 48 in the stochastic sea. Inside an island KAM curves form a structure which possesses all the features of the initial rotational KAM set. So they possess

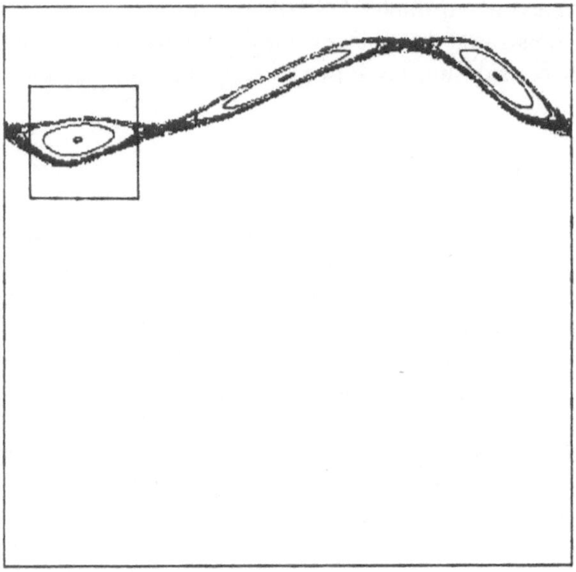

Fig. 40. A chain of islands surrounding an elliptic periodic orbit of period 3 for the standard map (17.4) with $k=0.95$. This chain is surrounded in turn by a stochastic layer formed by separatrices of the accompanying hyperbolic periodic orbit of period 3

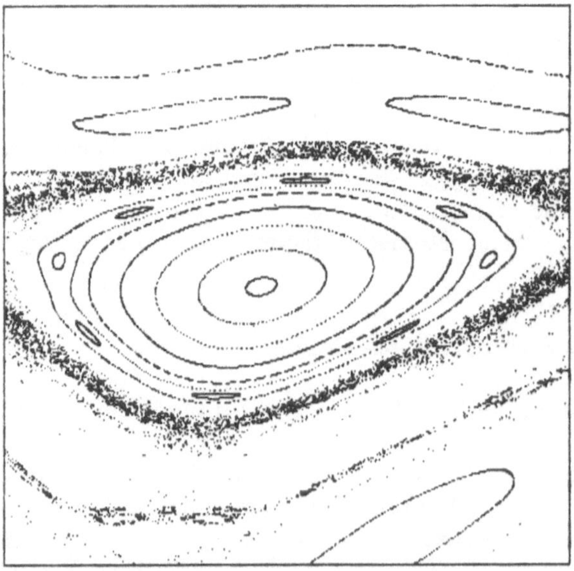

Fig. 41. The enlarged square in the left upper corner of Fig. 40. The components of chains of KAM curves and islands of higher period are visible

gaps, and the islands within these gaps, etc. A sort of hierarchy arises. This is a typical picture in Hamiltonian systems with two degrees of freedom.

Hyperbolic Periodic and Homoclinical Trajectories. Hyperbolic Cantor Sets of Zero Measure

The previous pictures (Figs. 38, 40) show that, typically, hyperbolic periodic orbits (Example 3.12) exist in gaps together with elliptic ones. Also one may naturally expect that there are a lot of such orbits in stochastic layers (see below). The investigation of a motion generated by a map f near a periodic orbit $\{z_0, z_1, \ldots, z_{q-1}\}$ of period q is reduced to that of the qth power of f near one of the points of the orbit, say z_0. As was mentioned in 7.17, this map, f^q, reads, in specially chosen smooth (not necessarily symplectic) coordinates (X_1, X_2) as $(X_1, X_2) \longmapsto (\Lambda \cdot X_1, \Lambda^{-1} \cdot X_2)$ with $\Lambda = \lambda_1 + c(X_1 X_2)^m, c \in \mathbb{R}, m \in \mathbb{Z}_+$, λ being the largest eigenvalue of $T_{z_0} f^q$. The equations $X_i = 0$ determine two smooth curves, the stable $W^s(z_0)$ and the unstable $W^u(z_0)$ manifolds of the point z_0. The images $f^k(W^{s(u)}(z_0))$, are the stable (unstable) manifolds of $z_k, 0 \leqslant k \leqslant q-1$. Their union is the stable (unstable) manifold of the orbit. Also, the manifolds $W^{s,u}(z_k)$ are called *separatrices*.

As mentioned in the preceding section, the separatrices $W^{s,u}$ of a periodic hyperbolic trajectory have intersections, homoclinic and heteroclinic trajectories which appear as an essential ingredient of stochastic phenomena. For example, a few homoclinic intersections for the fixed hyperbolic point $(-1/2, 0) = (1/2, 0)$ of the map (17.4) with $k = 1.5$ are shown in Fig. 43a–c. They form a kind of an intricate net. The existence of at least one transversal homoclinic trajectory results in the birth of infinitely many other such trajectories. Another consequence of the presence of a transversal homoclinic trajectory is the existence of compact Cantor like invariant sets of zero Lesbegue measure, which carry a uniform hyperbolic structure (see the definition further in this section). The mechanism of creating such sets is known as the Smale' horseshoe. The motion on such an invariant set is topologically conjugated to a so-called topological Markov' chain and has positive topological entropy. Since such sets have zero Letesgue measure, they are "invisible" in numerical experiments. In view of formula (0.7), there no use can be made of such sets for semiclassical asymptotics.

Birkhoffs' Theorem, Boundary Curves and the Instability Zone

To investigate to what extent the rotational KAM curves approach the region occupied by hyperbolic periodic points and their separatrices in the resonant gap corresponding to the rational value of the frequency (17.5), we shall use the well known Birkhoffs' theorem. Denote $A = \mathbb{T}^1 \times [0, +\infty[$, and consider a diffeomorphism $f: A \longrightarrow A$. Denote by $C_{\mathrm{I}}(\varepsilon) \subset T_z A = \mathbb{R}^2, z \in A$, the sector consisting of rays which form the ray $\{x\,\mathrm{ort}_1, x \leqq 0\}$ with the angle α satisfying the inequality $\pi/2 - \varepsilon \leqq \alpha \leqq \pi/2 + \varepsilon$. Let $C_{\mathrm{II}}(\varepsilon)$ be the opposite sector, $C_{\mathrm{I}}(\varepsilon) = -C_{\mathrm{II}}(\varepsilon)$. We say f *deviates the vertical to the right* if there is an $\varepsilon > 0$ such that $T_z f(\mathrm{ort}_2) \in C_{\mathrm{I}}(\varepsilon)$ and $T_z f^{-1}(\mathrm{ort}_2) \in C_{\mathrm{II}}(\varepsilon)$ for all $z \in A$.

Theorem 17.6 (Birkhoff). *Let $f: A \longrightarrow A$ preserve a measure with positive density. Let $U \subset A$ be a subset satisfying the following conditions:*

(1) *the closure of U in A is compact;*
(2) *U is diffeomorphic to $\mathbb{T}^2 \times \mathbb{R}_+$;*
(3) *there exists $\delta > 0$ such that $\mathbb{T}^1 \times [0, \delta] \subset U$;*
(4) *U is the interior of its closure;*
(5) *U is invariant under f, i.e. $f(U) = U$.*

Then the border $\bar{U} \backslash U$ of U in A is the graph of a Lipschitz function $\phi: \mathbb{T}^1 \longrightarrow \mathbb{R}_+$.

To apply Theorem 17.6 to the map (17.2), it is sufficient to extend K_0 onto $\mathbb{R}_+$ (making a shift in the y variable if necessary), to multiply h by an appropriate switching function, and to extend in onto A by zero. It is possible to make such changes in such a manner that the behavior of the map does not change in a neighbourhood of the rational gap (17.5). It follows from Theorem 17.6 that

Corollary 17.7. *Each rotational KAM curve of the map* (17.2) *is the graph of a smooth function.*

Consider the set of all rotational KAM curves of (17.2) which lie above a hyperbolic periodic trajectory belonging to the gap (17.5). Taking the union of all the open domains which these curves bound from above, we obtain a set U satisfying the conditions of Theorem 17.6. It follows that there exists the *upper KAM boundary curve* Γ_+^{KAM} to the gap (17.5). Analogously, one proves the existence

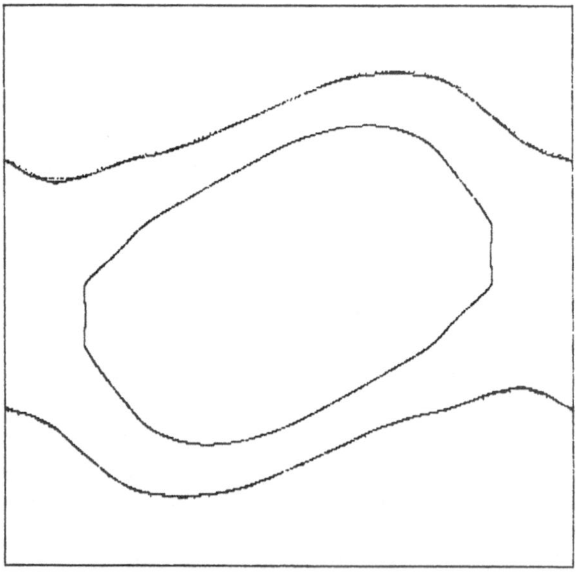

Fig. 42. Instability zone for the standard map (17.4) with $k = 0.9$ corresponding to the main resonance $\omega_0(y) = 0$ (near the separatrices of the fixed hyperbolic point

of the *lower KAM boundary curve*, $\Gamma_{-}^{\mathrm{KAM}}$. As a matter of fact, in the general situation there exists the third KAM boundary curve (or chain of curves), namely the *inner one*, Γ_0^{KAM}, which may be obtained by the analogous procedure applied to the KAM curves surrounding the points of an elliptic periodic trajectory belonging to the gap in question (those of period q, discussed in the subsection on *Periodic Points*). The inner KAM boundary curve has several components, each homeomorphic to the circle, their number is equal to the number of the corresponding elliptic trajectories multiplied by the period q. These are examples of critical curves. They are not KAM curves, since the latter possesses neighbouring KAM curves from both sides, as it follows from §12. Birkhoff's theorem asserts that $\Gamma_{\pm,0}^{\mathrm{KAM}}$ are Lipschitz. Now consider all the invariant curves which are graphs of Lipschitz functions in coordinates (x, y) and are situated above the gap. The same consideration yields the existence of the nearest curve to that gap, the *upper boundary curve*, Γ_+, to the gap. Analogously, there exists Γ_-, the *lower boundary curve*, and Γ_0, the *inner boundary curve* to the gap. The natural conjecture is that generically $\Gamma_\alpha^{\mathrm{KAM}} = \Gamma_\alpha$, $\alpha = \pm, 0$, but these curves may differ for individual maps.

Three boundary curves bound the so-called *instability zone* in the resonant gap. One deduces that, given two neighbourhoods W_1, W_2 of any of these three curves, there exist points $z_i, i = 1, 2$, and a positive integer N such that $f^N(z_1) = z_2$. A typical instability zone is visible in Fig. 42.

Stochastic Layer

Numerical experiments gives evidence that the instability zone in the rational gap has holes which are, for example, the islands around the points of elliptic periodic trajectories. So the closure of the union $\bigcup_{j=0}^{q-1}(W^s(z_j) \cup W^u(z_j))$ where $z_j, 0 \leqq j \leqq q-1$, are the points of a hyperbolic periodic trajectory belonging to the gap, does not coincide with the instability zone. We call this closure the *stochastic layer* generated by a periodic hyperbolic trajectory $\gamma = \langle z_j, 0 \leqq j \leqq q-1 \rangle$. It is possible that different periodic hyperbolic trajectories in a given rational gap near (17.5) generate different stochastic layers, but probably, in a typical case, there exists one "primary" stochastic layer in each instability zone. Of course, plenty of smaller stochastic layers may exist there.

Typically a stochastic layer looks like a very complicated closed set possessing on infinite number of holes. The reader can see stochastic layers in Figs. 1, 40, 41, 46, 47.

To retrace the formation of a stochastic layer let us consider that of the fixed point $(-1/2, 0) = (1/2, 0)$ of the same standard maps as (17.4) with $k = 1.5$. This map may be considered as a map of the torus $\mathbb{T}^2$ onto itself. In the following Figs. 43a–c the unfolding of the torus $\mathbb{T}^2$ is represented as the unit square $\langle(\phi, I)\colon -1/2 \leqq \phi \leqq 1/2, -1/2 \leqq I \leqq 1/2\rangle$. The said fixed point γ is represented by two points $(-1/2, 0)$ and $(1/2, 0)$ on the edges of the square. To obtain the picture of separatrices we used the following method. One can easily calculate the Taylor series of the separatrices $W^{s,u}(\gamma)$ near their origin γ and to use these series to calculate the initial small pieces of $W^s(\gamma)$ and $W^u(\gamma)$ near γ with high precision.

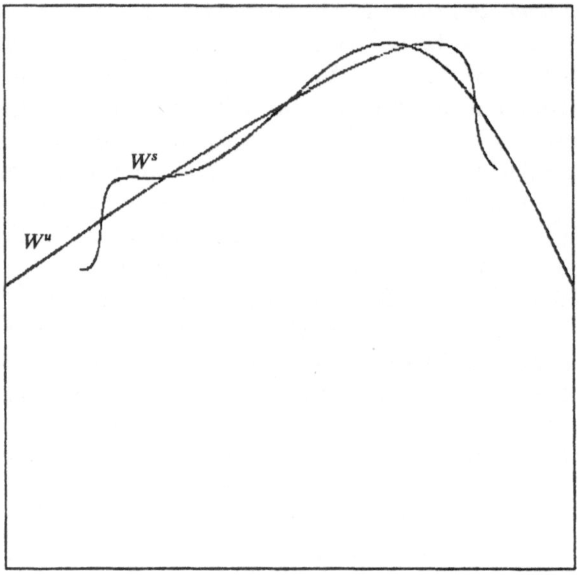

a

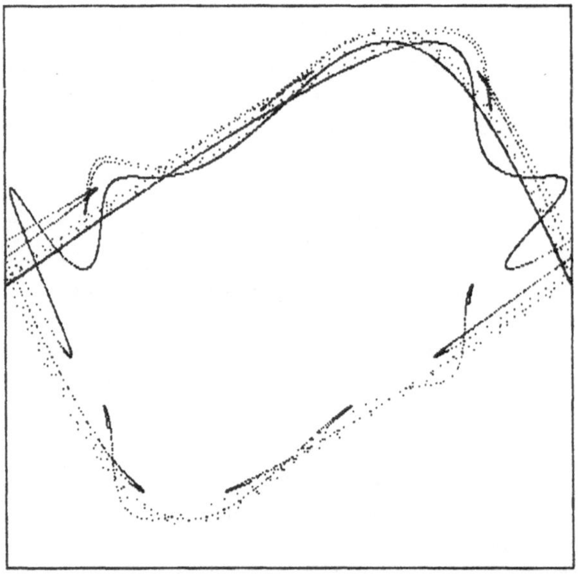

b

Fig. 43. The separatrices W^u and W^s of the hyperbolic fixed point $(-1/2, 0) = (1/2, 0)$ of the standard map (17.4) with $k = 1.5$: (a) few homoclinic intersections, (b) more homoclinic intersections, (c) large pieces of W^u and W^s

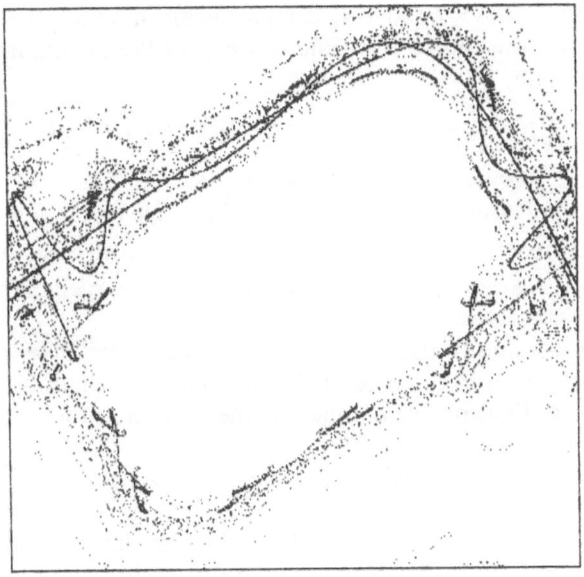

c

Fig. 43. (*Cont.*)

Then, to obtain their extension, it is sufficient to apply the iterations of the map, or its inverse, to the said pieces. The results of these calculations performed by use of a computer are presented consequently on Figs. 43a–c. Not quite so long pieces of $W^s(\gamma)$ and $W^u(\gamma)$ are shown in Fig. 43a. Transversal homoclinical points are visible. More long pieces are presented in Fig. 43b, and one can see the result of the "infinite continuation" of the pieces in Fig. 43c. A single transversal homoclinical intersection results in a catastrophic multiplication of numerous subsequent intersections. We chose such a value $k = 1.5$ in order to clearly demonstrate the picture of transversal intersections. As a matter of fact, the separatrices give rise to a "stochastic sea" for this value of k, since there are no rotational invariant curves. For smaller values of $k (k < k_c$ given by (17.17) this catastrophe does not speard far from the initial region of Fig. 43a because of the presence of the boundary curves which limit the creaping of separatrices. The presence of transversal homoclinical intersections, which leads to the creation of stochastic layers, is therefore a strong indication of nonintegrability.

It is interesting to notice that the said transversal homoclinical intersection takes place for arbitrarily small positive values of the parameter k entering the formulae (17.4) while $k = 0$ corresponds to the integrable map $(\phi, I) \longmapsto (\phi + I, I)$. The following asymptotic formula for the angle α of the intersection of the initial pieces of separatrices $W^s(\gamma)$ and $W^u(\gamma)$ at the point with $\phi = 1/2$ (see Fig. 41a) is valid as $k \longrightarrow 0$:

$$\alpha = \frac{\pi|\Theta_1|}{k}\exp\{-\pi^2/\sqrt{k}\}[I + \mathcal{O}(k^{(1/8)-\delta})]. \tag{17.8}$$

Here $|\Theta_1| = 1118.82770595$, $\delta \in]0, 1/8[$ is an arbitrary number, the constant in the estimate depends on the choice of δ. The width D of the instability zone (see Fig. 42) admits an estimate

$$c_1 k^{-1} \exp\{-\pi^2/\sqrt{k}\} \leqq D \leqq c_2 k^{-1} \exp\{-\pi^2/\sqrt{k}\}, \tag{17.9}$$

where $c_1 < c_2$ are positive constants. Since the amplitude of oscillations of separatrices is proportional to the product of the angle (17.8) and the elementary step of the size of order $\sqrt{k}$ we find the ratio of the width and the amplitude to be of the order $1/\sqrt{k}$ as $k \longrightarrow \infty$, i.e. the width is much larger than the amplitude of oscillations.

Little is known about the structure of a stochastic layer. The following questions are open, even in the case of the one-parameter family of the standard maps (17.4). The cardinal problem concerns the measure of a stochastic layer.

17.10. *Is the Lebesgue measure of a stochastic layer positive?*

The second question is how the system behaves in a stochastic layer. One naturally supposes that this behavior is hyperbolic. We say an invariant with respect to a $\{f^t\}$ set Λ possesses a *hyperbolic structure* if there are two linearly independent sections of $T_\Lambda x$, v^s and v^u, on Λ, not necessarily continuous, such that the following is true:

(i) v^s and v^u are invariant under f;

(ii) given Euclidean metrics $|\cdot|$ on X, there are positive measurable functions $\lambda, \varepsilon, C, K$ defined on Λ, λ, ε being invariant under f and satisfying the inequality $\lambda < 1 - \varepsilon$, C and K satisfying the inequalities for all $t \in \mathbb{Z}$ and $x \in A$:

$$C(f^t(x)) \leqq C(x) \exp\{\varepsilon(x)|t|\},$$
$$K(f^t(x)) \geqq K(x) \exp\{-\varepsilon(x)|t|\},$$

such that for all $t > 0$ and $x \in \Lambda$

$$|Tf^t v^S(x)| \leqq C(x) \lambda^t(x) |v^s(x)|,$$
$$|Tf^t v^u(x)| \geqq C^{-1}(x) \lambda^{-1}(x) |v^u(x)|,$$

and the angle $\alpha(x)$ between $v^s(x)$ and $v^u(x)$ admits the estimate:

$$\alpha(x) \geqq K(x).$$

If one puts in $\varepsilon = 0$ this definition and requires $C(x)$, $\lambda(x)$, and $K(x)$ to be independent of x, then one obtains the definition of a *uniform hyperbolic structure*. Note that in these definitions $\lambda, \varepsilon, C, K$, and α depend on the choice of Euclidean metrics, but the very fact of their existence is independent of that choice.

17.11. *Does a stochastic layer possess a hyperbolic structure?*

Another question arises about geometric structure of stochastic layers. Looking at Figs. 46 and 47, one sees the holes in the body of the stochastic layer. These are islands containing KAM curves surrounding periodic points.

17.12. *Do the holes form a dense set in a stochastic layer? In other words, is a stochastic layer nowhere dense?*

Note that the answer to this question is affirmative if the measure of a stochastic layer is zero.

A glance the phase portrait in Fig. 1, and in Figs. 41, 46 reveals many stochastic layers placed between KAM curves. One can naturally think that hyperbolic periodic points and stochastic layers are situated between quasiperiodic motions like rational numbers are between irrational ones. So we formulate our last question.

17.13. *Generically is the union of all stochastic layers dense in the phase space of a system?*

What happens to stochastic layers when one varies the system in question, say the parameter k in the standard family for example? It turns out that stochastic layers may join together or dissipate into smaller stochastic layers. A typical picture is when a very thin stochastic layer joins to or splits off from a larger stochastic layer. These phenomena are connected with the problem of destruction of KAM curves, and to describe them in more detail one needs to consider a new notion which will be expounded in the next subsection.

Generalized Quasiperiodic Motion. Cantori

Let $\mathbb{T}^1 = \mathbb{R}^1/\mathbb{Z}^1$ denote a one-dimensional torus, and $\pi:\mathbb{R} \longrightarrow \mathbb{T}^1$ the canonical projection. A map $s:\mathbb{T}^1 \longrightarrow \mathbb{T}^1$ is said to be *locally strictly increasing* if there exists a covering map $\tilde{s}:\mathbb{R} \longrightarrow \mathbb{R}$ such that (1) $\pi \circ \tilde{s} = s \circ \pi$, (2) $\tilde{s}$ strictly increases: $x < x' \Longrightarrow \tilde{s}(x) < \tilde{s}(x') \forall x, x' \in \mathbb{R}$, (3) $\tilde{s}(x+1) = \tilde{s}(x) + 1 \, \forall x \in \mathbb{R}$. The s is said to be continuous on the left if $\tilde{s}$ is also.

Let $\omega \in \mathbb{T}^1$ be an irrational number modulo 1. Consider the standard quasiperiodic cascade $g^t_\omega : \mathbb{T}^1 \longrightarrow \mathbb{T}^1$ with ω as a frequency: $g^t_\omega(x) = x + \omega t \pmod 1$, $t \in \mathbb{Z}$. Choose a finite (possibly empty) or denumerable set D of trajectories of $\{g^t_\omega\}$ and denote by A the union of all trajectories entering D. Define a locally strictly increasing, continuous on the left map $s:\mathbb{T}^1 \longrightarrow \mathbb{T}^1$ whose set of discontinuities is A. Denote $C = \overline{s(\mathbb{T}^1)}$, the closure of the image of s, and define a cascade $g^t_{\omega,D} : C \longrightarrow C$ as the image of $\{g^t_\omega\}$ under s. More precisely, we set $g^t_{\omega,D}|_{s(\mathbb{T}^1)} = s \circ g^t_\omega \circ s^{-1}$. It is easy to prove that these maps have only continuous extensions onto C, these forming a cascade. It will be called an *abstract generalized quasiperiodic motion.* There exists a continuous left inverse map k to $s (k \circ s = \mathrm{id})$ which semiconjugates g^t_ω and $g^t_{\omega,D}$, i.e. the following diagram commutes for all $t \in \mathbb{Z}$:

$$\begin{array}{ccc} C & \xrightarrow{\;g^t_{\omega,D}\;} & C \\ \Big\downarrow k & & \Big\downarrow k \\ \mathbb{T}^1 & \xrightarrow{\;g^t_\omega\;} & \mathbb{T}^1 \end{array}$$

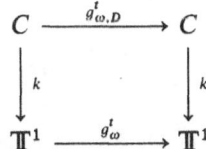

Obviously, the preimage of a point $x \in \mathbb{T}^1$ under k is either a single point if $x \notin A$, or a two-point set if $x \in A$. If $D = \emptyset, k$ is a homeomorphism, and $g^t_{\omega,D}$ is topologically isomorphic to g^t_ω via k. If D is nonempty, C is homeomorphic to the well known Cantor set, and $g^t_{\omega,D}$ turns out to be a nontrivial extension of g^t_ω. We call the former an *abstract cantorus*.

The preceding construction does not depend on the choice of s, only the positions of its jumps play a role. The following Proposition follows immediately from 9.5.

Proposition 17.14. *Two abstract generalized quasiperiodic motions* $g^t_{\omega,D}: C \longrightarrow C$ *and* $g^t_{\omega',D'}: C' \longrightarrow C'$ *are topologically isomorphic (i.e. there exists a homeomorphism* $h: C \longrightarrow C$ *which conjugates them) if and only if* (a) $\omega' = \pm\omega$, (b) *there is an* $a \in \mathbb{T}^1$ *such that the map* $x \longmapsto x + a$ *carries D onto D' bijectively.*

Now let $f^t: X \longrightarrow X$, $t \in \mathbb{Z}$, be a symplectic cascade on a 2-dimensional symplectic manifold X. We say a subset $\mathcal{T} \subset X$ bears a *generalized quasiperiodic motion* if

(GQM1) $\mathcal{T}$ is an invariant set with complete dynamics;

(GQM2) *there exists an abstract generalized quasiperiodic motion* $g^t_{\omega,D}: C \longrightarrow C$, *and a topological embedding* (i.e. an *injective continuous map*) $\chi: C \longrightarrow X$ *such that* $\mathcal{T} = \chi(C)$, *and* χ *conjugates* $\{g^t_{\omega,D}\}$ *and* $\{f^t|_{\mathcal{T}}\}$.

The number $\omega \in \mathbb{T}^1$ is called the *frequency* of $\mathcal{T}$, the map χ, a *conjugating map.*

If $D \neq \emptyset$, $\mathcal{T}$ is called a *cantorus.* If $D = \emptyset$, $\mathcal{T}$ is an *invariant circle.* One-dimensional KAM tori (KAM curves) are all particular cases of invariant circles. An invariant circle which is not a KAM curve is called a *critical curve.*

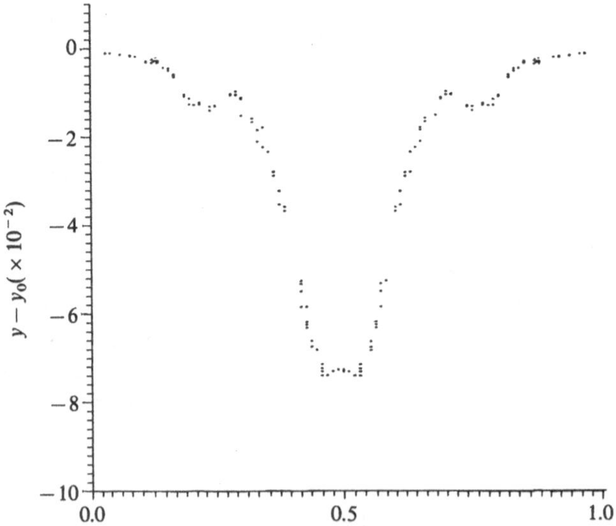

Fig. 44. Portrait of a cantorus (by MacKay, Meiss and Percival (1984)). (with permission)

It turns out that cantori and critical circles exist in the gaps between KAM curves. Let $\mathscr{T} \subset A_{[a,b]}$ be a set which bears a generalized quasiperiodic motion, and let $\chi: C \longrightarrow \mathscr{T}$ be its conjugating map. We say *$\mathscr{T}$ is projected properly onto the x-axis* if the map $p_1 \circ \chi: C \longrightarrow \mathbb{T}^1$ does not change the mutual ordering of points, i.e. there is a continuous extension of $p_1 \circ \chi$ onto $\mathbb{T}^1$ which is a locally strictly increasing map. Let $h(x, y_1)$ in (17.1) vanish at $y_1 = a$ and $y_1 = b$, and let the map (17.2) deviate the vertical to the right. Then one may assert that

Theorem 17.15 (Aubry–Mather). *For any $\omega \in [\omega_0(a), \omega_0(b)]/\mathbb{Z}$, he map* (17.2) *has only the invariant set bearing generalized quasiperiodic motion with frequency ω which projects properly on to the x-axis.*

We call the generalized quasiperiodic motion of Theorem 17.15 a *rotational generalized quasiperiodic motion.*

A detailed investigation of the situation in a rational gap near (17.5) results in the statement that there is no rotational KAM or critical curves with frequency sufficiently close to p/q. So, all generalized quasiperiodic motions of Theorem 17.15 with such frequencies are cantori. A typical cantorus is drawn in Fig. 44.

Turnstiles and Transport

There is a lot of evidence that generically cantori bear a uniform hyperbolic structure. So, the general stable–unstable manifolds theory yields the existence of stable and unstale manifolds to a cantorus. A typical picture of these manifolds may be described as follows. Let $\mathscr{T} \subset X$ be a cantorus with conjugating map $\chi: C \longrightarrow X$. Since C is a compact perfect subset of $\mathbb{T}^1$, its complement $\mathbb{T}^1 \backslash C$ may be represented as the union of open nonintersecting intervals. From these intervals take those which correspond to one element of the set D (see definition of generalized quasiperiodic motion above) and denote them $\Delta_j =]\alpha, \beta[$, $j \in \mathbb{Z}$, arranging the numeration so that $g^t_{\omega,D}(\Delta_j) = \Delta_{j+1}$. Denote $a = \chi(\alpha_j)$ and $b_j = \chi(\beta_j)$. Then one may choose the part of $W^s(\mathscr{T}) \overset{\text{def}}{=} \{x \in X\colon f^t(x) \longrightarrow \mathscr{T} \text{ if } t \longrightarrow +\infty\}$ corresponding to the chosen element of D, the said part being the union of smooth arcs σ^s_j joining a_j with b_j. The whole stable manifold $W^s(\mathscr{T})$ is the union of such parts corresponding to all elements of D. The unstable manifold $W^u(\mathscr{T})$ admits the analogous description, giving rise to the sequences of smooth arcs σ^u_j. Typically σ^s_j and σ^u_j intersect transversely forming a sort of figures of eight (see Fig. 45).

If the cantorus bears a uniform hyperbolic structure, then the set D is finite (R. S. MacKay 1983). For simplicity let D consist of one element. Then it is not a difficult problem to find out that the areas of the loops A_j (see Fig. 45) are equal to each other, and this common area is equal to the area which f carries through $\mathscr{T}$ per one iteration in the direction from the stable to the unstable subarcs which form the loops A_j. The analogous quantity, which is equal to the common area of the loops B_j, describes the flux in the opposite direction. This is the reason why these objects are called *turnstiles*. They are responsible for the transport of the

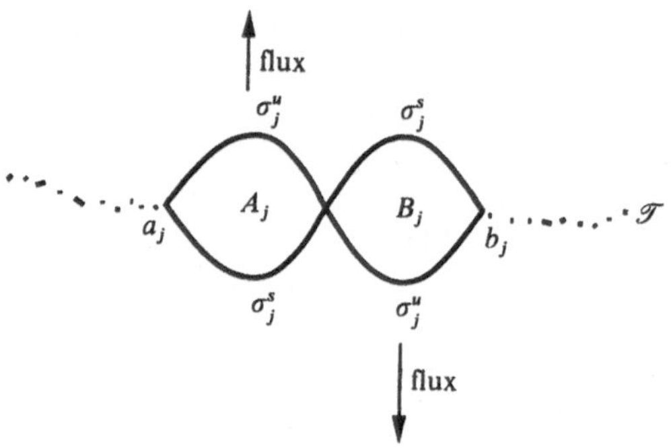

Fig. 45. Stable σ_j^s and unstable σ_j^u arcs to a gap with the ends a_j, b_j of a hyperbolic cantorus $\mathscr{T}$. The areas of the loops A_j and B_j are equal to the fluxes of the area per iteration

area in the phase space. If $\mathscr{T}$ is very close to the closed invariant curve, then the sizes of the holes in $\mathscr{T}$ and the areas of A_j, B_j are very small, and so the flux through $\mathscr{T}$ is slow. In the limiting case when $A_j = B_j = 0$ and $\mathscr{T}$ is an invariant curve, there is no flux. If flux is very small, one may regard $\mathscr{T}$ as a kind of semipenetrable barrier which puts an obstacle in the way of a trajectory in the phase space. Such semipenetrable barriers occur near the boundary of islands. This is the reason why a trajectory dwells there for a long time, as shown by numerical experiments.

Destruction of KAM Tori. Last Curves

Consider a family f_k, $k \in]\alpha, \beta[\subset \mathbb{R}$, of area preserving selfmaps of a two-dimensional symplectic annulus, each being of the form (17.2). As follows from Theorem 17.15, generalized quasiperiodic motions persist if one changes the parameter, but they may change their type and vary from being KAM curves to being cantori. Such a transition is shown in Figs. 46, and 47 a very interesting question is to expose the mechanism of this transition. This problem is closely connected to that of stability–instability transition. Indeed, as mentioned in the previous subsection, cantori serve as penetrable barriers while KAM curves give absolute obstacles to the transport in the y-direction. If there were no rotational KAM curves and no rotational critical curves (only cantori exist), then a point of a trajectory could travel in the y-direction limitlessly far, i.e. we had the instability regime. Otherwise, if at least one rotational closed invariant curve exists, there is no transport in the y-direction, and we have the stable regime.

Consider, as an example, the standard family (17.4), where k ranges over $\mathbb{R}$. If $k=0$, we have an integrable case, all curves $y=$ const are invariant. For positive k, when small enough the bulk of rotational generalized quasiperiodic motions, namely those with Diophantum frequencies with suitable σ and γ, remain KAM curves. If one enlarges k, KAM curves successively go to ruin, becoming cantori. Numerical experiments show that the most stable curve is that with $\omega = \omega_g = (\sqrt{5}-1)/2 \pmod 1$, the so-called "golden curve". This is due to the

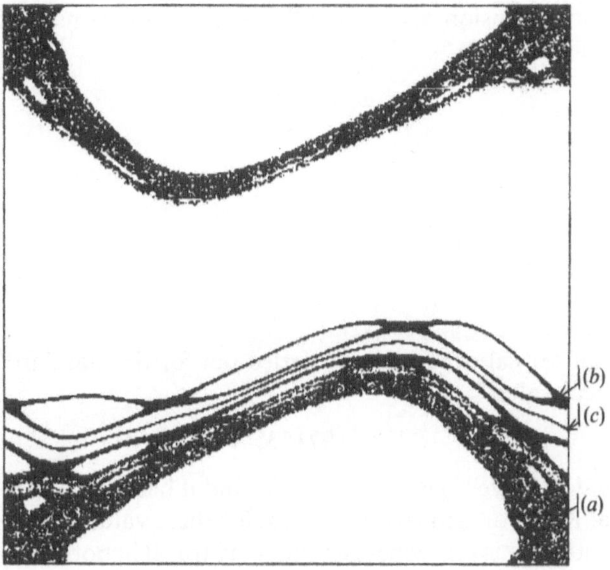

Fig. 46. Stochastic layers for the standard map (17.4) with $k = 0{,}9$, (a) the biggest one, corresponding to the fixed hyperbolic point, (b) the smaller one corresponding to a hyperbolic periodic orbit of period 3. A KAM curve (c) which separates these stochastic layers is visible

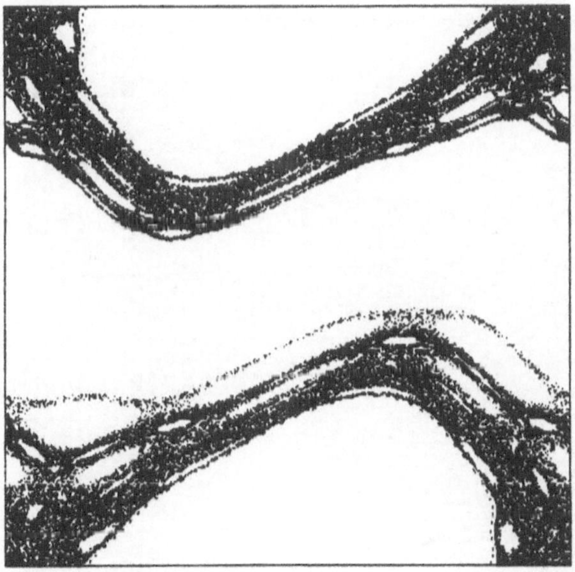

Fig. 47. Stochastic layer for the standard map (17.4) with $k = 1$. The former stochastic layers of Fig. 44 have now fused together

continuous fraction expansion of ω_g being the best one and consisting of the units. The ω_g satisfies the equation

(17.16) $$\omega_g = (1 + \omega_g)^{-1},$$

and so

$$\omega_g = \cfrac{1}{1 + \cfrac{1}{1 + \cfrac{1}{1 + \cdots}}}$$

Following from numerical experiments, carried out for the standard family (17.4), there is the "critical value"

(17.17) $$k_c \approx 0.97163540631\ldots$$

such that the "golden curve" exists if $0 < k < k_c$, and it becomes a cantorus if $k > k_c$, and no rotational KAM or critical curves exist for these values of k. The stochastic layer becomes a "stochastic sea" where a point may travel in both ϕ and I directions (Fig. 48). If $k = k_c$, there is only the rotational closed invariant curve, the *last curve*. It has the "golden" frequency ω_g (mod 1). It is natural to suppose that the bifurcation just described may serve as a model for the fusion of two stochastic layers in the general case.

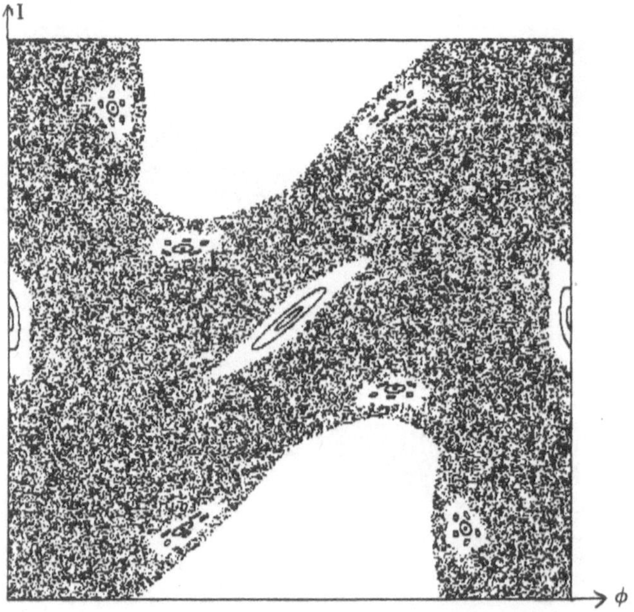

Fig. 48. Stochastic sea for the standard map (17.4) with $k = 1.7$. Islands of different periods are situated in this sea

Renorm-group Approach

An interesting method which clarifies the process of destruction of a KAM curve is the so-called *renorm-group* approach which has come from statistical physics. There is no room here to enter the detailed account of this method. So we restrict ourselves with a short sketch referring the reader to the literature cited in Notes to this chapter. Also, we restrict ourselves to the simplest case of destruction of the "golden curve" with frequency $\omega = \omega_g \pmod 1$.

Consider a map f of the form (17.2) near the "golden curve" $\mathcal{T}$ which is very close to the "moment of destruction" or is already destroyed, i.e. is a cantorus, but with very small holes. Fix a point $z_0 \in \mathcal{T}$, take a transversal segment of a line L and consider a domain D restricted by L and $f(L)$ within a small neighbourhood of $\mathcal{T}$. The map f generates the so-called *derivative map* $f': D' \longrightarrow D$ which is defined as follows: $D' = \{z \in D\colon \exists n > 0$ *such that* $f^n(z) \in D\}$, $f'(z) = f^{n(z)}(z)$, where $n(z) = \min\{n > 0\colon f^n(z) \in D\}$. Let us glue D by the map f identifying points $z \in L$ and $f(z) \in f(L)$. The glued domain D_g is nothing more than another symplectic annulus, and the glued map $f': D'_g \longrightarrow D_g$, D'_g being the glued image of D', has the glued image $\mathcal{T}'_g$ of $\mathcal{T} \cap D$ as an invariant curve (or cantorus). One may deduce, using the equation (17.16), that the frequency of $\mathcal{T}'_g$ is the same $\omega_g \pmod 1$. We have obtained a transformation of the pair $(f, \mathcal{T})$ to the other pair $(f'_g, \mathcal{T}'_g)$, where f, f'_g are symplectic maps of symplectic annuli with zero Calabi invariant, and $\mathcal{T}, \mathcal{T}'_g$ are rotational invariant sets which bear quasiperiodic motions with golden frequency $\omega_g \pmod 1$. One may consider this transformation $\mathcal{R}$ as a transformation

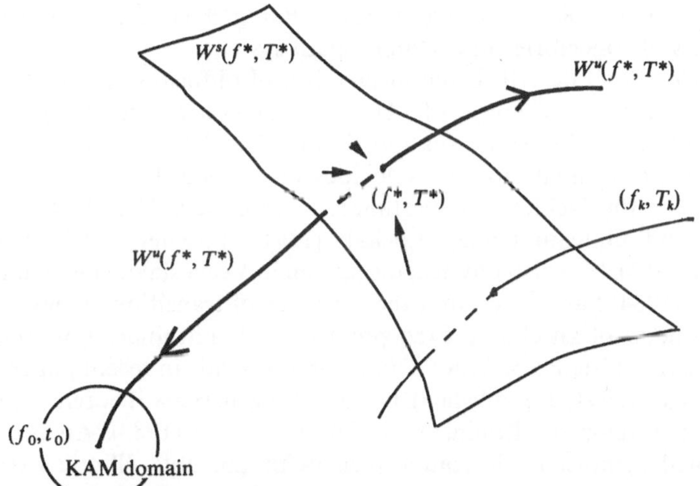

Fig. 49. Stable $W^s(f^*, \mathcal{T}^*)$ and unstable $W^u(f^*, \mathcal{T}^*)$ manifolds of the "universal" golden pair $(f^*, \mathcal{T}^*)$ in the space of all pairs $(f, \mathcal{T})$ (conventional picture). The renorm transform has two fixed points: the attracting trivial one $(f_0, \mathcal{T}_0)$ surrounded by the KAM neighbourhood and the nontrivial hyperbolic fixed point $(f^*, \mathcal{T}^*)$ with one-dimensional unstable manifold

on the space of all selfmaps of the form (17.2) (we omit technical details). $\mathcal{R}$ is called the *renorm transform*. Renorm-group approach suggests that the consider iterations of $\mathcal{R}$. A plausible behavior of these iterations is the following one:

(i) if $\mathcal{T}$ is a KAM curve, then the sequence of iterations $\mathcal{R}^n$, $n = 1, 2, \ldots$, applied to the initial pair $(f, \mathcal{T})$, *converges to a pair* $(f_0, \mathcal{T}_0)$, where f_0 is the integrable map $(x, y) \longmapsto (x + \omega(y), y)$ and $\mathcal{T}_0$ is the straightened golden KAM curve $\omega(y) = \omega_g$;

(ii) if $\mathcal{T}$ is a cantorus, then $\mathcal{R}^n(f, \mathcal{T})$ diverges;

(iii) if $\mathcal{T}$ is a critical curve (last curve) then $\mathcal{R}^n(f, \mathcal{T})$ converges to the "universal" pair $(f^*, \mathcal{T}^*)$, the latter serving as a universal model for a last curve.

The situation may be illustrated in Fig. 49 where the infinite-dimensional phase space, where $\mathcal{R}$ acts, is drawn symbolically in the two-dimensional picture. The point $(f_0, \mathcal{T}_0)$ appears to be an attractive fixed point of $\mathcal{R}$, while the point $(f^*, \mathcal{T}^*)$ appears to be a hyperbolic fixed point with one-dimensional unstable manifold $W^u(f^*, \mathcal{T}^*)$ and infinite-dimensional stable manifold $W^s(f^*, \mathcal{T}^*)$. A family $(f_k, \mathcal{T}_k)$ being represented in this picture as a line which intersects $W^s(f^*, \mathcal{T}^*)$ transversely. The moment of intersection corresponds to the destruction of the KAM curve $\mathcal{T}_k$. The line $W^u(f^*, \mathcal{T}^*)$ itself is the "universal" family which serves as a model for such a destruction. Naturally, other models will be used for the destruction of KAM curves with other frequencies.

Notes to Chapter III

Poincaré (1892) understood well the significance of stable and unstable manifolds and their homoclinical connection for creating a complicated behaviour and instabilities of trajectories in a Hamiltonian system.

Arnol'd (1964) described the mechanism of diffusion in gaps between KAM tori and presented an example of a Hamiltonian system with 3 degrees of freedom which displays the diffusion. Nekhoroshev (1977, 1979) gave in the analytical case an exponential estimate for the velocities of Arnol'd diffusion. There are many papers where the Nekhoroshev estimate is improved. The most recent ones are Giorgilli and Galgani (1985), Lochak (1991), Pöschel (1991), Delshams and Gutiérres (1991). (See also physical discussion in Vecheslavov and Chirikov (1989).) Delshams (1984) has proven that the existence of transition chains is generic in a neighbourhood of an elliptic fixed point of a Hamiltonian flow. Important step to proving the existence of Arnol'd diffusion was made in recent paper of Gallavotti and Chierchia (1991, unpublished). For general genericity theorems in Hamiltonian dynamical systems see Robinson (1970), Newhouse (1983). An example of a 4-dimensional symplectic diffeomorphism with unstable elliptic fixed point, the instability being due to Arnol'd diffusion, is given in R. Douady and Le Calvez (1983). R. Douady (1988a) proved that in dimension $2n$, $n \geqq 2$, a general elliptic fixed point of a C^∞ symplectic diffeomorphism may be stable or unstable regardless of the Taylor coefficients of the mapping at that point.

The existence of infinitely many transversal homoclinical points near an elliptic fixed point of a generic area-preserving diffeomorphism is established in Zehnder (1973).

A nice proof of Birkhoff's theorem about invariant curves (Theorem 17.6 here) is contained in Herman (1983) with exact references to the original Birkhoff papers. There is no rigorous theory of a stochastic layer. The reader can find physical comments in Chirikov (1979). The paper Greene, MacKay, Stark (1986) contains very interesting numerical data on boundary curves.

The formula (17.8) for the angle of the splitting of the separatrices of the standard map has been derived in Lazutkin (1984), the value of the coefficient $|\Theta_1|$ in that formula was calculated in Lazutkin, Schachmannski, Tabanov (1989). The estimate (17.9) for the width of the instability zone is obtained in Lazutkin (1990). The analogous problem for the perturbed time dependent pendulum equation was considered in Holmes, Marsden, Scheurle (1988).

Neishtadt (1984) proved in a very elegant way an exponential upper bound for the separatrices splittings in the analytical case.

There exists an interesting phenomenon of a "stochastic web" which is connected with separatrices splitting: Zaslavski, Sagdeev, Usikov, Chernikov (1988).

Cantori were discovered by Percival (1980). The existence of cantori is proved by Aubry (1983) (see also Aubry and Le Daeron (1983)) and Mather (1982). Further references on cantori are R. Douady (1988b). Le Calvez (1988), MacKay, Meiss, Percival (1984), MacKay (1987), Goroff (1985), Katok (1982, 1982a), Sinai and Khanin (1987), Veerman and Tangerman (1990).

The problem of the last curve was considered in Greene (1979), MacKay and Percival (1985).

Mather (1988) proved that an arbitrarily small C^∞ perturbation can destroy a given invariant circle with Liouville winding number (frequency).

For the renormalisation group approach the main references are MacKay (1983), Escande (1985), Khanin and Sinai (1986), Sinai and Khanin (1987).

Chapter IV. Proof of the Main Theorem

This chapter contains the detailed proof of Theorem 11.6. First, we prove the continuous-time version of the Theorem (§§18–27) and then, in §28, we deduce the discrete-time theorem. In §18 we show that one may set $\gamma = 1$, and pass to a suitable formulation in angle-frequency variables. Some technical devices are exposed in §19. The description of an iterative process which yields the desired objects is placed in §20. Other sections (§§21–26) contain routines checking the inequalities necessary for the iterative process to occur correctly and their simple consequences. In §27 we summarize the results and establish that the objects obtained satisfy the conditions of the conclusion of Theorem 11.6.

Further we use the notations stated in §11.

§18. Two Reductions

I. Removing γ

It is sufficient to prove Theorem 11.6 for the case $\gamma = 1$. To be convinced of that let us make the following change which corresponds to blowing up in the variable direction I with coefficient γ^{-1}:

$$\hat{K}_0 = \gamma^{-2} K_0 \circ s_\gamma, \qquad \hat{K} = \gamma^{-2} K \circ s_\gamma, \qquad \hat{D} = s_\gamma^{-1}(S), \qquad \hat{D}_i = s_\gamma^{-1}(D_{i\gamma}),$$
$$\hat{H}_1 = \gamma^{-2} H_1 \circ \tilde{s}_\gamma, \qquad \hat{\Sigma} = \tilde{s}_\gamma^{-1} \circ \Sigma \circ \tilde{s}_\gamma,$$

where $s_\gamma : \hat{D} \longrightarrow D$, and $\tilde{s}_\gamma : \mathbb{R}^n \times \hat{D} \longrightarrow \mathbb{R}^n \times D$ are the *stretching maps.* $I \longmapsto \gamma I$ and $(\phi, I) \longmapsto (\phi, \gamma I)$, respectively.

It is easy to verify that all the inequalities and inclusions of the theorem are equivalent to those for these new objects with the same constants R and c, and with γ replaced by 1. The map $\hat{\Sigma}$ appears to be symplectic also. Of course, the set $\mathscr{E}$ must be changed in an appropriate manner. We recall that we consider here only the case of continuous time.

II. Angle–Frequency Variables

The angle–action variables (ϕ, I) are not quite suitable here for manipulating with. It is more convenient to use the *angle–frequency* ones $(x, y) = (x^1, x^2, \ldots, x^n, y^1, y^2, \ldots, y^n)$ instead, the coordinate x being the same as ϕ and y being obtained from I by applying the frequency map. Let $\omega_0 = \operatorname{grad} K_0 : D \longrightarrow \mathbb{R}^n$ be the frequency map of the unperturbed system, and $B = \omega_0(D)$ be its image. Recall that the twist condition is no more than the requirement for ω_0 to be a diffeomorphism onto B. The inverse map $\omega_0^{-1} : B \longrightarrow D$ can also be expressed in gradient form $\omega_0^{-1}(y) = \operatorname{grad} F^0(y)$, where $F^0 : B \longrightarrow \mathbb{R}$ *is the Legendre transform* of K_0:

$$F^0(y) = \langle y, I \rangle - K_0(I), \qquad I = \omega_0^{-1}(y). \tag{18.1}$$

Verification is left to the reader.

Define a diffeomorphism $\Psi_0 : \mathbb{R}^n \times B \longrightarrow \mathbb{R}^n \times D$ by setting $\Psi_0(x, y) = (x, \omega_0^{-1}(y))$. The following formulae for objects carried by Ψ_0 are immediate consequences of the above definitions. The standard symplectic structure on $\mathbb{R}^n \times D$ is transformed to

$$\Psi_0^* \Omega_{st} = \Omega_{F^0} \overset{\text{def}}{=} \sum_{i=1}^{n} dx^i \wedge d\left(\frac{\partial F^0}{\partial y^i}\right), \tag{18.2}$$

the generating field of unperturbed system becomes

$$(T\Psi_0)^{-1} \circ \left(\sum_{i=1}^{n} \omega_0^i \frac{\partial}{\partial \phi^i} \right) \circ \Phi_0 = \partial \overset{\text{def}}{=} \sum_{i=1}^{n} y^i \frac{\partial}{\partial x^i}, \tag{18.3}$$

the Hamiltonian K_0 is transformed to

$$(\omega_0^{-1})^* K_0 = \mathfrak{L} F^0 \tag{18.4}$$

where the symbol $\mathfrak{L}$ denotes the operation

$$(\mathfrak{L}F)(y) \overset{\text{def}}{=} \sum_{i=1}^{n} y^i \frac{\partial F(y)}{\partial y^i} - F(y), \tag{18.5}$$

and, finally, the image of the perturbed Hamiltonian is

$$\Psi_0^* H = G + (\mathfrak{L} F^0) \circ \tilde{p}_2, \tag{18.6}$$

where $G = \Psi_0^* H_1$, and $\tilde{p}_2$ represents the second projection $p_2 : \mathbb{R}^n \times B \longrightarrow B$.

The Hamiltonian K of the perturbed model system, which we seek, also defines, like K_0, the transformation $\Psi : \mathbb{R}^n \times B \longrightarrow \mathbb{R}^n \times D$ by the same formulae as those defining Ψ_0 with the index 0 omitted. It carries Ω_{st} into the new structure

$$\Psi^* \Omega_{st} = \Omega_F \overset{\text{def}}{=} \sum_{i=1}^{n} dx^i \wedge d\left(\frac{\partial F}{\partial y^i}\right), \tag{18.7}$$

where F is the Legendre transform of K (omit 0 in (18.1)). The unknown conjugating diffeomorphism Σ must be changed by a diffeomorphism Φ of $\mathbb{R}^n \times B$ such that the diagram

$$\begin{array}{ccc} \mathbb{R}^n \times D & \xleftarrow{\;\Psi_0\;} & \mathbb{R}^n \times B \\ \uparrow{\scriptstyle \Sigma} & & \uparrow{\scriptstyle \Phi} \\ \mathbb{R}^n \times D & \xleftarrow{\;\Psi\;} & \mathbb{R}^n \times B \end{array} \tag{18.8}$$

commutes. Since Σ is supposed to be symplectic, our requirement is for Φ to preserve the symplectic structures carried by Ψ_0 and Ψ, i.e.

(18.9) $$\Phi^*\Omega_{F^0} = \Omega_F.$$

Now we reformulate Theorem 11.6 for the case of continuous time as follows.

Theorem 18.10. *Given an integer $n \geqq 2$, real $\sigma > n-1$ and real $R > 1$, there exist positive numbers $\delta < n^{-1}$ and c such that the following is true.*

Let $B \subset \mathbb{R}^n$ be a non-empty open bounded convex set with $\operatorname{diam} B < R$, *and let $\gamma \in]0, 1[$. Consider a smooth function $F^0 : B \longrightarrow \mathbb{R}$ such that* $\operatorname{grad} F^0$ *maps B diffeomorphically onto a bounded set $D \subset \mathbb{R}^n$, and a smooth function $G : \mathbb{R}^n \times B \longrightarrow \mathbb{R}$, which is periodic in its x-variable: $G(x + \mathrm{ort}_i, y) = G(x, y)\ \forall\ 1 \leqq i \leqq n$, $(x, y) \in \mathbb{R}^n \times B$. If the estimates*

(18.10a) $$\left\| \frac{\partial^2 F^0}{\partial y^i\, \partial y^k} \right\|_{r_0} \leqq R, \qquad \left| \det \left\{ \frac{\partial^2 F^0(y)}{\partial y^i\, \partial y^k} \right\} \right| \geqq R^{-1},$$

(18.10b) $$\| G \|_{r_0} \leqq \delta,$$

r_0 being given by (11.1), *and the inclusion*

(18.10c) $$\operatorname{supp} G \subset \mathbb{R}^n \times [B]_2$$

are satisfied, then there exist: (1) *a smooth function $F : B \longrightarrow \mathbb{R}$ such that* $\operatorname{grad} F$ *maps B diffeomorphically onto the same domain D as* $\operatorname{grad} F^0$, *and* (2) *a diffeomorphism $\Phi : \mathbb{R}^n \times B \longrightarrow \mathbb{R}^n \times B$, such that*

(i) *Φ commutes with the action of the group Γ generated by the shifts $(x, y) \longmapsto (x + \mathrm{ort}_i, y)$, $1 \leqq i \leqq n$, which is homotopic to the identity map* id *of $\mathbb{R}^n \times B$, and preserves the symplectic structures i.e.* (18.9) *holds;*

(ii) *the jet of infinite order of the function*

(18.10d) $$(G + (\mathfrak{L}F^0) \circ \tilde{p}_2) \circ \Phi - (\mathfrak{L}F) \circ \tilde{p}_2$$

vanishes at points of $\mathbb{R}^n \times (\mathscr{E}_{\sigma,1} \cap B)$;

(iii) *the following estimates*

(18.10e) $$\| F - F^0 \|_2 \leqq c \| G \|_{r_0}^{7/15},$$

(18.10f) $$\| \Phi - \mathrm{id} \|_1 \leqq c \| G \|_{r_0}^{4/15},$$

and inclusions

(18.10g) $$\operatorname{supp}(F - F^0) \subset [B]_1,$$

(18.10h) $$\operatorname{supp}(\Phi - \mathrm{id}) \subset \mathbb{R}^n \times [B]_1,$$

(18.10j) $$\Phi(\mathbb{R}^n \times (\mathscr{E}_{\sigma,1} \cap B)) \subset \mathbb{R}^n \times [B]_3$$

hold.

The Deduction: Theorem 18.10 ⇒ Theorem 11.6

Let $g_0^t: A_D \longrightarrow A_D$ and $H_1: \mathbb{R}^n \times D \longrightarrow \mathbb{R}$ satisfy the conditions of 11.6, the value of δ being fixed later. In particular, g_0^t is defined by the Hamiltonian $K_\sigma: D \longrightarrow \mathbb{R}$ such that $\omega_0 = \operatorname{grad} K_0$ maps D diffeomorphically onto a bounded set B. Consider the functions: $F^0: B \longrightarrow \mathbb{R}$ defined by (18.1), and $G = \Psi^* H_1$, where $\Psi_0(x, y) = (x, \omega_0^{-1}(y))$. The estimates (11.2) yield

$$|D_y^\rho \omega_0^{-1}(y)_i|, \qquad \|(D_y^\rho \Psi_0)_j\| \leqq \text{const}\,\gamma^{-|\rho|}, \tag{18.11}$$

$$1 \leqq |\rho| \leqq r_0 + 1, \quad 1 \leqq i \leqq n, \quad 1 \leqq j \leqq 2n,$$

where const depends only on n, r_0 and R, apply the differentiation rule to (18.1) and to $G = H_1 \circ \Psi_0$ one obtains, taking into account (18.11), (11.2) and (11.46)

$$\left\| \frac{\partial^2 F^0}{\partial y^i \partial y^k} \right\| \leqq R_1,$$

$$\left| \det \left\{ \frac{\partial^2 F^0(y)}{\partial y^i \partial y^k} \right\} \right| = \left| \det \left\{ \frac{\partial^2 K_0(I)}{\partial I_1 \partial I_j} \bigg|_{I = \omega_0^{-1}(y)} \right\} \right|^{-1} \geqq R_1^{-1}, \tag{18.12}$$

$$\|G\|_{r_0} \leqq R_1 \delta, \tag{18.13}$$

where R_1 depends only on n, r_0 and R.

Let δ_1 be the value of the constant δ of Theorem 18.10 calculated with R_1 standing for R. Set $\delta = \delta_1 / R_1$. Then the functions F_0 and G satisfy the conditions (18.10a) and (18.10b) with R_1 standing for R. The inclusion (18.10c) is also satisfied in virtue of (11.4a). Supposing Theorem 18.10 to be true, we obtain F and Φ satisfying (i), (ii), (iii) of 18.10. Then we take the Legendre transform K of F for the model Hamiltonian to the perturbed system and $\Sigma = \Psi_0 \circ \Phi \circ \Psi^{-1}$, (where $\Psi_9 x, y) = (x, \operatorname{grad} F(y)))$, for the conjugating diffeomorphism.

The assertions (iii) of 11.6 follow immediately from (iii) of 18.10, one has only to change the constant c. The application of $(\Psi^{-1})^*$ to (18.10d) yields $H \circ \tilde{\Sigma} - K \circ p_2$, as (ii) of 18.10 implies (ii) of (11.6) as well. Finally, the assertion (i) of Theorem 11.6 is a simple consequence of (ii) of that theorem. □

§19. Machinery

Here we describe three tools which we shall use in the process of proving the theorem. Two positive parameters M and N will be involved, they are assumed to be greater than 1. The variable r will range over the set $\mathbb{N}_{r_0} = \langle r \in \mathbb{Z}_+ : r \geqq r_0(\sigma, n) \rangle$ and will be the number of derivatives involved in the consideration. For the duration of the proof we define increasing functions $c_s: \mathbb{N}_{r_0} \longrightarrow \mathbb{R}_+ \backslash [0, 1]$, $1 \leqq s \leqq 11$ which will enter into the estimates. They will only depend on n and σ.

The Switching Multiplier

Given $N > 1$, define, for $i = 1, 2, 3$, the sets

(19.1) $$\mathscr{E}_N^{(i)} = \left\{ y \in [B]_4 : |\langle (m, y) \rangle| \geqq \frac{i}{3} |m|^{-\sigma} \textit{ for all } 0 \neq m \in \mathbb{Z}^n \textit{ such that } |m| \leqq N \right\}.$$

It is easy to see that $\mathscr{E}_N^{(i)}$ monotonously decreases as N increases $\lim_{N \to \infty} \mathscr{E}_N^{(3)}$, and $\mathscr{E}_N^{(i+1)} \subset \mathscr{E}_N^{(i)}$, $i = 1, 2$. We define the *switching multiplier* $\theta_{NM} \in C^\infty(\mathbb{R}^n \times B)$ by the formula

(19.2) $$\theta_{NM} = \chi_{A,\alpha} \circ \tilde{p}_2, \qquad A = \mathscr{E}_N^{(3)}, \qquad \alpha = M^{-1}.$$

Here $\chi_{A,\alpha}$ is the switching function of Lemma 11.13.

Proposition 19.3. *The switching multiplier θ_{NM} possesses the properties:*

(i) $0 \leqq \theta_{NM} \leqq 1$,
(ii) $\operatorname{supp} \theta_{NM} \subset \mathbb{R}^n \times \mathscr{E}_N^{(2)}$ *if* $M \geqq 6N^{\sigma+1}$,
(iii) $\theta_{NM}(x, y) = 1$ *if* $y \in \mathscr{E}_N^{(3)}$,
(iv) $|D_y^\rho \theta_{NM}(x, y)| \leqq c_1(r) M^{|\rho|}$, $1 \leqq |\rho| \leqq r$,
(v) *if* $M_+ \geqq 3M$, $N_+ \geqq N$, $|\Delta| \leqq M_+^{-1}$, *then:*

$(x, y) \in \operatorname{supp} \theta_{N_+ M_+} \Rightarrow \theta_{NM}(x, y + \Delta) = 1$ *and* $D_x^{\rho_1} D_y^{\rho_2} \theta_{NM}(x, y + \Delta) = 0$ *if* $|\rho_1| + |\rho_2| \geqq 1$.

Proof. The properties (i)–(iv) follow immediately from the corresponding properties of $\chi_{A,\alpha}$ (see Lemma 11.13). Let $(x, y) \in \operatorname{supp} \theta_{N_+ M_+}$. Then the property (ii) of Lemma 11.13 yields $\operatorname{dist}(y, \mathscr{E}_N^{(3)}) \leqq \operatorname{dist}(y, \mathscr{E}_{N_+}^{(3)}) \leqq 2M_+^{-1}$, and $\operatorname{dist}(y + \Delta, \mathscr{E}_N^{(3)}) \leqq 2M_+^{-1} + |\Delta| \leqq 3M_+^{-1} \leqq M^{-1}$. We obtain the conclusion of (v) due to (iii) of Lemma 11.13. □

The Smoothing Operator

Let $C_{\text{per},a}^\infty(\mathbb{R}^n \times B) \subset C_{\text{per}}^\infty(\mathbb{R}^n \times B)$ denote the subset consisting of functions $u \in C_{\text{per}}^\infty(\mathbb{R}^n \times B)$ with $\operatorname{supp} u \subset \mathbb{R}^n \times [B]_a$. We shall construct a sequence of smoothing operators $S_{NM}^{(r)}$ which carries functions in $C_{\text{per},a}^\infty(\mathbb{R}^n \times B)$ to $C_{\text{per},a-M^{-1}}^\infty(\mathbb{R}^n \times B)$ provided, of course, $M^{-1} < a$. The operator acts on a function h by the formula

(19.4) $$(S_{NM}^{(r)} h)(x, y) = \int_{\square^n \times B} X_{N,r}(x - \xi) Y_{M,r}(y - \eta) h(\xi, \eta) \, d\xi \, d\eta.$$

Here $\square^n \subset \mathbb{R}^n$ is the unit cube, $X_{N,r}(x)$ is a smooth periodic function with periods $\{\text{ort}_i, 1 \leqq i \leqq n\}$, $Y_{M,r}(y)$ is a smooth function with $\operatorname{supp} Y_{M,r}$ contained in the ball $\langle y : |y| \leqq M^{-1} \rangle$.

Construction of $Y_{M,r}(y)$.

Set

(19.5) $$Y_{M,r}(y) = M^n \varphi_r(My),$$

where $\varphi_r(t)$ is a smooth function with satisfies the conditions

(a) $$\varphi_r(t) = 0 \textit{ if } |t| \geqq 1,$$

(b) $$\int \varphi_r(t) t^\rho \, dt = \begin{cases} 1 \text{ if } \rho = 0, \\ 0 \text{ if } 1 \leqq |\rho| \leqq r-1, \end{cases} \qquad \rho = (\rho_1, \rho_2, \ldots \rho_n) \in \mathbb{Z}_+^n.$$

To obtain such a function $\varphi_r(t)$ fix a smooth function $\psi(t)$ which satisfies

(a′) $$\psi(t) = 0 \text{ if } |t| \geqq 1/2,$$

(b′) $$\psi(t) > 0 \text{ if } |t| < 1/2,$$

and look for a $\varphi_r(t)$ in the form

$$\varphi_r(t) = \sum_{0 \leqq |\lambda| \leqq r-1} c_\lambda \psi\left(t - \frac{\lambda}{2r}\right), \tag{19.6}$$

where $\lambda = (\lambda_1, \lambda_2, \ldots, \lambda_n) \in \mathbb{Z}_+^n$, c_λ are real numbers. It follows from (a′) that (19.6) satisfies the condition (a).

Proposition 19.7. *There are coefficients* c_λ *in* (19.6) *such that the condition* (b) *is satisfied.*

Proof. Condition (b) is equivalent to the linear system

$$\sum_{0 \leqq |\lambda| \leqq r-1} c_\lambda \int \psi\left(t - \frac{\lambda}{2r}\right) t^\rho \, dt = \begin{cases} 1, \text{if } \rho = 0, \\ 0, \text{if } 1 \leqq |\rho| \leqq r-1. \end{cases} \tag{19.7a}$$

Suppose that the determinant of (19.7a) is equal to zero. Then the rows of the matrix

$$\left\{ c_{\rho\lambda} = \int \psi(t - \lambda/2r) t^\rho \, dt \colon \rho, \lambda \in \mathbb{Z}_+, |\rho|, |\lambda| \leqq r-1 \right\}$$

are linearly independent. This means that there exists a collection of real numbers $\{p_\rho \colon \rho \in \mathbb{Z}_+, |\rho| \leqq r-1\}$, not all p_λ being zero, such that

$$\sum_\rho p_\rho c_{\lambda\rho} = P(\lambda/2\rho) = 0, \quad |\lambda| \leqq r-1.$$

Here

$$P(\tau) = \int \psi(t-\tau)\left(\sum_\rho p_\rho t^\rho\right) dt = \int \psi(t)\left(\sum_\rho p_\rho (t+\tau)^\rho\right) dt.$$

It is easy to see that $P(\tau)$ is a polynomial of degree at most $r-1$. It is zero at points of the form $\tau = \lambda/2r$, $\lambda \in \mathbb{Z}_+$, $|\lambda| \leqq r-1$. This condition yields $P(\tau) \equiv 0$, which contradicts the fact that $\{p_\rho\}$ contains elements not equal to zero. □

For completeness we prove the assertion we have just used.

Lemma 19.7a. *Let* $P(x)$, $x = (x^1, x^2, \ldots, x^n)$, *be a polynomial of degree at most* d. *If* $P(\rho) = 0$ *for all* $\rho \in \mathbb{Z}_+$, $|\rho| \leqq d$, *then* P *is identically zero.*

Proof. We use induction on d. The statement is obvious if $d = 0$. Let it valid for $d < p$ and let P be of degree at most p. Consider the polynomial of one variable

$Q_i(x) = P(x \cdot \mathrm{ort}_i)$. It vanishes at $p+1$ integer points $0, 1, 2, \ldots, p$. Hence Q_i is identically zero, $1 \leqq i \leqq n$. It follows that P may be divided by $x^1 x^2 \cdots x^n$ i.e. $P(x) = x^1 x^2 \cdots x^n P_1(x)$, where P_1 has degree at most $p - n$. If $n > p$, then $P_1 \equiv 0$. Let $n \geqq p$. The polynomial $P_2(x) = P_1(\sum_{i=1}^n \mathrm{ort}_i + x)$ satisfies the conditions of the Lemma with $d = p - n$. □

Construction of $X_{N,r}(x)$. Set $r_1 = (r+n)(r+n+1)$ and choose an integer $\tilde{N}$ in the interval $[(2\sqrt{nr_1})^{-1}, (\sqrt{nr_1})^{-1}]$. We suppose that N is greater than $2r_1$. Let us now construct the Dirichlet kernel

$$\mathscr{D}_{\tilde{N}}(x) = \sum_{|m_i| \leqq N, 1 \leqq i \leqq n} e^{i2\pi\langle m, x\rangle} = \prod_{i=1}^{n} \frac{\sin(2\pi N' x^i)}{\sin(\pi x^i)}, \tag{19.8}$$

where $N' = \tilde{N} + 1/2$. Consider the function

$$\psi_N(x) = c_N (N')^{-nr_1 + n} (\mathscr{D}_{\tilde{N}}(x))^{r_1}. \tag{19.9}$$

The constant c_N being found from the norming condition $\int_{\square^n} \psi_N(x)\,dx = 1$.

Lemma 19.10. $c_N = (\mathrm{Mes}\, D)^{-n} + \mathcal{O}(N^{-1})$, *where the domain* $D \subset \mathbb{R}^{r_1 - 1}$ *is defined by the inequalities* $|t_1 + t_2 + \cdots + t_{r_1 - 1}| \leqq 1$, $|t| \leqq 1$, $1 \leqq i \leqq r_1 - 1$.

Proof. We have

$$\int_{\square^n} (\mathscr{D}_{\tilde{N}}(x))^{r_1} dx = \left[\int_0^1 \left(\frac{\sin(2\pi N' x)}{\sin(\pi x)} \right)^{r_1} dx \right]^n,$$

and

$$\int_0^1 \left(\frac{\sin(2\pi N' x)}{\sin(\pi x)} \right)^{r_1} dx = \sum 1,$$

where the summation runs over the set of all r_1-tuples $(n_1, n_2, \ldots, n_{r_1})$ satisfying $\sum_{i=1}^n n_i = 0$ and $|n_i| \leqq \tilde{N}$. Let us introduce new variables $t_i = n_i(\tilde{N})^{-1}$. Then the above sum becomes equal to the number of points in D with rational coordinates of the form $n_i(\tilde{N})^{-1}$. This volume equals the sum of the $(r_1 - 1)$-volume of D divided by $\tilde{N}^{-r_1+1}$, the volume of the elementary $(r_1 - 1)$-cell, and of the error, the latter being estimated from the above by the $(r_1 - 2)$-volume of the elementary $(r_1 - 2)$-cell. □

In the Proposition which follows $\alpha = 1 - 1/(r+n)$. We also assume $N^{-\alpha} < 1/2$.

Proposition 19.11. (a) *If* $1/2 \geqq |x| > N^{-\alpha}$ *then*

$$\psi_N(x) = \mathcal{O}(N^{-r-1}), \tag{19.11a}$$

(b) *If* $|x| \leqq N^{-\alpha}$ *then*

$$\psi_N(x) = c_N (N')^n [x(N'x)]^{r_1} + \mathcal{O}(N^{n-1-\alpha}), \tag{19.11b}$$

where

$$\chi(t)=\prod_{i=1}^{n}\left[\frac{\sin 2\pi t^i}{\pi t^i}\right],\quad t=(t^1,t^2,\dots,t^n)\in\mathbb{R}^n.$$

Proof. (a) In this case there is at least one $i\in\{1,2,\dots,n\}$ such that $1/2\geqq x^i\geqq(\sqrt{n})^{-1}N^{-\alpha}$. For this value of i

$$\left|\frac{\sin(2\pi N'x^i)}{\sin(\pi x^i)}\right|\leqq|\sin\pi x^i|^{-1}\leqq 2^{-1}|x^i|^{-1}\leqq 2^{-1}\sqrt{n}N^\alpha$$

while the estimate

$$\left|\frac{\sin(2\pi N'x^i)}{\sin(\pi x^i)}\right|\leqq 2N'$$

holds for all other values x^j. Combining these we obtain:

$$|\psi_N(x)|\leqq c_N(N')^{n(r_1-1)}(2N')^{r_1(n-1)}(2^{-1}\sqrt{n}N^\alpha)^{r_1}\leqq\text{const}\,N^{n-(1-\alpha)r_1}=\text{const}\,N^{-r-1}.$$

(b) One immediately obtains that

$\left|\frac{1}{\sin\pi x^i}-\frac{1}{\pi x^i}\right|\leqq\frac{\pi^2}{12}|x^i|$ if $|x^i|\leqq 1/2$. Hence

$$\mathscr{D}_N(x)=(N')^n\chi(N'x)+\mathcal{O}(N^{n-1-\alpha})$$

for $|x^i|\leqq N^{-\alpha}$. Substituting this into (19.9) yields (19.11b). □

Let us continue the construction. Fix a set $\{b_k(x)\colon k\in\mathbb{Z}^n_+,\ 0\leqq|k|\leqq r-1\}$ of smooth, periodic with periods ort_i, $1\leqq i\leqq n$, functions, with $b_0(x)=1$ and $b_k(x)$, $|k|\geqq 1$, obeying the condition $b_k(x)=x^k$ if $|x|\leqq 1/4$. Now define the smothing kernel $X_{N,r}(x)$ by the formula

$$X_{N,r}(x)=\sum_{|\lambda|\leqq r-1,\ \lambda\in\mathbb{Z}^n_+}a_\lambda\psi_N\left(x-\frac{\lambda}{2rN'}\right),\tag{19.12}$$

where the coefficients a_λ are chosen so that

$$\int X_{N,r}(x)b_k(x)\,dx=\begin{cases}1 & \text{if } k=0,\\ 0 & \text{if } 1\leqq|k|\leqq r-1\end{cases}\tag{19.13}$$

holds.

Proposition 19.14. *There exist $c_2(r)$ and $c_3(r)$, $r\in\mathbb{N}_{r_0}$, such that, if $N>c_2(r)$, then there is a unique set of constants $\{a_\lambda\colon|\lambda|\leqq r-1\}$ which are the solutions to* (19.12), (19.13), *and satisfies the inequalities $|a_\lambda|\leqq c_3(r)$, $|\lambda|\leqq r-1$.*

Proof. The equations (19.12), (19.13) for determining a_λ may be written in the form

$$\sum_{|\lambda|\leqq r-1}c_{k\lambda}a_\lambda=\delta_{k0},\quad 0\leqq|k|\leqq r-1,\ k\in\mathbb{Z}^n_+,$$

where

$$c_{k\lambda}=(N')^{|k|}\int_{\square^n}\psi_N(x)b_k\left(x+\frac{\lambda}{2rN'}\right).$$

Using the estimates of Proposition 19.11, one obtains

$$c_{k\lambda}=\int_{|x|\leqq N^{-\varkappa}}c_N(N')^{|k|+n}[\chi(N'x)]^{r_1}\left(x+\frac{\lambda}{2rN'}\right)^k dx+\mathcal{O}(N^{-1}). \tag{19.14a}$$

Since

$$\int_{|t|\geqq N^{1-\varkappa}}[\chi(t)]^{r_1}\left(t+\frac{\lambda}{2r}\right)^k dt=\mathcal{O}(N^{-r-n}),$$

the integration in (19.14a) can be extended onto $\mathbb{R}^n$, the error estimate remaining the same. Hence

$$c_{k\lambda}=c^0_{k\lambda}+\mathcal{O}(N^{-1}), \tag{19.14b}$$

where

$$c^0_{k\lambda}=(\text{Mes}\,D)^{-n}\int_{\mathbb{R}^n}[\chi(t)]^{r_1}\left(t+\frac{\lambda}{2r}\right)^k \text{d}t.$$

The same considerations as were used when proving Proposition 19.7, yield $\det\{c^0_{k\lambda}: 0\leqq|k|, |\lambda|\leqq r-1\}\neq 0$. We have, in view of (19.14b), $\det\{c_{k\lambda}\}=\det\{c^0_{k\lambda}\}+\mathcal{O}(N^{-1})$, and $\det\{c_{k\lambda}\}$ is detached from zero provided N exceeds some positive constant depending on r. □

The next Proposition contains some properties of the smoothing operator $S^{(r)}_{NM}$ defined by (19.4) with $X_{N,r}$ and $Y_{M,r}(y)$ given by (19.12) and (19.5).

Proposition 19.15. (a) *Fourier coefficients of* $S^{(r)}_{NM}h$,

$$(S^{(r)}_{NM}h)_m(y)=\int_{\square^n}S^{(r)}_{NM}h(x,y)e^{-i2\pi\langle m,x\rangle}\,dx, \text{ are zero if } |m|>N.$$

(b) *If* $\operatorname{supp} h\subset\mathbb{R}^n\times[B]_a$ *then* $\operatorname{supp} S^{(r)}_{NM}h\subset\mathbb{R}^n\times[B]_{a-M^{-1}}$.
(c) *There is a function* $c_4(r)$, $r\in\mathbb{N}_{r_0}$, *such that*

$$|D^{\rho_1}_x D^{\rho_2}_y S^{(r)}_{NM}h|\leqq c_4(r)N^{|\rho_1|+n}M^{|\rho_2|}\max_{x,y}|h|,\quad 0\leqq\rho_1|+|\rho_2|\leqq r+4.$$

(d) *There is also a function* $c_5(r)$, $r\in\mathbb{N}_{r_0}$, *such that*

$$|h-S^{(r)}_{NM}h|\leqq c_5(r)\max_{\substack{|\rho_1|+|\rho_2|\leqq r':\, r'\leqq r\\ r'\leqq\min(r,x,y)}}N^{-|\rho_1|}M^{-|\rho_2|}|D^{\rho_1}_x D^{\rho_2}_y h(x,y)|.$$

Proof. The properties (a) and (b) immediately follow from the construction. To obtain (c) let us deduce the estimates for derivatives of the kernels from above.

Differentiating (19.9), (19.9), one obtains

$$D_x^{\rho_1}\psi_N(x) = c_N(N')^{-nr_1+n}\sum_\alpha C_\alpha \prod_{0\leqq s\leqq\rho_1} [D_x^s\mathcal{D}_{\tilde N}(x)]^{\alpha(s)}, \tag{19.15e}$$

where the summation runs over all functions $\alpha:\{s\in\mathbb{Z}_+^n: 0\leqq s\leqq\rho_1\}\longrightarrow\mathbb{Z}_+$ satisfying the equalities $\sum_{0\leqq s\leqq\rho_1} s\alpha(s)=\rho_1$ and $\sum_{0\leqq s\leqq\rho_1}\alpha(s)=r_1$, C_α being some constants. Taking into account the obvious estimates (see 19.8) $D_x^s\mathcal{D}_{\tilde N}(x)=\mathcal{O}(N^{|s|+n})$, we obtain from (19.15e) and (19.12) that

$$D_x^{\rho_1}X_{N,r}(x)=\mathcal{O}(N^{|\rho_1|+n}). \tag{19.15f}$$

The estimate

$$D_y^{\rho_2}Y_{M,r}(y)=\mathcal{O}(M^{|\rho_2|+n}) \tag{19.15g}$$

is obvious. Substituting (19.15f,g) into the expression for the derivative of (19.4), and taking into account $Y_{M,r}(y)=0$ if $|y|\geqq M^{-1}$, one obtains (19.15c)

To obtain (d), let us write

$$(S_{NM}^{(r)}h)(x,y)=\int X_{N,r}(-\xi)Y_{M,r}(-\eta)h(x+\xi,y+\eta)\,d\xi\,d\eta$$

and let us expand $h(x+\xi,y+\eta)$ into a Taylor series

$$h(x+\xi,y+\eta)=\sum_{|\rho_1|+|\rho_2|<r}(\rho_1!)^{-1}(\rho_2!)^{-1}D_x^{\rho_1}D_y^{\rho_2}h(x,y)b_{\rho_1}(\xi)\eta^{\rho_2}+\varphi_{r+1},$$

where φ_{r+1} admits the estimate

$$|\varphi_{r+1}|\leqq \text{const}\sum_{|\rho_1|+|\rho_2|=r}|\xi|^{\rho_1}|\eta|^{\rho_2}\max_{x,y}|D_x^{\rho_1}D_y^{\rho_2}h|.$$

Using (19.13) and the analogous property of $Y_{M,r}$ (the condition (b) in its construction), we obtain

$$|S_{NM}^{(r)}h-h|\leqq\max_{x,y}\int|X_{N,r}(-\xi)Y_{M,r}(-\eta)||\varphi_{r+1}(x,y\xi,\eta|\,d\xi\,d\eta \tag{19.15h}$$

$$\leqq\text{const}\max_{x,y,|\rho_1|+|\rho_2|=r}|D_x^{\rho_1}D_y^{\rho_2}h|\int_{|\xi^i|\leqq 1/2}|\xi^{\rho_1}X_{N,r}(-\xi)|\,d\xi\int|\eta^{\rho_2}Y_{M,r}(-\eta)|\,d\eta.$$

It remains for us to estimate the integrals in the right-hand side of (19.15h). The second one, in view of definition (19.5), admits the estimate $\mathcal{O}(M^{-|\rho_2|})$. As for the first onc, it is sufficient to estimate the expressions

$$\int_{\square^n}\left|\psi_N(\xi)\left(\xi+\frac{\lambda}{2rN'}\right)^{\rho_1}\right|d\xi\leqq\text{const}\sum_{0\leqq k\leqq\rho_1}\int_{\square^n}|\psi_N(\xi)\xi^k|\,d\xi\,N^{-|\rho_1-k|}. \tag{19.15i}$$

Using the estimates of Proposition 19.11, one obtains

$$\int_{\square^n}|\psi_N(\xi)\xi^k|\,d\xi=\int_{|\xi|\leqq N^{-\alpha}}c_N(N')^n[\chi(N'x)]^{r_1}|\xi^k|\,d\xi+\mathcal{O}(N^{n-1-\alpha})\int_{|\xi|\leqq N^{-\alpha}}|\xi^k|\,d\xi$$
$$+\mathcal{O}(N^{-r-1})=\mathcal{O}(N^{-|k|}).$$

Substituting this into (19.15i) we obtain that the first integral on the right-hand side of (19.15h) is less than const $N^{-|\rho_1|}$. This completes the proof of (d). □

The Homological Equation

This is the equation

$$\partial\Delta = h, \tag{19.16}$$

where $\partial = \sum_{i=1}^{n} y^i \partial/\partial x^i$. Our conditions on the right-hand side of (19.16) are the following:

(i) $\operatorname{supp} h \subset \mathbb{R}^n \times [B]_a$, a being a positive number.

The next condition deals with the Fourier coefficients

$$h_m(y) = \int_{\square^n} h(x, y) e^{-i2\pi\langle m, y\rangle} \, dx.$$

(ii) $h_m = 0$ if either $m = 0$ or $|m| > N$, N being a positive number.

(iii) $\operatorname{supp} h_m \subset \mathscr{E}_N^{(2)}$, $|m| \leqq N$.

Under the said conditions, one can write out a solution to (19.16) as follows:

$$\Delta(x, y) = \sum_{|m| \leqq N} \frac{h_m(y)}{i2\pi\langle m, y\rangle} e^{i2\pi\langle m, x\rangle}. \tag{19.17}$$

This solution is unique subject to the condition $\int_{\square^n} \Delta(x, y) \, dx = 0$.

Proposition 19.18. *Let h satisfy* (i), (ii), (iii), *then* (19.17) *admits the estimate*:

$$|D_x^{\rho_1} D_y^{\rho_2} \Delta| \leqq c_6(r) N^{n+|\rho_1|} \max_{0 \leqq k \leqq \rho_2} (|D_y^k h| N^{|\rho_2 - k|(\sigma+1)+\sigma}). \tag{19.18a}$$

Proof. The formula $D_y^{\rho}\langle m, y\rangle^{-1} = |\rho|!(-1)^{|\rho|} m^{\rho} \langle m, y\rangle^{-|\rho|-1}$ implies the estimate $|D_y^{\rho}\langle m, y\rangle^{-1}| \leqq \text{const}\, N^{|\rho|(\sigma+1)+\sigma}$. The estimate of the derivative of a typical term in (19.17) is

$$\left| D_x^{\rho_1} D_y^{\rho_2} \frac{h_m(y)}{i2\pi\langle m, y\rangle} e^{i2\pi\langle m, x\rangle} \right| \leqq \text{const} \max_{0 \leqq k \leqq \rho_2} N^{|\rho_1|} |D_y^k h_m(y)| N^{|\rho_2 - k|(\sigma+1)+\sigma}. \tag{19.18b}$$

The number of such terms in (19.17) is of order N^n. This, together with (19.18b), implies (19.18a). □

§20. Description of the Iterative Process

The function F and the diffeomorphism Φ, which we are trying to find, will be obtained as the limits $F = \lim_{i\to\infty} F^i$, $\Phi = \lim_{i\to\infty} \Phi_i$ where $\Phi_0 = \text{id}$, $\Phi_i = \Phi_{i-1} \circ \Psi_i$. The functions F^i and the diffeomorphism Ψ_i will be defined at the ith step of the iterative process.

The starting data for the iterative process are the functions $F^0 \in C^\infty(B)$ and $G^0 = G \in C^\infty_{\text{per},2}(\mathbb{R}^n \times B)$, satisfying the estimates (18.10a,b) with $\gamma = 1$. The constant δ in (18.9b) will be defined later. We suppose $\operatorname{grad} F^0$ maps the domain B diffeomorphically onto a bounded open subset $D \subset \mathbb{R}^n$.

At the ith step of the iterative process we define a triple (F^i, G^i, Ψ_i) where $F^i \in C^\infty(B)$, $\operatorname{grad} F^i : B \longrightarrow D$ is a diffeomorphism, $G^i = G \in C^\infty_{\text{per},a_i}(\mathbb{R}^n \times B)$, $\Psi_i \in \operatorname{Diff}_{\text{per},a_i}(\mathbb{R}^n = B)$, the following estimates being fulfilled:

(20.1i) $$|D^\rho_y F^i| \leqq (2 - 2^{-i}) R \cdot M_i^{|\rho| - 2}, \quad 2 \leqq |\rho| \leqq r_i + 3,$$

(20.2i) $$\left| \det \left\{ \frac{\partial^2 F^i(\eta)}{\partial \eta^k \partial \eta^l} \right\} \right|^{-1} \leqq (2 - 2^{-i}) R,$$

(20.3i) $$|D^{\rho_1}_x D^{\rho_2}_y (G^i + \mathfrak{L} F^i)| \leqq N_i^{|\rho_1|} M_i^{|\rho_2|}, \quad 1 \leqq |\rho_1| + |\rho_2| \leqq r_i,$$

(20.4i) $$|G^i| < \delta_i \quad \text{at points of } \operatorname{supp} \theta_{N_{i+1}, N_i}.$$

The constants entering these estimates satisfy the relations:

(20.5) $$M_i = N_i^\nu, \qquad \nu = \tfrac{4}{3}\sigma' + \varkappa',$$

(20.6) $$\delta_i = N_i^{-\mu_i},$$

(20.7) $$N_{i+1} = N_i^{4/3}.$$

Here $\sigma' = \sigma + 1$, the numbers $\varkappa' \in]0, 1/20[$ and $\mu_0 > 0$ are defined by the equalities where r_0 is a positive integer:

(20.8) $$r_0 = \tilde{r}_0 + 20\varkappa', \qquad \tilde{r}_0 = 4\tilde{\mu}_0,$$
$$\tilde{\mu}_0 = \tfrac{20}{3}\sigma' + \tfrac{16}{3} n, \qquad \mu_0 = \tilde{\mu}_0 + 4\varkappa'.$$

The positive integers r_i form a slowly increasing sequence starting with r_0.

We shall distinguish between different kinds of steps in our iterative process: the ith step is *regular* if $r_i = r_{i-1}$, it is *singular* if $r_i = r_{i-1} + 1$. No other kinds of steps will occur. The number μ_i will jump by 1/4 at the step which follows a singular one and remains constant for all other cases. The majority of steps will be regular ones. Sometimes we sporadically switch on a singular step in order to force r_i to tend to infinity. The rules which govern the onset of singular steps will be stated further.

The constants a_i are defined by the relations

(20.9) $$a_0 = 2 \quad \text{and} \quad a_{i+1} = a_i - M_{i+1}^{-1}.$$

The requirement $\Phi \in \operatorname{Diff}_{\text{per},1}(\mathbb{R}^n \times B)$ impels us to impose the restriction

(20.10) $$\sum_{i=0}^{\infty} M_i^{-1} < 1.$$

The only free constant that remains is that $N_0 > 1$ which we shall choose later sufficiently large for the inequalities (Ck_0), $1 \leqq k \leqq 9$, and CIV to be fulfilled (see §§21,23,24).

Let (F^j, G^j, Ψ_j), $0 \leqq j \leqq i$, be already defined, all of the above estimates and conditions being true. To construct $(F^{i+1}, G^{i+1}, \Psi_{i+1})$ we first define the function

(20.11) $$h_{i+1} = S^{(r_i)}_{N_{i+1}M_{i+1}} \theta_{N_{i+1}M_i} G^i,$$

and set

(20.12) $$F^{i+1} = F^i - [\![h_{i+1}]\!].$$

Here $[\![\cdot]\!]$ denotes the *averaging operation* over the x-variables:

(29.13) $$[\![h]\!](y) = \int_{\Box^n} h(x, y)\, dx$$

Note that $[\![h]\!](y)$ coincides with the zero Fourier coefficient $h_0(y)$ of h.

It follows from 19.15c and (20.9) that

(20.14) $$\operatorname{supp} h_{i+1} \subset \mathbb{R}^n \times [B]_{a_{i+1}}.$$

So we have

(20.15) $$\operatorname{supp}(F^{i+1} - F^i) \subset [B]_{a_{i+1}}.$$

To determine the last two terms of the triple $(F^{i+1}, G^{i+2}, \Psi_{i+1})$ consider the function Δ^{i+1} which is the solution to the equation

(20.16) $$\partial \Delta^{i+1} = h_{i+1} - [\![h_{i+1}]\!]$$

given by Proposition 19.18. Then the transformation $\Psi_{i+1} : (\xi, \eta) \longmapsto (x, y)$ is defined by the equality

(20.17) $$\langle \operatorname{grad}_\eta F^{i+1}(\eta) - \operatorname{grad}_y F^i(y), dx \rangle + \langle x - \xi, d \operatorname{grad}_\eta F^{i+1}(\eta) \rangle = d\Delta^{i+1}(x, \eta).$$

Equation (20.17) ensures that Ψ_{i+1} preserves symplectic structures:

(20.18) $$\Psi^*_{i+1} \Omega_{F^i} = \Omega_{F^{i+1}},$$

where Ω_F is defined as the right-hand side of (18.7). To obtain (20.18) it is sufficient to apply the d-operation to (20.17).

Once Ψ_{i+1} and F^{i+1} are found, G^{i+1} will be defined by the formula

(20.19) $$G^{i+1} + \mathfrak{Q} F^{i+1} = (G^i + \mathfrak{Q} F^i) \circ \Psi_{i+1}.$$

§21. Reproduction of (20.1i) and (20.2i). Convergence of F^i

Proposition 19.3(i), 19.15(c), and estimate (20.4i) involve the estimate for h_{i+1} given by (20.11):

(21.1) $$|D_x^{\rho_1} D_y^{\rho_2} h_{i+1}| \leqq c_4(r_i) N_{i+1}^{|\rho_1|+n} M_{i+1}^{|\rho_2|} \delta_i, \quad 1 \leqq |\rho_1| + |\rho_2| \leqq r_i + 4.$$

This leads (see (20.12)) to

(21.2) $$|D_y^{\rho}(F^{i+1} - F^i)| \leqq c_4(r_i) N_{i+1}^n M_{i+1}^{|\rho|} \delta_i, \quad |\rho| \leqq r_i + 4.$$

Let (20.1i) and (20.2i) hold. We are interested in whether they involve (20.1i + 1) and (20.2i + 1), provided $r_{i+1} \leqq r_i + 1$. It is convenient to consider the second derivatives ($|\rho| = 2$) separately. Using (21.2), we have

$$|D_y^\rho F^i| \leqq (2 - 2^{-i-1})R - 2^{-i-1}R + c_4(r_i)N_i^{(4/3)n+(8/3)\nu-\mu_i}, \quad |\rho| = 2. \tag{21.3}$$

Since $\frac{4}{3}n + \frac{8}{3}\nu - \mu_i \leqq \frac{4}{3}n + \frac{8}{3}\nu - \mu_0 \leqq -1$, the sum of the last two terms in (21.3) becomes negative if

$$c_4(r_i)N_i^{-1}2^{i+1}R^{-1} < 1. \tag{C1i}$$

This implies $|F^{i+1}|_2 \leqq 2R$, and (20.2i), (21.2) yield

$$\begin{aligned}\left|\det\left\{\frac{\partial^2 F^{i+1}}{\partial y^k \partial y^l}\right\}\right| &\geqq \left|\det\left\{\frac{\partial^2 F^i}{\partial y^k \partial y^l}\right\}\right| - d_1 c_4(r_i) N_{i+1}^n M_{i+1}^2 \delta_i \\ &\geqq (2 - 2^{-i-1})^{-1}R^{-1} + 4^{-1}2^{-i-1}R^{-1} - d_1 c_4(r_i)N_i^{-1}.\end{aligned} \tag{21.4}$$

The sum of the last two terms in (21.4) will be positive if

$$4d_1 c_4(r_i)N_i^{-1}2^{i+1}R < 1, \tag{C2i}$$

d_1 being a constant depending only on n, σ and R.

Let us turn our attention to higher derivatives. Denote $\gamma_i^{(\rho)} = \max_y |D_y^\rho F^i| \cdot M_i^{-|\rho|+2}$. Then (21.2) implies

$$\gamma_{i+1}^{(\rho)} \leqq (M_i/M_{i+1})\gamma_i^{(\rho)} + c_4(r_i)N_{i+1}^n M_i^2 \delta_i, \quad |\rho| \leqq r_i + 4. \tag{21.5}$$

If (20.1i) holds and $|\rho| \leqq r_i + 3$ then $\gamma_i^{(\rho)} \leqq (2 - 2^{-i})R < R$. We obtain (20.1$i$ + 1) for these values of ρ provided

$$2M_i/M_{i+1} + c_4(r_i)RN_{i+1}^n M_i^2 \delta_i < 1. \tag{C3i}$$

this constraint guarantees the reproduction of (20.1i) at a regular step when $r_{i+1} = r_i$.

The reproduction of (20.1i) at a singular step involves obtaining an estimate of the additional derivatives of order $r_i + 4$. The inequality (21.5) is valid for $|\rho| = r_i + 4$ too, but $\gamma_i^{(\rho)}$ is not obliged to satisfy any regular estimate in this case. However, after a large number of regular steps it will be as small as one wishes. This is due to the following lemma which is left to the reader to prove.

Lemma 21.6. *Let $\{\gamma_i\}$ and $\{\varepsilon_i\}$ be two sequences of nonnegative numbers satisfying $\gamma_{i+1} < \varepsilon_i(\gamma_i + 1)$. Then if $\sum \varepsilon_i$ converges, then so does $\sum \gamma_i$.*

Of course, for each positive integer r, the sequence $\varepsilon_i = \max\{M_i/M_{i+1}, c_4(r)N_{i+1}^n M_i^2 \delta_i\}$ satisfies the condition of the Lemma. We formulate the constraint

(CI) The $(i + 1)$*th step may be singular only if we have* $\gamma_{i+1}^{(\rho)} < R$ *for* $|\rho| = r_i + 4$.

Lemma 21.6 shows that a sufficiently long finite sequence of regular steps results in the correct conditions to do the singular step. We may set $r_{i+1} = r_i + 1$, and (20.1i + 1) will hold.

Do the F^i and its derivatives converge if i goes to infinity? Fix $r>0$, and consider the behavior of $D^\rho F^i$, $|\rho|=r$. The estimate (21.2) ensures the convergence of $D^\rho F^i$ provided $\sum_i c_4(r_i)N^n_{i+1}M^r_{i+1}\delta_i$ converges. Since μ_i grows, there is an $i_0=i_0(r)$ such that $N^n_{i+1}M^r_{i+1}\delta_i<N_i^{-\varkappa}<2^{-(4/3)^i\varkappa}$ for all $i>i_0$. The following is obvious.

21.7. *There is a nondecreasing sequence of positive integers $\{r_i^0\}$, $\lim r_i^0=\infty$, such that the series*

(21.7a) $$\sum_{i=0}^{\infty} c_\alpha(r_i^0)2^{-(4/3)^i\varkappa}, \quad 1\leqq\alpha\leqq 6.$$

converges.

We fix some sequence $\{r_i^0\}$ satisfying 21.7 and impose on $\{r_i\}$ the constraint

(CII) $$r_i\leqq r_i^0, \quad i=0,1,2,\dots$$

Obviously, the series (21.7a) preserves the convergence if one replaces r_i^0 by r_i satisfying (CII), and the corresponding sums do not become larger. The condition (CII) implies the convergence of F^i together with all of its derivatives.

Let us now obtain estimates for F^i-F^0 in the norm $\|\cdot\|_2$. As follows from (21.2)

$$\|F^i-F^0\|_2\leqq\sum_{j=1}^{i-1}c_4(r_j)N^n_{j+1}M^2_{j+1}\delta_j\leqq\sum_{j=1}^{i-1}c_4(r_j)N_0^{-(4/3)^j\alpha},$$

where

$$\alpha=\mu_0-\tfrac{4}{3}n-\tfrac{8}{3}\nu=\tfrac{28}{3}\sigma'+4n+\tfrac{4}{3}\varkappa\geqq\tfrac{7}{15}\mu_0+\varkappa.$$

In view of the latter inequality we have

(21.8) $$\|F^i-F^0\|_2\leqq\delta_0^{7/15}\sum_{j=1}^{\infty}c_4(r_j)N_0^{-(4/3)^j\varkappa}\leqq c_{\mathrm{II}}\delta_0^{7/15},$$

where c_{II} is the maximum of the sums (21.7a). It follows from (21.8) that the limiting function F satisfies the same inequality

(21.9) $$\|F-F^0\|_2\leqq c_{\mathrm{II}}\delta_0^{7/15}.$$

Note that the constant c_{II} depends only on n and σ. Our next constraint is

(CIII) $$c_{\mathrm{II}}\delta_0^{7/15}<\min\{n^{-1},\tilde{R}^{-1}n^{-2}\},$$

where $\tilde{R}=\tilde{R}(R,n)$ is the constant which bounds from above the elements of the inverse matrix to $\{\partial^2F^0/\partial y^k\,\partial y^l\}$ (see (18.10a)). Since $\operatorname{supp}(F^i-F^0)\subset[B]_1$, (20.15), Lemma A11.22 asserts that $\operatorname{grad}F^i$ as well as $\operatorname{grad}F$ map the domain B diffeomorphically onto D (take into account (21.18), (21.9) and (CIII)). It also follows that $\operatorname{supp}(F-F^0)\subset[B]_1$, i.e. (18.10g) holds.

§22. Estimates of Ψ_{i+1}

Note that $h_i-[\![h_i]\!]$ satisfies the conditions (i)–(iii) of subsection entitled "The Homological Equation' of §19 with $N=N_{i+1}$. The condition (iii), for instance, follows from 19.3(ii) provided $6N^{\sigma+1}\leqq M_i$. Our choice of ν (20.5) guarantees the latter inequality if N_0 is sufficiently large: (CIV) $N_0^\varkappa>6$.

So the equation (20.16) only has the solution $\Delta^{i+1} \in C^{\infty}_{\text{per},\alpha_{i+1}}(\mathbb{R}^n \times B)$ with $[\![\Delta^{i+1}]\!] = 0$. Combining Proposition 19.18 and the estimate 21.1 yields

$$(22.1) \qquad |D_x^{\rho_1} D_y^{\rho_2} \Delta^{i+1}| \leqq c_6(r_i) c_4(r_i) N_{i+1}^{2n+\sigma+|\rho_1|} M_{i+1}^{|\rho_2|} \delta_i, \quad 0 \leqq |\rho_1| + |\rho_2| \leqq r_i + 2.$$

Write $\Psi_{i+1}(\xi, \eta) = (x, y)$. We have to determine x and y as functions of ξ and η by solving equation (20.17). The latter being reduced to the following two equations:

$$(22.2) \qquad x - \xi = Q^{i+1}(\eta) \cdot \operatorname{grad}_\eta \Delta^{i+1}(x, \eta),$$

$$(22.3) \qquad \operatorname{grad} F^i(y) = \operatorname{grad} F^i(\eta) - \operatorname{grad}_x \Delta^{i+1}(x, \eta) + \operatorname{grad}(F^{i+1}(\eta) - F^i(\eta)).$$

In (22.2) $Q^{i+1}(\eta) = \{Q_{kl}^{i+1}(\eta)\}$ denotes the inverse matrix to $\{\partial^2 F^{i+1}(\eta)/\partial\eta^k \partial\eta^l\}$.

Proposition 22.4. *There are a number d_2 and function $c_7(r)$ such that if $N_0 > d_2$ and* (20.1*i*)–(20.4*i*), (20.1*i* + 1) and (20.2*i* + 1) *hold then the system* (22.2), (22.3) *has a solution $x(\xi, \eta)$, $y(\xi, \eta)$ such that $x - \xi$ and $y - \eta$ belong to $C^{\infty}_{\text{per},\alpha_{i+1}}$, and the following estimates are true*:

$$(22.4a) \qquad |D_x^{\rho_1} D_y^{\rho_2}(x - \xi)| \leqq c_7(r_i) N_{i+1}^{2n+\sigma+|\rho_1|} M_{i+1}^{|\rho_2|+1} \delta_i,$$

$$(22.4b) \qquad |D_x^{\rho_1} D_y^{\rho_2}(y - \eta)| \leqq c_7(r_i)[N_{i+1}^{2n+\sigma+|\rho_1|+1} M_{i+1}^{|\rho_2|} + N_{i+1}^{|\rho_1|+n} M_{i+1}^{|\rho_2|+1}]\delta_i, \quad 0 \leqq |\rho_1| + |\rho_2| \leqq r_i + 1.$$

Proof. Let us make the change of variables: $\hat{x} = N_{i+1}x$, $\hat{y} = M_{i+1}y$, $\hat{\xi} = N_{i+1}\xi$, $\hat{\eta} = M_{i+1}\eta$. Denote $\tilde{u}(x, y) = Q^{i+1}(y) \cdot \operatorname{grad}_y \Delta^{i=1}(x, y)$, and let $u(\hat{x}, \hat{y}) = N_{i+1}\tilde{u}(N_{i+1}^{-1}\hat{x}, M_{i+1}^{-1}\hat{y})$. Then the map $g: (\hat{x}, \hat{\eta}) \longmapsto (\hat{x} + u(\hat{x}, \hat{\eta}), \hat{\eta})$ is of the form considered in Lemma AII.24, and u satisfies the estimate $\|u\|_{r_i+2} \leqq c_8(r_i) N_{i+1}^{2n+\sigma+1} M_{i+1}\delta_i$. Since $\mu_i \geqq \mu_0 \geqq \frac{4}{3}(2n + v + \sigma + 1) + \varkappa$, the conditions of Lemmas AII.4 and AII.6 are fulfilled if

$$(C3i) \qquad c_8(r_i) N_i^{\varkappa} < \min\{n^{-1/2}; d \text{ of Lemma AII.26}\}.$$

If it is true then there is an inverse map g^{-1} of the form $(\hat{\xi}, \hat{\eta}) \longmapsto (\hat{\xi} + V(\hat{\xi}, \hat{\eta}), \hat{\eta})$ satisfying the estimate of Lemma A11.26. Returning to the old variables gives the estimate (22.4a) for the function $x(\xi, \eta) = \xi + N_{i+1}^{-1} V(\hat{\xi}, M_{i+1}\eta)$ which is a solution to (22.2). Consider the equation (22.3). As above, it is convenient to use the stretched variables $\hat{x}$, $\hat{y}$, $\hat{\xi}$, and $\hat{\eta}$. Then solving (22.3) with respect to y is nothing more than inverting of the map $y \longmapsto \operatorname{grad} F^i(y)$. Denote this inverse by $z \longmapsto \Gamma(z)$. By the induction hypotheses (20.1*i*) and (20.2*i*), the derivatives of order up to $r_i + 2$ of Γ in stretched variables are estimated by constants depending but on R, n, and r_i. The preceding consideration showed us that $x(\xi, \eta)$ has the derivatives of orders up to $r_i + 2$ (again in the stretched variables) smaller than 1. It follows that the r-norm of $y(\xi, \eta)$ defined from (22.3) admits the same estimate as the right side of this equation, probably one must enlarge the values of the constants. Returning to the old variables and taking into account (22.1) and (21.2), one obtains (22.4). □

Remark 22.5. Let R_{i+1} denote the stretching map $(x, y) \longmapsto (N_{i+1}x, M_{i+1}y)$, and let $\hat{\Psi}_{i+1} = R_{i+1} \circ \Psi_{i+1} \circ R_{i+1}^{-1} = \mathrm{id} + \hat{U}_{i+1}$. The above proof yields the estimate

$$(22.5a) \qquad \|\hat{U}_{i+1}\|_{r_{i+1}} \leqq c_7(r_i) N_{i+1}^{2n+\sigma+1} M_{i+1}\delta_i \leqq c_7(r_i) N_i^{-\varkappa} \leqq 1$$

provided N_0 is sufficiently small and $\{r_i\}$ increases sufficiently slowly (see the constraint (C4i) below.

The formula $(\xi,\eta)\longmapsto(x(\xi,\eta), y(\xi,\eta))$ defines a smooth map $\Psi_{i+1}:\mathbb{R}^n\times B\longrightarrow \mathbb{R}^n\times B$. Since $\Delta^{i+1}\in C^{\infty}_{\mathrm{per},a_{i+1}}(\mathbb{R}^n\times B)$, so does $\Psi_{i+1}-\mathrm{id}$. To be persuaded of $\Psi_{i+1}\in \mathrm{Diff}^{\infty}_{\mathrm{per},a_{i+1}}(\mathbb{R}^n\times B)$ it is sufficient to check that Ψ_{i+1} is a diffeomorphism. To do that we use Lemma AII.21 and the estimate

$$\|\Psi_{i+1}-\mathrm{id}\|_1\leqq 2c_7(r_i)N_{i+1}^{2n+\sigma}M_{i+1}^2\delta_i\leqq 2c_7(r_i)N_i^{-\varkappa},$$

which follows from (22.4a, b) and from our relations between constrants. Imposing the constraint

(C4i) $$2c_7(r_i)N_i^{-\varkappa}<(2n)^{-1},$$

we have $\Psi_{i+1}-\mathrm{id}$ satisfies the condition of AII.21. So Ψ_{i+1} is a diffeomorphism.

§23. Reproduction of (20.3i)

If we define R_{i+1} and $\hat{\Psi}_{i+1}$ as in Remark 22.5 and set $\hat{H}^i=(G^i+\mathfrak{L}F^i)\circ R_{i+1}^{-1}$ then (20.3$i+1$) becomes equivalent to

(23.1) $$|\hat{H}^i\circ\hat{\Psi}_{i+1}|_r\leqq 1,\quad 1\leqq r\leqq r_{i+1},$$

(see (20.19)) whilst (20.3i) gives

(23.2) $$|\hat{H}^i|_r\leqq (N_i/N_{i=1})^r,\quad 1\leqq r\leqq r_i.$$

To obtain (23.1), let us apply the estimate of AII.19(a):

(23.3) $$|\hat{H}^i\circ\hat{\Psi}_{i+1}|_r\leqq|\hat{H}^i|_r+C\sum_{l=1}^{r}|\hat{H}^i|_l(r+|\hat{u}|_1)^{l-1}|\hat{u}|_{r-l+1},\quad 1\leqq r\leqq r_i+1.$$

In this estimate $\hat{\Psi}_{i+1}=\mathrm{id}+\hat{u}$, and C only depends on r_i and n. Taking into account (23.2) and (22.5a), we obtain

$$|\hat{H}^i\circ\hat{\Psi}_{i+1}|_r\leqq c_8(r_i)N_i^{-1/3},\quad 1\leqq r\leqq r_i,$$

which leads to (20.31) for the case of regular step if we impose

(C5i) $$c_8(r_i)N_i^{-1/3}\leqq 1.$$

Now consider a possibility for making a singular step. Set $r=r_i+1$, and let $j\geqq i$. We have

$$\hat{H}^{j+1}=\hat{H}^j\circ R_{j+1}\circ\Psi_{j+1}\circ R_{j+2}^{-1}=\hat{H}^j\circ\hat{\Psi}_{j+1}\circ R_{j+1}\circ R_{j+2}^{-1}.$$

Then (23.3) and (22.5a) give us

$$|\hat{H}^{j+1}|_r\leqq\left(\frac{N_{j+1}}{N_{j+2}}\right)^r|\hat{H}^j\circ\hat{\Psi}_{j+1}|_r\leqq\left(\frac{N_{j+1}}{N_{j+2}}\right)^r\cdot\left(|\hat{H}^j|_r+\mathrm{const}\frac{N_j}{N_{j+1}}\right).$$

Denote $\gamma_j=|\hat{H}^j|_r$ and $\varepsilon_j=\max\left\{\left(\frac{N_{j+1}}{N_{j+2}}\right)^r,\mathrm{const}\left(\frac{N_{j+1}}{N_{j+2}}\right)^r\frac{N_j}{N_{j+1}}\right\}$, $j\geqq i$. Then the above inequality may be reduced to $\gamma_{j+1} \leqq \varepsilon_j(\gamma_j + 1)$ where $\sum_j \varepsilon_j < \infty$. Applying Lemma 21.6 yields $\gamma_j \longrightarrow 0$ as $j \longrightarrow \infty$. It then follows from (23.3) that

$$|\hat{H}^j \circ \hat{\Psi}_{j+1}|_r \leqq c_9(r_j + 1)(\gamma_j + N_j^{-1/3}). \tag{23.4}$$

Our next constraint which governs the possibility of making a singular step is

(CIV) *If $i = 0$, or i is a number of a singular step (that is $r_i = r_{i-1}$), then $j + 1, j \geqq i$ may be a number of a singular step if the right-hand side of* (23.4) *is less than* 1.

§24. Reproduction of (20.4*i*)

Setting $\Psi_{i+1}(\xi, \eta) = (x, y)$, one may write

$$G^{i+1}(\xi, \eta) = \mathrm{I} + \mathrm{II} + \mathrm{III}, \tag{24.1}$$

where

$$\mathrm{I} = G^i(x, y) - (S^{(r_i)}_{N_{i+1}M_{i+1}} \theta_{N_{i+1}M_i} G^i)(x, y),$$
$$\mathrm{II} = (S^{(r_i)}_{N_{i+1}M_{i+1}} \theta_{N_{i+1}M_i} G^i)(x, y) - (S^{(r_i)}_{N_{i+1}M_{i+1}} \theta_{N_{i+1}M_i} G^i)(x, y),$$
$$\mathrm{III} = F^i(\eta) - F^i(y) - \langle \eta - y, \operatorname{grad} F^i(y) \rangle.$$

These are immediate consequences of our formulae (20.11), (20.12), (20.16), (20.17), and (20.19) (see also (18.5) for the definition of $\mathfrak{L}$).

Estimation of I. At first we need to be persuaded that one can insert the multiplier $\theta_{N_{i+1}M_i}(x, y)$ at the first term in I also.

24.2. *If $(\xi, \eta) \in \operatorname{supp} \theta_{N_{i+2}M_{i+1}}$ and $(x, y) = \Psi_{i+1}(\xi, \eta)$, then*

$$\theta_{N_{i+1}M_i}(x, y) = 1 \text{ and } D_x^{\rho_1} D_y^{\rho_2} \theta_{N_{i+1}M_i}(x, y) = 0 \text{ for all}$$
$$(\rho_1, \rho_2) \in \mathbb{Z}_+^n \times \mathbb{Z}_+^n, \qquad |\rho_1| + |\rho_2| \geqq 1.$$

Proof. The (22.4b) implies

$$|y - \eta| \leqq c_7(r_i)[N_{i+1}^{2n+\sigma+1} + N_{i+1}^{n+\nu}]\delta_i < M_{i+1}^{-1}$$

provided N_0 and $\{r_i\}$ are chosen in an appropriate manner. The conclusion follows from 19.3(*v*) if we impose

$$M_{i+1} \geqq 3M. \tag{C7i}$$

The inequality $N_0 \geqq 27$ is sufficient. □

Let $\eta \in \mathscr{E}^{(2)}_{N_{i+2}}$. By use of (19.15d) and 19.3(*iv*) we have

$$|\mathrm{I}| \leqq c_5(r_i) \max_{|\rho_1| + |\rho_2| = r_i} N_{i+1}^{-|\rho_1|} M_{i+1}^{-|\rho_2|} |D_x^{\rho_1} D_y^{\rho_2} \theta_{N_{i+1}M_i} G^i|$$
$$\leqq c_{10}(r_i) \max_{|\rho_1| + |\rho_2| = r_i} (N_i/N_{i+1})^{-|\rho_1|} (M_i/M_{i+1})^{-|\rho_2|} = c_{10}(r_i) N_i^{-1/3r_i} < \tfrac{1}{3}\delta_{i+1}$$

provided

(24.3) $$\mu_{i+1} \leqq \tfrac{1}{4} r_i - \varkappa',$$

and N_0 and $\{r_i\}$ are chosen so that

(C8i) $$c_{10}(r_i) N_i^{-1/3\varkappa'} < \tfrac{1}{3}.$$

Estimation of II.

$|\mathrm{II}| \leqq \max_{|\rho|=1} |D_y^\rho S_{N_{i+1}M_{i+1}}^{\langle r_i\rangle} \theta_{N_{i+1}M_i} G^i| \cdot |y-\eta| \leqq$ (use 19.15c and (22.4b)) $c_4(r_i) N_{i+1}^n \cdot M_{i+1} \delta_i c_7(r_i)(N_{i+1}^{2n+\sigma} + N_{i+1}^n M_{i+1})\delta_i < \frac{1}{3}\delta_{i+1}$ provided

(24.4) $$\mu_{i+1} \leqq \tfrac{3}{2}\mu_i - \max\langle \sigma + 1 + 3n + v, 2n + 2v\rangle - \varkappa',$$

and N_0 and $\{r_i\}$ are such that

(C9i) $$2c_4(r_i)c_7(r_i)N_i^{-4/3\varkappa'} < \tfrac{1}{3}.$$

Estimation of III. By Taylor's formula

(24.5) $$\mathrm{III} = \int_0^1 \left\langle \frac{\partial^2 F^i}{\partial y^2}(y + (\eta - y)t)\cdot(\eta - y), (\eta - y)\right\rangle (1-t)\, dt.$$

Estimates (20.1i) and (22.4b) yield

$$|\mathrm{III}| \leqq c_{11}(r_i)[N_{i+1}^{\sigma+2n+1} + N_{i+1}^2 M_{i+1}]^2 \delta_i^2 < \tfrac{1}{3}\delta_{i+1}$$

provided

(24.46) $$\mu_{i+1} \leqq \tfrac{3}{2}\mu_i - \max\langle 2\sigma + 4n + 2, 2n + 2v\rangle - \varkappa',$$

and N_0 and $\langle r_i\rangle$ secure

(C10i) $$2c_{11}(r_i)N_i^{-(4/3)\varkappa'} < \tfrac{1}{3}.$$

The estimates stated above imply (20.4i + 1). It remains for us to check (24.3), (24.4), (24.6), and (C7i)–(C10i). These impose constraints on $\{\mu_i\}$ and $\{r_i\}$ the slowly growing sequences. The inequality (24.3) holds for $i = 0$ if $\mu_1 = \mu_0$ due to our choice of r_0 and μ_0 (see (20.8)). The growth of r_i will enable us to enlarge μ_i. At a step which does not follow a singular one we take $\mu_i = \mu_{i-1}$, and we enforce μ_i to jump by 1/4 if r_{i-1} jumps by 1. Both the inequalities (24.4) and (24.6) are true for $i = 0$ if $\mu_1 = \mu_0$. Since $\mu_i \geqq \mu_0 \geqq 1$, (24.4) and (24.6) permit μ_{i+1} to be greater than $\mu_i + 1/4$ at any step of our iterative process. The conditions (Cki) are restrictions on the choice of N_0 and a sequence $\{r_i\}$.

§25. Convergence of the Process and the Estimate of $\|\Phi - \mathrm{id}\|$

Denote $\Phi_i = \mathrm{id} + W_i$. The equation $\Phi_{i+1} = \Phi_i \circ \Psi_{i+1}$ leads to a recurrence relation

(25.1) $$W_{i+1} = u_{i+1} + W_i \circ (\mathrm{id} + u_{i+1}),$$

where we set $\Psi_{i+1} = \mathrm{id} + u_{i+1}$. Let $\Phi_i \in \mathrm{Diff}^\infty_{\mathrm{per},a_i}(\mathbb{R}^n \times B)$. Since $u_{i+1} \in C^\infty_{\mathrm{per},a_{i+1}}(\mathbb{R}^n \times B)$, so does W_{i+1}, and hence $\Phi_{i+1} \in \mathrm{Diff}^\infty_{\mathrm{per},a_{i+1}}(\mathbb{R}^n \times B)$. The limiting diffeomorphism

Φ, if exists, would belong to $\mathrm{Diff}^{\infty}_{\mathrm{per},1}(\mathbb{R}^n \times B)$, i.e. (18.10h) and the commutation with Γ would hold. Also, (20.18) implies $\Phi_i^*\Omega_{F0} = \Omega_{Fi}$. Passing to the limit $i \longrightarrow \infty$ in the latter equality yields (18.9)

Proposition 25.2. *Given an integer $r > 0$, the series $\sum_i \|u_i\|_r$ converges. The sum $\sum_{i=1}^{\infty} \|u_i\|_2$ is bounded from above by a constant depending only on R, n, σ. The sum corresponding to $r = 1$ admits an estimate*

$$\sum_{i=1}^{\infty} \|u_i\|_1 \leqq \mathrm{const}\, \delta_0^{4/15}, \tag{25.2a}$$

where const *depends only on R, n, σ.*

Proof. It follows from (22.4a, b) that, if $r_i \geqq r - 1$,

$$\|u_{i+1}\|_r \leqq 2c_7(r_i) N_{i+1}^{\sigma+2n} M_{i=1}^{r+1} \delta_i. \tag{25.2b}$$

Starting with some $i_0 = i_0(r)$, the right-hand side of (25.2b) becomes less than $2c_7(r_i)N_i^{-\varkappa}$ (this is due to the growth of μ_i). The convergence follows then from (CII). If $r = 2$, one may take $i_0 = 0$, as it follows from the relations (20.5)–(20.8). So we can bound $\sum \|u_i\|_2$ from above by twice the sum (21.7a). The estimate (25.2b) for the case $r = 1$ gives us

$$\begin{aligned} \|u_{i+1}\|_1 &\leqq 2c_7(r_i) N_i^{-[\mu_0 - (4/3)(\sigma + 2n + 2\nu)]} \leqq 2c_7(r_i) N_i^{-(4/15)\mu_0 - \varkappa} \\ &\leqq \delta_0^{4/15} 2c_7(r_i) 2^{-(4/3)^i \varkappa}. \end{aligned}$$

Again (CII) and (21.7) yield the desired estimate, namely (25.2a). □

Proposition 25.3. *Given an integer $r \geqq 0$, there is a positive constant A_r which depends only on n, σ, R, N_0, such that $\|W_i\|_r \leqq A_r$. The constant A_2 may be chosen to be independent of N_0.*

Proof. Applying Lemma AII.15 to (25.1) gives us

$$\|W_{i+1}\|_r \leqq \|u_{i+1}\|_r + \|W_i\|_r (1 + \mathrm{const}(\|u_{i+1}\|_r + \|u_{i+1}\|_{r-1}^r)).$$

Denote $\varepsilon_i = \max\langle \|u_{i+1}\|_r, \mathrm{const}(\|u_{i+1}\|_r + \|u_{i+1}\|_{r-1}^r)\rangle$. Proposition 25.2 ensures $\sum \varepsilon_i$ to be convergent and, in the case $r = 2$, to be bounded by a constant independent of N_0. Use Lemma 25.4 below. □

Lemma 25.4. *Let $\{\gamma_i\}, \{\varepsilon_i\}, i \in \mathbb{Z}_+$, be two sequences of nonnegative numbers satisfying the sequence of inequalities $\gamma_{i+1} \leqq \gamma_i + \varepsilon_i \gamma_i + \varepsilon_i$, $z \in \mathbb{Z}_+$. If $\sum \varepsilon_i$ convergences then γ_i is bounded. Moreover*

$$\gamma_i \leqq \left(\prod_{i=0}^{\infty} (1 + \varepsilon_i)\right)\left(\gamma_0 + \sum_{i=0}^{\infty} \varepsilon_i\right).$$

Proof. Define x_i, $i \geqq 0$, by setting $x_0 = \gamma_0$ and $x_{i+1} = x_i + \varepsilon_i x_i + \varepsilon_i$. Obviously $\gamma_i \leqq x_i$. One immediately checks

$$x_i = \gamma_0 \prod_{k=0}^{i-1} (1 + \varepsilon_k) + \sum_{j=0}^{i-1} \varepsilon_j \prod_{k=j+1}^{i-1} (1 + \varepsilon_k).$$ □

Now we are able to establish the convergence of Φ_i. It is sufficient to prove that the series $\sum_i \| W_{i+1} - W_i \|_r$ converges for each $r \geqq 0$. Applying Lemma AII.15 to (25.1), one obtains

$$\| W_{i+1} - W_i \|_r \leqq \| u_{i+1} \|_r + \text{const} \, \| W_i \|_{r+1} (\| u_{i+1} \|_r + \| u_{i+1} \|_{r-1}^r), \tag{25.5}$$

where const only depends on r, R, σ, n. Since $\| W_i \|_{r+1}$ is bounded due to Proposition 25.3, and $\| u_{i+1} \|_r$ tends to zero due to Proposition 25.2, the estimate (25.5) reduces to

$$\| W_{i+1} - W_i \|_r \leqq \text{const} \, \| u_{i+1} \|_r, \tag{25.6}$$

where const only depends on r, R, σ, n, N_0. In view of 25.2, (25.6) yields the convergence of W_i in the C^r-norm. There exists $W = \lim_{i \to \infty} W_i \in C^\infty_{\text{per},1}(\mathbb{R}^n \times B)$.

Let us obtain estimates for $\| W_i \|_1$. First, we notice that const in (25.6) does not depend on N_0 if $r = 1$ (take into account the assertion 25.3 about A_2 when deriving (25.6) from (25.5)). Since $W_0 = 0$, we have, using (25.2a),

$$\| W_i \|_1 \leqq \sum_{j=1}^{\infty} \| W_{i+1} - W_i \|_1 \leqq \text{const} \sum_{j=1}^{\infty} \| u_{i+1} \|_1 \leqq c \delta_0^{4/15},$$

where c only depends on R, n, σ. Taking a limit $t \longrightarrow \infty$, we obtain the estimate

$$\| \Phi - \text{id} \|_1 = | W |_1 \leqq c \delta_0^{4/15}. \tag{25.7}$$

If

$$c \delta_0^{4/15} < (2n)^{-1}. \tag{CV}$$

then Lemma AII.21 ensures that $\Phi = \text{id} + W$ is a diffeomorphism.

§26. Derivatives of G at points of $\mathbb{R}^n \times \mathscr{E}$

To establish 18.10(*ii*) we are going to estimate higher derivatives of G^i at points of $\mathbb{R}^n \times \mathscr{E}$.

Proposition 26.1. *Given multiindices $\rho_1, \rho_2 \in \mathbb{Z}^n_+$,*

$$\lim_{i \to \infty} \sup_{(x,y) \in \mathbb{R}^n \times \mathscr{E}} | D_x^{\sigma_1} D_y^{\rho_2} G^i(x, y) | = 0.$$

Proof. We will use the representation (24.1) and consequently will estimate the derivatives of I, II, III.

Estimate of $D_\xi^{\rho_1} D_\eta^{\rho_2}$I. Let $\eta \in \mathscr{E}$. As in §24 we may insert $\theta_{N_{i+1} M_i}(x, y)$ as a multiplier of $G_j^i(x, y)$. We have, by use of Lemma AII3,

$$| D_\xi^{\rho_1} D_\eta^{\rho_2} \text{I} | \leqq \| A \circ \Psi_{i+1} \|_r \leqq \| A \|_r (1 + \text{const}(\| U_{i+i} \|_r + \| U_{i+1} \|_{r-1}^r)), \tag{26.1a}$$

where $r = |\rho_1| + |\rho_2|$ and

$$A(x, y) = \theta_{N_{i+1} M_i}(x, y) G^i(x, y) - (S^{(r_i)}_{N_{i+1} M_{i+1}} \theta_{N_{i+1} M_i} G^i)(x, y).$$

Proposition 19.15(d) and estimates (20.1i), (20.3i) give us

$$\| A \|_r \leqq c_5(r_i) 2R^2 \max_{|\rho_1|+|\rho_2|=r_i-r} (N_i/N_{i+1})^{-|\rho_1|}(M_i/M_{i+1})^{-|\rho_2|} M_i^r$$

$$\leqq 2R^2 c_5(r_i) N_i^{-(1/3)(r_i-r)+\nu r} \leqq 2R^2 c_5(r_i) N_i^{-\varkappa}$$

provided i is sufficiently large so that $r_i > r(1+3\nu)+3\varkappa$. The condition (CII) ensures that $\| A \|_r \longrightarrow 0$ as $i \longrightarrow \infty$. Taking into account Proposition 25.2 we conclude that so does the right-hand side of (26.1a).

Estimate of $D_\xi^{\rho_1} D_\eta^{\rho_2}$II. Using the notation (20.11), we have, starting with some $i_0 = i_0(r)$, $r = |\rho_1| + |\rho_2|$, $|D_\xi^{\rho_1} D_\eta^{\rho_2} \mathrm{II}| = |D_\xi^{\rho_1} D_\eta^{\rho_2} h_{i+1}(x, y) - D_x^{\rho_2} D_\eta^{\rho_2} h_{i+1}(x, y)| \leqq$ (use Lemma AII16) const $\| h_{i+1} \|_{r+1} \| y - \eta \|_r \leqq$ (use 21.1 and 22.4b) const $c_4(r_i) M_{i+1}^{r+1} \times N_{i+1}^{\eta} \delta_i 2c_7(r_i) N_{i+1}^{\sigma+2\eta} M_{i+1}^{r+1} \delta_i \leqq$ const $c_4(r_i) c_7(r_i) N_i^{-\varkappa}$, where const only depends on r, n and $\max_i \| u_i \|_r$ (see Proposition 25.2). Our constraint CII forces the right-hand side of (26.16) to go to zero as $i \longrightarrow \infty$.

Estimate of $D_\xi^{\rho_1} D_\eta^{\rho_2}$III. Differentiating with respect to the variables (ξ, η) of the right-hand side of (24.5), we see that $D_\xi^{\rho_1} D_\eta^{\rho_2}$III may be represented as a finite sum, the typical term being an integral of a product of $D_y^s F^i$, $|s| \leqq |\rho_1| + |\rho_2|$, some multipliers of the form $D_\xi^{k_1} D_\eta^{k_2}(y(\xi, \eta) - \eta)$, or $D_\xi^{k_1} D_\eta^{k_2}(y + (\eta - y)t)$, and a bounded function. At least one of these multipliers tends uniformly to zero as $i \longrightarrow \infty$ (see Proposition 25.2) while the others are bounded ((20.1i) and (25.2)). The derivatives of y being bounded follows from 25.2, that of derivatives of F^i is a consequence of their convergence, the latter being proven in §21. □

Now consider the expression (18.10d). Its derivatives can be computed by applying differentiation to and taking the limit in the equality

$$G^{i+1} = (G^0 + \mathfrak{L} F^0) \circ \Phi_{i+1} - \mathfrak{L} F^{i+1}. \tag{26.2}$$

One can obtain (26.2) by the iteration of (20.19) and using the relation $\Phi_{i+1} = \Phi_i \circ \Psi_{i+1}$. Proposition 26.1 ensures that the mentioned derivatives vanish at points of $\mathbb{R}^n \times \mathscr{E}$. So the limiting F and Φ satisfy (*ii*) of Theorem 18.10.

§27. The End of the Proof of Theorem 18.10

To finish the proof we have to choose suitable values of the constants N_0 (or equivalently $\delta_0 = N_0^{-\mu_0}$) and δ, to choose the sequences $\{r_i\}$ and $\{\mu_i\}$ so that the iterative process can be carried out without hindrance, the necessary estimates being fulfilled, and to check that the limiting objects satisfy the conclusion of the theorem.

The choice of δ_0. We set $\delta_0 = \| G \|_{r_0}$ and will impose $\delta_0 \leqq \delta$.

The choice of δ. This is the maximal value of the quantity δ_0 which is sufficient to secure the validity of the above constraints (Ck_0), $1 \leqq k \leqq 9$, (CIII), (CV), (CVI), and the initial inequality (20.30). Such a δ depends only on n, σ, and R.

The choice of $\{r_i\}$. The value of r_0 is given by (20.8). At a regular step $r_i = r_{i-1}$. The rules for the appearance of singular steps, when $r_i = r_{i-1} + 1$, are limited by demanding (Ck_i), $1 \leqq k \leqq 9$, and (CI), (CII), (CIV). The growth of N_i enables us to occasionally switch on a singular step. We shall do this infinitely often. The sequence $\{r_i\}$ such constructed tends to infinity.

The choice of $\{\mu_i\}$. If ith step is singular, we set $\mu_{i+1} = \mu_i + 1/4$. Otherwise $\mu_{i+1} = \mu_i$. This rule ensures that (24.3), (24.4), and (24.6) are satisfied.

Thus, all elements of our construction are fixed, and the process occurs without obstruction. As was stated in §21 and §25, the sequences F^i and Φ_i converge respectively to a function F, such that grad F maps D diffeomorphically onto B, and a diffeomorphism $\Phi: \mathbb{R}^n \times B \longrightarrow \mathbb{R}^n \times B$. The conclusion (i) of the Theorem is stated in the beginning of §25, (ii) in §26, inclusions and estimates of (iii) in §§21, 25. We have obtained that F and Φ satisfy all the conditions of the conclusion of Theorem 18.10.

§28. Deduction of the Theorem for Discrete Time from That of Continuous Time

It is convenient here to use the letter n for denoting the number of degrees of freedom. So we consider a Hamiltonian diffeomorphism $f: \mathbb{R}^{n-1} \times D \longrightarrow \mathbb{R}^{n-1} \times D$ which commutes with the action of the group Γ' generated by shifts $z \longrightarrow z + \mathrm{ort}_i$, $1 \leqq i \leqq n-1$, and which possesses a Hamiltonian

$$H(\phi', \tilde{I}') = K_0(\tilde{I}') + H_1(\phi', \tilde{I}'). \tag{28.1}$$

Here $\phi' = (\phi^1, \phi^2, \ldots, \phi^n)$, $\tilde{I}' = (\tilde{I}_1, \tilde{I}_2, \ldots, \tilde{I}_n)$ are the old angles and new momenta, prime denoting that the number of variables equals $n-1$, an incomplete quantity. The transformation $f: (\phi', I') \longmapsto (\tilde{\phi}', \tilde{I}')$ is defined by $H(\phi', \tilde{I}')$ via the formulae (8.4). We suppose that K_0 and H_1 satisfy conditions (11.2), (11.4a, b) with δ to be determined later. For the duration of the proof, the symbol const denotes constants only depending on n, R and σ. The words "δ is sufficiently small" mean δ is less than a constant only depending on n, R and σ.

To prove the theorem for the system such defined, we will apply the suspension construction described in §6 in order to obtain a system with continuous time. Then the theorem for that system will yield, after some efforts, the desired result for the discrete time case.

Suspension

The construction we use differs formally from that of §6, but it is equivalent to the latter if one takes return time function $\tau(x, p) = (1 + 2\varepsilon p)^{-1/2}$. Let us add the nth angle and momentum (ϕ^n, I_n) to the list of coordinates and consider the product $Z = (\mathbb{R}^{n-1} \times D) \times (\mathbb{R} \times]-1, 1[)$ of symplectic spaces, the $2(n-1)$-dimensional one and the 2-dimensional one. We use the following notations for the coordinates in Z: $Z \ni z = (\phi, I) = (\phi', I', \phi^n, I_n)$ where $(\phi', I') \in \mathbb{R}^{n-1} \times D$ and $(\phi^n, I^n) \in \mathbb{R} \times]-1, 1[$, while $\phi = (\phi', \phi^n) = (\phi^1, \phi^2, \ldots, \phi^n)$ and $I = (I_1, I_2, \ldots, I_n)$. I hope the reader will forgive me for such an abuse of notations and will understand them correctly.

Consider Hamiltonian flow on Z defined by Hamiltonian function $h^{\text{flow}} = I_n + (\varepsilon/2)I_n^2$, where ε is a small positive number to be defined later.

Denote by $\hat{\Gamma}$ the transformation group in Z, generated by the following transformations: the shifts $(\phi, I) \longmapsto (\phi + \text{ort}_i, I)$, $1 \leqq i \leqq n-1$, and the special transformation

$$\hat{g}: (\phi', I', \phi^n, I_n) \longmapsto (f(\phi', I'), \phi^n - 1, I_n).$$

One verifies easy $\hat{\Gamma}$ consists of symplectic diffeomorphisms and acts in properly disconnected manner.

Let $p: Z \longrightarrow Z/\hat{\Gamma}$ be the quotient space projection. Since $\hat{\Gamma}$ leaves h^{flow} invariant, the flow on Z generated by h flow projects onto a Hamiltonian flow on $Z/\hat{\Gamma}$. The set $p(S)$, where $S \subset Z$ is defined by the equations $\phi^n = 0, I_n = 0$, is a symplectic submanifold and forms a cross section in $Z/\hat{\Gamma}$ to that flow (see §6). Obviously, the Poincaré map of this cross section is isomorphic to f via $p \circ i$ where $i: (\phi', I') \longmapsto (\phi', I', 0, 0)$ is the inclusion map of $\mathbb{R}^{n-1} \times D$ into Z.

The Untwisting Map

The space $Z/\hat{\Gamma}$ is, of course, isomorphic to a symplectic annulus, but the coordinates used are not suitable for applying the flow KAM theorem. It is necessary to untwist the coordinates. All calculations will be made in covering spaces. We pass from the coordinates (ϕ, I) to those denoted

$$(\psi, J) = (\psi', J', \psi^n, J_n), \quad \psi = (\psi', \psi^n), \quad J = (J', J_n),$$
$$\psi' = (\psi^1, \psi^2, \ldots, \psi^{n-1}) \in \mathbb{R}^{n-1}, \quad \psi^n \in \mathbb{R}^1, \quad J' = (J_1, J_2, \ldots, J_{n-1}).$$

The (J', J_n) ranges over the domain D^{flow} which is the image of $D \times]-1, 1[$ under the map $(I', I_n) \longmapsto (I', I_n - K_0(I'))$.

The untwisting map

$$u: (\mathbb{R}^{n-1} \times D) \times (\mathbb{R} \times]-1, 1[) \longrightarrow \mathbb{R}^n \times D^{\text{flow}}$$

will be symplectic and will carry the action of $\hat{\Gamma}$ to that of the "standard" group Γ generated by the unit shifts of the first n coordinates ψ^i. It will act slicewise, carrying the slice $\phi^n = \text{const}$ to the slice $\psi^n = \phi^n = \text{const}$. At first we define u on the slices $\phi^n \in]-1/3, 1]$. The old coordinates (ϕ, I) and the new ones (ψ, J) are linked via a Hamiltonian function

$$\Phi(\phi', J', \phi^n, J_n) = \phi^n K_0(J') + \theta(\phi^n) H_1(\phi', J'). \tag{28.2}$$

Here $\theta:]-1/3, 1[\longrightarrow [0, 1]$ is a smooth function such that $\theta(x) = 0$ if $x \in]-1/3, 1/3[$, and $\theta(x) = 1$ if $x \in]2/3, 1[$. One can obtain explicit formulae linking (ϕ, I) and (ψ, J) using (8.4). Note that Φ does not contain J_n which implies the above slice-preserving property. Proposition (8.17), being applied to each slice $\psi^n = \phi^n = \text{const}$, yields the existence and the uniqueness of u provided $\|H_1\|_{2,\gamma}$ is sufficiently small ($<\min\langle 1/(n-1), 1/2\rangle$).

Let g denote the shift $(\psi, J)\longmapsto(\psi - \mathrm{ort}_n, J)$ of the space $\mathbb{R}^n \times D^{\mathrm{flow}}$. One immediately obtains that

$$u\circ\hat{g} = g\circ u \tag{28.3}$$

on the common domain, i.e. at points with $\phi^n \in]2/3, 1[$. We extend to the right and to the left, and demand that (28.3) be satisfied. Such an extension uniquely restores the map $u:(\mathbb{R}^{n-1} \times D) \times (\mathbb{R} \times]-1, 1[) \longrightarrow \mathbb{R}^n \times D^{\mathrm{flow}}$ which is evidently a symplectic diffeomorphism. Since Φ is invariant under shifts belonging to Γ', and since u carries $\hat{g}$ to g, the image of the lattice $\hat{\Gamma}$ under the map u is Γ, the standard lattice generated by unit shifts of the first n variables.

Application of the KAM Theorem

The image under u of the flow in question is also Hamiltonian, its Hamiltonian function being

$$\tilde{H}^{\mathrm{flow}}(\psi, J) = (u^{-1})^* h^{\mathrm{flow}}(\psi, J) = K_0^{\mathrm{flow}}(J) + \tilde{H}_1^{\mathrm{flow}}(\psi, J). \tag{28.4}$$

The functions in the right-hand side of (28.4) have explicit expressions:

$$K_0^{\mathrm{flow}}(J) = J_n + K_0(J') + \frac{\varepsilon}{2}(J_n + K_0(J'))^2, \tag{28.5}$$

$$\tilde{H}_1^{\mathrm{flow}}(\psi, J) = \theta'(\psi^n) H_1(\phi'(\psi, J), J')[1 + \varepsilon(J_n + K_0(J')) + \frac{\varepsilon}{2}\theta'(\psi^n) H_1(\phi'(\psi, J), J')] \tag{28.6}$$

One must prolong $\theta'(\psi^n)$ periodically for (28.6) to have sense for all values of variables. The Hamiltonian (28.4), although being in the form required for KAM theorem, does not satisfy the condition (11.4a). In view of that, we replace $\tilde{H}^{\mathrm{flow}}(\psi, J)$ by more suitable Hamiltonian

$$H^{\mathrm{flow}}(\psi, J) = K_0^{\mathrm{flow}}(J) + H_1^{\mathrm{flow}}(\psi, J), \tag{28.7}$$

where

$$H_1^{\mathrm{flow}}(\psi, J) = \tilde{H}_1^{\mathrm{flow}}(\psi, J)\chi(J_n + K_0(J')), \tag{28.8}$$

and $\chi:[-1, 1] \longrightarrow [0, 1]$ is a smooth function with $\operatorname{supp}\chi \subset [-2/3, 2/3]$ and such that $\chi(x) = 1$ if $x \in [-1/3, 1/3]$. Such a spoiling will cause no trouble since the KAM tori in question will be situated close to the set $J_0 + K_0(J') = 0$ where χ is identically equal to 1. Let us check the remaining conditions of Theorem 11.6.

We are going to apply the continuous time version of Theorem 11.6 to the system defined by the Hamiltonian (28.7), (28.8). In order to distinguish between constants and functions concerning the continuous time case from those of the discrete time case, we shall supply the formers with the upper indices "flow". Thus the constant $\delta = \delta(n, \sigma, R)$ of the flow case of Theorem 11.6 will be denoted as δ^{flow}. Checking the conditions of the flow KAM theorem, we shall investigate the properties of the domain D^{flow} and the frequency map $\omega_0^{\text{flow}} : D^{\text{flow}} \longrightarrow B^{\text{flow}}$ where

$$\omega_0^{\text{flow}}(J) = \operatorname{grad} K_0^{\text{flow}}(J) = [1 + \varepsilon(J_n + K_0(J')](\omega_0(J'), 1). \tag{28.9}$$

Obviously D^{flow}, defined above in this section, is non-empty, open, bounded, and simply connected. Since the derivatives of (28.5) are bounded, the frequency domain $B^{\text{flow}} = \omega_0^{\text{flow}}(D^{\text{flow}})$ is also bounded.

Proposition 28.10. B^{flow} is convex.

Proof. The domain B^{flow} may be represented as the image of $B \times]-1, 1[$ under the map $(\omega', I_n) \longmapsto ((1 + \varepsilon I_n)\omega', 1 + \varepsilon I_n)$. So it has the form of a basket (see Fig. 50) whose bottom and cover are linear images of B. Let $\omega_i \in B^{\text{flow}}$, $i = 1, 2$. Then $\omega_i = (q_i\omega_i', q_i), \omega_i' \in B, q_i \in]1 - \varepsilon, 1 + \varepsilon[$. Given $\tau \in [0, 1]$,

$$\tau\omega_1 + (1 - \tau)\omega_2 = (Q[\tau'\omega_1' + (1 - \tau')\omega_2'], Q), \tag{28.10a}$$

where $Q = \tau q_1 + (1 - \tau)q_2 \in]1 - \varepsilon, - + \varepsilon[$ and $\tau' = \tau q_1 Q^{-1} \in [0, 1]$. Since B is convex, $[\tau'\omega_1' + (1 - \tau')\omega_2' \in B$, and the right-hand side of (28.10a) belongs to the basket □

Proposition 28.11. *There exists $\varepsilon = \varepsilon(R, n)$ and $R_1 = R_1(R, n)$, such that, if K_0 satisfies* (11.2), *then K_0^{flow} satisfies* (11.2) *with R replaced by R_1, and $\omega_0^{\text{flow}} : D^{\text{flow}} \longrightarrow B^{\text{flow}}$, defined by* (28.9), *is a diffeomorphism.*

Proof. One can easily compute that the map (28.9) is single valued provided $0 \leqq \varepsilon \leqq 1$. On the other hand, (28.5) yield

$$\left\{\frac{\partial^2 K_0^{\text{flow}}}{\partial J_i \partial J_k}\right\} = \begin{array}{|c|c|} \hline \dfrac{\partial^2 K_0}{\partial J_i \partial J_k} & 0 \\ \hline 0 & \varepsilon \\ \hline \end{array} + \varepsilon\mathscr{R}(J) \tag{28.11a}$$

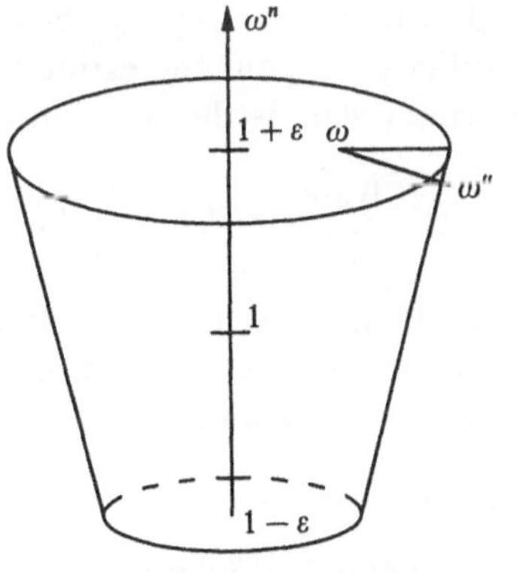

Fig. 50

where the $n \times n$ matrix $\mathscr{R}(J)$ contains the derivatives of K_0 up to second order and has zero at the lower right corner. It follows from (28.11a) that

$$\det\left\{\frac{\partial^2 K_0^{\text{flow}}}{\partial J_i \partial J_k}\right\} = \varepsilon \det\left\{\frac{\partial^2 K_0}{\partial J_i \partial J_k}\right\} + \mathcal{O}(\varepsilon^2). \tag{28.11b}$$

If ε is sufficiently small then ω_0^{flow} becomes a diffeomorphism, (28.11a, b) yielding (11.2) for K_0^{flow} with R replaced by an $R_1 = R_1(R, n)$. □

In particular, 28.11 involves B^{flow} being open and bounded for this value of ε. Fixing ε we proceed with our construction. Set

$$\gamma^{\text{flow}} = \min\{(1-\varepsilon)(1+R^2)^{-1/2}, \sigma^{-1}\varepsilon\}\cdot\gamma.$$

Proposition 28.12. $\operatorname{Supp} H_1^{\text{flow}} \subset \mathbb{R}^n \times (\omega_0^{\text{flow}})^{-1}([B^{\text{flow}}]_{2\gamma^{\text{flow}}})$, i.e. *the* H_0^{flow} *satisfies the corresponding form of the condition* (11.4a).

Proof. Let $\omega = \omega_0^{\text{flow}}(J) = ((1+\varepsilon I_n)\omega', 1+\varepsilon I_n) \notin [B^{\text{flow}}]_{2\gamma^{\text{flow}}}$. Then either it belongs to the $2\gamma^{\text{flow}}$-neighbourhood of the bottom, or to that of the cover, or to that of the side of the basket (see Fig. 50). In the first two cases $H_1^{\text{flow}}(\psi, J) = 0$ due to the presence of the multiplier $\chi(I_n)$ in (28.8) since $\varepsilon\gamma^{\text{flow}}$ is less than a sixth of *the height of the basket*. Consider the final case. Let $\tilde{\omega} = ((1+\varepsilon\tilde{I}_n)\tilde{\omega}', 1+\varepsilon\tilde{I}_n)$ be the nearest point to ω at the side of the basket, and let $\hat{\omega} = ((1+\varepsilon I_n)\tilde{\omega}', 1+\varepsilon I_n)$. The points $\omega, \tilde{\omega}$, and $\hat{\omega}$ form a right-angled triangle parallel to the bottom, whose hypotenuse lies in a slice of the level I_n (Fig. 50). The angle α of this triangle belonging to the vertex ω has $\tan\alpha = |\omega'|$. We have

$$|\omega - \hat{\omega}| = |\omega - \tilde{\omega}|(1+|\omega'|^2)^{1/2} \leqq 2\gamma^{\text{flow}}(1+R^2)^{1/2}.$$

On the other hand $|\omega - \hat{\omega}| = (1+\varepsilon I_n)|\omega' - \tilde{\omega}'|$. It follows that $|\omega' - \tilde{\omega}'| \leqq 2\gamma^{\text{flow}}(1 + R^2)^{1/2}(1-\varepsilon)^{-1} \leqq 2\gamma$. Recall that $\omega' = \omega_0(J')$ and $\tilde{\omega}'$ belongs to the boundary of B. The condition (11.4a) yields $H_1(\phi', J') = 0$. So does $H_1^{\text{flow}}(\psi, J', J_n)$ (see (28.6)). □

We will apply the flow theorem 11.2 with γ^{flow} standing for γ. To be convinced of its applicability one must check the remaining condition (11.4b) where all the symbols must be supplied with upper indices "flow", where $r_0^{\text{flow}} = r_0 - 1$. It is sufficient, in view of periodicity, to gain the estimate on the interval $0 \leqq \psi^n \leqq 1$. The function $\phi'(\psi, J)$, entering (28.6), is the solution of the equation

$$\psi' = \phi' + \psi^n \frac{\partial K_0(J')}{\partial J'} + \theta(\psi^n)\frac{\partial H_1}{\partial J'}(\phi', J')\chi(J_n + K_0(J')). \tag{28.13}$$

It follows from (28.13) and (11.4b) that the derivatives of $\phi'(\psi, J)$ of order up to r_0^{flow} are bounded by constants only depending on n and R. Taking this into account, one obtains the following estimate for H_1^{flow} given by (28.8), (28.6):

$$\| H_1^{\text{flow}} \|_{r_0^{\text{flow}}, \gamma^{\text{flow}}} \leqq \text{const} \| H_1 \|_{r_0, \gamma}, \tag{28.14}$$

where const depends only on n and R. If we take $\delta = \text{const}^1 \min\{(1-\varepsilon)^2(1+R^2)^{-1}, \varepsilon^2/36\}\delta^{\text{flow}}$ where δ^{flow} is the δ of the flow version of Theorem 11.6, then (11.4b) gives the desired estimate $(\gamma^{\text{flow}})^2\delta^{\text{flow}}$ for (28.14) from above. Hence we may assert the conclusion of the flow version of Theorem 11.6 that there exists an integrable system in $A_{D^{\text{flow}}}$ with Hamiltonian function $K^{\text{flow}}: D^{\text{flow}} \longrightarrow \mathbb{R}$ which satisfies

$$\|K^{\text{flow}}\circ\tilde{p}_2 - K_0^{\text{flow}}\circ\tilde{p}_2\|_{2,\gamma^{\text{flow}}} \leqq c^{\text{flow}}(\gamma^{\text{flow}})^{16/15}\|H_1^{\text{flow}}\|^{7/15}_{r_0^{\text{flow}},\gamma^{\text{flow}}}, \tag{28.15}$$

$$\operatorname{supp}(K^{\text{flow}} - K_0^{\text{flow}}) \subset D^{\text{flow}}_{\gamma^{\text{flow}}}, \tag{28.16}$$

and a map $\Sigma^{\text{flow}} \in \operatorname{Diff}^{\text{symp}}_{0,\gamma^{\text{flow}}}(A_{D^{\text{flow}}})$ such that

$$\|\Sigma^{\text{flow}} - \mathrm{id}\|_{1,\gamma^{\text{flow}}} \leqq c^{\text{flow}}(\gamma^{\text{flow}})^{-8/15}\|H_1^{\text{flow}}\|^{4/15}_{r_0^{\text{flow}},\gamma^{\text{flow}}}, \tag{28.17}$$

$$\Sigma^{\text{flow}}(\Xi^{\text{flow}}) \subset D^{\text{flow}}_{3\gamma^{\text{flow}}}, \tag{28.18}$$

$$\operatorname{supp}(\tilde{\Sigma}^{\text{flow}} - \mathrm{id}) \subset \mathbb{R}^n \times D^{\text{flow}}_{\gamma^{\text{flow}}}, \tag{28.19}$$

and

28.20. *the jets of infinite order of* $H^{\text{flow}}\circ\tilde{\Sigma}^{\text{flow}}$ *and* $K^{\text{flow}}\circ\tilde{p}_2$ *coincide at points of* $\tilde{\Xi}^{\text{flow}}$.

Here $\tilde{\Sigma}^{\text{flow}}$ and $\tilde{\Xi}^{\text{flow}}$ are the lifts of Σ^{flow} and Ξ^{flow}, $\Xi^{\text{flow}} = \mathbb{T}^n \times (\omega^{\text{flow}})^{-1}(\mathscr{E}^{\text{flow}})$, $\omega^{\text{flow}} = \operatorname{grad} K^{\text{flow}}$,

$$\mathscr{E}^{\text{flow}} = \mathscr{E}_{\sigma,\gamma^{\text{flow}}} \cap [B^{\text{flow}}]_{4\gamma^{\text{flow}}}.$$

The lift $\tilde{\Sigma}^{\text{flow}}: \mathbb{R}^n \times D^{\text{flow}} \longrightarrow \mathbb{R}^n \times D^{\text{flow}}$ of the map Σ^{flow} may be written in our notations for coordinates as

$$(\bar{\psi}', \bar{\psi}^n, \bar{J}', \bar{J}_n) \longmapsto (\psi', \psi^n, J', J_n), \psi' = \bar{\psi}' + a(\bar{\psi}, \bar{J}),$$
$$\psi^n = \bar{\psi}^n + b(\bar{\psi}, \bar{J}), J' = \bar{J}' + c(\bar{\psi}, 0\bar{J}), J_n = \bar{J}_n + d(\bar{\psi}, 0\bar{J}),$$

where a, b, c, d satisfy the relations which follow from (28.17)–(28.19). We assume that

28.21. δ is sufficiently small so that the image under $\tilde{\Sigma}^{\text{flow}}$ of the hypersurface $\bar{\psi}^n = 0$ is wholly contained in the domain where $\tilde{H}_1^{\text{flow}} = 0$ (see (28.6)). Then we have $\tilde{H}_1^{\text{flow}}\circ\tilde{\Sigma}^{\text{flow}}|_{\bar{\psi}^n=0} = m + (\varepsilon/2)m^2$, where

$$m(\bar{\psi}', \bar{J}', \bar{J}_n) = \bar{J}_n + d(\bar{\psi}', 0, \bar{J}', \bar{J}_n) + K_0(\bar{J}' + c(\bar{\psi}', 0, \bar{J}', \bar{J}_n)). \tag{28.22}$$

The condition (28.20) may be expressed as follows

28.23. for each $\bar{\psi}'$ the jets of infinite order to functions $m + (\varepsilon/2)m^2$ and K^{flow} as functions of the variables $\bar{J}', \bar{J}_n$ only coincide on the set $(\omega^{\text{flow}})^{-1}(\mathscr{E}^{\text{flow}})$.

The K for a Cascade

To derive K from that of the assertion of the discrete time theorem, let us first define $K: D \longrightarrow \mathbb{R}$ as the solution to the equation

$$K^{\text{flow}}(J', -K(J')) = 0. \tag{28.24}$$

Writing $K^{\text{flow}} = K_0^{\text{flow}} + \varkappa^{\text{flow}}$ and $K = K_0 + \varkappa$, one obtains (use (28.5)) the equation

$$\varkappa = (1 - (1 - 2\varepsilon\varkappa^{\text{flow}})^{1/2})/\varepsilon, \quad \varkappa^{\text{flow}} = \varkappa^{\text{flow}}(J', -K_0(J') - \varkappa) \tag{28.25}$$

for determining $\varkappa$. Note that $\|\varkappa^{\text{flow}}\|_{2,\gamma} \leqq \text{const}\,\gamma^{46/15}\delta$. The contraction principle, applied to equation (28.25) in the space $C^0(D)$, gives us the existence and uniqueness of $\varkappa(J')$ provided $\delta \leqq$ const, const only depending on c^{flow}. Then the implicit function theorem yields the existence, and the estimates for, the derivatives of $\varkappa$ up to second order:

$$\|\varkappa\|_{2,\gamma} \leqq \text{const}\,\|\varkappa^{\text{flow}}\|_{2,\gamma} \leqq \text{const}\,\gamma^{16/15}\|H_1\|_{r_0,\gamma}^{7/15}. \tag{28.26}$$

The (28.26) coincides with (11.7).

To localize the support of $\varkappa$, let us set $\gamma' = \gamma^{\text{flow}}(1+\varepsilon)^{-1}$. If $\operatorname{dist}(\omega_0(J'), \partial B) \leqq \gamma'$ then, as follows from (28.9), $\operatorname{dist}(\omega_0^{\text{flow}}(J', J_n), \partial B^{\text{flow}}) \leqq \gamma^{\text{flow}}$ and $\varkappa^{\text{flow}}(J', J_n) = 0$ (28.16). Hence $\varkappa(J') = 0$, and we have the weaker form of the inclusion (11.9) with γ replaced by γ':

$$\operatorname{supp}(K - K_0) \subset D_{\gamma'}. \tag{28.27}$$

At the end of the proof we shall say how to improve the statement.

Embeddings Θ and Θ_{00}

Let us define a symplectic embedding $\Theta: \mathbb{R}^{n-1} \times D \longrightarrow \mathbb{R}^n \times D^{\text{flow}}$ by the formulae $(\phi', I') \longmapsto (\psi, J)$ where

$$\psi' = \phi', \quad \psi^n = 0, \quad J' = I', \quad J_n = -K(I') + \alpha(\phi', I'). \tag{28.28}$$

A function α will be chosen so that the image of Θ is contained in the hypersurface $H^{\text{flow}} = 0$:

$$\begin{aligned}\alpha(\phi', I') = K(I') &- K_0(I' + c(\phi', 0, I', -K(I') + \alpha(\phi', I')))\\ &- d(\phi', 0, I', -K(I') + \alpha(\phi', I'))\end{aligned} \tag{28.29}$$

(use the condition (28.21)).

Let us consider (28.29) as the equation of the form $\alpha = A(\alpha)$ in the ball $B = \{\alpha : \|\alpha\|_0 \leqq 1/3\}$ of the Banach space $C^0_{\text{per},\gamma'}(\mathbb{R}^{n-1} \times D)$. Taking into account (28.15)–(28.19), one may assert that $A(B) \subset B$ and A is a contraction provided δ is sufficiently small. So there exists a unique $\alpha \in B$, a solution to (28.29). Applying the implicit function theorem yields $\alpha \in C^\infty$. We have by the construction

$$\operatorname{supp}\alpha \subset \mathbb{R}^{n-1} \times D_{\gamma'}. \tag{28.30}$$

Differentiating (28.29) gives us, in view of (28.15) and (28.17),

$$\|\alpha\|_{1,\gamma} \leqq \text{const}\,\gamma^{7/15}\|H_1\|_{r_0^{\text{flow}}}^{4/15} \leqq \text{const}\,\gamma\delta^{4/15}. \tag{28.31}$$

Recall that the set Ξ consists, by definition, of all $(\phi', I') \in \mathbb{R}^{n-1} \times D$ such that $\omega(I') \in \mathscr{E}$, $\omega = \operatorname{grad} K$, $\mathscr{E} = \mathscr{E}_{\sigma,\gamma} \cap [B]_{4\gamma}$.

Proposition 28.32. *The jet of infinite order of the function α vanishes at points of Ξ.*

Proof. Let $J' \in \omega^{-1}(\mathscr{E})$. Since K was defined by the equation (28.24).

(28.32a) $$\omega(J') = \omega'_f(J', J_n)/\omega^n_f(J', J_n), \quad J_n = -K(J'), \quad \omega_f = (\omega'_f, \omega^n_f) = \operatorname{grad} K^{\text{flow}}.$$

It is easy to prove, using (28.15) that, if

$$c^{\text{flow}} \| H_1^{\text{flow}} \|^{7/15}_{r_0^{\text{flow}}, \gamma^{\text{flow}}} < \varepsilon, \quad \text{then}$$

(28.32b) $$\omega'_f/\omega^n_f \in \mathscr{E}_{\sigma,\gamma} \Longrightarrow (\omega'_f, \omega^n_f) \in \mathscr{E}_{\sigma,\gamma^{\text{flow}}}.$$

The equation (28.29) for determining α may be written in the form $m(\phi', I', -K(I') + \alpha) = 0$, where m is defined by (28.22). On the other hand, in view of condition (28.23), the function K at points of $\omega^{-1}(\mathscr{E})$ may be defined together with all of its derivatives as the solution of the equation $m(\phi', I', -K(I')) = 0$. Since the solution α of (28.29) is unique, we have α vanishes at $\tilde{\Xi}$ and so do all its variables. □

Define also the symplectic embedding $\Theta_{00}: (\phi', I') \longmapsto (\psi, J)$ by the formulae: $\psi' = \phi', \psi^n = 0, J' = I', J_n = -K_0(I')$. It follows from (28.31) that

(28.33) $$\| \Theta - \Theta_{00} \|_{1,\gamma} \leqq \text{const}\, \gamma^{7/15} \| H_1 \|^{4/15}_{r_0^{\text{flow}}} \leqq \text{const}\, \gamma \delta^{4/15}.$$

Then (28.27) and (28.30) yield

(28.34) $$\operatorname{supp}(\Theta - \Theta_{00}) \subset \mathbb{R}^{n-1} \times D_{\gamma'}.$$

Obviously Θ and Θ_{00} commute with the action of Γ' (see the beginning of the section).

The Σ for a Cascade

We will construct the lift of the map Σ for a cascade by taking a superposition.

(28.35) $$\tilde{\Sigma} = \pi \circ u^{-1} \circ \tilde{\Sigma}^{\text{flow}} \circ \Theta,$$

where $\pi: (\mathbb{R}^{n-1} \times D) \times (\mathbb{R}^1 \times]-1, 1[) \longrightarrow \mathbb{R}^{n-1} \times D$ is the projection $(\phi', I', \phi^n, I_n) \longmapsto (\phi', I')$. One immediately finds $\text{id} = \pi \circ u^{-1} \circ \text{id} \circ \Theta_{00}$. This formula permits us to compare $\tilde{\Sigma}$ and id:

(28.36) $$\begin{aligned} \| \tilde{\Sigma} - \text{id} \|_{1,\gamma} &\leqq \text{const}\, \| \tilde{\Sigma}^{\text{flow}} \circ \Theta - \Theta_{00} \|_{1,\gamma} \leqq \text{const}\, \| \tilde{\Sigma}^{\text{flow}} \circ \Theta - \Theta \|_{1,\gamma} \\ &+ \text{const}\, \| \Theta - \Theta_{00} \|_{1,\gamma} \leqq \text{const}\, \| \tilde{\Sigma}^{\text{flow}} - \text{id} \|_{1,\gamma} \cdot | \Theta |_{1,\gamma} \\ &+ \text{const}\, \| \Theta - \Theta_{00} \|_{1,\gamma} \leqq \text{const}\, \gamma^{-8/15} \| H_1 \|^{4/15}_{r,\gamma} \leqq \text{const}\, \delta^{4/15}. \end{aligned}$$

The inclusion (28.34) shows that $\tilde{\Sigma}^{\text{flow}} \circ \Theta$ coincides with $\tilde{\Sigma}^{\text{flow}} \circ \Theta_{00}$ on $\mathbb{R}^{n-1} \times (D \backslash D_{\gamma'})$. Since Θ_{00} maps the latter set into $\mathbb{R}^n \times (D^{\text{flow}} D^{\text{flow}}_{\gamma'})$, and $\tilde{\Sigma}^{\text{flow}}$ coincides with id there, we have

(28.37) $$\operatorname{supp}(\tilde{\Sigma} - \text{id}) \subset \mathbb{R}^{n-1} \times D_{\gamma'}.$$

It follows from (28.36), (28.37), and Lemma AII.21, that $\tilde{\Sigma}$ is a diffeomorphism provided δ is sufficiently small.

Proposition 28.38. *$\tilde{\Sigma}$ is symplectic.*

Proof. Let us represent $\tilde{\Sigma}$ as the superposition of diffeomorphisms:

(28.38a) $$\tilde{\Sigma} = \pi' \circ (u^{-1})' \circ (\tilde{\Sigma}^{\text{flow}})' \circ \Theta,$$

where the prime denotes the restriction on the image of the precedent from the right. All images mentioned are symplectic submanifolds. We are going to show that the members of decomposition (28.38a) are symplectic. The Θ is symplectic because it is a symplectic embedding (see formulae (28.28)). As for the two middle terms in (28.38a), they are obviously symplectic since they arose from symplectic diffeomorphisms of surrounding spaces. Let us prove that $\pi': S' \longrightarrow \mathbb{R}^{n-1} \times D$ are symplectic where $S' = \text{im}(u^{-1} \circ \tilde{\Sigma}^{\text{flow}} \circ \Theta)$. It may be represented as a superposition $\pi' = \pi \circ i$ where $i: S' \hookrightarrow (\mathbb{R}^{n-1} \times D) \times (\mathbb{R}^1 \times]-1, 1[)$ is the inclusion. Denote $\Omega' = d\phi' \wedge dI'$, the symplectic structure on $\mathbb{R}^{n-1} \times D$, and let $\Omega = d\phi' \wedge dI' + d\phi^n \wedge dI'_n$ be that of $(\mathbb{R}^{n-1} \times D) \times (\mathbb{R}^1 \times]-1, 1[)$. The symplectic structure on S' is $i^*\Omega$. We are to show the equality $I^*\Omega = i^*\pi^*\Omega'$. But $\Omega = \pi^*\Omega' + d\phi^n \wedge dI'_n$. So it is sufficient to prove $i^*d\phi^n \wedge dI'_n = 0$. This is evident since the surface S' satisfies the equation $I_n = 0$. □

The $\Sigma^{-1} \circ f^t \circ \Sigma$

It is sufficient to investigate the titled map at $t = 1$. By the construction $f = f^t$ coincides with the Poincaré map of the flow on $Z/\tilde{\Gamma}$ with Hamiltonian $I_n + (\varepsilon/2)I_n^2$, if one takes the quotient projection of the submanifold $S = \{\phi^n = 0, I_n = 0\}$ as the surface of section. Since $\pi': S' \longrightarrow S$ is a symplectic diffeomorphism which follows the lines of the vector field of our Hamiltonian, this Poincaré map is isomorphic (via π') to that generated on the quotient projection of S'. Applying the symplectic diffeomorphism $\tilde{\Sigma}^{\text{flow}} \circ u^{-1}$ carries this picture into $\mathbb{R}^n \times D^{\text{flow}}$. Under that map the surface S' goes diffeomorphically into $\hat{S} = \Theta(\mathbb{R}^{n-1} \times D)$, and the flow with Hamiltonian $I_n + (\varepsilon/2)I_n^2$ into the flow $\{F^t, t \in \mathbb{R}\}$ generated by the Hamiltonian $\tilde{H}^{\text{flow}}$. Near the surface $\hat{S}$ it does not differ from that of the Hamiltonian H^{flow}. The map $\Sigma^{-1} \circ f^t \circ \Sigma$ may be constructed as follows. Let S_0 be a surface in $\mathbb{R}^n \times D^{\text{flow}}$ defined by the equation $\bar{\psi}^n = 0$, and S_1 be that defined by the equation $\bar{\psi}^n = 1$. Since $\tilde{H}^{\text{flow}}$ is close to K_0^{flow}, every orbit started at a point $(\bar{\psi}', 0, \bar{J}', \bar{J}_n) \in S_0$ reaches S_1 at time $\tau(\bar{\psi}', 0, \bar{J}', \bar{J}_n)$, and the orbit intersects S_1 transversely. Denote $\hat{F} = \hat{g} \circ F^\tau: S_0 \longrightarrow S_0$ ($\hat{g}$ is the special generator of $\hat{\Gamma}$ defined above). Obviously, $\hat{F}$ is a diffeomorphism. Since a flow preserves its Hamiltonian, and $\hat{S} = S_0 \cap \{H^{\text{flow}} = 0\}$, we have $\hat{F}(\hat{S}) = \hat{S}$. It follows from our construction that

$$\tilde{F} \stackrel{\text{def}}{=} \tilde{\Sigma}^{-1} \circ \tilde{f} \circ \tilde{\Sigma} = \Theta^{-1} \circ \hat{F}|_{\hat{S}} \circ \Theta.$$

Side by side with $\tilde{F}$ we construct the map $\tilde{F}_0 = \Theta_0^{-1} \circ \hat{F}_0|_{\hat{S}_0} \circ \Theta_0: \mathbb{R}^{n-1} \times D \longrightarrow \mathbb{R}^{n-1} \times D$ where Θ_0 is defined by the relations $(\phi', I') \longmapsto (\psi', \psi^n, J', J_n)$, $\psi' = \phi'$, $\psi^n = 0$, $J' = I'$, $J_n = -K(I')$, $\hat{S}$ is the image of Θ_0, and $\hat{F}_0$ is defined in the same manner as $\hat{F}$ with $\hat{S}_0$ standing for $\hat{S}$ and with the flow of the Hamiltonian K^{flow} instead of that of H^{flow}. As one can easily compute, $\tilde{F}_0(\phi', I') = (\phi' + \omega'(I'), I')$ where ω' is given by (28.32a). The estimates (28.15), (28.14), (28.31), and inclusions (28.30),

(28.27) result in

(28.39) $$\|\tilde{F}-\tilde{F}_0\|_{1,\gamma} \leqq \text{const}\,\gamma\delta^{4/15},$$

(28.40) $$\operatorname{supp}(\tilde{F}-\tilde{F}_0) \subset \mathbb{R}^{n-1} \times D_{\gamma'}.$$

Applying Lemma AII.21 yields that F is a factorizible diffeomorphism (see §8) provided δ is sufficiently small. Due to 28.32, (28.32a, b), and 28.20, we have:

28.41. *$\tilde{F}$ coincides with $\tilde{F}_0$ together with all of its derivatives at points of $\hat{\Sigma} = \mathbb{R}^{n-1} \times \omega^{-1}(\mathscr{E})$.*

In particular, the factor F of $\tilde{F}$ with respect to the action of Γ' possesses invariant tori. So the Calabi invariant of F is zero (8.2), and F possesses a Hamiltonian $\mathbf{H} \in C^{\infty}_{\text{per},0}(\mathbb{R}^{n-1} \times D)$ (8.6). We normalize $\mathbf{H}$ by demanding that

(28.42) $$\mathbf{H}(\phi'_0, I'_0) = K(I'_0)$$

at one chosen point $I'_0 \in \omega^{-1}(\mathscr{E})$. It appears that this is sufficient for $\tilde{F}$ to coincide with $\tilde{F}_0$ together with all their derivatives at all points of $\omega^{-1}(\mathscr{E})$.

Proposition 28.43. $\mathbf{H}$ *coincides with $K \circ \tilde{p}_2$ together with all of its variables at points of $\tilde{\Sigma}$.*

Proof. Since the derivatives of $\mathbf{H}$ and $K \circ \tilde{p}_2$ are the components of generated diffeomorphism, their coincidence is a consequence of (28.41). It remains to check the equality $\mathbf{H} = K \circ \tilde{p}_2$ at points of $\tilde{\Sigma}$. Let $A, B \in \tilde{\Sigma}$, and $A = (\phi'_0, I'_0)$ be the point where the equality was satisfied by definition (see (28.42)). Denote $\tilde{A} = \Theta(A)$, $\tilde{B} = \Theta(B)$, and let l be a path joining $\tilde{A}$ with $\tilde{B}$ and lying in the image of Θ. Consider the film $\mathscr{F}$ generated by the segments of trajectories of $\{F^t\}$ which start at points of l and finish at points with $\psi^n = 1$ (see Fig. 51). The path l goes along these trajectories to the final path $\bar{l}$ with the ends $\bar{A}, \bar{B}$. The $\mathscr{F}$ lies wholly in the energy hypersurface $H^{\text{flow}} = 0$, and it is easy to derive from this that $\mathscr{F}$ is isotropic. Hence $\int_{\partial \mathscr{F}} \psi \, dJ = 0$. Since $A, B \in \tilde{\Sigma}$, the moment J is constant along the trajectories joining

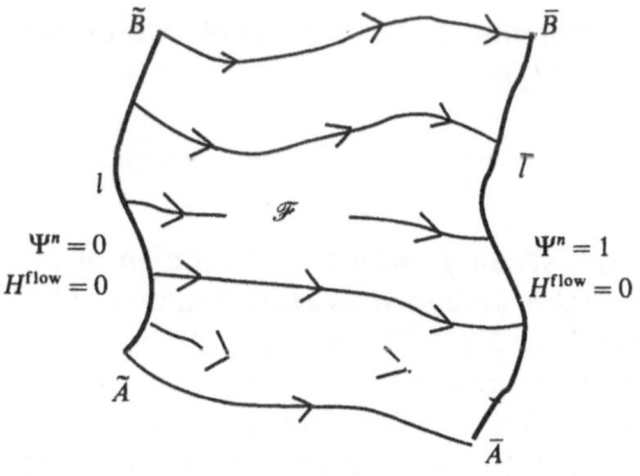

Fig. 51

$\tilde{A}$ with $\bar{A}$ and $\tilde{B}$ with $\bar{B}$ (see 28.41), i.e. $\int \psi \, dJ = 0$ along these segments of trajectories. Hence

(28.43a)
$$\int_l \psi \, dJ = \int_{\bar{l}} \psi \, dJ.$$

We have for the variables in (28.43a): $\psi = (\psi', \psi^n)$, $J = (J', J_n)$, $\psi^n = 0$ on l, $\psi^n = 1$ on $\bar{l}$. The values of ψ' and J' at points of $\bar{l}$ are functions of those on l at starting points of trajectories: $\psi'|_{\bar{l}} = \bar{\psi}'(\psi', J')$, $J'|_{\bar{l}} = \bar{J}'(\psi', J')$. Since the flow F^t generates a cascade $\hat{F}^t$ on $\hat{S}$ whose image under Θ is one with Hamiltonian $\mathbf{H}$, the functions $\bar{\psi}', \bar{J}'$ satisfy the equations (cf. 8.4):

(28.43b)
$$J' = \bar{J}' + \partial \mathbf{H}(\psi', \bar{J}')/\partial \psi', \qquad \bar{\psi}' = \psi' + \partial \mathbf{H}(\psi', \bar{J}')/\partial \bar{J}'.$$

Substituting (28.43b) into (28.43a) and taking into account the mentioned values of ψ^n at points of l and $\bar{l}$, yields

(28.43c)
$$\int_{\tilde{A}}^{\tilde{B}} \psi' \, dJ' = \int_{\tilde{A}}^{\tilde{B}} \psi' \, d\bar{J}'(\psi', J') + \int_{\tilde{A}}^{\tilde{B}} \psi' \, d(\partial \mathbf{H}(\psi', \bar{J}')/\partial \psi')$$

(*the left-hand side of* (28.43a)) $= \int_{\tilde{A}}^{\tilde{B}} \psi' \, d\bar{J}'(\psi', J') + \int \partial \mathbf{H}(\psi', \bar{J}')/\partial \bar{J}' \, d\bar{J}' + J_n(\bar{B}) - J_n(\bar{A})$ (*the right-hand side of* (28.43a)). Note that $\partial \mathbf{H}/\partial \psi' = 0$ at the end points of paths, since they lie on KAM tori. The formula of integration by parts then yields

$$\int_{\tilde{A}}^{\tilde{B}} \psi' \, d(\partial \mathbf{H}(\psi', \bar{J}')/\partial \psi') = -\int_{\tilde{A}}^{\tilde{B}} \partial \mathbf{H}(\psi', \bar{J}')/\partial \psi' \, d\psi'.$$

Combining these facts, we obtain from (28.43c)

(28.43d)
$$\mathbf{H}(\psi'_A, J'_A) - \mathbf{H}(\psi'_B, J'_B) = -\int \partial \mathbf{H}/\partial \psi' \, d\psi' - \int \partial \mathbf{H}/\partial J' \, dJ'$$
$$= J_n(\bar{B}) - J_n(\bar{A}).$$

Here $(\psi'_A, J'_A) = (\phi'_0, I'_0)$, (ψ'_B, J'_B) denote the coordinates of the points A and B. The norming condition (28.42) gives us

$$\mathbf{H}(\phi'_0, I'_0) = \mathbf{H}(\psi'_A, J'_A) = K(I'_0) = -J_n(A).$$

Cancelling the corresponding terms in (28.43d), we obtain the desired equality $\mathbf{H}(\psi'_B, J'_B) = -J'(B) = K(J'_B)$. □

The End of the Proof

Given a discrete time system f^t with a Hamiltonian of the form (28.1), we have constructed: an integrable system with Hamiltonian K (28.24) and a conjugating symplectic diffeomorphism Σ (28.35), such that $\Sigma^{-1} \circ f^t \circ \Sigma$ possesses a Hamiltonian $\mathbf{H}$ whose jets of infinite order coincide with those of $K \circ \tilde{p}_2$ at points of $\tilde{\Xi} = \mathbb{R}^{-1} \times \omega^{-1}(\mathscr{E})$ (Proposition 28.43). So the condition (ii) in the conclusion of the Theorem is satisfied. Since (i) is a consequence of (ii), it is also satisfied. The validity of estimates (11.7) and (11.8) is established in (28.26) and (28.36). The inclusions

(28.27) and (28.37) differ from (11.9) and (11.10) by replacing γ by $\gamma' = \text{const}\,\gamma, 0 < \text{const} < 1$. To remove this disagreement, let us change D for $\hat{D} = D_\gamma$ and γ for $\hat{\gamma} = \gamma/2$, and let us then fulfill our preceding construction for the changed position. The functions obtained will satisfy (11.9), (11.10) being extended by zero onto the initial domains.

Note, finally, that all constructed objects behave properly with respect to the action of the group of integer translations Γ'.

Notes to Chapter IV

The proof given here is nothing more than a development of that given in the original paper by Moser (1962), but the formal scheme is adopted from Pöschel (1982). I followed the ideas of R. Douady (1982) in proving that the discrete time version of the theorem is a consequence of that of continuous time (§28).

Part II. Eigenfunctions Asymptotics

Chapter V. Laplace–Beltrami–Schrödinger Operator and Quasimodes

This chapter plays an auxiliary role in our exposition. It contains necessary definitions and facts concerning operators, spectra, and quasimodes.

§29. Basic Facts about Self-Adjoint Operators and Spectra

Here we recall the main definitions about linear operators in Hilbert spaces and their spectra. We refer the reader to [Birman & Solomjak (1987)] for a detailed account. The letter $\mathfrak{H}$ will denote a Hilbert space supplied with an inner product $(\!(.,.)\!)$, a positively defined sesquilinear form, being complete with respect to the norm $\|\cdot\|$, where for $u \in \mathfrak{H}$, $\|u\|^2 = (\!(u, u)\!)$.

Let $\boldsymbol{L}: \mathscr{D}(\boldsymbol{L}) \longrightarrow \mathfrak{H}$ be a linear map, $\mathscr{D}(\boldsymbol{L})$, the domain of $\boldsymbol{L}$, being a linear subset of $\mathfrak{H}$. The *resolvent set of* $\boldsymbol{L}$ is a subset of $\mathbb{C}$ consisting of all complex numbers z such that there exists $(\boldsymbol{L} - \lambda \mathrm{id})^{-1}$, bounded and defined on the whole space $\mathfrak{H}$. The complement to the resolvent set is called the *spectrum of* $\boldsymbol{L}$. Particularly, *eigenvalues*, i.e. numbers λ such that the equation $\boldsymbol{L}u = \lambda u$ has solutions with $u \neq 0$, called *eigenvectors*, belong to the spectrum and form the *point spectrum*. The *multiplicity* of an eigenvalue λ is the dimension of its *eigenspace*, the subspace spanned by all eigenvectors corresponding the said eigenvalue λ.

Let $\mathscr{D}(\boldsymbol{L})$, the domain of $\boldsymbol{L}$, be a dense linear subset of $\mathfrak{H}$. Then one can define the *adjoint map* $\boldsymbol{L}^*: \mathscr{D}(\boldsymbol{L}^*) \longrightarrow \mathfrak{H}$, as follows: $u \in \mathscr{D}(\boldsymbol{L}^*)$ if and only if there exists a $w \in \mathfrak{H}$ such that for all $v \in \mathscr{D}(\boldsymbol{L})$ $(\!(\boldsymbol{L}v, u)\!) = (\!(v, w)\!)$; then, by definition, $\boldsymbol{L}^*u = w$. A densely defined linear map $\boldsymbol{L}$ is called *self-adjoint*, or a *self-adjoint* operator, if $\boldsymbol{L} = \boldsymbol{L}^*$.

The spectrum of a self-adjoint operator is contained in the real axis. The *discrete spectrum* of a self-adjoint $\boldsymbol{L}$ consists of all isolated eigenvalues of finite multiplicity (one can prove that any isolated point of the spectrum of a self-adjoint operator is an eigenvalue). All the remaining points of the spectrum form the *continuous spectrum*. There are analogous definitions for the part of the spectrum lying in a given interval of the real axis. For instance, we say the spectrum of $\boldsymbol{L}$ in the interval $]a, b[$ is discrete if each point of the spectrum, which belongs to $]a, b[$, is an isolated eigenvalue of finite multiplicity. The spectral theorem asserts that, given a self-adjoint operator $\boldsymbol{L}$, there exists an operator valued measure $\boldsymbol{E}$ defined on the set of all Borel subsets of $\mathbb{R}$. The values of $\boldsymbol{E}$ are self-adjoint projections in $\mathfrak{H}$, such that

$$L = \int_{\mathbb{R}} \lambda E(d\lambda) \tag{29.1}$$

and $\mathscr{D}(L) = \{u \in \mathfrak{H}: \int \lambda^2 \langle\!\langle E(d\lambda)u, u\rangle\!\rangle < \infty\}$. We call E the *spectral measure* of L. The support of E coincides with the spectrum of L. The eigenspace corresponding to an eigenvalue λ is $E(\{\lambda\})$. A number λ is an eigenvalue if and only if $E(\{\lambda\}) \neq 0$. If δ is a Borel subset of $\mathbb{R}$ with no intersection with the continuous spectrum, then $E(\delta) = \sum_{\lambda \in \delta} E(\{\lambda\})$, and $E(\{\lambda\})u = \sum_{k=1}^{N} \langle\!\langle u, e_k\rangle\!\rangle e_k$, where N is the multiplicity of an eigenvalue λ, and e_k, $1 \leqq k \leqq N$ is a vector base in the eigenspace corresponding to λ.

The self-adjoint property is difficult to check. The weaker-one, the *symmetry property*, is easier in that respect. A linear map $L: \mathscr{D}(L) \longrightarrow \mathfrak{H}, \mathscr{D}(L) \subset \mathfrak{H}$, is called *symmetric* if (i) $\mathscr{D}(L)$ is a linear and dense subset of $\mathfrak{H}$, and (ii) $\langle\!\langle Lu, v\rangle\!\rangle = \langle\!\langle u, Lv\rangle\!\rangle$ for all $u, v \in \mathscr{D}(L)$. Any self-adjoint operator is symmetric, but the converse is false in the case of infinite-dimensional $\mathfrak{H}$. Let L be a symmetric linear map. A symmetric linear map L' is called a *symmetric extension* of L if $\mathscr{D}(L') \supset \mathscr{D}(L)$ and $L'|\mathscr{D}(L) = L$. One of the problems in operator theory is to construct a self-adjoint extension of a given symmetric operator L. In general, this problem may have no solution, or have many solutions. There is one important case where this problem admits a regular procedure for solving; namely the case of semibounded symmetric operators. A symmetric linear map L is called *semibounded* from below if there is a real constant C such that $\langle\!\langle Lu, u\rangle\!\rangle \geqq C\|u\|^2$ for all $u \in \mathscr{D}(L)$. If $C > 0$ L is called *positively defined.* A semibounded operator can be made positively defined by adding the id multiplied by a sufficiently large positive constant. So it is sufficient to discuss only positively defined operators. If L is positively defined, then $\mathscr{D}(L)$ supplied with the new inner product $\langle\!\langle .,. \rangle\!\rangle_L$ defined as $\langle\!\langle u, v\rangle\!\rangle_l = \langle\!\langle Lu, v\rangle\!\rangle$ is a pre-Hilbert space. This can be extended to a Hilbert space $\tilde{\mathscr{D}} \supset \mathscr{D}(L)$ so that $\tilde{\mathscr{D}} \subset \mathfrak{H}$. The inner product $\langle\!\langle .,. \rangle\!\rangle_1$ of $\tilde{\mathscr{D}}$ can then be written as $\langle\!\langle u, v\rangle\!\rangle_1 = \langle\!\langle \tilde{L}^{1/2}u, \tilde{L}^{1/2}v\rangle\!\rangle$, where $\tilde{L}$ is a positively defined self-adjoint operator, and the domain $\tilde{\mathscr{D}}$ then coincides with $\mathscr{D}(L^{1/2})$. The uniquely constructed operator $\tilde{L}$ is called the *Friedrichs extension* of L. As mentioned above, this procedure can be applied to semibounded symmetric operators; one must add $C \cdot id$ with $C > 0$ large enough, apply the above procedure, and then $C \cdot id$. The result is then independent of C.

§30. Laplace–Beltrami–Schrödinger Operator

Let (Q, g) be a Riemannian manifold, that is let Q be a manifold and $g: Q \longrightarrow S^2T^*Q$ be a smooth section of the bundle of symmetric bilinear forms on Q (see AI.39, 40) whose values are positively defined forms. The manifold Q may contain the boundary ∂Q, and it may or may not be compact. The letter n will denote the dimension of Q. We start with the definition of one special measure on Q.

Measure μ

The Riemannian metric $\boldsymbol{g}$ has the following expression in local coordinates $(q^1, q^2, \ldots, q^n)$ of a smooth chart $\alpha : U \longrightarrow \mathbb{R}^n_{()}$:

$$\boldsymbol{g}|_U = \sum_{i,j=1}^{n} g_{ij}\, dq^i \odot dq^j \tag{30.1}$$

(see (6.1a) for the definition of the symmetrized product $\odot$). in (30.1) the g_{ij} are smooth real-valued functions defined on U. Denote g by the determinant of the matrix $\{g_{ij}\}$. Since $\boldsymbol{g}$ is positively defined, g is a positive function, but depends on the choice of the chart (U,α). We call g the *determinant of the metric in a chart* (U,α). Let $\alpha' : U' \longrightarrow \mathbb{R}^n_{(+)}$ be another smooth chart, and g' be the determinant of the corresponding matrix $\{g'_{kl}\}$ of $\boldsymbol{g}$ in the coordinates $(q'^1, q'^2, \ldots, q'^n)$ of (U', α'). If $U \cap U' \neq \varnothing$ then one obtains, using the bilinearity of the $\odot$-product, that

$$g'_{kl} = \sum_{i,j=1}^{n} g_{ij} \frac{\partial q^i}{\partial q'^k} \cdot \frac{\partial q^j}{\partial q'^l} \quad \text{on } U \cap U' \tag{30.2}$$

and

$$g' = g\left(\det\left\{\frac{\partial q^i}{\partial q'^k}\right\}\right)^2 \tag{30.3}$$

For $\phi \in C_c(Q, \mathbb{C})$, we define the integral $\int \phi\, d\mu$ by setting

$$\int \phi\, d\mu = \sum_k \int_{U_k} (\chi_k \phi) \circ \alpha_k^{-1}(q^1, q^2, \ldots, q^n) \sqrt{g_k}\, dq^1 dq^2 \cdots dq^n, \tag{30.4}$$

where $\{(U_k, \alpha_k)\}$ is a finite collection of charts whose domains U_k cover $\operatorname{supp} \phi$, $(q^1, q^2, \ldots, q^n)$ and g_k are respectively the coordinates and the determinant of the metric in (U_k, α_k), $\{\chi_k\}$ is a partition of unity subordinated to the covering $\{U_k\}$ such that $\operatorname{supp} \chi_k \subset U_k$, and $\sum_k \chi_k = 1$. It follows from (30.3) that the right-hand side of (30.4) is independent of the choice of (U_k, α_k, χ_k).

Thus the form (30.4) defines a measure $\mu = \mu_g$ on Q whose coordinate expression in a chart (U, α) with coordinates $(q^1, q^2, \ldots, q^n)$ is

$$d\mu|_U = \sqrt{g}\, dq^1 dq^2 \cdots dq^n \tag{30.5}$$

Remark 30.6. If Q is orientable, then one defines a volume form $d\hat{\mu}$ which is represented in the oriented chart (U, α) as

$$d\hat{\mu} = \sqrt{g}\, dq^1 \wedge dq^2 \wedge \cdots \wedge dq^n \tag{30.7}$$

In that case the integral (30.4) is nothing more than the integral of the n-form $\phi\, d\hat{\mu}$ over the oriented manifold Q. In our treatment of the subject we need not confine ourselves to the orientable case. So we use the measure (30.5) instead of the volume form (30.7).

Since the measure (30.5) and the integral (30.4) ar defined, one introduces the Hilbert space $L_2(Q, \mu)$ as follows. The elements of $L_2(Q, \mu)$ are classes of measurable functions $\phi : Q \longrightarrow \mathbb{C}$. Two functions belong to the same class if their difference vanishes almost everywhere with respect to the measure μ, such that

$$\|\phi\|^2 = \int |\phi|^2 \, d\mu < \infty. \tag{30.8}$$

The right-hand side of (30.8) is independent of the choice of the representative of a class. The square root of (30.8) is, by definition, the L_2-norm of the class of ϕ. The operations of sum and multiplication by a complex number acting on functions satisfying (30.8), results in functions also satisfying (30.8), and the class of the result again is independent of the choice of the representatives; so the said operations are naturally defined on $L_2(Q, \mu)$, and $L_2(Q, \mu)$ becomes a vector space over $\mathbb{C}$. The inner product in $L_2(Q, \mu)$ is defined as

$$(\!(\phi, \psi)\!) = \int \phi \bar{\psi} \, d\mu \tag{30.9}$$

(the top line denotes complex conjugation). If ϕ and ψ satisfy (30.8) then the right-hand side of (30.9) exists, is finite, and is independent of the choice of ϕ and ψ in their classes. Further we shall not distinguish between the classes and their representatives, so ϕ and ψ in (30.9) may be regarded as elements of $L_2(Q, \mu)$. The formula (30.9) defines a positively defined sesquilinear form which turns $L_2(Q, \mu)$ into a Hilbert space.

It will be of use for us to distinguish the following linear subsets of $L_2(Q, \mu)$:

$C_c^\infty(Q)$ consists of all smooth complex valued functions on Q with compact support;

$C_0^\infty(Q)$ consists of all $\phi \in C_c^\infty(Q)$ which vanish on the boundary: $\phi|_{\partial Q} = 0$;

$C_{00}^\infty(Q)$ consists of those functions whose supports have no intersection with the boundary.

Evidently

$$C_{00}^\infty(Q) \subset C_0^\infty(Q) \subset C_c^\infty(Q) \subset C_c(Q) \subset L_2(Q, \mu). \tag{30.10}$$

It is well known that all the sets in (30.10) are dense in $L_2(Q, \mu)$ with the topology defined by the norm (30.8).

Laplace–Beltrami Operator

Consider the sesquilinear Dirichlet form $\mathscr{D}$ defined on the functions $\phi, \psi \in C_c^\infty(Q)$ by the formula

$$\mathscr{D}(\phi, \psi) = \int \langle d\phi, d\psi \rangle \, d\mu, \tag{30.11}$$

where $\langle .,. \rangle$ denotes the inner product on the fibers of the cotangent bundle T^*Q defined by (6.3).

Proposition 30.12. *There is uniquely defined linear map $\Delta : C_c^\infty(Q) \longrightarrow C_c^\infty(Q)$ such that for any $\phi \in C_c^\infty(Q)$ and any $\psi \in C_0^\infty(Q)$*

(a) $\mathscr{D}(\phi, \psi) = -\langle\!\langle \Delta\phi, \psi \rangle\!\rangle$
(b) $\operatorname{supp} \Delta\phi \subset \operatorname{supp} \phi$

and in the local coordinates $(q^1, q^2, \ldots, q^n)$ *of a chart* (U, α)

$$\text{(c)}\quad \Delta \cdot|_U = \alpha^* \circ \left[\frac{1}{\sqrt{g}} \sum_{i,j=1}^{n} \frac{\partial}{\partial q^i} \left(\sqrt{g} g^{ij} \frac{\partial}{\partial q^j} \right) \right] \circ \alpha^{*-1}$$

Here $\{g^{ij}\}$ *is the inverse matrix to the matrix* $\{g_{ij}\}$ *of* $\mathbf{g}$.

The operator Δ is called a Laplace–Beltrami operator.

Proof. Given open $U \subset Q$, let $C^\infty_{0,U}(Q)$ $(C^\infty_{c,U}(Q))$ denote the subset of $C^\infty_0(Q)(C^\infty_c(Q)$ correspondingly) consisting of functions ϕ with $\operatorname{supp}\phi \subset U$. Let $\phi \in C^\infty_{c,U}(Q)$ and $\psi \in C^\infty_{0,U}(Q)$, where U is the domain of a chart with coordinates $(q^1, q^2, \ldots, q^n)$. We now obtain an explicit expression for the form $\mathscr{D}$ in these coordinates. Applying (6.6b), we have

$$\text{(d)}\qquad (d\phi)^{\#(g)} = \sum_{i,j=1}^{n} g^{ij} \frac{\partial \phi}{\partial q^i} \cdot \frac{\partial}{\partial q^j}$$

Since $dq^i \odot dq^j(\partial/\partial q^j, \partial/\partial q^j) = \frac{1}{2}(\delta^i_k \delta^j_l + \delta^i_l \delta^j_k)$, immediate calculation yields $\langle d\phi, d\bar{\psi} \rangle = \sum_{i,j=1}^{n} g^{ij} (\partial\phi/\partial q^i) \cdot (\partial\bar{\psi}/\partial q^j)$. So (30.11) becomes

$$\text{(e)}\qquad \mathscr{D}(\phi, \psi) = \int \sum_{i,j=1}^{n} g^{ij} \frac{\partial \phi}{\partial q^i} \cdot \frac{\partial \bar{\psi}}{\partial q^j} \sqrt{g}\, dq.$$

We apply the formula of integration by parts to the right-hand side of (e), taking into account that the boundary terms vanish due to the nullity of ψ on ∂Q, to give

$$\text{(f)}\qquad \mathscr{D}(\phi, \psi) = \int \left[-\frac{1}{\sqrt{g}} \sum_{i,j=1}^{n} \frac{\partial}{\partial q^j} \left(\sqrt{g} g^{ij} \frac{\partial \phi}{\partial q^i} \right) \right] \bar{\psi} \sqrt{g}\, dq = -\langle\!\langle \Delta_{U,\alpha}\phi, \psi \rangle\!\rangle,$$

where $\Delta_{U,\alpha}$ denotes the operator $\Delta_{U,\alpha} : C^\infty_{c,U}(Q) \longrightarrow C^\infty_{c,U}(Q)$, defined by the right-hand side of (c).

Lemma (g) *Let* U *be an open subset of* Q, *then a linear map* $A : C^\infty_{c,U}(Q) \longrightarrow C^\infty_{c,U}(Q)$ *such that* $\langle\!\langle A\phi, \psi \rangle\!\rangle = \mathscr{D}(\phi, \psi)$ *for any* $\phi \in C^\infty_{c,U}(Q), \psi \subset C^\infty_{0,U}(Q)$ *is unique if it exists.*

Proof. Suppose there is another such map A'. Then $\langle\!\langle (A - A')\phi, \psi \rangle\!\rangle = 0$ for all $\psi \in C^\infty_{0,U}(Q)$. Since $C^\infty_{0,U}(Q)$ is dense in $L_2(U)$, we have $A\phi = A'\phi$, i.e. $A = A'$. □

(h) **Corollary.** *The uniqueness assertion of the Proposition. (Apply Lemma to the case* $U = Q$.)

(i) **Corollary.** *Let* (U, α) *and* (V, β) *be two charts, and let* $\phi \in C^\infty_{c,U}(Q) \cap C^\infty_{c,V}(Q)$. *Then* $\Delta_{U,\alpha}\phi = \Delta_{V,\beta}\phi$.

To define the global operator Δ on a function $\phi \in C_c^\infty(Q)$, given a point $x \in Q$, take a chart (U, α), $U \ni x$, fix a smooth function $\chi \in C_{0,U}^\infty(Q)$ such that $\chi|_v = 0$ for some open set $x \in V \subset U$, and set

$$\Delta\phi(x) = \Delta_{U,\alpha}(\chi \cdot \phi)(x). \tag{j}$$

The explicit formula (c) for $\Delta_{U,\alpha}$ implies $\Delta\phi(x)$ is independent of the choice of χ. Corollary (i) implies the independence of the choice of the chart. Clearly $\Delta\phi \in C_c^\infty(Q)$. If $x \notin \operatorname{supp} \phi$ then $\Delta\phi(x) = 0$. This proves (b). The expression (c) follows from the definition (j). □

The Definition of the Laplace–Beltrami–Schrödinger Operator

Here we define our main object, the Laplace–Beltrami–Schrödinger operator (LBS), which is, in fact, a family $\{\mathscr{H}_\hbar, \hbar \in]0, 1]\}$ of self-adjoint operators in $L_2(Q, \mu)$. To define the LBS we need one thing apart from the Riemannian metric, namely the *potential*. This is a smooth family $\{V(.,\hbar), \hbar \in [0, 1]\}$ of smooth real-valued functions defined on Q, i.e. a function $V\colon Q \times [0, 1] \longrightarrow \mathbb{R}$. The *pre-LBS* operator $\mathscr{H}_\hbar^{\text{pre}}\colon C_c^\infty(Q) \longrightarrow C_c^\infty(Q)$ is defined by the formula

$$\mathscr{H}_\hbar^{\text{pre}}\phi(x) = -\frac{\hbar^2}{2}\Delta\phi(x) + V(x, \hbar)\cdot\phi(x). \tag{30.13}$$

Let $\mathscr{H}_{\hbar,0}^{\text{pre}}$ denote the restriction of $\mathscr{H}_\hbar^{\text{pre}}$ onto $C_0^\infty(Q)$. The essential feature of $\mathscr{H}_\hbar^{\text{pre}}$ is its symmetry:

$$(\!(\mathscr{H}_{\hbar,0}^{\text{pre}}\phi, \psi)\!) = (\!(\phi, \mathscr{H}_{\hbar,0}^{\text{pre}}\psi)\!), \quad \forall \phi, \psi \in C_0^\infty(Q), \tag{30.14}$$

which follows immediately from Proposition 30.12. The operator $\mathscr{H}_{\hbar,0}^{\text{pre}}$ is not self-adjoint. One needs to apply the extension procedure to obtain a more satisfactory mathematical object, i.e. a self-adjoint operator. The convenient procedure of extension is the one by Friederichs described in §29. To be able to apply this we impose the following semiboundness condition on the potential V: there exists a constant C such that

$$\int V|\phi|^2\, d\mu \geqq C\|\phi\|^2 \tag{30.15}$$

for all $\phi \in C_0^\infty(Q)$. The condition (30.15) is obeyed, for instance, if V is semibounded from below: i.e. $V(x) \geqq C \;\forall\; x \in Q$. If (30.15) is fulfilled, the Friederichs procedure gives rise to a uniquely defined self-adjoint operator $\mathscr{H}_\hbar$ which is an extension of $\mathscr{H}_{\hbar,0}^{\text{pre}}$. We call $\mathscr{H}_\hbar$ the *Laplace–Beltrami–Schrödinger operator*, or, in more detail, the *Laplace–Beltrami–Schrödinger operator with Dirichlet boundary conditions.*

Remark 30.16. Given a smooth function $\sigma\colon \partial\Omega \longrightarrow \mathbb{R}$, consider the operator $\mathscr{H}_{\hbar,\sigma}^{\text{pre}} = \mathscr{H}_\hbar^{\text{pre}}|_{\mathscr{D}^\sigma}$, where $\mathscr{D}^\sigma$ consists of all $\phi \in C_c^\infty(Q)$ satisfying the *mixed boundary condition*

$$\left.\left(\frac{\partial\phi}{\partial n} - \sigma\phi\right)\right|_{\partial Q} = 0. \tag{30.16a}$$

Here $\partial/\partial n$ is the normal derivative to ∂Q, it has the following expression in local coordinates (see (6.9b) and (6.6a, b)):

$$\frac{\partial}{\partial n} = -\frac{1}{\sqrt{g^{11}}} \sum_{i=1}^{n} g^{1i} \frac{\partial}{\partial q^i} \tag{30.16b}$$

The set $\mathscr{D}^\sigma$ is dense in $L_2(Q, \mu)$ since it contains $C_0^\infty(Q)$. Let σ satisfy the inequality

$$\int \sigma |\phi|^2 \, d\mu_{\partial Q} \geqq C_2 \|\phi\|^2. \tag{30.16c}$$

In (30.16c) $d\mu_{\partial Q}$ is the measure on ∂Q generated by $\boldsymbol{g}|_{T\partial Q}$. It has the following local coordinate expression:

$$d\mu_{\partial Q} = \sqrt{g^{11}}\sqrt{g}\, dq^2 \cdots dq^n. \tag{30.16d}$$

The Friederichs extension procedure may be applied to $\mathscr{H}_\hbar^{\text{pre}}$ to obtain the *LBS-operator with mixed boundary conditions.* If $\sigma = 0$, we have the *LBS-operator* with *Neumann boundary conditions.* The semiclassical theory for the LBS with mixed boundary conditions is quite similar to that for the LBS with Dirichlet conditions. So, in further considerations, we restrict ourselves to the Dirichlet boundary conditions, the mixed case will be mentioned in remarks or left to the reader. The term *LBS-operator* and the notation $\mathscr{H}_\hbar$ will correspond to the operator with Dirichlet conditions.

Semiclassical Counterpart

The common belief is that the behaviour of the spectrum of the LBS-operator when $\hbar \longrightarrow 0$ is governed by a definite classical dynamical system, namely the Generalized Geodesic Flow constructed as in §6 for the Riemannian manifold $(Q, \boldsymbol{g})$ with $V(.,0)$ in the role of the potential. Recall that the Generalized Geodesic Flow is a Hamiltonian dynamical system defined on a glued phase space $Z \xleftarrow{\rho} T^*Q \backslash \Sigma$. Its Hamiltonian H satisfies the relation $H \circ \rho = \tilde{H}$ where the pre-Hamiltonian $\tilde{H}$ is given by (6.4) with $V(q) = V(q, 0)$. Note that the systems, classical counterpart is the same for all the types of boundary conditions: the Dirichlet one and mixed ones with different σ. The change of boundary conditions results in the changing of the secondary terms of eigenfunctions asymptotics only.

Semiclassical Formula for the Number of Eigenvalues

Consider an interval $[E_1, E_2] \subset \mathbb{R}$. Suppose the part of the spectrum of $\mathscr{H}_\hbar$, which lies in $[E_1, E_2]$, is discrete, that is all the points of the spectrum which enter $[E_1, E_2]$ are isolated eigenvalues of finite multiplicity. Let this property hold for all $\hbar \in]0, 1]$. Denote by $N(E_1, E_2, \hbar)$ the number of eigenvalues of $\mathscr{H}_\hbar$, entering $[E_1, E_2]$, each eigenvalue being counted as many times as its multiplicity. How do we estimate $N(E_1, E_2, \hbar)$ provided $\hbar$ tends to zero? To obtain the answer, one appeals to the general heuristic concept of a "quantum cell" which occupies a volume $(2\pi\hbar)^n$ in the phase space and corresponds to one quantum state (= eigenfunction) (see Landau and Livschitz [1] formula (48.7)). The cells which correspond to eigenvalues of the interval $[E_1, E_2]$ must lie in the domain $H^{-1}([E_1, E_2])$ of the phase space. Dividing the volume of this domain by that of the elementary cell, one obtains the following semiclassical estimate

$$N(E_1, E_2, \hbar) \sim \frac{\operatorname{Vol}\{z \in Z: E_1 \leqq H(z) \leqq E_2\}}{(2\pi\hbar)^n} \tag{30.17}$$

There are many cases for which the formula (30.17) is proved to be valid asymptotically when $\hbar \longrightarrow 0$ with the error of order $\mathcal{O}(\hbar^{1-n})$ (see Notes in the end of the chapter), but the author does not know if (30.17) is true in the general setting, with the error of order $\mathcal{O}(\hbar^{1-n})$ or of another order less than $\hbar^{-n}$.

§31. Particular Cases

We shall distinguish as particular cases the Schrödinger operator, the Laplace–Beltrami operator on a compact manifold, and the wave equation. It is convenient to modify slightly the setting of the eigenvalue problems in the last two cases by introducing another spectral parameter tending to infinity and playing the role of the large quantity instead of $\hbar^{-1}$.

Schrödinger Operator

The Schrödinger Operator $\mathscr{H}_\hbar$ is the LBS-operator with $(\mathbb{R}^n, g_{st})$, where $g_{st} = \sum_{k=1}^n dq^k \odot dq^k$, for (Q, g) and with the potential V independent of $\hbar$. The equation for eigenfunctions

$$\mathscr{H}_\hbar u \overset{\text{def}}{=\!=} -\frac{\hbar^2}{2}\Delta u + Vu = Eu \tag{31.1}$$

is known as the Schrödinger equation; this is the main equation of nonrelativistic quantum mechanics. Its eigenvalues E are the values of energy of the quantum system; the corresponding eigenfunctions are quantum states with the given energy. The important quantum mechanical problem is to calculate the eigenvalues when V is given. The nature of the spectrum of $\mathscr{H}_\hbar$ depends essentially on properties of the potential V. It is clear that all the spectrum is situated above the constant C of (30.15). In the general case the spectrum of $\mathscr{H}_\hbar$ may contain both discrete and continuous components, but there is a zone where one can guarantee the discreteness of the spectrum.

Proposition 31.2. *Let* $\lim_{R\to\infty}(\inf_{\|q\| \geqq R} V(q)) \geqq a$. *Then the spectrum of* $\mathscr{H}_\hbar$ *in the interval* $]-\infty, a[$ *is discrete. In particular, if* $V(q) \longrightarrow \infty$ *as* $\|q\| \longrightarrow \infty$ *uniformly in the angle variable, then all the spectrum of* $\mathscr{H}_\hbar$ *is discrete.*

This proposition is proved in [Beresin and Shubin 1983] Chapter 3, §3 Theorem 3.1. The exponential estimate for the decreasing rate of the corresponding eigenfunctions is also proven.

The Laplace–Beltrami Operator

The Laplace–Beltrami operator corresponds to the case of a compact manifold Q and the potential $V=0$. We put $\hbar^2/2=1$ and use the latter λ for denoting the spectral parameter. So the equation for eigenfunctions u_j and eigenvalues λ_j reads

$$-\Delta u_j = \lambda_j u_j \tag{31.3}$$

The spectrum of $-\Delta$ is discrete (independently of the boundary conditions if $\partial Q \neq 0$). The problem which we are interested in is to investigate the high lying eigenvalues λ_j of the Laplace–Beltrami operator and the corresponding eigenfunctions. This problem is semiclassical in nature, because large λ_j in (31.3) correspond to small $\hbar$ in (31.1). The classical counterpart here is the geodesic flow on (Q, g), with reflections if $\partial Q \neq \emptyset$. One may reduce the problem of finding the $\lambda \longrightarrow \infty$ asymptotics to that of finding the $\hbar \longrightarrow 0$ asymptotics by multiplying $-\Delta$ by $\hbar^2/2$ and considering the interval $[0, 1]$ for the new spectral parameter $E = (\hbar^2/2)\lambda$. Thus, the asymptotic formula (30.17) gives us the leading term in the following well-known formula for $N(\lambda)$, the number of eigenvalues of $-\Delta$ (counting multiplicities) less than λ:

$$N(\lambda) = (2\pi)^{-n}\lambda^{n/2}\Omega_n \mu(\Omega) + \mathcal{O}(\lambda^{(n-1)/2}) \tag{31.4}$$

Here $\Omega_n = 2\pi^{n/2}/\Gamma(n/2)$ is the $(n-1)$-volume of the unit sphere in $\mathbb{R}^n$.

Wave Equation

Let Q be a bounded domain in $\mathbb{R}^n$ with a smooth boundary ∂Q, and let, $c: Q \longrightarrow \mathbb{R}$ be a smooth positive function, the *velocity*. The *wave equation* in Q reads

$$\Delta u + \frac{\omega^2}{c^2} u = 0. \tag{31.5}$$

Here $\Delta = \sum_{k=1}^{n} (\partial/\partial q^k)^2$ is the Laplacian, and $\omega \in \mathbb{R}$ is the spectral parameter called *frequency*. We add to (31.5) the boundary condition

$$u|_{\partial Q} = 0. \tag{31.6}$$

(or mixed condition). The problem (31.5), (31.6) has a nontrivial solution u except for the discrete set $\{\omega_j\}$ of values of the parameter ω, $\omega_j \longrightarrow \infty$ and $j \longrightarrow \infty$. The numbers ω_j are called *eigenfrequencies* of the domain Q with velocity distribution c. This is a mathematical setting of the problem of sound vibrations in the hole filled by (inhomogeneous) media. The solutions (u_j, ω_j) of (31.5), (31.6) are amplitudes u_j and frequencies ω_j of small acoustical vibrations in Q. As a partial case we note that a vibrating homogeneous membrane with fixed edges occurs if $n=2$ and $c=\text{const}$.

The stated problem may also be interpreted as that of finding eigenvalues and eigenfunctions of a self-adjoint operator. Consider the Hilbert space $L_2(Q,\mu)$, where μ is the measure which has the density $1/c^2(q)$ with respect to the lebesgue measure: $\mu(A)=\int_A c^{-2}(q)\,dq$, and the *wave operator* L^{wave} which equals minus the Laplacian multiplied by c^2: its value at a function $u\in C_0^\infty(Q)$ is

$$L^{\text{wave}}u = -c^2\Delta u. \tag{31.7}$$

L^{wave} is densely defined, linear, symmetric and positive and $\langle\!\langle L^{\text{wave}}u,u\rangle\!\rangle \geqq 0$. We attach the same name and the same notation to its Friederichs extension. The spectrum of L^{wave} is descrete and clearly concides with the set of the square of eigenfrequencies of the problem (31.5), (31.6).

The subject of our interest is the eigenvalues and eigenfunctions asymptotics when the number j of an eigenvalue tends to infinity. We are going to show how this problem is reduced to that of the particular case of the LBS operator. Consider another Hilbert space, namely $L_2(Q,\mu_2)$ where the measure μ_1 has the density $1/c^n(q)$, and let $S:L_2(Q,\mu_1)\longrightarrow L_2(Q,\mu)$ be an isometry which acts by the formula

$$(Su)(q)=c(q)^{-n/2+1}V(q). \tag{31.8}$$

Then clearly $S(C_0^\infty(Q))=C_0^\infty(Q)$ and, as one may easily checks,

$$S^{-1}L^{\text{wave}}S=L_1^{\text{wave}}. \tag{31.9}$$

where for functions $u\in C_0^\infty(Q)$

$$nL_1^{\text{wave}}u=-\Delta_g u+v\cdot u. \tag{31.10}$$

Here $\Delta_g=c^n(q)\sum_{k=1}^n(\partial/\partial q^k)c^{2-n}(q)\partial/\partial q^k$ is the Laplace-Beltrami operator constructed with the Riemannian metric

$$g=c^{-2}(q)\sum_{k=1}^n dq^k\odot dq^k. \tag{31.11}$$

and the potential

$$v=\frac{n-2}{2}c\Delta c-\frac{n(n-2)}{4}(\operatorname{grad} c)^2. \tag{31.12}$$

Here we use the letter Δ for denoting the Laplace operator corresponding to the standard Euclidean metric.

We obtained L^{wave} as the unitary equivalent of the LBS operator with $\hbar/2=1$. Multiplying by $\hbar^2/2$ and rescaling of the spectral parameter $(\hbar^2/2)\omega=E$ reduces the problem to that of finding the asymptotics as $\hbar\longrightarrow 0$ of the spectrum in a given interval of on LBS operator. The latter is defined in the domain Q with g given by (3.11) and $V=(\hbar^2/2)v$, where v is given by (31.12).

Returning to our initial problem and taking into account that the $\hbar^2$ terms of V do not contribute to the leading terms, yields the following asymptotic formula for the number $N(\omega)$ of eigenfrequencies ω_j less than ω:

(31.13) $$N(\omega) = (2\pi)^{-n}\omega^n\Omega_n\int_\Omega c^{-n}(q)\,dq + \mathcal{O}(\omega^{n-1}).$$

§32. Quasimodes

In this section we shall consider a self-adjoint operator L in a Hilbert space $\mathfrak{H}$. The domain of definition of L will be denoted $D(L)$. We use the notations $(\!(\cdot,\cdot)\!)$ and $\|\cdot\|$ for the inner product and the norm in $\mathfrak{H}$ respectively.

Definition of an Individual Quasimode

Let ε be a nonnegative number. A *quasimode* of L with discrepancy ε is a pair (u, λ) where $u \in D(L)$, $\|u\| = 1$, $\lambda \in \mathbb{R}$, such that $\|Lu - \lambda u\| \leqq \varepsilon$. Thus, quasimodes with discrepancy 0 are nothing more than the eigenvectors and eigenvalues of L. Suppose that, instead of knowing the exact eigenvector and eigenvalue, one knows an approximate solution to the eigenvalue problem, i.e. a quasimode with discrepancy ε. What conclusion can one make about the spectrum of L? The following proposition gives the answer.

Proposition 32.1. *Let (u, λ) be a quasimode with discrepancy $\varepsilon > 0$ for the operator L, and let there be no continuous spectra of L in the interval $[\lambda - \varepsilon, \lambda + \varepsilon]$. Then this interval contains an eigenvalue of L.*

Proof. Suppose λ does not belong to the spectrum. The distance d_λ between λ and the spectrum of L is estimated as

$$d_\lambda^{-1} = \|(L - \lambda\cdot\mathrm{id})^{-1}\| = \sup_{v \neq 0} \frac{\|(L - \lambda\cdot\mathrm{id})^{-1}v\|}{\|v\|} \geqq \frac{\|(L - \lambda\cdot\mathrm{id})^{-1}v'\|}{\|v'\|},$$

where v' is an arbitrary nonzero vector in $\mathfrak{H}$. Substituting $v' = (L - \lambda\cdot\mathrm{id})u$, one obtains $d_\lambda \leqq \varepsilon$. □

Note that one can say nothing about the proximity of u to an exact eigenvector. The following though, is true. Let $E(\Delta)$ be the spectral projection of L (see §29) corresponding to the interval $\Delta = [\lambda - \mu, \lambda + \mu]$, μ being a positive number. Then the inequality

(32.2) $$\|E(\Delta)u - u\| \leqq \varepsilon\mu^{-1}$$

holds for the quasimode (u, λ) with discrepancy ε. Particularly, if there is exactly one eigenvalue λ^* of L in the interval Δ with eigenvector u^*, $\|u^*\| = 1$, then one can assert that, for some $\alpha \in \mathbb{R}$,

(32.3) $$\|u - e^{i\alpha}u^*\| \leqq 2\varepsilon\mu^{-1}.$$

The estimate (32.2) is a simple consequence of the spectral theorem (29.1) and the definition of a quasimode, (32.3) follows immediately from (32.2) if one puts $E(\Delta)u = (\!(u, u^*)\!)u^*$.

A Family of Quasimodes

The semiclassical method usually provides us with a lot of quasimodes, and the problem arises: what part of the actual spectrum of L is approximated by quasimodes? In this situation, the important characteristic concerns the proximity of the quasimodes to the orthonormal system. Let ε and δ be nonnegative numbers. A *family of quasimodes* with discrepancy ε and deviation from orthogonality δ is a family $\langle (u_l, \lambda_l), l \in \Lambda \rangle$ where Λ is a finite set of indices, $u_l \in D(L)$, $\lambda_l \in \mathbb{R}$, such that for all $l, k \in \Lambda$

(i) $\|u_l\| = r$,

(ii) $\|Lu_l - \lambda_l u_l\| \leqq \varepsilon$,

(iii) $|(\!(u_k, u_l)\!)| \leqq \delta$ if $k \neq l$.

Suppose that, given such a family of quasimodes, the set $\bigcup_{l\in\Lambda}[\lambda_l - \varepsilon, \lambda_l + \varepsilon]$ contains only the discrete spectrum of L. Then, from proposition 32.1, for each $l \in \Lambda$, the interval $[\lambda_l - \varepsilon, \lambda_l + \varepsilon]$ contains at least one eigenvalue λ^* of the operator L. Fix a positive number μ and set $\mathscr{E}_\mu = \bigcup_{l\in\Lambda}[\lambda_l - \mu, \lambda_l + \mu]$. Denote by $E(\mathscr{E}_\mu)$ the spectral projection of L corresponding to the set $\mathscr{E}_\mu$, by $\mathscr{L}\{(u_l, \lambda_l), l \in \Lambda\}$ the linear shell of the system $\{(u_l, \lambda_l), l \in \Lambda\}$, and by $N^*(\mathscr{E}_\mu)$ the dimension of the subspace $E(\mathscr{E}_\mu)\mathscr{L}\{(u_l, \lambda_l), l \in \Lambda\}$. It is natural to call $N^*(\mathscr{E}_\mu)$ the *total multiplicity* of the part of the spectrum of L approximated by the family of quasimodes. Denote by $|\Lambda|$ the number of elements of the set Λ. It is useful to know when the total multiplicity of the part of the spectrum approximated by quasimodes coincides with the number of quasimodes, i.e. with $|\Lambda|$.

Proposition 32.4. *Let $\{(u_l, \lambda_l), l \in \Lambda\}$ be a family of quasimodes of L with discrepancy ε and deviation from orthogonality δ. If $\varepsilon\mu^{-1} + \delta < |\Lambda|^{-1}$ then $N^*(\mathscr{E}_\mu) = |\Lambda|$.*

Proof. It is clear that $N^*(\mathscr{E}_\mu) \leqq |\Lambda|$. The equality holds if and only if the vectors $\{Pu_k, k \in \Lambda\}$, where $P = E(\mathscr{E}_\mu)$, from a linearly independent family. This is true if and only if the matrix $(\!(Pu_k, Pu_l)\!)$ has nonzero determinant. Let $(\!(Pu_k, Pu_l)\!) = \delta_{kl} + a_{kl}$. Taking into account the estimates: (iii) of the definition of a family of quasimodes, and (32.2), and obtains

$$|a_{kl}| = |(\!(u_k, u_l)\!) - \delta_{kl} + (\!(Pu_k - u_k, u_l)\!)| \leqq \delta + \varepsilon\mu^{-1} < |\Lambda|^{-1}.$$

Applying Lemma AII.20 to the matrix $A = \{a_{kl}\}$ yields the desired result. □

§33. Degenerated Quasimodes

This section contains an example which shows that quasimodes do not necessarily approximate the true eigenfunctions, even when constructed with the use of semiclassical asymptotic techniques.

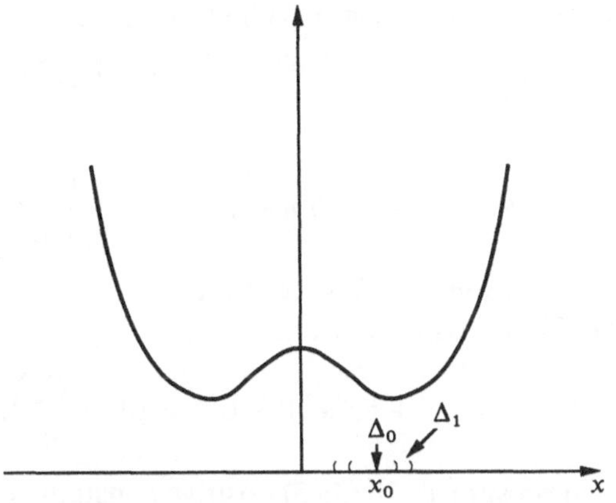

Fig. 52. Symmetric potential with two minima

Consider the Schrödinger one-dimensional operator

(33.1) $$\mathscr{H}_\hbar = -\frac{1}{2}\hbar^2\frac{d^2}{dx^2} + V,$$

where the potential V is smooth, real valued, and possesses the symmetry: $V(x) = V(-x)$. We suppose that V has two strict minima at points $\pm x_0$ ($V'(x_0) = 0$, $V''(x_0) > 0$ and $V(x) > V(x_0)$ if $x \neq x_0$) and tends to infinity as $x \longrightarrow \pm\infty$ (see Fig. 52). The typical example is $V_0(x) = (1 - x^2)^2$. The operator, defined originally by the differential expression (33.1) on $C_c^\infty(\mathbb{R})$, is essentially self-adjoint. This means it has the only self-adjoint extension, the basic Hilbert space being $L_2(\mathbb{R})$ with respect to the Lebesgue measure. We keep the same notation $\mathscr{H}_\hbar$ for the extended operator.

It is well-known that the spectrum of $\mathscr{H}_\hbar$ is discrete (see Proposition 31.2) and simple. The latter means that each eigenvalue is of multiplicity one. The spectrum is bounded from below by min V.

The symmetry condition implies $\mathscr{H}_\hbar$ commutes with the reflection map $(Ru)(x) = u(-x)$. So the eigenfunctions are of two kinds: symmetric $Ru = u$ and antisymmetric $Ru = -u$, the former kind satisfying $u'(0) = 0$, the latter $u(0) = 0$.

Let $\hbar$ tend to zero. Let us first construct quasimodes which correspond to the lowest eigenvalues and are concentrated near one of the minima. Take for instance x_0. Choose two open neighbourhoods Δ_0, Δ_1 of x such that $\bar{\Delta}_0 \subset \Delta_1$ and $\bar{\Delta}_1$ lies in positive axis and is separated from zero. Fix a smooth real-valued function χ which equals 1 on Δ_0 and has supp $\chi \subset \Delta_1$. We seek for an approximate solution to the equation $\mathscr{H}_\hbar u = Eu$ in the form

(33.2) $$u^+ = c_n(\psi')^{-1/2}D_n(\hbar^{-1/2}\psi)\cdot\chi,$$

where $D_n(t) = e^{-t^2/2}H_n(t)$, $H_n(t) = e^{t^2}(d/dt)^n e^{-t^2}$ is the Hermitian polynomial, $n \in \mathbb{Z}_+$ being fixed, $\chi \in C_0^\infty(\mathbb{R})$, c_n is the norming constant. Substituting (33.2) in the eigenfunction equation (note that $D_n(t)$ obeys the equation $D_n'' = (t^2 - 2n - 1)D_n$)

yields the following expression for the discrepancy:

$$\mathscr{H}_\hbar u^+ - Eu^+ = \delta\cdot u^+ - \hbar^2(\chi' u^{+\prime} + \tfrac{1}{2}\chi'' u^+), \tag{33.3}$$

where

$$\delta = -\tfrac{1}{2}\psi'^2\psi^2 + V - E + \hbar(n+\tfrac{1}{2})\psi'^2 + \hbar^2\left(\frac{1}{4}\cdot\frac{\psi'''}{\psi'} - \frac{3}{8}\cdot\frac{\psi''^2}{\psi'^2}\right). \tag{33.4}$$

Here prime denotes derivation with respect to x.

The last term in (33.3) satisfies the estimation

$$|\hbar^2(\chi' u^{+\prime} + \tfrac{1}{2}\chi'' u^+)| \leqq C_1\hbar\exp\left(-\frac{C_2}{\hbar}\right), \tag{33.5}$$

provided ψ is bounded on $\operatorname{supp}\psi$. In (33.5) positive constants C_1, C_2 depends only on ψ, χ, and n.

To obtain quasimodes, we solve the equation $\delta = 0$ approximately by substitutions

$$E = \sum_{s=1}^{N} \hbar^s\lambda_s, \qquad \psi = \sum_{s=1}^{N} \hbar^s\psi_s, \tag{33.6}$$

where N is a positive integer, ψ_s are smooth real-valued functions defined on an interval which contains $\operatorname{supp}\chi$, $\psi_0' > 0$, and λ_s are real numbers.

Substituting (33.6) into the right-hand side of (33.4), re-expanding the latter into a series of powers of $\hbar$, and requiring the terms for $\hbar^s$, $0 \leqq s \leqq N$, in the obtained expansion to be zero, yields the sequence of equations which determine ψ_s and λ_s uniquely. The equations are of the form:

$$-\tfrac{1}{2}(\psi_0')^2\psi_0^2 + V - \lambda_0 = 0, \tag{33.70}$$

$$-\psi_0\psi_0'(\psi_1\psi_0)' - \lambda_1 + (n+\tfrac{1}{2})(\psi_0')^2 = 0, \tag{33.71}$$

$$s \geqq 2, \quad -\psi_0\psi_0'(\psi_s\psi_0)' - \lambda_s + f_s = 0, \tag{33.7s}$$

where f_s depends on ψ_k, λ_k, $k < s$. If ψ_k, $k < s$, are smooth, so is f_s. The unique solution of (33.70) satisfying the condition for ψ_0 to be smooth and defined on Δ_1 is

$$\lambda_0 = V(x_0), \qquad \psi_0(x) = \sqrt{2\int_{x_0}^{x}\sqrt{2(V(y)-\lambda_0)}\,dy}. \tag{33.8}$$

Let λ_k, ψ_k, $k < s$, be already defined. Then the unique solution to (33.7s) defined on Δ_1 and subjected to the above mentioned condition is given by the formulae

$$\lambda_s = f_s(x_0), \qquad \psi_s(x) = \left(\int_{x_0}^{x}\frac{f_s(y)-\lambda_s}{(\psi\psi')(y)}\,dy\right)\cdot(\psi_0(x))^{-1}. \tag{33.9}$$

Particularly, as it follows from (33.71),

$$\lambda_1 = (n+\tfrac{1}{2})(\psi_0'(x_0))^2 = (n+\tfrac{1}{2})\sqrt{V''(x_0)}. \tag{33.10}$$

Since the terms of orders $\hbar^s$, $s \leqq N$, in (33.4) vanish, δ satisfies the estimate $|\delta| \leqq \text{const}\, \hbar^{N+1}$, const depending only on N, V, and Δ_1. Let us summarize the result as follows.

Proposition 33.11. *Under the above stated conditions on the potential, the equations* (33.7*s*), $0 \leqq s \leqq N$, *have unique solutions* $E_s \in \mathbb{R}$, $\psi_s \in C^\infty(\Delta_1)$, *which are defined by* (33.8), (33.9) *and* (33.10). *The pair* (ψ, E), *given by* (33.6), *is a quasimode with discrepancy of order* $\hbar^{N+1}$.

To obtain the quasimodes u^- concentrated near $-x_0$, it is sufficient to apply the symmetry R to the above formulae. The support of u^- is contained in the set $-\Delta_1$ and has no intersection with the support of u^+. Hence u^+ and u^- are mutually orthogonal. So, the pairs $\{(u^i, E^i), i \in (+,-)\}$ form a family of quasimodes in the sense of §32. The set Λ consists of two elements, the discrepancy $\varepsilon = \mathcal{O}(\hbar^{N+1})$, and the deviation from orthogonality $\delta = 0$.

Note that the eigenvalue expansion (33.6) remains the same for u^-: $E^+ = E^-$. Applying Proposition 32.1 and 32.4 we may formulate

Proposition 33.12. *Given a potential V obeying the above stated conditions, positive integers n and N, and E defined by* (33.6), (33.8), (33.9) *and* (33.10). *Then there is an eigenvalue E^* of $\mathcal{H}_\hbar$ such that*

$$|E - E^*| \leqq c\hbar^{N+1},$$

where the positive constant c depends only on V, n, N. Moreover, the interval $[E - 2c\hbar^{N+1}, E + 2c\hbar^{N+1}]$ contains at least two eigenvalues of $\mathcal{H}_\hbar$.

Information 33.13. The low lying spectrum of $\mathcal{H}_\hbar$ consists of pairs (E_n^-, E_n^+), $n = 0, 1, 2, \ldots$, of simple eigenvalues which have the same power asymptotics (33.6) and are divided by the space

$$\Delta E_n = E_n^+ - E_n^- = a e^{-b/\hbar}(1 + \mathcal{O}(\hbar)),$$

where

$$a = 2^{n+1}(n!)^{-1}\psi_0'(x_0)|\psi_0(0)|^{2n+1}\pi^{-1/2}$$
$$\times \exp\langle (2n+1)\int_0^{x_0} [\psi_0'(x)^2(2(V(x)-\lambda_0))^{-1/2} + \psi_0^{-1/2}(x)\psi_0'(x)]\,dx,$$

$$b = \int_{-x_0}^{x_0} \sqrt{2(V(x)-\lambda_0)}\,dx,$$

ψ_0 and λ_0 are given by (33.8).

The constructed quasimodes $u^\pm$ are concentrated near the points $\pm x_0$ correspondingly. So they do not possess any symmetry property with respect to the reflection R. Since the spectrum of $\mathcal{H}_\hbar$ is simple and each eigenfunction is either symmetric or antisymmetric, the quasimodes obtained do not approximate the individual eigenfunctions. The true asymptotics of an eigenfunctions, $(1/\sqrt{2})$ $(u^+ + u^-)$ for the symmetric one, and $(1/\sqrt{2})\,(u^+ - u^-)$ for the antisymmetric one, contain the sum of two quasimodes corresponding to two different invariant sets (fixed points $(\pm x_0, 0)$ of the vector field ξ in $T^*\mathbb{R}$ in our case) of the classical counterpart.

Remark 33.14. The methods developed in Chapter VII, being applied to (33.1), permit us to construct quasimodes corresponding to large quantum numbers n. The essential condition is that the interval $[a, b]$, which the eigenvalues E_n range over, contains no critical values of the potential. Of course, the classical counterpart in this case is integrable, and the construction becomes much more simpler. Consider for example the potential $V_0(x) = (1 - x^2)^2$. If $a > 1$, then a quasimode with $E_n \in [a, b]$ occupies the region $S = \{x\colon E_n - V(x) > 0\}$, and goes rapidly to zero outside of S. If $[a, b] \subset]0, 1[$, then the region S consists of two non-intersecting intervals S_+ and S_-, and the reflection R permutes $S_\pm$. There are two quasimodes $(u^\pm, E)$ concentrated in $S_\pm$ with the same E.

Notes to Chapter V

The reader can find a detailed account of notions and theorems of the theory of self-adjoint operators in Hilbert space in Birman and Solomjak (1987). The theory of Friederichs extension is expounded in the same place. The original paper is by Friederichs (1934).

A great deal of facts concerning the Schrödinger operator are contained in Beresin and Shubin (1983).

The semiclassical estimates for the number of eigenvalues (30.17) is well known to physicists. See e.g. Landau and Lifschitz (1963) formula (48.7). The rigorous mathematical proof is the subject of numerous investigations, and it has been tested in many particular cases. The early papers until 1975 are covered by survey of Birman and Solomjak (1977). We give here some references with no pretension to completeness. The most recent result with error of order $\hbar^{n-1}$ in the case of a compact manifold is formulated by Ivrii (1984). The λ-asymptotics of the form (31.4) are proven by Ivrii (1980) and for more general operators by Vasil'ev (1987). Ivrii and Vasil'ev in the cited papers investigated also (and it was their main goal) two-term asymptotics of the form.

$$N(\lambda) = c_0 \lambda^{n/2} + b\lambda^{(n-1)/2} + \mathcal{O}(\lambda^{(n-1)/2}).$$

They found that the second term existed if the measure of periodic trajectories of the classical counterpart is zero. Earlier, the corresponding result for the case of a compact manifold without boundary was obtained by Duistermaat and Guillemin (1975).

Some results about the formula (30.17) for the case of the Schrödinger operator in $\mathbb{R}^n$ are contained in papers by Helfer and Robert (1981), (1983) and by Ivrii (1986a, b, c).

The segments of semiclassical expansions (33.6) for any N were constructed in Slavjanov (1989). The exponential estimates and asymptotic formulae for the width of splitting ΔE_n with different conditions on the potential V can be found in Fedoryuk (1966), Alenitsyn (1982), Simon (1984), Harrel (1980), Jona-Lasinio, Martinelli, Scoppola (1981), and Pankratova (1984).

Chapter VI. Maslov's Canonical Operator

We adopt the canonical operator invented by Maslov (1972). It is an effective tool for constructing the global high frequency asymptotic to the solutions of the Laplace–Beltrami–Schrödinger equation. Maslov's canonical operator is attached to an invariant Lagrangian submanifold in the phase space of the corresponding classical dynamical system (the generalized geodesic flow in our case). It carries functions defined on the Lagrangian submanifold in the glued cotangent bundle to those defined on the coordinate space.

§34. Assumptions

In more detail, the situation we shall discuss is the following. A Lagrangian (boundaryless) submanifold $\mathcal{T}$ is given in the phase space Z of the generalized geodesic flow (§6) of an n-dimensional Riemannian manifold (Q, g). It is useful to recall the notations attached to the construction of Z by writing out the following commutative diagram:

(34.1)
$$\begin{array}{ccccc} \tilde{\mathcal{T}} & \subset & T^*Q\backslash\Sigma & \subset & T^*Q \\ \downarrow{\scriptstyle \rho|\tilde{\mathcal{T}}} & & \downarrow{\scriptstyle \rho} & & \downarrow{\scriptstyle \pi} \\ \mathcal{T} & \subset & Z & \xrightarrow{\pi'} & Q \end{array}$$

Here π is the cotangent bundle projection, and ρ is the gluing projection. The map π' is continuous and its restriction onto $(\pi')^{-1}(Q\backslash\partial Q) = Z\backslash\Pi$ is a submersion. The letter $\tilde{\mathcal{T}}$ denotes the pre-image of $\mathcal{T}$ under ρ, $\tilde{\mathcal{T}}$ is a manifold with boundary if $\mathcal{T} \cap \Pi \neq \varnothing$.

Our first assumption is

34.2. *$\mathcal{T}$ is contained in some submanifold of constant energy $H^{-1}(E)$, $E \in \mathbb{R}$.*

Here H is the Hamiltonian function of the generalized geodesic flow. This condition implies the invariance of $\mathcal{T}$ with respect to the flow: for each $z \in \mathcal{T}$, $\xi_z \in T_z\mathcal{T}$, ξ being the vector field of the generalized geodesic flow.

Also we impose two topological conditions:

34.3. *$\mathcal{T}$ is connected,*

34.4. *the cohomology group $H^1(\mathscr{T}, \mathbb{Z})$ is finitely generated and has no torsion.*

It follows that there is natural inclusion $H^1(\mathscr{T};\mathbb{Z}) \subset H^1(\mathscr{T};\mathbb{R})$, the latter being a finite dimensional real vector space. Due to de Rhamm's theorem, $H^1(\mathscr{T};\mathbb{R})$ may be also identified with the space of closed 1-forms modulo exact forms. We suppose that

34.5. *there are smooth circles $C_k \subset \mathscr{T}$ and closed 1-forms $\mathbf{e}_k \in \Lambda^1(\mathscr{T})$, $1 \leqq k \leqq N$, such that $\int_{C_k} \mathbf{e}_s = \delta_{ks}$, $1 \leqq k, s \leqq N$, and the cohomology classes $[\mathbf{e}_k]$, $r \leqq k \leqq N$, form a vector base in $H^1(\mathscr{T};\mathbb{R})$.*

Each integer class in $H^1(\mathscr{T};\mathbb{R})$ is of the form $\sum_{k=1}^{N} m_k[\mathbf{e}_k]$, $(m_1, m_2, \ldots, m_N) \in \mathbb{Z}^N$. So, the map $(x^1, x^2, \ldots, x^N) \longmapsto \Sigma x^k[\mathbf{e}_k]$ is an isomorphism between $\mathbb{R}^N$ and $H^1(\mathscr{T};\mathbb{R})$, which carries the integer lattice $\mathbb{Z}^N$ onto $H^1(\mathscr{T};\mathbb{Z})$: We adopt the notation $\mathbf{e} = (\mathbf{e}_1, \mathbf{e}_2, \ldots, \mathbf{e}_N)$ for the corresponding vector valued 1-form.

The last assumption is the existence of an invariant volume:

34.6. *there is a volume form σ' on $\mathscr{T}$ invariant under the generalized geodesic flow.*

Recall that a volume form on an n-dimensional manifold is an n-form which nowhere vanishes. The invariance means that $f^{t*}\sigma' = \sigma'|_U$ for each local flow $f^t: U \longrightarrow Z$ of ξ. This is equivalent, as one can obtain by applying the differentiation formula (AI.68), to the transport equation

$$d(\xi \lrcorner \sigma') = 0. \tag{34.7}$$

The form $\sigma = \rho^*\sigma'$ appears to be a volume form on $\rho^{-1}(\mathscr{T})$. The presence of a volume form enables us to define an inner product in $C_c^\infty(\rho^{-1}(\mathscr{T}))$ by setting

$$(\!(u, v)\!) = \int_{\rho^{-1}(\mathscr{T})} u\bar{v}\sigma, \tag{34.8}$$

the manifold $\rho^{-1}(\mathscr{T})$ being oriented by σ. The corresponding norm $\sqrt{(\!(u,u)\!)}$ will be denoted $\|u\|_{L_2}$ or simply $\|u\|$.

Under the said assumptions we shall construct a linear map

$$\mathfrak{M}: C_c^\infty(\rho^{-1}(\mathscr{T})) \longrightarrow C_0^\infty(Q),$$

Maslov's canonical operator, and shall investigate its properties. This construction will be used for making quasimodes in Chapter VII. Note that in this case it appears to be preferable to deal with the Lagrangian manifold $\rho^{-1}(\mathscr{T})$ in the cotangent bundle instead of $\mathscr{T}$ itself contrary to the problems in dynamics such as the existence of KAM tori considered in the first part of this book.

§35. The Local Canonical Operator

The local canonical operator $\mathfrak{M}_{z_0}^{\text{loc}}$, up to an appropriate multiplier, gives the expression for $\mathfrak{M}$ restricted on local functions, i.e. on functions whose support is contained in a sufficiently small neighbourhood of the pre-image $\rho^{-1}(z_0)$ of a given point $z_0 \in \mathcal{T}$. To build $\mathfrak{M}^{\text{loc}}_{z_0}$ we need to construct some objects attached to z_0. In the following subsection we shall give the list of these objects, establish their properties, and give the proofs of their existences if necessary.

The Arrangement in the Neighbourhood of a Point

We shall distinguish two different cases: $\pi'(z_0) \notin \partial Q$ (interior case), and $\pi'(z_0) \in \partial Q$ (boundary case) (see diagram (34.1)). Our construction will be slightly more complicated in the boundary case. The pre-image $\rho^{-1}(z_0)$ under the gluing projection consists of one point x_0 in the interior case and of two points $x_0^{\text{in}} \in C_{\text{in}}$ and $x_0^{\text{out}} \in C_{\text{out}}$ in the boundary case.

A Chart (V, α). Choose a chart (V, α) in Q such that

(i) $\pi'(z_0) \in V$, *and in the interior case we suppose additionally that* $V \cap \partial Q = \varnothing$.

Domains U *and* $W_{(\text{in},\text{out})}$. Pick an open neighbourhood U of z_0 in Z possessing the properties:

(ii) *U is diffeomorphic to an open ball,*

(iii) *in the interior case the set $\rho^{-1}(U)$ is contained, together with its closure, in the domain T_v^*Q of the associated chart $(T_v^*Q, T^{*-1}\alpha)$ of (V, α) in T^*Q* (see A.I.60); *in the boundary case $\rho^{-1}(U)$ consists of two connected components W_{in} and W_{out}, neighbourhoods of x_0^{in} and x_0^{out} respectively, such that $\rho^{-1}(U) = W_{\text{in}} \cup W_{\text{out}} \subset T_v^*Q$, $\partial W_{\text{in}} \subset C_{\text{in}}$, $\partial W_{\text{out}} \subset C_{\text{out}}$ and the triple $(U, U_{\text{in}} = \rho(W_{\text{in}}), U_{\text{out}} = \rho(W_{\text{out}}))$ is diffeomor-*

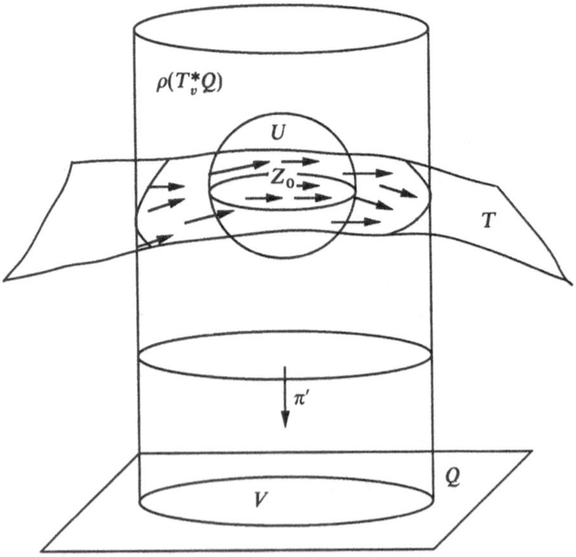

Fig. 53. Arrangement at a point $z_0 \in \mathcal{T}$ (interior case)

phic to the standard triple $B_1=\{x\in\mathbb{R}^{2n}:|x|<I\}$, $B_1^-=\{x\in B_1:x^1\leqq 0\}$, $B_1^+=\{x\in B:x^1\geqq 0\}$. *In both cases we shall use the brief notation* $W=\rho^{-1}(U)$.

Note that in the boundary case ρ maps W_ι diffeomorphically onto U_ι, $\iota\in\{\text{in},\text{out}\}$, and $U_{\text{in}}\cup U_{\text{out}}=U$, $U_{\text{in}}\cap U_{\text{out}}=\partial U_{\text{in}}=\partial U_{\text{out}}=U_{\text{in}}\cap\Pi=U_{\text{out}}\cap\Pi$, In the interior case ρ maps W diffeomorphically onto U. The situations in the interior and boundary cases are shown in Fig. 53 and Fig. 54 respectively.

Attached charts: the chart $(W,\beta)=(W,T^{*-1}\alpha)|_W\circ(\rho|_W)^{-1})$ in the interior case, and the charts $(W_\iota,\beta_\iota)=(U_\iota,(T^{*-1}\alpha)_{W_\iota}\circ(\rho|_{W_\iota})^{-1})$, $\iota\in\{\text{in},\text{out}\}$, defined in the halves of the chosen neighbourhood U in the boundary case.

Sheet(s). In the interior case the intersection $\rho^{-1}(\mathscr{T})\cap W$ will be called the sheet of $\mathscr{T}$. In the boundary case we have two sheets: $\rho^{-1}(\mathscr{T})\cap W_\iota$, $\iota\in(\text{in},\text{out})$, we suppose

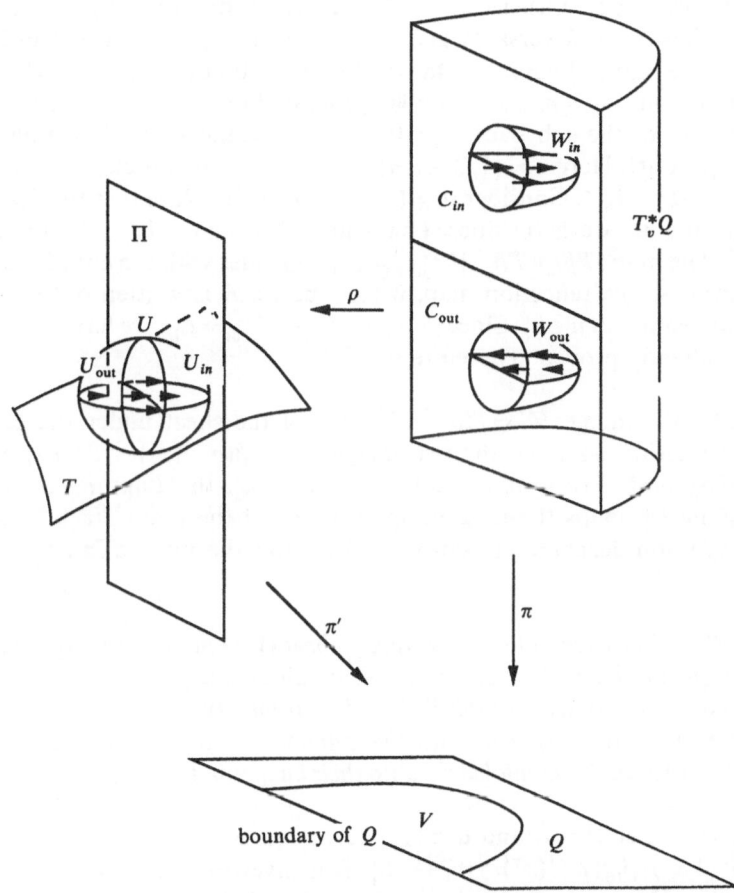

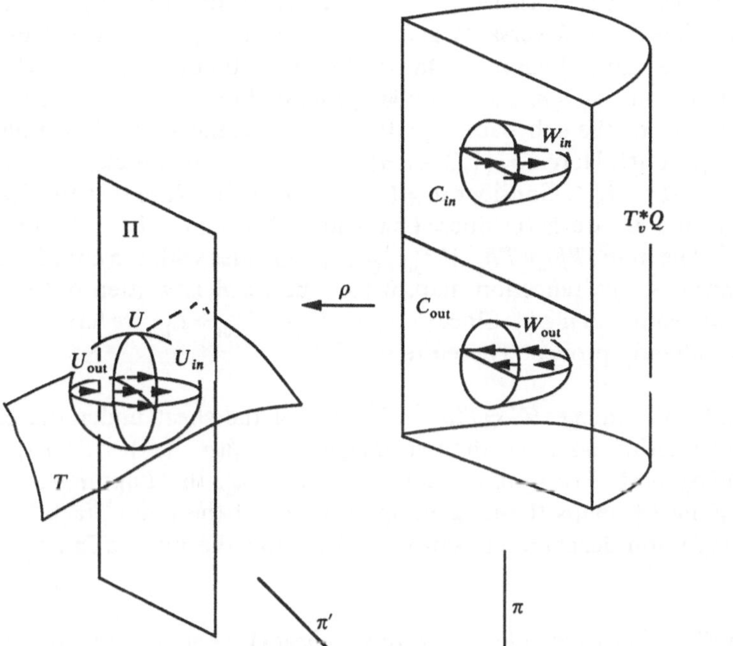

Fig. 54. Arrangement at a point $z_0\in\mathscr{T}$ (boundary case)

(iv) *in the interior case the sheet* $\rho^{-1}(\mathscr{T})$ *is diffeomorphic to on open n-ball; in the boundary case each of the two sheet* $\rho^{-1}(\mathscr{T})\cap W_\iota$ *is diffeomorphic to a half of an open n-ball* $\{x\in\mathbb{R}^n\colon |x|<1,\ x^1\geqq 0\}$

Denote by L_z the tangent space to the sheet(s) at a point $z\in\rho^{-1}(\mathscr{T})\cap W_{(\iota)}$.

The set $\mathfrak{a}$. We pick a subset $\mathfrak{a}\subset(1,2,\dots,n\}$ satisfying the conditions of the following proposition.

Proposition 35.1. *There is a subset* $\mathfrak{a}\subset\{1,2,\dots,n\}$ *such that the coordinate Lagrangian subspace L of* $(\mathbb{R}^{2n},\Omega_{st})$ *(see Example 1.16) is transversal to* $T\beta(L_{z_0})$ *in the interior case, and both are to* $T\beta_\iota(L_{z_0})$ *in the boundary case. In the latter case* $\mathfrak{a}$ *may be chosen so that* $1\notin\mathfrak{a}$.

Proof. In the interior case the Proposition is equivalent to 1.14. We need to prove 35.1 for the remaining boundary case. Denote $L'_\iota=T\beta_\iota(\xi_{z_0})\subset(\mathbb{R}^{2n},\Omega_{st})$, $\iota\in\{\text{in},\text{out}\}$. Since $\mathscr{T}$ is invariant under ξ, the vectors $\xi_\iota=T\beta_\iota(\xi_{z_0})$ belong to L'_ι. From this it follows that $\text{ort}_{n+1}\notin L'_\iota$. Otherwise $\Omega_{st}(\xi'_\iota,\text{ort}_{n+1})=0$ due to L'_ι being Lagrangian. This contradicts the transversality of ξ_{z_0} to Π: $\xi'_\iota=c(\partial/\partial q^1)+\cdots,c\neq 0$. Following a similar proof to that of Lemma 1.14a we obtain a subset $\bar{\mathfrak{a}}\subset\{1,2,\dots,n\}$, $1\in\bar{\mathfrak{a}}$, such that the coordinate subspace $K\subset 0\times\mathbb{R}^n$ generated by $\{\text{ort}_{n+i},i\in\bar{\mathfrak{a}}\}$ is the complement in $0\times\mathbb{R}^n$ to the subspace $L'_{\text{in}}\cap 0\times\mathbb{R}^n$. We claim that $L^{\bar{\mathfrak{a}}}_{\mathfrak{a}}$ is transversal to both L'_ι, $\iota\in\{\text{in},\text{out}\}$. Here $\mathfrak{a}=\{1,2,\dots,n\}\backslash\bar{\mathfrak{a}}$. As for L'_{in}, this is clear from the proof of 1.14. Let $u\in L^{\bar{\mathfrak{a}}}_{\mathfrak{a}}\cap L'_{\text{out}}$. Consider $u'=(T\beta_{\text{in}})(T\beta_{\text{out}})^{-1}u$. Note that the first components of u and u' are zero since $1\notin\mathfrak{a}$ and $T\beta_\iota|_{z_0}$ maps $T_{z_0}C_\iota$ bijectively onto $0\times\mathbb{R}^{2n-1}$. The map $(T\beta_{\text{in}})(T\beta_{\text{out}})^{-1}|_{0\times\mathbb{R}^{2n-1}}$ coincides with the coordinates image of the tangent to the reflection map, whose matrix R is written out in 6.11. The latter shows again that $u'\in L^{\bar{\mathfrak{a}}}_{\mathfrak{a}}$. Since $(T\beta_{\text{in}})(T\beta_{\text{out}})^{-1}L'_{\text{out}}=L'_{\text{in}}$, we have $u'\in L^{\bar{\mathfrak{a}}}_{\mathfrak{a}}\cap L'_{\text{in}}=\{0\}$ as was already proven. Therefore $u=0$. □

Consider the image $\mathscr{T}'=\beta(\rho^{-1}(\mathscr{T})\cap W)$ of the sheet under the coordinate map in the interior case, and the two images $\mathscr{T}'_\iota=\beta(\rho^{-1}(\mathscr{T})\cap W_\iota)$, $\iota\in\{\text{in},\text{out}\}$, in the boundary case. Proposition 35.1 shows that $\pi^{\bar{\mathfrak{a}}}_{\mathfrak{a}}$, the Lagrangian projection onto $L^{\bar{\mathfrak{a}}}_{\mathfrak{a}}$ along $L^{\mathfrak{a}}_{\bar{\mathfrak{a}}}$, maps the tangent space to $\mathscr{T}'$ at the point $\beta(z_0)$ (to $\mathscr{T}'_\iota$ at the points $\beta_\iota(z_0)$) non-degenerately onto $L^{\bar{\mathfrak{a}}}_{\mathfrak{a}}$. This and the inverse function theorem yield

Corollary 35.2. *For each point* $z_0\in\mathscr{T}$ *and a chart* (V,α) *in* Q *satisfying* (i) *there are: an open neighbourhood* U *of* z_0 *in* Z *and a subset* $\mathfrak{a}\subset\{1,2,\dots,n\}$ *such that the previous conditions* (ii)–(iv) *are fulfilled, and additionally*

(v) $\pi^{\bar{\mathfrak{a}}}_{\mathfrak{a}}$ *maps diffeomorphically* (a) *the manifold* $\mathscr{T}'$ *onto its image in* $L^{\bar{\mathfrak{a}}}_{\mathfrak{a}}$ *in the interior case,* (b) *both the manifolds* $\mathscr{T}'_\iota$ *onto their images in* $L^{\bar{\mathfrak{a}}}_{\mathfrak{a}}$ *in the boundary case.*

The map(s) $\gamma_{(\iota)}$. Let $U\subset Z$ and $\mathfrak{a}\subset\{1,2,\dots,n\}$ be as in Corollary 35.2. Consider the map $(\pi^{\bar{\mathfrak{a}}}_{\mathfrak{a}}|_{\mathscr{T}'_{(\iota)}})\circ\beta_{(\iota)}\colon\rho^{-1}(\mathscr{T})\cap W_\iota\longrightarrow L^{\bar{\mathfrak{a}}}_{\mathfrak{a}}$. It is invertible because so are the maps $\beta_{(\iota)}$ and $\pi^{\bar{\mathfrak{a}}}_{\mathfrak{a}}|_{\mathscr{T}'}$. Denote by $\gamma_{(\iota)}$ the inverse of $(\pi^{\bar{\mathfrak{a}}}_{\mathfrak{a}}|_{\mathscr{T}'_{(\iota)}})\circ\beta_{(\iota)}$. It is defined on an open set $G_{(\iota)}\subset L^{\bar{\mathfrak{a}}}_{\mathfrak{a}}$.

The Generating Function(s). The manifold $\mathscr{T}'_{(l)}$ possesses the generating function $S_{(l)}(q^{\bar{a}}, p_a)$ defined on $G_{(l)}$, $q^{\bar{a}}, p_a$ denoting $\{q^i, i \in \bar{a}, p_j, j \in a\}$ (see the end of §2, (2.33)). The equations of $\mathscr{T}'_{(l)}$ are the following:

$$
(35.3) \qquad \begin{aligned} p_i &= p_i^s(q^{\bar{a}}, q_a) \overset{\text{def}}{=} \frac{\partial S_{(l)}(q^{\bar{a}}, p_a)}{\partial q^i}, \quad i \in \bar{a}, \\ q^i &= q_s^i(q^{\bar{a}}, q_a) \overset{\text{def}}{=} -\frac{\partial S_{(l)}(a^{\bar{a}}, p_a)}{\partial p_j}, \quad j \in a. \end{aligned}
$$

Generating functions are defined up to additive constants. It is convenient to impose on $S_{(l)}$ the following norming condition:

$$
(35.4) \qquad \left(S_{(l)}(q^{\bar{a}}, p_a) + \sum_{j \in a} q_s^j(q^{\bar{a}}, p_a) p_j \right) \Bigg|_{\gamma_{(l)}(q^{\bar{a}}, p_a) = z_0} = 0.
$$

In the boundary case the condition (35.4) yields

$$
(35.5) \qquad S_{\text{in}}|_{q^1=0} = S_{\text{out}}|_{q^1=0}.
$$

Indeed, the explicit formulae (6.9a) for the reflection map imply that p_i, q^i, $1 \leqq i \leqq n$, are continuous if one moves on $\mathscr{T}$ crossing Π. Hence, the first partial derivatives of $S_{\text{in}}, S_{\text{out}}$ coincide if $q^1 = 0$ except $\partial S_{(l)}/\partial q^1$. Another consequence of (35.4) is the formula

$$
(35.6) \qquad S_{(l)}(q^{\bar{a}}, p_a) + \sum_{j \in a} q_s^j(q^{\bar{a}}, p_a) p_j = \int_{x_0}^{\gamma(q^{\bar{a}}, p_a)} \Theta_Q.
$$

Both the sides of (35.6) coincide at $\gamma(q^{\bar{a}}, p_a)$ and have the same partial derivatives. The result follows as $\mathscr{T}_{(l)} \cap U$ and $G_{(l)}$ are simply connected.

Switching function. Pick a function $\chi: \mathbb{R}^n \longrightarrow \mathbb{R}$ satisfying

$$
(\text{vi}) \qquad \operatorname{supp} \chi \subset \alpha(V), \quad \chi = 1 \text{ on } \alpha \circ \pi'(U).
$$

An *arrangement* at a point z_0 is a collection $\{(V, \alpha), U, a, \chi\}$ satisfying the above condition (i)–(vi). Once an arrangement is given, all the other elements listed above, attached chart(s), generating function(s), etc., are determined uniquely.

1/ħ-Fourier Transform

Given a constant $\hbar > 0$, we define the *1/ħ-Fourier transform* $F: \mathscr{T}(\mathbb{R}^n) \longrightarrow \mathscr{T}(\mathbb{R}^n)$ by the formula

$$
(35.7) \qquad (Fv)(p) = (2\pi\hbar)^{-n/2} \int e^{(i/\hbar)\langle q, p\rangle} v(q)\, dq,
$$

where $p = (p_1, p_2, \ldots, p_n)$, $q = (q^1, q^2, \ldots, q^n) \in \mathbb{R}^n$, $\langle q, p\rangle = \sum_{k=1}^n q^k p_k$ and $dq = \prod_{k=1}^n dq^k$.

Here $\mathscr{T}(\mathbb{R}^n)$ is the Schwartz space of complex-valued smooth functions on $\mathbb{R}^n$ which tend to zero together with all their derivatives faster than any positive power of the norm of the variable when the latter tends to infinity. We also shall write $F_{p \leftarrow q} v(q)$ to denote the right-hand side of (35.7). We shall refer to F also as the *complete Fourier transform.*

Given a subset $\mathfrak{a} \subset \{1, 2, \ldots, n\}$ we also define the *partial* $1/\hbar$-Fourier transform in the $\mathfrak{a}$-direction, $F_{\mathfrak{a}}$, by the formula

$$(35.8) \qquad (F_{\mathfrak{a}} u)(p_{\mathfrak{a}}, q^{\bar{\mathfrak{a}}}) = F_{p_{\mathfrak{a}} \leftarrow q} \mathfrak{a} v(q) = (2\pi\hbar)^{-|\mathfrak{a}|/2} \int \left[\exp(i/h) \sum_{k \in \mathfrak{a}} q^k p_k \right] v(q) \prod_{k \in \mathfrak{a}} dq^k,$$

with $|\mathfrak{a}|$ denoting the number of elements of $\mathfrak{a}$ and $\bar{\mathfrak{a}} = \{1, 2, \ldots, n\} \setminus \mathfrak{a}$. It is clear that $F = F_{\langle 1, 2, \ldots n \rangle}$.

It is well known that both the partial and the complete Fourier transforms map $\mathcal{T}(\mathbb{R}^n)$ onto itself linearly and bijectively (see for instance Birman and Solomjak (1987)). The inverse to $F_{\mathfrak{a}}$ has the expression

$$(35.9) \quad (F_{\mathfrak{a}}^{-1} v)(q) = F_{p_{\mathfrak{a}} \leftarrow q}^{-1} \mathfrak{a} v(p_{\mathfrak{a}}, q^{\bar{\mathfrak{a}}}) = (2\pi\hbar)^{-|\mathfrak{a}|/2} \int \left[\exp(i/h) \sum_{k \in \mathfrak{a}} q^k p_k \right] v(p_{\mathfrak{a}}, q^{\bar{\mathfrak{a}}}) \prod_{k \in \mathfrak{a}} dp_k.$$

All the transformations (35.7)–(35.9) preserve the L_2-norm $\| \phi \|_2 = (\int |\phi(q)|^2)^{1/2}$.

Local Expression for the Canonical Operator

Consider firstly an interior case, i.e. $z_0 \notin \Pi$. Given an arrangement at a point z_0, we define a linear map $\mathfrak{M}_{z_0}^{\mathrm{loc}} : C_W^\infty(\rho^{-1}(\mathcal{T})) \longrightarrow C_V^\infty(Q)$ by the formula

$$(35.10) \qquad \mathfrak{M}_{z_0}^{\mathrm{loc}} v = \alpha^* \frac{\chi}{\sqrt[4]{g}} F_{\mathfrak{a}}^{-1} \left(\sqrt{\left| \frac{\gamma^* \sigma}{V_{\mathfrak{a}}} \right|} e^{(i/\hbar) S} \gamma^* v \right).$$

Here we used the notations of the previous subsection, and the following additional notations: g is the determinant of the matrix of the Riemannian structure $\boldsymbol{g}$ in the chart (V, α),

$$(35.11) \qquad V_{\mathfrak{a}} = \left(\bigwedge_{i \in \bar{\mathfrak{a}}} dq^i \right) \wedge \left(\bigwedge_{j \in \mathfrak{a}} dp_j \right)$$

is the standard volume from on $L_{\mathfrak{a}}^{\bar{\mathfrak{a}}}$. The function $\gamma^* v$ defined originally on $G \subset L_{\mathfrak{a}}^{\bar{\mathfrak{a}}}$ is extended by the zero function onto the whole space $L_{\mathfrak{a}}^{\bar{\mathfrak{a}}}$. In the boundary case we define

$$(35.12) \qquad \mathfrak{M}_{z_0}^{\mathrm{loc}} = \mathfrak{M}_{z_0, \mathrm{out}}^{\mathrm{loc}} - \mathfrak{M}_{z_0, \mathrm{in}}^{\mathrm{loc}},$$

where $\mathfrak{M}_{z_0, \iota}^{\mathrm{loc}}$, $\iota \in \{\mathrm{in}, \mathrm{out}\}$, are given by the same formula as (35.10) with S_ι and γ_ι standing for S and γ respectively. Since in the boundary case $1 \notin \mathfrak{a}$, the Fourier transform does not affect the coordinate q^1. Taking into account (35.5) yields

35.13. *If* $v \in C_{V,\rho}^\infty(\rho^{-1}(\mathcal{T}))$ *then* $\mathfrak{M}_z^{\mathrm{loc}} v|_{\partial Q} = 0$

Recall that the subscript ρ on the symbol of a function space implies that the functions in question have the same value at pairs of points glued by ρ.

To indicate the dependence of $\mathfrak{M}^{loc}_{z_0}$ on $\mathscr{T}$, we shall write $\mathfrak{M}^{loc}_{z_0, \mathscr{T}}$.

The $\mathcal{O}(\hbar^\infty, (x)_\infty, (y)_b)$

This symbol will be useful in further calculations. Let $f(\hbar, x, y, z)$ be a complex valued function of variables $\hbar \in \,]0, 1[$ and $(x, y, z) \in G \subset \mathbb{R}^{n_1} \times \mathbb{R}^{n_2} \times \mathbb{R}^{n_3}$. We write $f = \mathcal{O}(\hbar^\infty, (x)_\infty, (y)_b)$ in G if (a) there is a positive constant C such that $f(\hbar, x, y, z) = 0$ if $|y| > C$; (b) for each positive integer k there is a positive constant C_k such that $|f(\hbar, x, y, z)| \leq C_k \hbar^k (1 + |x|)^{-k}$.

Isometry Property

The multiplier $1/\sqrt[4]{g}$ in (35.10) implies the preservation of the inner products (30.9), (34.8) up to the terms of order $\hbar^2$.

Proposition 35.14.

$$(35.14a) \qquad (\!(\mathfrak{M}^{loc}_{z_0} u, \mathfrak{M}^{loc}_{z_0} v)\!) = (\!(u, v)\!) + \mathcal{O}(\hbar^2).$$

Proof. It is sufficient to check the preservation modulo $\mathcal{O}(\hbar^2)$ of the L_2-norm with respect to the measure σ. At first, we consider the interior case. We have

$$(35.14b) \qquad \|\mathfrak{M}^{loc}_{z_0} u\|^2 = \int \left| F^{-1} \sqrt{\left| \frac{\gamma^* \sigma}{V} \right|} e^{(i/\hbar)S} \gamma^* u \right|^2 dq$$
$$+ \int (1 - \chi^2) \left| F^{-1} \sqrt{\left| \frac{\gamma^* \sigma}{V} \right|} e^{(i/\hbar)S} \gamma^* u \right|^2 dq.$$

Here the subscription $\mathfrak{a}$ of V and $F_\mathfrak{a}^{-1}$ are omitted for brevity. The first term on the right-hand side of (31.14b) equals $\|u\|^2$ since the Fourier transform preserves the L_2-norm, the latter being calculated with respect to the Euclidean measure dq (or V) in $\mathbb{R}^n$ (or $L_\mathfrak{a}^{\bar{\mathfrak{a}}}$), and

$$\left(\sqrt{\left| \frac{\gamma^* \sigma}{V} \right|} \right)^2 V = \gamma^* \sigma.$$

To get the estimate for the second term in (35.14b), let us consider the integrand $|\cdot|^2$ as a stationary phase integral over $\prod_{k \in \mathfrak{a}} dp_k \prod_{k \in \mathfrak{a}} dp'_k$, the phase function being

$$(35.14c) \qquad \Phi(q, p_\mathfrak{a}, p'_\mathfrak{a}) = \sum_{j \in \mathfrak{a}} (p_j - p'_j)q^j + S(q^{\bar{\mathfrak{a}}}, q_\mathfrak{a}) - S(q^{\bar{\mathfrak{a}}}, p_\mathfrak{a}).$$

The explicit calculation yields that there are no stationary points of (35.14c) on the support of the integrand of $q \in \sup(1 - \chi^2)$ (use (vi)). Moreover, there is a positive constant c such that $|\mathrm{grad}_{p_\mathfrak{a}, p'_\mathfrak{a}} \Phi| \geq c(1 + |q^\mathfrak{a}|)$ at these points. Integrating by parts as in Lemma AII.5, yields the estimate for the integrand in (35.14b):

(35.14d) $$|\cdot|^2 = \mathcal{O}(\hbar^\infty, (q^{\mathfrak{a}})_\infty, (\bar{q})_b).$$

The estimate $\mathcal{O}(\hbar^\infty)$ for the second term in (35.14b) follows from (35.14d).

In the boundary case, the integral to be evaluated is

(35.14e) $$\|\mathfrak{M}^{\mathrm{loc}}_{z_0,\mathrm{in}}u - \mathfrak{M}^{\mathrm{loc}}_{z_0,\mathrm{out}}u\|^2 = \|\mathfrak{M}^{\mathrm{loc}}_{z_0,\mathrm{in}}u\|^2 + \mathfrak{M}^{\mathrm{loc}}_{z_0,\mathrm{out}}u\|^2 - 2\,\mathrm{Re}(\!(\mathfrak{M}^{\mathrm{loc}}_{z_0,\mathrm{in}}u, \mathfrak{M}^{\mathrm{loc}}_{z_0,\mathrm{out}}u)\!).$$

The first two terms in the right-hand side of (35.14e) admit the same treatment as in the interior case. So they are $\mathcal{O}(\hbar^\infty)$. Analogous arguments show also that one can remove the Fourier transform in the last term, the error being of order $\hbar^\infty$:

(35.14f) $$-2\,\mathrm{Re}(\!(\mathfrak{M}^{\mathrm{loc}}_{z_0,\mathrm{in}}u, \mathfrak{M}^{\mathrm{loc}}_{z_0,\mathrm{out}}u)\!) = -2\int_{(L^{\bar{\mathfrak{a}}}_{\mathfrak{a}})_+} \cos\left(\frac{1}{\hbar}(S_{\mathrm{in}} - S_{\mathrm{out}})\right) \times A(q^{\bar{\mathfrak{a}}}, p_{\mathfrak{a}}) \prod_{i\in\bar{\mathfrak{a}},\, i\in\mathfrak{a}} dq^i dp_j + \mathcal{O}(\hbar^\infty).$$

Here A is a smooth real-valued function with compact support. The half-space $(L^{\bar{\mathfrak{a}}}_{\mathfrak{a}})_+$ is defined by the inequality $q^1 \geqq 0$. Recall that $1 \notin \mathfrak{a}$. The gradient of the phase function of the last integal nowhere vanishes and is transversal to the boundary of $(L^{\bar{\mathfrak{a}}}_{\mathfrak{a}})_+$. Indeed, the reflection formulas (6.9a) imply the value of the said gradient at a boundary point to be equal to

(35.14g) $$p_1^{\mathrm{in}} - p_1^{\mathrm{out}} = -2\left(\sum_{j=1}^n g^{j1}p_j\right)(g^{11})^{-1}.$$

The right-hand side of (35.14g), up to a nonzero multiplier, coincides with the q^1-component of the vector field ξ at a point on $L \cap \Pi$ (see (6.6d)). Since ξ meets Π transversally, (35.14g) does not vanish. We may insert an appropriate partition of unity in (35.14f) to reduce the problem of estimation to that for an integral over a small region, if necessary. If the support of the integrand has no intersection with $\partial(L^{\bar{\mathfrak{a}}}_{\mathfrak{a}})_+$ then the corresponding term is $\mathcal{O}(\hbar^\infty)$. Otherwise, integrating twice by parts yields the estimate $\mathcal{O}(\hbar^2)$. One must take into account that, due to (35.5), $\sin((1/\hbar)(S_{\mathrm{in}} - S_{\mathrm{out}})) = 0$ on the boundary. □

Remark 35.15. In the interior case above the proof yields the preservation of the products up to the elements of order $\hbar^\infty$.

Agreement on Intersections

Consider two points $z_0, \tilde{z}_0 \in \mathcal{T}\setminus\Pi$, supplied with the arrangements $\{(V,\alpha), U, \mathfrak{a}, S, \chi\}$ and $\{(\tilde{V},\tilde{\alpha}), \tilde{U}, \mathfrak{b}, \tilde{S}, \tilde{\chi}\}$ respectively. We shall use the conventional notations assigning $\sim$ to the objects attached to $\tilde{z}_0$ with one exception for the set $\mathfrak{a}$ which is to be replaced by $\mathfrak{b}$ in the second arrangement. Suppose that $U \cap \tilde{U} \cap \mathcal{T}$ is non-empty and simply connected. The domains of $\mathfrak{M}^{\mathrm{loc}}_{z_0}$ and $\mathfrak{M}^{\mathrm{loc}}_{\tilde{z}_0}$, constructed with the use of the above arrangements, have the non-empty intersection $C^\infty_{W\cap\tilde{W}}(\rho^{-1}(\mathcal{T}))$. Our goal here is to compare the values of these operators on a function v belonging to the said intersection. The natural way to do this is to apply the pertinent Fourier transform. Instead of comparing $\mathfrak{M}^{\text{loc}}_{z_0} v$ with $\mathfrak{M}^{\text{loc}}_{\tilde{z}_0} v$ we shall compare

$$u = F_{\mathfrak{b}} \sqrt[4]{\tilde{g}} (\tilde{\alpha}^{-1})^* \mathfrak{M}^{\text{loc}}_{z_0} v, \tag{35.16}$$

$$\tilde{u} = F_{\mathfrak{b}} \sqrt[4]{\tilde{g}} (\tilde{\alpha}^{-1})^* \mathfrak{M}^{\text{loc}}_{\tilde{z}_0} v = \sqrt{\left|\frac{\tilde{\gamma}^* \sigma}{V_{\mathfrak{b}}}\right|} e^{(i/\hbar)\tilde{S}} \tilde{\gamma}^* v + \mathcal{O}(\hbar^\infty). \tag{35.17}$$

In these formulae $\tilde{g}$ stands for the determinant of the matrix of the Riemannian structure in the chart $(\tilde{V}, \tilde{\alpha})$. The asymptotic estimate in the right-hand side of (35.17) is obtained in the same way as in the proof of Proposition (35.14).

To calculate u we will also apply the stationary phase method. The latter, failing to evaluate the original values of Maslov's operator, turns out to be applicable here due to the fine position of $\beta(\rho^{-1}(\mathcal{T}) \cap W \cap \tilde{W})$ and $\tilde{\beta}(\rho^{-1}(\mathcal{T}) \cap W \cap \tilde{W})$ with respect to the corresponding coordinate Lagrangian subspaces. At first, we write out the integral (35.16) in the form

$$u(\tilde{q}^{\bar{\mathfrak{b}}}, \tilde{p}_{\mathfrak{b}}) = \int A(p_{\mathfrak{a}}, \tilde{q}) \exp[(i/\hbar)\Phi(p_{\mathfrak{a}}, \tilde{q}, \tilde{p}_{\mathfrak{b}})] \prod_{j \in \mathfrak{a}, \hat{j} \in \mathfrak{b}} dp_j \, d\tilde{q}_{\hat{j}}, \tag{35.18}$$

where

$$\Phi(p_{\mathfrak{a}}, \tilde{q}, \tilde{p}_{\mathfrak{b}}) = S(q^{\bar{\mathfrak{a}}}(\tilde{q}), p_{\mathfrak{a}}) + \sum_{j \in \mathfrak{a}} p_j q^j(\tilde{q}) - \sum_{j \in \mathfrak{b}} \tilde{p}_j \tilde{q}^j \tag{35.19}$$

and

$$A(p_{\mathfrak{a}}, \tilde{q}) = (2\pi\hbar)^{-(|\mathfrak{a}|+|\mathfrak{b}|)/2} \sqrt[4]{\tilde{g}'(\tilde{q})} \chi(q(\tilde{q})) \frac{1}{\sqrt[4]{g(q(\tilde{q}))}} \left(\sqrt{\left|\frac{\gamma^*, \sigma}{V_{\mathfrak{a}}}\right|} \tilde{\gamma}^* v\right)(q^{\bar{\mathfrak{a}}}(\tilde{q}), p_{\mathfrak{a}}). \tag{35.20}$$

Here $q^{\bar{\mathfrak{a}}}(q) = \{q^i(\tilde{q}^1, \tilde{q}^2, \ldots, \tilde{q}^n),\ i \in \bar{\mathfrak{a}}\}$ are the functions which represent the $\bar{\mathfrak{a}}$-components of the coordinate transform $\alpha \circ \tilde{\alpha}^{-1}$. We shall use analogous notations also for the remaining components of $\alpha \circ \tilde{\alpha}^{-1}$. The equations (AIII.4) for finding a stationary point $(\rho_{\mathfrak{a}}, \tilde{q}^{\mathfrak{b}})$ read in our case:

$$\begin{gathered} q^j(\tilde{q}) - q^j_s(q^{\bar{\mathfrak{a}}}(\tilde{q}), p_{\mathfrak{a}}) = 0, \quad j \in \mathfrak{a}, \\ \sum_{i \in \bar{\mathfrak{a}}} p^s_i(q^{\bar{\mathfrak{a}}}(\tilde{q}), p_{\mathfrak{a}}) \frac{\partial q^i(\tilde{q})}{\partial \tilde{q}^k} + \sum_{j' \in \mathfrak{a}} p_{j'} \frac{\partial q^i(\tilde{q})}{\partial \tilde{q}^k} - \tilde{p}_k = 0, \quad k \in \mathfrak{b}. \end{gathered} \tag{35.21}$$

The main result of this subsection can be formulated as follows.

Proposition 35.22. *Under the conditions and notations of this subsection and in the interior case,*

(i) *if* $(\tilde{q}^{\bar{\mathfrak{b}}}, \tilde{p}_{\mathfrak{b}}) \notin \tilde{G}'' = \pi^{\bar{\mathfrak{b}}}_{\mathfrak{b}}(\tilde{\mathcal{T}}'')$, *where* $\tilde{\mathcal{T}}'' = \beta(\rho^{-1}(\mathcal{T}) \cap W \cap \tilde{W})$ *(see Fig. 55), then there is no stationary point of the integral* (35.18) *in the support of* A, *and so*

$$u(\tilde{q}^{\bar{\mathfrak{b}}}, \tilde{p}_{\mathfrak{b}}) = \mathcal{O}(\hbar^\infty, (\tilde{p}_{\mathfrak{b}}), (\tilde{q}^{\mathfrak{B}})_{\mathfrak{b}}) \quad \text{on } L^{\bar{\mathfrak{b}}}_{\mathfrak{b}} \backslash \tilde{G}''. \tag{35.22a}$$

(ii) *if* $(\tilde{q}^{\bar{\mathfrak{b}}}, \tilde{p}_{\mathfrak{b}}) \in \tilde{G}''$ *then there is unique solution* $(p^{st}_{\mathfrak{a}}, \tilde{q}^{\mathfrak{b}}_{st})$ *of* (35.21) *which can be found by the following recipe: let*

$$x_{st} = \tilde{\beta}^{-1} \circ (\pi^{\bar{\mathfrak{b}}}_{\mathfrak{b}}|_{\mathcal{T}''})^{-1} (\tilde{q}^{\bar{\mathfrak{b}}}, \tilde{p}_{\mathfrak{b}}). \tag{35.22b}$$

Then (see Fig. 55)

$$\text{(35.22c)}\qquad \begin{aligned} p^{st}_{\mathfrak{a}} &= \pi_{\mathfrak{a}}(A), \quad A = \beta(x_{st}) \in \mathscr{T}'', \\ \tilde{q}^{\mathfrak{b}}_{st} &= \pi^{\mathfrak{b}}(B), \quad B = \tilde{\beta}(x_{st}) \in \tilde{\mathscr{T}}, \end{aligned}$$

where $\pi_{\mathfrak{a}}$ *and* $\pi^{\mathfrak{b}}$ *are the corresponding coordinate projections.* $(p^{st}_{\mathfrak{a}}, \tilde{q}^{\mathfrak{b}}_{st})$ *is a non-degenerate stationary point, all the conditions of Theorem AII.7 being fulfilled, and, uniformly on* $(\tilde{q}^{\bar{\mathfrak{b}}}, \tilde{p}_{\mathfrak{b}}) \in \tilde{g}''$,

$$\text{(35.22d)}\qquad u \sim \exp\left[i\frac{\pi}{4} \operatorname{sign} M + (i/\hbar) L(z_0, \tilde{z}_0) \right] \sqrt{\left| \frac{\tilde{\gamma}^* \sigma}{V_{\mathfrak{b}}} \right|} e^{(i/\hbar)\tilde{S}} \tilde{\gamma}^* \left\{ v + \sum_{k=1}^{\infty} (i\hbar)^k T_k v \right\}.$$

Here $T_k: C^{\infty}_{W \cap \tilde{W}}(\rho^{-1}(\mathscr{T})) \longrightarrow C^{\infty}_{W \cap \tilde{W}}(\rho^{-1}(\mathscr{T}))$, $k = 1, 2, \ldots$, *is a linear differential operator of order* $2k$,

$$\text{(35.22e)}\qquad L(z_0, \tilde{z}_0) = \int_{z_0}^{\tilde{z}_0} \Theta,$$

the path of integration lying in $\mathscr{T} \cap U \cap \tilde{U}$, *M is the matrix of second derivatives of the left-hand sides of* (35.21) *with respect to* $(p_{\mathfrak{a}}, \tilde{q}^{\mathfrak{b}})$ *being evaluated at the stationary point: the signature of this matrix can be calculated also as follows:*

$$\text{(35.22f)}\qquad \operatorname{sign} M = \operatorname{sign}\left\{ \frac{\partial^2 S}{\partial p_i \, \partial p_j}, i, j \in \mathfrak{a} \right\} - \operatorname{sign}\left\{ \frac{\partial^2 \tilde{S}}{\partial \tilde{p}_i \, \partial \tilde{p}_j}, i, j \in \mathfrak{b} \right\},$$

the matrices on the right-hand side of (35.22f) *being calculated at the points* $(q^{\bar{\mathfrak{a}}}, p_{\mathfrak{a}}) = \pi^{\bar{\mathfrak{a}}}_{\mathfrak{a}}(A)$ *and* $(\tilde{q}^{\bar{\mathfrak{b}}}, \tilde{p}_{\mathfrak{b}})$ *correspondingly (see Fig. 55).*

Thus, comparing (35.17) and (35.22d), one observes that their leading terms differ by a simple constant multiplier $\exp\{(i\pi/4) \operatorname{sign} M - (i/\hbar) L(z_0, \tilde{z}_0)\}$. Note that both the terms on the right-hand side of (35.21f) depend on the point separately, while their difference, $\operatorname{sign} M$, is constant when z_{st} runs over $\mathscr{T} \cap U \cap \tilde{U}$ as it is simply connected.

Remark 35.23. When one or both of the operators in question belong to the boundary case, the statement of Proposition 35.21 and the subsequent proof remain valid, being applied to in and out components separately. One need only attach the indices in and out in appropriate places in the formulae.

Proof of Proposition 35.22. given $(\tilde{q}^{\bar{\mathfrak{b}}}, \tilde{p}_{\mathfrak{b}}) \in L^{\bar{\mathfrak{b}}}_{\mathfrak{b}}$, consider the transformation $\mathscr{F}: (p_{\mathfrak{a}}, \tilde{q}^{\mathfrak{b}}) \longmapsto (\mu^{\mathfrak{a}}, \tilde{v}_{\mathfrak{b}})$ which is defined by the equations

$$\text{(g)}\qquad \begin{aligned} & q^j(\tilde{q}) - q^j_s(q^{\bar{\mathfrak{a}}}(\tilde{q}), p_{\mathfrak{a}}) = \mu^j, \quad j \in \mathfrak{a}, \\ & \sum_{i \in \bar{\mathfrak{a}}} p^s_i(q^{\bar{\mathfrak{a}}}(\tilde{q}), p_{\mathfrak{a}}) \frac{\partial q^i(\tilde{q})}{\partial \tilde{q}^k} + \sum_{j' \in \mathfrak{a}} p_{j'} \frac{\partial q^i(\tilde{q})}{\partial \tilde{q}^k} - \tilde{p}_k = v_k, \quad k \in \mathfrak{b}. \end{aligned}$$

The domain of definition of $\mathscr{F}$ is the maximal one which is consistent with those of q_s, p^s and $\alpha \circ \tilde{\alpha}^{-1}$, the map $\mathscr{F}$ depending on $(\tilde{q}^{\bar{\mathfrak{b}}}, \tilde{p}_{\mathfrak{b}})$ as parameters. We are to prove the unique reversibility of $\mathscr{F}$ at $(\mu^{\mathfrak{a}}, \tilde{v}_{\mathfrak{b}}) = (0, 0)$, and to evaluate the matrix

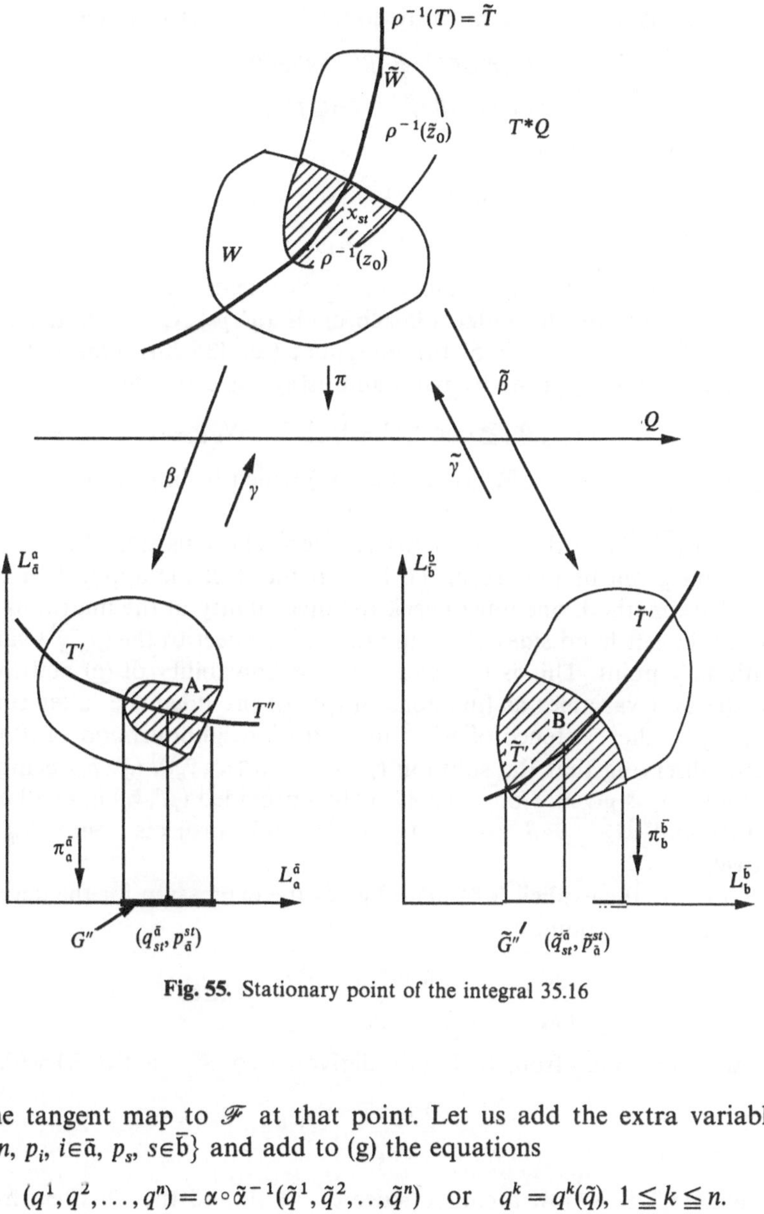

Fig. 55. Stationary point of the integral 35.16

M of the tangent map to $\mathscr{F}$ at that point. Let us add the extra variables $\{q^k, 1 \leqq k \leqq n, p_i, i \in \bar{\mathfrak{a}}, p_s, s \in \bar{\mathfrak{b}}\}$ and add to (g) the equations

$$(q^1, q^2, \dots, q^n) = \alpha \circ \tilde{\alpha}^{-1}(\tilde{q}^1, \tilde{q}^2, \dots, \tilde{q}^n) \quad \text{or} \quad q^k = q^k(\tilde{q}),\ 1 \leqq k \leqq n.$$

$$p_i = p_i^s(q^{\bar{\mathfrak{a}}}, p_{\mathfrak{a}}) = \frac{\partial S}{\partial q^i}(q^{\bar{\mathfrak{a}}}, p_{\mathfrak{a}}), \quad i \in \bar{\mathfrak{a}},$$

$$\tilde{p}_s = \sum_{k=1}^{n} p_k \frac{\partial q^k(\tilde{q})}{\partial q^s}, \quad s \in \bar{\mathfrak{b}}.$$

Denote correspondingly by s_μ and $\tilde{s}_{\tilde{\nu}}$ the shifts which act in $\mathbb{R}^{2n}$ as follows $(p, q^{\mathfrak{a}}, q^{\bar{\mathfrak{a}}}) \longmapsto (p, q^{\mathfrak{a}} + \mu^{\mathfrak{a}}, q^{\bar{\bar{\mathfrak{a}}}})$, and $(\tilde{p}_{\mathfrak{b}}, \tilde{p}_{\bar{\mathfrak{b}}}, \tilde{q}) \longmapsto (\tilde{p}_{\mathfrak{b}} + \tilde{\nu}_{\mathfrak{b}}, \tilde{p}_{\bar{\mathfrak{b}}}, \tilde{q})$. One can observe that the above equations are equivalent to the following geometric statements:

(h)
$$(\tilde{q},\tilde{p})\in\tilde{s}_{\tilde{v}}^{-1}\circ\tilde{\beta}\circ\beta^{-1}\circ s_\mu(\mathscr{T}'),$$
$$(q,p)=\beta\circ\tilde{\beta}^{-1}\circ\tilde{s}_{\tilde{v}}(\tilde{q},\tilde{p}).$$

Let $(\mu^{\mathfrak{a}},\tilde{v}_{\mathfrak{b}})=(0,0)$. Then (h) read

(h_0)
$$(\tilde{q},\tilde{p})\in\tilde{\beta}\circ\beta^{-1}\circ s_\mu(\mathscr{T}')=\tilde{\mathscr{T}},$$
$$(q,p)=\beta\circ\tilde{\beta}^{-1}(\tilde{q},\tilde{p}).$$

The latter, in virtue of the invertibility of $\pi_{\mathfrak{b}}^{\bar{\mathfrak{b}}}|_{\tilde{\mathscr{T}}'}$, yields the uniqueness of the stationary point and the formulae (35.22b, c). If $(\tilde{q}^{\bar{\mathfrak{b}}},\tilde{p}_{\mathfrak{b}})\notin\tilde{G}''$, then there is no stationary point in the support of the integrand (see (35.20)). One obtains the assertion (35.22a) by integrating by parts and using the estimates

$$|\operatorname{grad}_{p_{\mathfrak{a}},\tilde{q}^{\mathfrak{b}}}\Phi|\geqq \mathrm{const}_k(1+|\tilde{p}_{\mathfrak{b}}|)^{-k},\quad \forall_k\geqq 0,$$

which are uniform on $(\tilde{q}^{\bar{\mathfrak{b}}},\tilde{p}_{\mathfrak{b}})\in\mathbb{R}^n\setminus\tilde{G}''$ and $(\tilde{q}^{\mathfrak{b}},p_{\mathfrak{a}})$ which belongs to the support of the integrand.

Let $(\tilde{q}^{\bar{\mathfrak{b}}},\tilde{p}_{\mathfrak{b}})\in\tilde{G}''$. Then the above considerations show us that there is unique stationary point given by (35.21b, c). To be convinced of the applicability of the stationary phase method, one must check the invertibility of the matrix M of the derivatives of the left-hand sides of (g) and (h) with respect to the $(p_{\mathfrak{a}},\tilde{q}^{\mathfrak{b}})$-variables at the stationary point. This is equivalent to the solvability of (g) and (h) with respect to the said variables as functions of $(\mu^{\mathfrak{a}},\tilde{v}_{\mathfrak{b}})$ provided the latter run near the origin, i.e. to the existence of $\mathscr{F}^{-1}$ in a small neighbourhood of (0,0). We gain this by observing that the solution $(p_{\mathfrak{a}}(\mu^{\mathfrak{a}},\tilde{v}_{\mathfrak{b}}),\tilde{q}^{\mathfrak{b}}(\mu^{\mathfrak{a}},\tilde{v}_{\mathfrak{b}}))$, $(\tilde{q}^{\bar{\mathfrak{b}}},\tilde{p}_{\mathfrak{b}})$ being fixed, which we look for, is given geometrically by (h_0) provided $(\mu^{\mathfrak{a}},\tilde{v}_{\mathfrak{b}})$ is small enough so the manifold $\tilde{s}_{\tilde{v}}^{-1}\circ\tilde{\beta}\circ\beta^{-1}\circ s_\mu(\mathscr{T}')\approx\sim\tilde{\mathscr{T}}'$ still projects onto $L_{\mathfrak{b}}^{\bar{\mathfrak{b}}}$ non-degeneratively.

Differentiating the explicit formula (g) gives the expression for the derivatives of $\mathscr{F}$ at the stationary point:

(k)
$$\frac{\partial\mu^j}{\partial p_k}=-\frac{\partial q_s^j(q^{\bar{\mathfrak{a}}}(\tilde{q}_s^{\bar{\mathfrak{b}}},\tilde{q}_{st}^{\mathfrak{b}}),p_{\mathfrak{a}}^{st})}{\partial p_k},\quad j,k\in\mathfrak{a},$$

and it follows immediately from (h) that the derivatives of $\mathscr{F}^{-1}$ at $(\mu^{\mathfrak{a}},\tilde{v}_{\mathfrak{b}})=(0,0)$ are

(l)
$$\frac{\partial\tilde{q}^j}{\partial\tilde{v}^k}=\frac{\partial\tilde{q}_{\bar{s}}^j(\tilde{q}^{\bar{\mathfrak{b}}},\tilde{p}_{\mathfrak{b}})}{\partial\tilde{p}_k},\quad j,k\in\mathfrak{b}.$$

The formula (35.22f) is now a consequence of (k), (l) and the Theorem from the reference Lazutkin (1988).

The stationary phase method will provide us with the formula (35.22d) if we check the values of the phase and the amplitude. Substituting $p_{\mathfrak{a}}^{st}(\tilde{q}^{\bar{\mathfrak{b}}},\tilde{p}_{\mathfrak{b}})$ and $\tilde{q}_{st}^{\mathfrak{b}}(\tilde{q}^{\bar{\mathfrak{b}}},\tilde{p}_{\mathfrak{b}})$ into (35.19), one obtains a function of variables $(\tilde{q}^{\mathfrak{b}},\tilde{p}_{\mathfrak{b}})$. Its differential is equal to

$$\text{(m)}\quad d\Phi=\sum_{h\in\bar{\mathfrak{b}}}\left(\sum_{i\in\bar{\mathfrak{a}}}p_i^s(q^{\bar{\mathfrak{a}}}(\tilde{q}),p_{\mathfrak{a}})\frac{\partial q^i(\tilde{q})}{\partial\tilde{q}^h}+\sum_{j\in\mathfrak{a}}p_j^{st}\frac{\partial q^i(\tilde{q})}{\partial\tilde{q}^h}\right)d\tilde{q}^h-\sum_{k\in\mathfrak{b}}\tilde{q}_{st}^k\,d\tilde{p}_j,\quad \tilde{q}=(\tilde{q}^{\bar{\mathfrak{b}}},\tilde{q}_{st}^{\mathfrak{b}}),$$

as one can immediately calculate, taking into account the equations (35.21) which define the stationary point. The formulae (35.22c) imply

$$\tilde{q}^j_{st} = -\frac{\partial \tilde{S}(\tilde{q}^{\bar{\mathfrak{b}}}, \tilde{p}_{\mathfrak{b}})}{\partial \tilde{p}_j},$$

$$\sum_{i \in \bar{\mathfrak{a}}} p^s_i(q^{\bar{\mathfrak{a}}}(\tilde{q}), p_{\mathfrak{a}}) \frac{\partial q^i(\tilde{q})}{\partial \tilde{q}^h} + \sum_{j \in \mathfrak{a}} p^{st}_j \frac{\partial q^i(\tilde{q})}{\partial \tilde{q}^h} = \frac{\partial \tilde{S}(\tilde{q}^{\bar{\mathfrak{b}}}, \tilde{p}_{\mathfrak{b}})}{\partial \tilde{p}_j},$$

so the right-hand side of (m) coincides with $d\tilde{S}$. The norming condition (35.4) yields the value of the phase in (35.22d).

To obtain the formula for the amplitude, let us consider the map $\mathscr{F}_1 = \mathscr{F} \times \mathrm{id}: (p_{\mathfrak{a}}, \tilde{q}^{\mathfrak{b}}, \tilde{q}^{\bar{\mathfrak{b}}}) \longmapsto (\mu^{\mathfrak{a}}, \tilde{\nu}_{\mathfrak{b}}, \tilde{q}^{\bar{\mathfrak{b}}})$ by adding to (g), (h) the equation $\tilde{a}^{\bar{\mathfrak{b}}} = \tilde{q}^{\bar{\mathfrak{b}}}$. Evidently, the determinant of the matrix of the, tangent to $\mathscr{F}_1$ at $(p^{st}_{\mathfrak{a}}, \tilde{q}^{\mathfrak{b}}_{st}, \tilde{q}^{\bar{\mathfrak{b}}})$ equals M. Consider also the superposition

$$\mathscr{F}_1 \circ \mathscr{F}_2 \circ \mathscr{F}_3 : (\mu^{\mathfrak{a}}, q^{\bar{\mathfrak{a}}}, p_{\mathfrak{a}}) \longmapsto (\mu^{\mathfrak{a}}, \tilde{\nu}_{\mathfrak{b}}, \tilde{q}^{\bar{\mathfrak{b}}}),$$

where $\mathscr{F}_2 : (p_{\mathfrak{a}}, q) \longmapsto (p_{\mathfrak{a}}, \tilde{q})$ equals $\mathrm{id} = \alpha \circ \alpha^{-1}$ and $\mathscr{F}_3 : (\mu^{\mathfrak{a}}, q^{\mathfrak{a}}, p_{\mathfrak{a}}) \longmapsto (p_{\mathfrak{a}}, \tilde{q}^{\mathfrak{a}}, \tilde{q}^{\bar{\mathfrak{a}}})$ is given by $p_{\mathfrak{a}} = p_{\mathfrak{a}}$, $q^{\bar{\mathfrak{a}}} = q^{\bar{\mathfrak{a}}}$, and $q^{\mathfrak{a}} = q^a_s(q^{\mathfrak{a}}, p_{\mathfrak{a}}) + \mu^{\mathfrak{a}}$, one can easily check that the matrix of the tangent to $\mathscr{F}_1 \circ \mathscr{F}_2 \circ \mathscr{F}_3$ at the pre-image of $(0, 0, \tilde{q}^{\bar{\mathfrak{b}}})$ has the form

$$\begin{array}{c|c|c|c|}
 & \mu^{\mathfrak{a}} & q^{\bar{\mathfrak{a}}} & p_{\mathfrak{a}} \\
\hline
\mu^{\mathfrak{a}} & 1 & 0 & 0 \\
\hline
\tilde{q}^{\bar{\mathfrak{b}}} & & \multicolumn{2}{c|}{} \\
 & & \multicolumn{2}{c|}{\tilde{M}_1} \\
\tilde{\nu}_{\mathfrak{b}} & & \multicolumn{2}{c|}{} \\
\hline
\end{array}$$

where $\tilde{M}_1$ is the $n \times n$-matrix of the tangent to the map $\tilde{\gamma}^{-1} \circ \gamma$ at $\pi^{\bar{\mathfrak{a}}}_{\mathfrak{a}}(A)$. Since its determinant equals the product of the determinants of three matrices corresponding to $\mathscr{F}_i$, $i = 1, 2, 3$, we have, evaluating at the corresponding points

$$\det(T\tilde{\gamma}^{-1} \circ \gamma) = \det(T\tilde{\alpha} \circ \alpha^{-1}) \cdot \det M.$$

Taking into account the equations

$$(\tilde{\gamma}^{-1} \circ \gamma)^* V_{\mathfrak{b}} = [\det T(\tilde{\gamma}^{-1} \circ \gamma)]^{-1} V_{\mathfrak{a}},$$

$$\det T(\tilde{\alpha} \circ \alpha^{-1}) = \sqrt{g/\tilde{g} \circ \tilde{\alpha} \circ \alpha^{-1}},$$

one obtains

$$\frac{1}{\sqrt{|\det M|}} \cdot \frac{1}{\sqrt[4]{g \circ \alpha \circ \tilde{\alpha}^{-1}}} \sqrt{\left| \frac{\tilde{\gamma}^* \sigma}{(\gamma^{-1} \circ \tilde{\gamma})^* V_{\mathfrak{a}}} \right|} = \frac{1}{\sqrt[4]{\tilde{g}}} \sqrt{\left| \frac{\tilde{\gamma}^* \sigma}{V_{\mathfrak{b}}} \right|},$$

i.e. the quoted value of the amplitude. □

§36. The Commutation Rule

Here we consider the question of now the LBS-operator $\mathscr{H}_\hbar - \frac{1}{2}\hbar^2\Delta + \mathscr{V}$, defined in §30, acts on functions of the form $\mathfrak{M}^{\text{loc}}_{z0}v$. It will be convenient, in view of further use, to discuss the action of a more general pseudo-differential a–b-operator

The a–b-Operator

The construction of the a–b-operator is attached to a chart (V, α) in Q which enters some chosen arrangements. Here we suppose $V \cap \partial Q = \varnothing$, i.e. we are working in the interior case. Let us introduce the special space of symbols $C^\infty_{\text{symb}}(\mathbb{R}^{2n})$. We shall denote here the variable running over $\mathbb{R}^{2n}$ as (q, p) where $q = (q^1, q^2, \ldots, q^n)$, $p = (p_1, p_2, \ldots, p_n)$ run over $\mathbb{R}^n$. A function $a \in C^\infty(\mathbb{R}^{2n})$ if and only if

(a1) there is a compact set $K' \subset \alpha(V)$ such that for all $p \in \mathbb{R}^n$ supp $a(.,p) \subset K'$,

(a2) there is a positive number m such that for each multi-index $k = (k_1, k_2, \ldots, k_b)$ the inequality

$$\left|\left(\frac{\partial}{\partial q}\right)^k a(q,p)\right| \leqq c_k |p|^m$$

holds with some positive c_k for all $(q, p) \in \mathbb{R}^{2n}$.

Given two functions $a, b \in C^\infty_{\text{symb}}(\mathbb{R}^{2n})$ define the operator $\mathfrak{O}_{a,b}: C^\infty_V(Q) \longrightarrow C^\infty_V(Q)$ by the formula

$$\mathfrak{O}_{a,b} u \circ \alpha^{-1}(q) = F^{-1}_{q \leftarrow p} a(q,p) F_{p \leftarrow q'} b(q',p) u \circ \alpha^{-1}(q'), \tag{36.1}$$

where F and F^{-1} stand for the Fourier transform and its inverse (see the notations (35.7, 8, 9)). Our conditions (a1), (a2) imply the convergence of the integrals on the right-hand side of (36.1) and the existence of all q-derivatives, and so $\mathfrak{O}_{a,b}$ is correctly defined.

We shall consider also the case when a and b depend on an extra variable $\hbar \in [0, 1]$. In that case we assume that either they are polynomials in $i\hbar$ or they depend smoothly on $\hbar^2$. The above properties will be supposed to be fulfilled uniformly on $\hbar$. Such functions can be regarded as families of elements of $C^\infty_{\text{symb}}(\mathbb{R}^{2n})$, the formula (36.1) defining the family of operators $\{u_\hbar\colon \hbar \in [0,1]\} \in C^\infty_V(Q)$ of functions rather than isolated functions. We adopt the following notation for a "negligible" family: $u_\hbar = \mathcal{O}_V(\hbar^\infty)$ means that (i) there is a compact set $K \subset V$ such that supp $u_\hbar \subset K$ for all $\hbar$, (ii) for each multi-index $k = (k_1, k_2, \ldots, k_n)$ and a positive number m there is a constant $c_{k,m}$ such that

$$\left|\left(\frac{\partial}{\partial q}\right)^k u_\hbar \circ \alpha^{-1}(q)\right| \leqq c_{k,m} \hbar^m \quad \forall (q, \hbar) \in \alpha(V) \times [0, 1].$$

The following assertion is evident:

36.2. $$u_\hbar = \mathcal{O}_V(\hbar^\infty) \;\Rightarrow\; \mathfrak{O}_{a,b} u_\hbar = \mathcal{O}_V(\hbar^\infty),$$

Nonintersecting Supports

Let us consider how $\mathfrak{D}_{a,b}$ acts on a function of the form $\mathfrak{M}^{\mathrm{loc}}_{zo}v$. The (V,α) of $\mathfrak{D}_{a,b}$ do not necessarily enter the arrangement $\{(\tilde{V},\tilde{\alpha}), U, \mathfrak{a}, \chi\}$ of $\mathfrak{M}^{\mathrm{loc}}_{zo}, v$. The case $\tilde{V}\cap\partial Q\neq\varnothing$ is not excluded.

Proposition 36.3. *Let* $v\in C^{\infty}_{U}(\mathcal{T})$ *and* $\beta = T^{*-1}\alpha$. *If* $\operatorname{supp}\beta^{*}a\cap\operatorname{supp}\beta^{*}b\cap\operatorname{supp}v = \varnothing$, *then* $\mathfrak{D}_{a,b}\mathfrak{M}^{\mathrm{loc}}_{0}v = \mathcal{O}_{V}(\hbar^{\infty})$.

Proof. $(\partial/\partial q)^{k}\mathfrak{D}_{a,b}\mathfrak{M}^{\mathrm{loc}}_{zo}v$ may be represented as an integral containing a rapidly oscillating integrand. We use the notations q, p, q' for the variables concerning $\mathfrak{D}_{a,b}$ and $\tilde{q},\tilde{p}$ for those of $\mathfrak{M}^{\mathrm{loc}}_{zo}v$. It is sufficient, in view of AIII to show that the gradient of the phase function

$$\text{(36.3a)}\quad \Phi(q,p,q';\tilde{p}_{\mathfrak{a}}) = \langle q-q',p\rangle + \sum_{j\in\mathfrak{a}}\tilde{q}^{j}\tilde{p}_{j} + S(\tilde{q}^{\bar{\mathfrak{a}}},\tilde{p}_{\mathfrak{a}}),\quad \tilde{q} = \tilde{\alpha}\cdot\alpha^{-1}(q'),$$

with respect to the variables $(p,q',\tilde{p}_{\mathfrak{a}})$ does not vanish on the support of the integrand. Immediate calculation yields $\operatorname{grad}\Phi = 0$ to be equivalent to the equations

$$\text{(36.3b)}\quad q = q',\qquad p_{k} = \sum_{j\in\mathfrak{a}}\frac{\partial\tilde{q}^{j}}{\partial q'^{k}}\tilde{p}_{j} + \sum_{i\in\bar{\mathfrak{a}}}\frac{\partial\tilde{q}^{i}}{\partial q'^{k}}p^{s}_{i}(\tilde{q}^{\bar{\mathfrak{a}}},\tilde{p}_{\mathfrak{a}}),\quad \tilde{q}^{j} = q^{j}_{s}(\tilde{q}^{\bar{\mathfrak{a}}},\tilde{p}_{\mathfrak{a}}),\quad j\in\mathfrak{a}.$$

Given q, let $p,q',\tilde{p}_{\mathfrak{a}}$ obey (36.3b). Set $\tilde{q}^{i} = (\tilde{\alpha}\circ\alpha^{-1}(q'))^{i}$, $i\in\bar{\mathfrak{a}}$, $\tilde{q} = (\tilde{q}^{\bar{\mathfrak{a}}}, q^{\mathfrak{a}}_{s}(\tilde{q}^{\bar{\mathfrak{a}}},\tilde{p}_{\mathfrak{a}}))$, $\tilde{p} = (p^{s}_{\bar{\mathfrak{a}}}(\tilde{q}^{\bar{\mathfrak{a}}},\tilde{p}_{\mathfrak{a}}),\tilde{p}_{\mathfrak{a}})$. Then (36.3b) implies

$$\text{(36.3c)}\qquad (q',p) = (q,p) = (T^{*-1}\alpha\circ\tilde{\alpha}^{-1})(\tilde{q},\tilde{p}).$$

Suppose that $(p,q',\tilde{p}_{\mathfrak{a}})$ belongs to the support of the integrand. The latter is contained in

$$\{(p,q',\tilde{p}_{\mathfrak{a}})\colon (q,p)\in\operatorname{supp}a,\ (q',p)\in\operatorname{supp}b,\ \tilde{\gamma}(\tilde{q}^{\bar{\mathfrak{a}}},\tilde{p}_{\mathfrak{a}})\in\operatorname{supp}v\}$$

(here $\tilde{\gamma}$ is of $\mathfrak{M}^{\mathrm{loc}}_{zo}$). The equality (36.3c) implies that in this case the point $z = (T^{*-1}\tilde{\alpha}^{-1})(\tilde{q},\tilde{p})$ belongs simultaneously to $\operatorname{supp}\beta^{*}a$, $\operatorname{supp}\beta^{*}b$ and $\operatorname{supp}v$ which contradicts the condition of the Proposition. □

The Commutation Rule

Now we pass to the case when the chart (V,α) in the definition of $\mathfrak{D}_{a,b}$ belongs to the arrangement $\{(V,\alpha), U, \mathfrak{a}, \chi\}$ of $\mathfrak{M}^{\mathrm{loc}}_{zo}$. Fix an open set $\tilde{U}\supset\bar{U}$ which is sufficiently close to U so that all the properties hold when $\tilde{U}$ replaces U. Denote $W = \rho^{-1}(\tilde{U})$. The operator $\mathfrak{M}^{\mathrm{loc}}_{zo}$ may be extended onto $C^{\infty}_{c,\tilde{W}}(\rho^{-1}(\mathcal{T}))$ by the same formulae.

Proposition 36.4. *There is a linear map*

$$\mathfrak{D}^{\square}_{a,b}\colon C^{\infty}_{c,W}(\rho^{-1}(\mathcal{T}))\longrightarrow C^{\infty}_{c,\tilde{W}}(\rho^{-1}(\mathcal{T}))$$

such that

(36.4a) $$\mathfrak{D}_{a,b}\mathfrak{M}^{\mathrm{loc}}_{zo}v = \mathfrak{M}^{\mathrm{loc}}_{zo}\mathfrak{D}^{\square}_{a,b}v + \mathcal{O}_V(\hbar^\infty).$$

There operator $\mathfrak{D}^{\square}_{a,b}$ *has the following asymptotic expansion:*

(36.4b) $$\mathfrak{D}^{\square}_{a,b} \sim \sum_{k=0}^{\infty} (i\hbar)^k \mathfrak{D}^{\square}_{a,b,k},$$

where, for each k, $\mathfrak{D}^{\square}_{a,b,k}$ *is a differential operator of order not exceeding* k *with smooth coefficients. More particularly,* $\mathfrak{D}^{\square}_{a,b,0}$ *is the operator of multiplication by the function* $\beta^*(a\cdot b)|_{\rho^{-1}(U)\cap\tilde{W}}$, $\mathfrak{D}^{\square}_{a,b,1}$ *is given by the formulae (g), (h) below.*

The estimate in (36.4a) holds in $L_2(Q)$. *The expansion (36.4b) means that for any positive integer* N *and for any* $v \in C^\infty_{c,W}(\rho^{-1}(\mathcal{T}))$,

(36.4c) $$\left|\mathfrak{D}^{\square}_{a,b}v(z) - \sum_{k=0}^{N}(i\hbar)^k\mathfrak{D}^{\square}_{a,b,k}v(z)\right| = \mathcal{O}(\hbar^{N+1})$$

uniformly on $z \in \tilde{W}\cap\rho^{-1}(\mathcal{T})$.

Proof. Fx a smooth function $\tilde{\chi}:\mathcal{T}\longrightarrow\mathbb{R}$ with compact support such that $\operatorname{supp}\tilde{\chi}\subset\tilde{U}$, $\tilde{\chi}=1$ on U, and set

(d) $$\mathfrak{D}^{\square}_{a,b}v = \tilde{\chi}\gamma^{*-1}e^{-(i/\hbar)S}\lambda^{-1/2}w,$$

where

(36.4e) $$\lambda = \left|\frac{\gamma^*\sigma}{V_\alpha}\right|$$

and

(f) $$w(q^{\bar{\alpha}}, p_\alpha) = F_{q_\alpha\leftarrow p^\alpha}F^{-1}_{q\leftarrow p'}\sqrt[4]{g(q)}a(q,p')F_{p'\leftarrow q'}\frac{\chi(q')}{\sqrt[4]{g(q)}}b(q',p')F^{-1}_{(q')^\alpha\leftarrow p''_\alpha}$$
$$\times\left(\sqrt{\lambda}\gamma^*v(q'^{\bar{\alpha}},p''_\alpha)e^{(i/\hbar)S(q'^{\bar{\alpha}},P''_\alpha)}\right).$$

To obtain the expansion (36.4b) we use the following

The ABC-Lemma

Let $A, B\in C^\infty_{\mathrm{symb}}(\mathbb{R}^{2n})$, $C\in C^\infty_G(L^{\bar{\alpha}}_\alpha)$. Consider the integral of the form

$$I(A,B,C)(q^{\bar{\alpha}},p_\alpha)$$
$$= F_{q_\alpha\leftarrow p^\alpha}F^{-1}_{q\leftarrow p'}A(q,p')F_{p'\leftarrow q'}B(q',p')F^{-1}_{(q')^\alpha\leftarrow p''_\alpha}C(q'^{\bar{\alpha}},p''_\alpha)\exp\langle(i/\hbar)S(q'^{\bar{\alpha}},p''_\alpha)\rangle.$$

Define a special tensor product

$$C^\infty_{\mathrm{symb}}(\mathbb{R}^{2n})\times C^\infty_{\mathrm{symb}}(\mathbb{R}^{2n})\times C^\infty_G(L^{\bar{\alpha}}_\alpha)\xrightarrow{\cdot\overset{p}{\otimes}\cdot\overset{q}{\otimes}\cdot} C^\infty(\mathbb{R}^{3n+|\alpha|})$$

as

$$A \overset{p}{\otimes} B \overset{q}{\otimes} C(q, q', p', p''_{\mathrm{a}}) = A(q, p')B(q', p')C(q'^{\bar{\mathrm{a}}}, p''_{\mathrm{a}}), \quad q' = (q'^{\bar{\mathrm{a}}}, q'^{\mathrm{a}}),$$

and a map $\delta: G \longrightarrow \mathbb{R}^{3n+|\mathrm{a}|}$ by the formula

$$\delta(q^{\bar{\mathrm{a}}}, p_{\mathrm{a}}) = (q, q', p', p''_{\mathrm{a}}), \tag{36.5}$$

where $q = q' = (q^{\bar{\mathrm{a}}}, q^{\mathrm{a}}_s(q^{\bar{\mathrm{a}}}, p_{\mathrm{a}}))$, $p' = (p^s_{\bar{\mathrm{a}}}(q^{\bar{\mathrm{a}}}, p_{\mathrm{a}}), p_{\mathrm{a}})$, $p''_{\mathrm{a}} = p_{\dot{\mathrm{a}}}$.

Lemma 36.6 (ABC-Lemma). *There is a sequence*

$$I_s: C^\infty_{\mathrm{symb}}(\mathbb{R}^{2n}) \times C^\infty_{\mathrm{symb}}(\mathbb{R}^{2n}) \times C^\infty_G(L^{\bar{\mathrm{a}}}_{\mathrm{a}}) \longrightarrow C^\infty_G(L^{\bar{\mathrm{a}}}_{\mathrm{a}}), \quad s = 0, 1, 2, \ldots$$

of differential trilinear operators containing the application of δ^* to a sum of the derivatives of $A \overset{p}{\otimes} B \overset{q}{\otimes} C$ multiplied by smooth functions, the order of I_s with respect to C not exceeding s, such that

$$I(A, B, C)|_G \sim e^{(i/\hbar)S} \sum_{s=1}^{\infty} (i\hbar)^s I_s(A, B, C)|_G, \tag{36.6a}$$

and, if $(q^{\bar{\mathrm{a}}}, p_{\mathrm{a}}) \notin G$, then, for any positive integer k

$$|I(A, B, C)(q^{\bar{\mathrm{a}}}, p_{\mathrm{a}})| \leq \mathrm{const}\, \hbar^k (1 + |p_{\mathrm{a}}|)^{-k}, \tag{36.6b}$$

where const > 0 *depends only on k, S, A, B, C.*

The explicit formulae for the first two operators I_s are:

$$I_0(A, B, C) = \delta^*(A \overset{p}{\otimes} B \overset{q}{\otimes} C), \tag{36.5c}$$

$$\begin{aligned} I_1(A, B, C) = \delta^*\Bigg[& \left(\sum_{j\in\mathrm{a}}\left(\frac{\partial^2}{\partial p''_j \partial q'^j} + \frac{\partial^2}{\partial p''_j \partial q^j} + \frac{\partial^2}{\partial p'_j \partial q^j}\right) - \sum_{i\in\bar{\mathrm{a}}} \frac{\partial^2}{\partial p''_j \partial q'^j}\right) \\ & \times A \overset{p}{\otimes} B \overset{q}{\otimes} C + \sum_{i\in\bar{\mathrm{a}}, j\in\mathrm{a}} \left(\frac{\partial^2}{\partial p'_i \partial q^j} + \frac{\partial^2}{\partial p'_i \partial q'^j}\right) \\ & \times A \overset{p}{\otimes} B \overset{q}{\otimes} \left(C \frac{\partial^2 S}{\partial p''_j \partial q'^i}\right) \\ & - \sum_{i,k\in\bar{\mathrm{a}}} \left(\frac{1}{2} \frac{\partial^2}{\partial q'^i \partial q'^k} + \frac{1}{2} \frac{\partial^2}{\partial q^j \partial q^k} + \frac{\partial^2}{\partial q^j \partial q'^k}\right) \\ & \cdot A \overset{p}{\otimes} B \overset{q}{\otimes} \left(C \times \frac{\partial^2 S}{\partial p''_j \partial p''_k}\right) - \frac{1}{2} \sum_{i,k\in\mathrm{a}} \frac{\partial^2}{\partial p'_i \partial p'_k} \\ & \cdot A \overset{p}{\otimes} B \overset{q}{\otimes} \left(C \frac{\partial^2 S}{\partial q''^i \partial q'^k}\right)\Bigg]. \end{aligned} \tag{35.6d}$$

We postpone the proof of the ABC-Lemma to the end of this subsection.

End of the Proof of Proposition 36.4

Applying the ABC-Lemma to (f) with

(g) $A = \sqrt[4]{g}a, \qquad B(\chi/\sqrt[4]{g})b, \qquad C = \sqrt{\lambda}\gamma^* v$

yields the expansion (36.4b). We have

(h) $\mathfrak{O}^{\square}_{a,b,1} v = \tilde{\chi}\gamma^{*-1}\lambda^{-1/2}I_1(A, B, C).$

To obtain the commutation rule we take into account (36.1), (f), the definition
(3.10) of $\mathfrak{M}^{loc}_{z_0}$, and write

(i) $\alpha^{*-1}\mathfrak{O}_{a,b}\mathfrak{M}^{loc}_{z_0} v = \dfrac{1}{\sqrt[4]{g}} F_a^{-1} w = \dfrac{\chi}{\sqrt[4]{g}} F_a^{-1}(\gamma^* \tilde{\chi})w + \dfrac{1-\chi}{\sqrt[4]{g}} F_a^{-1} w$

$$+ \frac{\chi}{\sqrt[4]{g}} F_a^{-1}\gamma^*(1 - \tilde{\chi})w.$$

The first term in the right-hand side of (i) is

$$\frac{\chi}{\sqrt[4]{g}} F_a^{-1}(\gamma^* \tilde{\chi})w = \frac{\chi}{\sqrt[4]{g}} F_a^{-1}\sqrt{\lambda}e^{(i/\hbar)S}\gamma^*[\tilde{\chi}\gamma^{*-1}(\lambda^{-1/2}e^{-(i/\hbar)S}w)] = \mathfrak{M}^{loc}_{z_0}\mathfrak{O}^{\square}_{a,b}v.$$

The second one, due to Proposition 36.3, is

$$\frac{1-\chi}{\sqrt[4]{g}} F_a^{-1}w = \mathfrak{O}_{(1-\chi)a,b}\mathfrak{M}^{loc}_{z_0}v = \mathcal{O}_V(\hbar^\infty).$$

since $\beta^* \chi = 1$ on supp v. The third one is also $\mathcal{O}_V(\hbar^\infty)$. The proof of the last assertion
is left to the reader, as it is quite similar to that of Proposition 36.3.   □

The Commutation Rule for the LBS Operator

The LBS-operator $\mathscr{H}_\hbar = -\frac{1}{2}\hbar^2\Delta + \mathscr{V}$ can be represented as in a–b-operator being
restricted on some subspaces of $C^\infty_0(Q)$. Let (V, α) be an interior chart in Q, and
$K \subset V$ be a compact subset. Fix a smooth function $\chi_1 : \mathbb{R}^n \longrightarrow \mathbb{R}$ with properties
supp $\chi \subset \alpha(V)$ and $\chi_1 = 1$ on $\alpha(K)$. Then $\mathscr{H}_\hbar$ restricted on $C^\infty_K(Q)$ coincides with
$\mathfrak{O}_{a,b}$ where

(36.7) $a(q,p) = \left\{ \displaystyle\sum_{k,l=1}^n \left(\frac{1}{2}g^{kl}(q)p_k p_l + \frac{i\hbar}{2} p_l \frac{1}{\sqrt{g(q)}} \right. \right.$

$$\left. \left. \times \frac{\partial}{\partial q^k}(g^{kl}(q)\sqrt{g(q)}) \right) + \mathscr{V}\circ\alpha^{-1}(q) \right\}\chi_1(q), \qquad b(q,p) = \chi_1(q).$$

If (V, α) enters the arrangement of $\mathfrak{M}^{loc}_{z_0}$, the commutation rule for $\mathscr{H}_\hbar$ and
$\mathfrak{M}^{loc}_{z_0}$ follows immediately from Proposition 36.4.

Now let (V, α) be a boundary chart, i.e. $V \cap \partial Q \neq \varnothing$, and let $K \subset V$ and χ_1 be
chosen in the same manner as above. The local nature of $\mathscr{H}_\hbar$ enables us to

represent $\mathscr{H}_\hbar$ restricted on $C_K^\infty(Q)$ also as an a–b-operator. The construction which follows utilizes an extension operator E.

Lemma 36.8. *There is a linear map* $E: C_c^\infty(\mathbb{R}_+^n) \longrightarrow C_c^\infty(\mathbb{R}^n)$ *such that*

(i) $R\circ E = \mathrm{id}$, *where* $R: C_c^\infty(\mathbb{R}^n) \longrightarrow C_c^\infty(\mathbb{R}_+^n)$ *is the restriction map, i.e.* $Ru = u|_{\mathbb{R}_+^n}$;

(ii) *for each multi-index k there is a constant C_k such that*

$$\left|\left(\frac{\partial}{\partial q}\right)^k Eu\right| \leq C_k \max_{q,l\leq |k|}\left|\left(\frac{\partial}{\partial q}\right)^l u(q)\right|.$$

Proof. (cf. Hamilton, p. 138). The function

$$\phi(t) = \frac{e^{2\sqrt{2}}}{\pi(1+t)} e^{-(t^{1/4}+t^{-1/4})}\sin(t^{1/4}-t^{-1/4})$$

obeys the equalities $\int_0^\infty t^n \phi(t)\,dt = (-1)^n$ for $n = 0, 1, 2, \ldots$. Fix another smooth function $\psi: \mathbb{R} \longrightarrow \mathbb{R}$ with compact support, such that $\psi|_{[-1,0]} = 1$. Set

$$(Eu)(q^1, q^2, \ldots, q^n) = \begin{cases} u(q^1, q^2, \ldots, q^n) & \text{if } q^1 \geqq 0, \\ \psi(q^1)\displaystyle\int_0^\infty \phi(t) u(-tq^1, q^2, \ldots, q^n) & \text{if } q^1 < 0. \end{cases}$$

Both the properties (i), (ii) are evident. □

Applying E to the coefficients in (36.7) and to a function $u\circ\alpha^{-1}(q)$ we obtain the extended functions $\tilde{a}, \tilde{b} \in C^\infty(\mathbb{R}^{2n})$ and $\tilde{u} \in C_c^\infty(\mathbb{R}^{2n})$. a, b obey the conditions (a2), the condition (a1) is also fulfilled for some compact set $\tilde{K}(\tilde{K}\cap\alpha(V) = \alpha(K))$ replacing K'. Define $\mathfrak{O}_{\tilde{a},\tilde{b}}$ by using the same formula (36.1) with $\tilde{a}, \tilde{b}$ and $\tilde{u}$ standing for $a, b, u\circ\alpha^{-1}$ correspondingly. Clearly $\mathfrak{O}_{\tilde{a},\tilde{b}}u = \mathscr{H}_\hbar u$, if $u \in C_K^\infty(Q)$, and this result does not depend on the choice of an extension $\tilde{u}$ of $u\circ\alpha^{-1}$. Particularly, we may extend the value $\alpha^{*-1}\mathfrak{M}_{z_0,t}^{\mathrm{loc}} v$ of a component of a local Maslov's operator in the boundary case by using appropriate continuations of $\lambda_t, S_t, \gamma^* v$ and other functions entering the definition of $\mathfrak{M}_{z_0,t}^{\mathrm{loc}}$. Then, acting in the same manner as in the interior case, one obtains the commutation rule for $\mathscr{H}_\hbar$ and $\mathfrak{M}_{z_0,t}^{\mathrm{loc}}$ with $\operatorname{supp}\chi$ in the role of a set K, χ belonging to the arrangement of $\mathfrak{M}_{z_0}^{\mathrm{loc}}$. Note that all the operators $\mathscr{H}_\hbar$ and $\mathfrak{O}_{\tilde{a},\tilde{b},s}$, $s = 0, 1, 2, \ldots$ are of a local nature, and hence they do not depend on the way the function was extended. So we may speak about $\mathscr{H}_{z_0,s}^{\square}: C_{c,W}^\infty(\rho^{-1}(\mathscr{T})) \longrightarrow C_{c,W}^\infty(\rho^{-1}(\mathscr{T}))$. Note, at last, that, due to the isometry property of $\mathfrak{M}_{z_0}^{\mathrm{loc}}$ (see 35.14) the estimate (36.5c) gives the corresponding discrepancy in the L_2-norm if one replaces $\mathfrak{O}_{a,b}^{\square}$ in (36.4a) by the partial sum of the series (36.4b). The result of these discussions can be formulated as follows.

Proposition 36.9. *Given an arrangement $\{(V,\alpha), U, a, \chi\}$ at a point $z_0 \in \mathscr{T}$, there is a sequence of differential operators $\mathscr{H}_{z_0,s}^{\square}: C_{c,W}^\infty(\rho^{-1}(\mathscr{T})) \longrightarrow C_{c,W}^\infty(\rho^{-1}(\mathscr{T}))$, $s = 0, 1, 2, \ldots$ with coefficients in $C^\infty(\rho^{-1}(\mathscr{T})\cap W)$, $W = \rho^{-1}(U)$, the order of $\mathscr{H}_{z_0,s}^{\square}$ not exceeding* *s, such that for each natural N, and for each* $v \in C^\infty_{c,W}(\rho^{-1}(\mathcal{T}))$

(36.9a) $$\mathcal{H}_h \mathfrak{M}^{\mathrm{loc}}_{z_0} v = \mathfrak{M}^{\mathrm{loc}}_{z_0} \sum_{s=0}^{N} (i\hbar)^s \mathcal{H}^{\square}_{z_0,s} v + \mathcal{O}(\hbar^{N+1}).$$

Here $\mathfrak{M}^{\mathrm{loc}}_{z_0}$ *is the local Maslov's operator constructed by the use of the said arrangement. The estimate* $\mathcal{O}(\hbar^{N+1})$ *in (36.9a) holds in* $L_2(Q)$.

The explicit expressions for the first two $\mathcal{H}^{\square}_{z_0,s}$ *are the following:*

(36.9b) $\mathcal{H}^{\square}_{z_0,0} = E$, *the operator of multiplication by the number* $E = H(\mathcal{T})$,

(36.9c) $$\mathcal{H}^{\square}_{z_0,1} = -\tilde{\xi}|_{W \cap \rho^{-1}(\mathcal{T})} = -(dH)^{\#}|_{W \cap \rho^{-1}(\mathcal{T})},$$

up to the sign the application of the pre-Hamiltonian vector field ξ *of generalized geodesic flow (see §6).*

Proof of the Formulae (36.9b) and (36.9c)

The first formula is a simple consequence of the assertion of Proposition 36.4 concerning $\mathfrak{D}^{\square}_{a,b,0}$ and the assumption (34.2). The formulae (g), (h), (36.7) yield the following expression for $\mathcal{H}^{\square}_{z_0,1}$:

(36.9d) $$\gamma^* \mathcal{H}^{\square}_{z_0,1} v = \lambda^{-1/2} I_1\left(\sqrt[4]{g}\, a_0, \frac{1}{\sqrt[4]{g}}, \sqrt{\lambda}\gamma^* v\right) + \lambda^{-1/2} I_0\left(\sqrt[4]{g}\, a_1, \frac{1}{\sqrt[4]{g}}, \sqrt{\lambda}\gamma^* v\right),$$

where

(36.9e) $$a_0(q,p) = \frac{1}{2}\sum_{k,l=1}^{n} g^{kl}(q) p_k p_l + \mathcal{V} \circ \alpha^{-1}(q) = \tilde{H} \circ \beta^{-1}(q,p),$$

(36.9f) $$a_1(q,p) = \frac{1}{2}\sum_{k,l=1}^{n} p_l \frac{1}{\sqrt[4]{g}} \frac{\partial}{\partial q^k}\left(g^{kl}(q)\sqrt{g(q)}\right).$$

The substitution of $A = \sqrt[4]{g}\, a_k$, $k = 0, 1$, $B = g^{-1/4}$, $C = \sqrt{\lambda}\gamma^* v$ into (36.6c) and (36.6d), and then into (36.9d) results, after some cancellations, in

$$\gamma^* \mathcal{H}^{\square}_{z_0,1} v = -\xi_1 \gamma^* v + \mathcal{G}\gamma^* v,$$

where

(36.9g) $$\xi_1 = \sum_{i \in \bar{a}} \delta_1^*\left(\frac{\partial a_0}{\partial p_i}\right)\frac{\partial}{\partial q^i} - \sum_{j \in a} \delta_1^*\left(\frac{\partial a_0}{\partial q^j}\right)\frac{\partial}{\partial p_j},$$

δ_1 standing for the map

(36.9h) $$\delta_1 : (q^{\bar{a}}, p_a) \longmapsto (q^{\bar{a}}, q^a_s(q^{\bar{a}}, p_a), p^s_{\bar{a}}(q^{\bar{a}}, q_a), p_a)$$

i.e. the inclusion of $\mathscr{T}'$ into $\mathbb{R}^{2n}$, $\mathscr{G}$ is the operator of multiplication by the function

$$\mathscr{G} = -(2\lambda)^{-1}\Bigg\{\xi_1 + \delta_1^*\Bigg(\sum_{i\in\bar{a}}\frac{\partial^2 a_0}{\partial p_i \partial q^i} - \sum_{j\in a}\frac{\partial^2 a_0}{\partial p_j \partial q^j}\Bigg) + \sum_{i,k\in a}\delta_1^*\Bigg(\frac{\partial^2 a_0}{\partial p_i \partial p_k}\Bigg)\frac{\partial^2 S}{\partial q^i \partial q^k} - 2\sum_{i\in\bar{a},j\in a}\delta_1^*\Bigg(\frac{\partial^2 a_0}{\partial q^j \partial p_i}\Bigg)\frac{\partial^2 S}{\partial p_j \partial q^i} + \sum_{j,l\in a}\delta_1^*\Bigg(\frac{\partial^2 a_0}{\partial p_j \partial p_l}\Bigg)\frac{\partial^2 S}{\partial p_j \partial p_l}\Bigg\}\lambda. \tag{36.9i}$$

The vector field ξ_1 is nothing more than the coordinate image of $\tilde{\xi}|_{p^{-1}(\mathscr{T})}$: $\xi_1 = \gamma_*^{-1}\tilde{\xi} \overset{\text{def}}{=} \gamma^* \circ \tilde{\xi} \circ \gamma^{*-1}$, a vector field being considered as a map which acts on functions. To be persuaded of the validity of (36.9g), let us expand its left-hand side, it being a vector field on $\mathscr{T}' = \beta(\mathscr{T} \cap U)$, into the natural basic vector fields:

$$\xi_1 = \gamma_*^{-1}\tilde{\xi} = \sum_{i\in\bar{a}} \xi^i \frac{\partial}{\partial q^i} + \sum_{j\in a} \xi_j \frac{\partial}{\partial p_j},$$

where

$$\xi^i = dq^i(\gamma_*^{-1}\tilde{\xi}) = (\gamma_*^{-1} dq^i)(\tilde{\xi}|_{\mathscr{T}}) = (\text{see (6.6c)}) \sum_k \delta_1^*(g^{ik})p_k = \delta_1^*\Bigg(\frac{\partial a_0}{\partial p_i}\Bigg), \tag{36.9j}$$

and analogously

$$\xi_j = -\delta_1^*\Bigg(\frac{\partial a_0}{\partial q^j}\Bigg). \tag{36.9k}$$

We shall finish the proof of formula (36.9c) by establishing that $\mathscr{G} = 0$. The transport equation, which is a consequence of the invariancy of σ with respect to ξ, reads (see (34.7)):

(36.9l) $0 = \gamma^* d(\xi \lrcorner \sigma) = d(\gamma^*(\xi \lrcorner \sigma)) = d(\gamma_*^{-1}\xi \lrcorner \gamma^*\sigma) = (\text{use (36.4e)})\ d(\lambda\gamma_*^{-1}\xi \lrcorner V_a) =$ (use (36.9j), (36.9k), and take into account the definition (35.11) of V_a)

$$\Bigg[\sum_{i\in\bar{a}}\frac{\partial}{\partial q^i}\Bigg(\lambda\delta_1^*\Bigg(\frac{\partial a_0}{\partial p_i}\Bigg)\Bigg) - \sum_{j\in a}\frac{\partial}{\partial p_j}\Bigg(\lambda\delta_1^*\Bigg(\frac{\partial a_0}{\partial q^j}\Bigg)\Bigg)\Bigg] V_a.$$

Let us make the differentiation operation in the right-hand side of (36.9l) and take into account (36.9h). Comparing the result with (36.9i) one observes the right-hand side of (36.9l) to be equal to $-2\lambda\mathscr{G}V_a$ hence $\mathscr{G} = 0$. □

Proof of the ABC Lemma

Let us represent the integral in question as

$$I(A,B,C)(q^{\bar{a}}, p_a) = \frac{1}{(2\pi\hbar)^{n+|a|}}\int \prod_{i\in a} dq^j \prod_{k=1}^{n} dp'_k \prod_{l=1}^{n} dq'^l \prod_{j\in a} dp''_j \times e^{(i/\hbar)\Phi(q,p_a,q',p',p''_a)} A(q,p')B(q',p')C(q'^{\bar{a}}, p''_a), \tag{36.10}$$

where the phase function is

(36.11) $$\Phi(q,p_{\mathfrak{a}},q',p',p''_{\mathfrak{a}}) = S(q'^{\bar{\mathfrak{a}}},p''_{\mathfrak{a}}) + \sum_{i\in\bar{\mathfrak{a}}} p'_i(q^i-(q')^i) + \sum_{j\in\mathfrak{a}} ((q')^j(p''_j-p'_j)+q^j(p'_j-p_j)).$$

If $(q^{\bar{\mathfrak{a}}},p_{\mathfrak{a}})\in G$ then the integral (36.10) has only one stationary-phase point, $(q^{\mathfrak{a}}_{st},q'_{st},p'_{st},p''_{\mathfrak{a},st})$, which is a smooth function of the variables $(q^{\bar{\mathfrak{a}}},p_{\mathfrak{a}})$. It follows immediately that

$$(q^{\bar{\mathfrak{a}}},q^{\mathfrak{a}}_{st},q'_{st},p'_{st},p''_{\mathfrak{a},st}) = \delta(q^{\bar{\mathfrak{a}}},p_{\mathfrak{a}})$$

(see (36.5)). One can easily write out the matrix M of the second derivatives of Φ with respect to $(q^{\mathfrak{a}},q',p',p''_{\mathfrak{a}})$ (see formulae (36.12), (36.13) below) and find the determinant of the latter to be equal to 1, using the immediate definition of the determinant (but one member of the sum over permutations does not vanish, it equals $1\cdot(-1)^{|\bar{\mathfrak{a}}|}\cdot(-1)^{|\bar{\mathfrak{a}}|}=1$).

To compute the signature of M, let us divide the variables involved into two groups, (1) and (2), and other them as follows:

$$(1) = (q'^{\bar{\mathfrak{a}}},p''_{\mathfrak{a}},p'_{\mathfrak{a}}), \qquad (2) = (q'^{\mathfrak{a}},p'_{\bar{\mathfrak{a}}},q^{\mathfrak{a}}).$$

Placing the cites in accordance with this partition and ordering, the matrix M reads

(36.12) $$M = \begin{array}{|c|c|}\hline \mathscr{A} & \mathscr{P} \\ \hline \mathscr{P}^T & 0 \\ \hline \end{array}\begin{array}{c}(1)\\(2)\end{array}$$
$$\qquad (1)\quad (2)$$

where $\mathscr{A},\mathscr{P}$ are $(n+|\mathfrak{a}|)\times(n+|\mathfrak{a}|)$ matrices:

(36.13) $\mathscr{A} =$

the matrix of second derivatives of $S(q'^{\bar{\mathfrak{a}}},p''_{\mathfrak{a}})$		0
		0
0	0	0

,

$\mathscr{P} =$

$I_{\lvert\mathfrak{a}\rvert}$	0	0
0	$-I_{\lvert\bar{\mathfrak{a}}\rvert}$	0
$-I_{\lvert\mathfrak{a}\rvert}$	0	$I_{\lvert\mathfrak{a}\rvert}$

,

(I_k stands for $k\times k$ unit matrix). Evidently

$$M^{-1} = \begin{array}{|c|c|}\hline 0 & (\mathscr{P}^{-1})^T \\ \hline \mathscr{P}^{-1} & \mathscr{B} \\ \hline \end{array}\begin{array}{c}(1)\\(2)\end{array}$$
$$\qquad (1)\quad (2)$$

where $\mathscr{B} = -\mathscr{P}^{-1}\mathscr{A}(\mathscr{P}^{-1})^T$. Since sign $\mathscr{B} = -$ sign $\mathscr{A}$, the identity (Lazutkin (1988))

$$\text{sign}\, M = \text{sign}\, \mathscr{A} + \text{sign}\, \mathscr{B} \tag{36.14}$$

implies sign $M = 0$.

Applying the stationary phase theorem AIII.7 yields the expansion of the form (36.6a) with the first term (36.6c). To obtain the formulae for the other terms, let us expand the product ABC in the Taylor series in the neighbourhood of the stationary point:

$$A(q, p')B(q', p')C(q'^{\bar{\mathfrak{a}}}, p''_{\mathfrak{a}}) \sim \sum_{k,l,r,t} \frac{1}{k!l!r!t!}\delta^*\left[\left(\frac{\partial}{\partial q^{\mathfrak{a}}}\right)^k\left(\frac{\partial}{\partial q'}\right)^l\left(\frac{\partial}{\partial p'}\right)^r\left(\frac{\partial}{\partial p''_{\mathfrak{a}}}\right)^t A^p \otimes B^q \otimes C\right]$$
$$\times (q^{\mathfrak{a}} - q^{\mathfrak{a}}_{st})^k (q' - q'_{st})^l (p' - p'_{st})^r (p'' - p''_{st})^t,$$

where $k = (k_j, j \in \mathfrak{a})$, $l = (l_k, 1 \leqq k \leqq n)$, $r = (r_k, \ 1 \leqq k \leqq n)$, $t = (t_j, \ j \in \mathfrak{a})$ denote multi-indices. Since the kth term of the expansion (AIII.7a) contains the derivatives of the integrand of the order not exceeding $2k$, the latters being evaluated at the stationary point, the k, l, r, t term of (36.15) gives rise to the terms of order $\hbar^s$ where $2s \geqq |k| + |l| + |r| + |t|$. The terms with $|k| + |l| < |t|$ give no contribution because of the identity

$$(p'' - p''_{st})e^{(i/\hbar)\Phi} = \left[\frac{\hbar}{i}\frac{\partial}{\partial \dot{q}^j} + \frac{\hbar}{i}\frac{\partial}{\partial q^j}\right]e^{(i/\hbar)\Phi}, \quad j \in \mathfrak{a} \tag{36.16}$$

(apply integration by parts). This implies the order of the derivation of C in $I_s(A, B, C)$ does not exceed s.

In order to derive the explicit formula for I_1, let us write out the remaining expressions, which are analogues of (36.16). In the following formulae we omit $\exp((i/\hbar)\Phi)$, and use the sign $\doteq$ for denoting the sameness of two operations when acting on this exponent. The index j runs $\mathfrak{a}, g$ runs $\bar{\mathfrak{a}}$.

$$\begin{aligned}
&q^j - q^j_{st} \doteq -i\hbar\frac{\partial}{\partial p'_j} - i\hbar\frac{\partial}{\partial p''_j} + q^j_s(q'^{\bar{\mathfrak{a}}}, p''_{\mathfrak{a}}) - q^j_s(q^{\bar{\mathfrak{a}}}, p_{\mathfrak{a}}),\\
&q'^g - q'^g_{st} \doteq i\hbar\frac{\partial}{\partial p'_g},\\
&q'^j - q'^j_{st} \doteq -i\hbar\frac{\partial}{\partial p''_j} + q^j_s(q'^{\bar{\mathfrak{a}}}, p''_{\mathfrak{a}}) - q^j_s(q^{\bar{\mathfrak{a}}}, p_{\mathfrak{a}}),\\
&p'_g - p'_{g,st} \doteq i\hbar\frac{\partial}{\partial q'^g} + p^s_g(q'^{\bar{\mathfrak{a}}}, p''_{\mathfrak{a}}) - p^s_g(q'^{\bar{\mathfrak{a}}}, p''_{\mathfrak{a}}),\\
&p'_j - p'_{j,st} \doteq -i\hbar\frac{\partial}{\partial q'^j}.
\end{aligned} \tag{36.17}$$

The linear terms of the expansion (36.15) give no contribution in $I_1(A, B, C)$. Indeed, looking at (36.10), (36.15)–(36.17), one observes that the terms containing the operation of differentiation vanish (apply integration by parts), as do the terms containing the multiplications by a function of the form $f(q'^{\bar{\mathfrak{a}}}, p''_{\mathfrak{a}}) - f(q^{\bar{\mathfrak{a}}}, p_{\mathfrak{a}})$, for the latter may be represented as $\sum_{i \in \bar{\mathfrak{a}}} \varphi_i(q^{\bar{\mathfrak{a}}}, p_{\mathfrak{a}}, q'^{\bar{\mathfrak{a}}}, p''_{\mathfrak{a}})(q'^i - q^i) + \sum_{j \in \mathfrak{a}} \psi_j(q^{\bar{\mathfrak{a}}}, p_{\mathfrak{a}}, q'^{\bar{\mathfrak{a}}}, p''_{\mathfrak{a}})(p''_j - p_j)$

(see AIII.8′). Again $(q'^i - q^i)$, $(p''_j - p_j)$ may be substituted by the differentiation ((36.16) and (36.17)), and integration by parts yields zero contribution. Therefore, only the quadratic terms of (36.15) are to be taken into account when computing $I_1(A, B, C)$. Integration by parts now gives the nonzero contribution. Let us consider, for instance, a term containing $q'^g - q'^g_{st}$, $g \in \mathfrak{a}$. Denote by $\square$ the other multiplier which forms together with $q'^g - q'^g_{st}$ the quadratic term in question. We have

$$\int \square\,(q'^g - q'^g_{st})e^{(i/\hbar)\Phi} = \int \square\; i\hbar \frac{\partial}{\partial p'_g} e^{(i/\hbar)\Phi} = \int -i\hbar\Big((\partial/\partial p'_g)\square\Big)e^{(i/\hbar)\Phi}. \tag{36.18}$$

(integrating by parts)

It follows that only

$$\square = \delta^*\left[\frac{\partial^2 A^p \otimes B^q \otimes C}{\partial q'^g \partial p'_g}\right](p'_g - p'_{g,st})$$

does not give zero contribution when inserted in (36.18). This term gives rise to a contribution $\delta^*\left[\dfrac{\partial^2 A^p \otimes B^q \otimes C}{\partial q'^g \partial p'_g}\right]$. Acting in the same manner one establishes consequently that the terms which contain $p'_j - p'_{j,st}$, $j \in \mathfrak{a}$, and do not contain $q'^g - q'^g_{st}$, $g \in \bar{\mathfrak{a}}$, give rise to $\delta^*[\partial^2 A^p \otimes B^q \otimes C/\partial p'_j \partial q^j]$, those containing $p'_j - p'_{j,st}$ and not containing these already considered, give rise to

$$\delta^*\left[\frac{\partial^2 A^p \otimes B^q \otimes C}{\partial p''_j \partial q'^j} + \frac{\partial^2 A^p \otimes B^q \otimes C}{\partial p'_j \partial q'^j}\right].$$

Considering the terms containing $q^j - q^j_{st}$, $j \in \mathfrak{a}$, and not containing previous ones, in consecutive order the terms with $q'^j - q'^j_{st}, p'_g - p'_{g,st}$, $j \in \mathfrak{a}$, $g \in \bar{\mathfrak{a}}$, gives rise to the contributions of another form:

$$\begin{aligned} \mathrm{I} &= \int \boxed{\mathrm{I}}\,(q^j_s(q'^{\bar{\mathfrak{a}}}, p''_{\mathfrak{a}}) - q^j_s(q^{\bar{\mathfrak{a}}}, p_{\mathfrak{a}}))e^{(i/\hbar)\Phi}, \\ \mathrm{II} &= \int \boxed{\mathrm{II}}\,(p^s_g q'^{\bar{\mathfrak{a}}}, p''_{\mathfrak{a}}) - p^s_g(q^{\bar{\mathfrak{a}}}, p_{\mathfrak{a}}))e^{(i/\hbar)\Phi}, \end{aligned} \tag{36.19}$$

where $\boxed{\mathrm{I}}$ comes from the terms of (36.15) containing $(q^j - q^j_{st})(q^l - q^l_{st})$, $(q^j - q^j_{st})(q'^l - q'^l_{st})$, $(q'^j - q^j_{st})(q'^l - \acute{q}^l_{st})$, $j, l \in \mathfrak{a}$, $(q^j - q^j_{st})(p'_g - p'_{g,st})$, $j \in \mathfrak{a}$, $g \in \bar{\mathfrak{a}}$, and $\boxed{\mathrm{II}}$ comes from those containing $(p'_g - p'_{g,st})(p'_h - p'_{h,st})$, $g, h \in \bar{\mathfrak{a}}$. Substituting the equivalences

$$q^j_s(q'^{\bar{\mathfrak{a}}}, p''_{\mathfrak{a}}) - q^j_s(q^{\bar{\mathfrak{a}}}, p_{\mathfrak{a}}) \doteq i\hbar\left[\sum_{l \in \mathfrak{a}} \frac{\partial^2 S}{\partial p_j \partial p_l}\bigg|_{\substack{\text{stat}\\\text{point}}}\left(\frac{\partial}{\partial q'^l} + \frac{\partial}{\partial q^l}\right) - \sum_{g \in \bar{\mathfrak{a}}} \frac{\partial^2 S}{\partial q^g \partial p_j}\bigg|_{\substack{\text{stat}\\\text{point}}} \frac{\partial}{\partial p'_g}\right] + \mathcal{O}(\hbar^2),$$

$$p^s_g(q'^{\bar{\mathfrak{a}}}, p''_{\mathfrak{a}}) - p^s_g(q^{\bar{\mathfrak{a}}}, p_{\mathfrak{a}}) \doteq i\hbar\left[\sum_{h \in \bar{\mathfrak{a}}} \frac{\partial^2 S}{\partial q^g \partial q^h}\bigg|_{\substack{\text{stat}\\\text{point}}} \frac{\partial}{\partial p'_h} - \sum_{j \in \mathfrak{a}} \frac{\partial^2 S}{\partial q^g \partial p_j}\bigg|_{\substack{\text{stat}\\\text{point}}}\left(\frac{\partial}{\partial q'^j} + \frac{\partial}{\partial q^j}\right)\right] + \mathcal{O}(\hbar^2)$$

into (36.19), making the differentiations by parts, and combining the results together, one obtains the resulting formula (36.6d).

To prove (36.6b) it is sufficient to note that the support of the integrand contains no stationary points in that case, as an immediate calculation shows. Moreover, we have the estimate $|\operatorname{grad}\Phi| \geqq \text{const}(1+|p_a|)$, const >0 depending only on S. Integrating by parts as in AIII.5 yields (36.6b). □

§37. Theory of Maslov's Indices

Here we continue the investigation of Lagrangian subspaces of a symplectic vector space (V,Ω). Ternary and binary Maslov's indices will be defined, the former for Lagrangian subspaces, the latter for the corresponding points in the universal covering $\hat{\mathscr{L}}(V)$ of the manifold $\mathscr{L}(V)$ of Lagrangian subspaces of V. Maslov's indices are the essential part of global Maslov's theory. They will be employed in sewing together the local expressions of §35.

The Ternary Maslov's Index

Given a triple L, M, N of Lagrangian subspaces of a symplectic vector space (V,Ω), $\dim V = 2n$, consider the subspace $K = (L+M)\cap N$. Its vectors are those $v\in N$ whcih are of the form $v = l + m$, $l\in L$ and $m\in M$. Let $v_i = l_i + m_i$, $i = 1, 2$, be two such vectors. Then the number

(37.1) $$b(v_1, v_2) = \Omega(m_1, v_2) = \Omega(m_1, l_2) = \Omega(v_1, l_2)$$

does not depend of the choice of expansion $v_i = l_i + m_i$ and defines a symmetric bilinear form on K. The said independence and the bilinearity follow from the obvious equalities (37.1). Those equalities yield also the symmetry: $b(v_1, v_2) - b(v_2, v_1) = \Omega(m_1, v_2) - \Omega(v_2, l_1) = \Omega(m_1, v_2) + \Omega(l_1, v_2) = \Omega(v_1, v_2) = 0$ since N is isotropic. We define the *ternary Maslov's index* $\tau(L, M, N)$ as

(37.2) $$\tau(L, M, N) = \operatorname{sign} b,$$

where sign b, the *signature of a form* b, equals the number of positive squares minus that of negative ones in the representation of b as a sum of squares.

The Meyer–Turaev Cocycle

We shall define the function $\varphi: \mathrm{Sp}(V) \times \mathrm{Sp}(V) \longrightarrow \mathbb{Z}$ which plays an important role in the theory of Maslov's indices. Here $\mathrm{Sp}(V)$ is the symplectic group of a symplectic vector space V, defined in §1. Given a pair $f, g\in\mathrm{Sp}(V)$, consider the subspace

$$M_{f,g} = (f-1)(V)\cap(g-1)(V),$$

where 1 stands for the identity map of V, id_V, for brevity. The maps $f-1$ and $g-1$ have multivalued inverses $(f-1)^{-1}$ and $(g-1)^{-1}$ defined on $M_{f,g}$.

Proposition 37.3. *The formula*

(37.3a) $$C(u,v)=\Omega((f-1)^{-1}(u)+(g-1)^{-1}(u)+u,v)$$

defines correctly a symmetric bilinear form on $M_{f,g}$.

Proof. First we shall establish that the right-hand side of (37.3a) does not depend on the choice of the values of $(f-1)^{-1}(u)$ and $(g-1)^{-1}(u)$. It is sufficient to prove the same assertion for the other expression:

(37.3b) $$(u,v)\longmapsto\Omega((g-1)^{-1}(u)+\tfrac{1}{2}u,v).$$

Let $u=(f-1)x$, $v=(f-1)y$, $x,y\in V$. Then

(37.3c) $$\begin{aligned}\Omega(x+u/2,v)&=\tfrac{1}{2}\Omega(f(x)+x,f(y)+y)\\&=\tfrac{1}{2}\Omega(x,f(y))+\tfrac{1}{2}\Omega(y,f(x))=\Omega(y+v/2,u)\end{aligned}$$

The symmetry with respect to the permutation of x and y gives us the last equality.

Comparing the ends of (37.3c) we find that (1) $\Omega(x+u/2,v)$ does not depend on the choice of x, and (2) it is linear both with respect to u and v. The symmetry of (37.3b) follows from the above mentioned symmetry of (37.3c) with respect to x, y. □

The *Meyer–Turaev cocycle* $\varphi(f,g)$ is the signature of the form (37.1a).

Proposition 37.4. φ *is a 2-cocycle on* $\mathrm{Sp}(V)$, *that is, the following identity is fulfilled for all* $f,g,h\in\mathrm{Sp}(V)$:

(37.4a) $$\varphi(f,g)+\varphi(fg,h)=\varphi(f,gh)+\varphi(g,h).$$

To prove of this proposition, we need some Lemmata.

Lemma 37.5. *Let* $f\in\mathrm{Sp}(V)$. *Then* $\ker(f-1)^{\perp(\Omega)}=\mathrm{im}(f-1)$.

Proof. For all $x,y\in V$, $\Omega(f(x)-x,y)=\Omega(f(x),y-f(y))$. Hence $f(x)-x\perp\ker(f-1)$, and we have $\mathrm{im}(f-1)\subset(\ker(f-1))^{\perp(\Omega)}$. The equality follows, since both the subspaces have the same dimension. □

If D is a bilinear symplectic form on a vector space, then sign D denotes the signature of D.

Lemma 37.6. *Let H be a vector space, D be a nonsingular symmetric bilinear form on H, and $H_0\subset H$ be a subspace. If* $H_0=H_0^{\perp(\Omega)}$ *then* $\operatorname{sign}D=0$.

Proof. Let $\dim H_0=n$. Take a vector base $\{e_i, 1\leqq i\leqq n\}$ in H_0. Let H_j be the linear hull of the set $\{e_i,1\leqq i\leqq n\}\backslash\{e_j\}$. Since $H_j\subsetneqq H_0\hookrightarrow$, we have $H_j^{\perp(D)}\supsetneqq H_0$. Hence there is $f_j\in H_j^{\perp(D)}\backslash H_0$. One may assume that $D(f_j,e_j)=1$ and $D(f_j,e_k)=0$, $j\neq k$. It follows $\{f_j,e_j,1\leqq k\leqq n\}$ is a linearly independent set. As a matter of fact this is a vector base in H because $\dim H=2n$ which follows from $H_0=H_0^{\perp(D)}$ and from the nonsingularity of D. Take another vector $\tilde{f}_j = f_j + \sum_{k=1}^{n} c_{jk} e_k$ to obtain $D(\tilde{f}_j, f_k) = 0$, and as before $D(\tilde{f}_j, e_j) = 1$, $D(\tilde{f}_j, e_k) = 0$, $j \neq k$. One easily finds c_{jk}. Let G_k be the two-dimensional subspace generated by $\{f_k, e_k\}$. We have $H = \sum_{k=1}^{n} \oplus G_k$ the orthogonal sum with respect to D, and $\operatorname{sign} D = \sum_{k=1}^{n} \operatorname{sign} D|_{G_k}$. Note that in the base $\{f_k, e_k\}$ the matrix of the form $D|_{G_k}$ is $\begin{pmatrix} 0 & 1 \\ 1 & 0 \end{pmatrix}$, and hence $\operatorname{sign} D|_{G_k} = 0$.

□

Lemma 37.7. *Let E be a vector space, D be a symmetric bilinear form on E, and E_i, $i = 1, 2$, be subspaces of E. If*

(3.7a) $$(E_1 \cap E_2)^{\perp(D)} = E_1 + E_2 \quad \text{and} \quad E_1 \perp^{(D)} E_2$$

then

$$\operatorname{sign} D = \operatorname{sign} D|_{E_1} + \operatorname{sign} D|_{E_2}.$$

Proof. First of all we shall reduce the problem to one with nonsingular D. Denote $N = E^{\perp(D)}$ and consider the canonical projection $p: E \longrightarrow \hat{E} = E/N$. The form D induces the nonsingular form $\hat{D}$ on $\hat{E}$. The subspaces $\hat{E}_i = p(E_i)$ satisfy both the conditions (37.7a) in $\hat{E}$ with respect to the form $\hat{D}$. Indeed, in $\hat{x} \in (\hat{E}_1 \cap \hat{E}_2)^{\perp(D)}$ then $\hat{x} = p(x)$ and $D(x, y_1) = 0$ for all $y_1 \in E_1$ such that there is a $y_2 \in E_2$ with the property $y_1 - y_2 \in N$. In particular $D(x, y) = 0$ if $y \in E_1 \cap E_2$. Then (37.7a) implies $x = x_1 + x_2$, $x_i \in E_i$, $i = 1, 2$, and $\hat{x} = p(x) = p(x_1) + p(x_2) \in \hat{E}_1 + \hat{E}_2$. Conversely, if $\hat{x} \in \hat{E}_1 + \hat{E}_2$ then $\hat{x} = p(x_1 + x_2)$, $x_i \in E_i$, $i = 1, 2$, and (37.7a) implies $x_1 + x_2 \perp E_1 \cap E_2$. Let $y_i \in E_i$, $i = 1, 2$ be such that $y_1 - y_2 \in N$, then $D(x_1 + x_2, y_1) = D(x_1, y_1) = D(x_1, y_2) = 0$. Here we took into account the second condition of (37.7a). This means $\hat{x} = p(x_1 + x_2) \in (\hat{E}_1 \cap \hat{E}_2)^{\perp(\hat{D})}$. The mutual orthogonality of $\hat{E}_i$ follows from that of E_i. Evidently $\operatorname{sign} D = \operatorname{sign} \hat{D}$ and $\operatorname{sign} D|_{E_i} = \operatorname{sign} \hat{D}|_{\hat{E}}$, $i = 1, 2$.

So we may assume D to be nonsingular. Denote $H_0 = E_1 \cap E_2$, and take (algebraic) complements E'_i to H_0 in E_i. We have

(i) $E_i = E'_i \oplus H$ and $\operatorname{sign} D|_{E'_i} = \operatorname{sign} D|_{E'_i}$ (D vanishes on H_0),

(ii) $E_1 + E_2 = E'_1 \oplus E'_2 \oplus H_0$.

The orthogonal complement

(iii) $H = (E'_1 \oplus E'_2)^{\perp(D)}$

has the properties

(iv) $H \supset H_0$,

(v) $E = E'_1 \oplus E'_2 \oplus H$,

(vi) $D' = D|_H$ is nonsingular in H,

(vii) $H_0^{\perp(D')} = H_0$, H_0 being regarded as a subspace of H.

Let us prove (v). If $x \in (E'_1 \oplus E'_2) \cap H$ then, by (37.7a) $x \perp^{(D)} H_0$, and by (ii) $x \perp^{(D)} E'_1 \oplus E'_2$. It follows from (ii) that $x \perp^{(D)} E_1 \oplus E_2$, so, due to (37.7a), $x \in H_0$. We have $x \in (E'_1 \oplus E'_2) \cap H_0 = \{0\}$, and so $(E'_1 \oplus E'_2) \cap H = \{0\}$. The equality (iv) follows from the fact that the dimension of H is complementary to that of $E'_1 \oplus E'_2$ (see (iii)). To prove (vi), let us note that otherwise, due to (iv), D would be singular in E. The property (vii) follows from $H_0^{\perp(D)} \supset H_0$ (see (37.7a)) and (vi).

To complete the proof of the Lemma, we write, using the orthogonal decomposition (v):

(viii) $$\operatorname{sign} D = \operatorname{sign} D|_{E'_1} + \operatorname{sign} D|_{E'_2} + \operatorname{sign} D|_H,$$

using (i) and Lemma 37.6. The latter, being applied to $H_0 \subset H$, yields the last term in the right side of (viii) to be zero. □

Proof of Proposition 37.4

Denote by E the subspace of $V \oplus V \oplus V$ consisting of all triples (a_1, a_2, a_3) with the properties $a_1 + a_2 + a_3 = 0$, $a_1 \in (f-1)(V)$, $a_2 \in (g-1)(V)$, $a_3 \in (h-1)(V)$. Let E_i, $1 \leqq i \leqq 4$, be the subspaces of E defined by the equations

$$\begin{aligned}
&E_1\colon\ a_1 = 0,\\
&E_2\colon\ a_2 = (gh-h)(x),\ a_3 = (h-1)(x),\ x \in V,\\
&E_3\colon\ a_3 = 0,\\
&E_4\colon\ a_1 = (fg-g)(x),\ a_2 = (g-1)(x),\ x \in V.
\end{aligned}$$

Define the map $D\colon E \times E \longrightarrow \mathbb{R}$ by the formula

(b) $$\begin{aligned} D((a_1,a_2,a_3),(b_1,b_2,b_3)) &= \Omega((f-1)^{-1}(a_1), b_1) + \Omega((g-1)^{-1}(a_2), b_2)\\ &\quad + \Omega((h-1)^{-1}(a_3), b_3) + \Omega((a_3, b_3) - \Omega((a_2, b_1). \end{aligned}$$

The last two terms of (b) may be transformed as

$$\begin{aligned} \Omega((a_3,b_3) - \Omega((a_2,b_1) = \tfrac{1}{2}(\Omega((a_1,b_1) &+ \Omega((a_2,b_2)\\ &+ \Omega((a_3,b_3) + \Omega((a_1,b_2) + \Omega((b_1,a)) \end{aligned}$$

in virtue of the definition of E. So, the right side of (b) may be represented as a sum of three forms like (37.3b) and a correctly defined symmetric form $\frac{1}{2}(\Omega((a_1,b_2) + \Omega((b_1,a_2))$.

The proof of 37.3 yields the correct definition, bilinearity, and symmetry of D. We claim:

(c) the signatures of the restrictions of D onto E_i, $1 \leqq i \leqq 4$, are equal respectively, and to $\varphi(g,h)$, $\varphi(f,gh)$, $\varphi(f,g)$, and $\varphi(fg,h)$;

(d) $$\begin{cases} (E_1 \cap E_2)^{\perp(D)} = E_1 + E_2,\ E_1 \perp^{(D)} E_2,\\ (E_3 \cap E_4)\perp^{(D)} = E_3 + E_4,\ E_3 \perp^{(D)} E. \end{cases}$$

It follows from (d) and Lemma 37.5 that

(e) $$\operatorname{sign} D = \operatorname{sign} D|_{E_1} + \operatorname{sign} D|_{E_2} = \operatorname{sign} D|_{E_3} + \operatorname{sign} D|_{E_4}.$$

The assertion of the proposition is the consequence of (c) and (e). It remains for us to prove (c) and (d).

Proof of (c). Denote the form (37.3a) on $M_{f,g}$ by $C_{f,g}$. Evidently the map $c \longmapsto (a, -a, 0)\colon M_{f,g} \longrightarrow E_3$ is an isomorphism carrying $C_{f,g}$ to $D|_{E_3}$. Hence

$\operatorname{sign} D|_{E_3} = \varphi(f,g)$. Analogously $\operatorname{sign} D|_{E_1} = \varphi(g,h)$. The image of E_2 under the projection $p_1:(a_1,a_2,a_3)\longmapsto a_1: E \longrightarrow V$, equals $M_{f,gh}$. Let $(a_1,a_2,a_3),(b_1,b_2,b_3)\in E_2$, and $a_1=(f-1)(x)$, $a_2=(gh-h)(y)$, $a_3=(h-1)(y)$, $x,y\in V$. Then, by definition (b):

$$\begin{aligned}D((a_1,a_2,a_3),(b_1,b_2,b_3)) &= \Omega(x,b_1)+\Omega(h(y),b_2)+\Omega((h-1)y,b_3)+\Omega(y,b_3)-\Omega(a_2,b_1)\\ &= \Omega(x,b_1)+\Omega(h(y),b_2+b_3)-\Omega(a_2,b_1)\\ &= \Omega(x-h(y)-a_2,b_1)=\Omega(x-y)+a_1,b_1)\\ &= C_{f,gh}(a_1,b_1),\ \text{i.e.}\end{aligned}$$

$$(p_1|_{E_2})^* C_{f,gh} = D|_{E_2}.$$

Hence, $\operatorname{sign} D|_{E_2} = \varphi(f,gh)$. Similarly, $\operatorname{sign} D|_{E_4} = \varphi(fg,h)$. □

Proof of (d). We shall prove the assertion for E_1, E_2. The proof for E_3, E_4 goes is analogous. Let $(a_1,a_2,a_3)\in E_1$ and $(b_1,b_2,b_3)\in E_2$. Then $a_1=0$. $a_2=-a_3=g(x)-x=h(y)-y$, $y\in V$ and $b_2=(gh-h)(z)$, $b_3=h(z)=z$, $b_1=z-gh(z)$. Substituting these into (b), one obtains

$$\begin{aligned}D((a_1,a_2,a_3),(b_1,b_2,b_3)) &= \Omega(x,(gh-h)(z))+\Omega(-y,h(z)-z)+\Omega(-h(y)+y,h(z)-z))\\ &\quad -\Omega(g(x)-x,z-gh(z)) = (\text{after cancellations})\\ &\quad -\Omega(x,h(z))-\Omega(h(y),h(z))+\Omega(h(y),z)+\Omega(g(x),gh(z))\\ &\quad +\Omega(x-g(x),z) = (\text{substituting } x-g(x)=y-h(y))\\ &\quad -\Omega(x,h(z))-\Omega(g(y),h(z))+\Omega(y,z)\\ &\quad +\Omega(g(x),gh(z))=0\end{aligned}$$

since g and h are symplectic. This proves the orthogonality in (d). To prove the equality, let us firstly note that

(f) $E_1\cap E_2=\{(a_1,a_2,a_3)\in E: a_1=0,\ a_2=(hg-h)(x),\ a_3=h(x)-x$ for some $x\in \ker(gh-1)\}$.

(g) $$E_1+E_2=\{(b_1,b_2,b_3)\in E: b_1\in \operatorname{im}(gh-1)\}.$$

Let $(a_1,a_2,a_3)\in E_1\cap E_2$. Then $D((a_1,a_2,a_3),(b_1,b_2,b_3)=\Omega(h(x),b_2)+\Omega(x,b_3)+\Omega(h(x)-x,b_3)-\Omega(gh(x)-h(x),b_1)=\Omega(h(x),b_2+b_3)+\Omega(h(x)-x,b_1)=-\Omega(x,b_1)$.

Therefore the condition $(b_1,b_2,b_3)\perp^{(D)}E_1\cap E_2$ is equivalent to $b_1\perp(\Omega)\ker(gh-1)$. Applying Lemma 37.5 and (g), we obtain this condition to be equivalent to $(b_1,b_2,b_3)\in E_1+E_2$. □

It is important to know the sets where $\varphi(f,g)$ is constant.

Proposition 37.8. *Given* $p,q,r,s,t\in\mathbb{Z}_+$, *the function* $\varphi:\operatorname{Sp}(V)\times\operatorname{Sp}(V)\longrightarrow\mathbb{Z}$ *is constant on the path components of the set*

(37.8a) $$\{(f,g)\in\operatorname{Sp}(V)\times\operatorname{Sp}(V): \dim\operatorname{im}(f-1)=p,\ \dim\operatorname{im}(g-1)=q,\ \dim\operatorname{im}(fg-1)=r,\ \dim\operatorname{im}((f-1)\cap\operatorname{im}(g-1))=s\}.$$

Lemma 37.9. $M_{f,g}^{\perp(C)} = (g-1)(\ker(fg-1))$.

Proof. Let $z \in (g-1)(\ker(fg-1))$. Then $z = g(x) - x$ and $fg(x) = x$. We have

$$(*) \qquad z = g(x) - fg(x) = (f-1)(-g(x)),$$

i.e. $z \in M_{f,g}$. The $b \in M_{f,g}$ and consider the value of C: $C(z,b) = \Omega((f-1)^{-1}(z) + (g-1)^{-1}(z) + z, b) = \Omega((f-1)^{-1}(z) + g(x), b) =$ (by using $(*)$)$\Omega(-g(x) + g(x), b) = 0$. We have proven $(g-1)(\ker(fg-1)) \subset M_{f,g}^{\perp(C)}$. To prove the converse inclusion, take $z \in M_{f,g}$, $z \perp^{(C)} M_{f,g}$, then $z = f(x) - x = g(y) - y$ for some $x, y \in V$. The condition $z \perp^{(C)} M_{f,g}$ is equivalent to $z + x + y \perp^{(\Omega)} M_{f,g}$. It follows from the definition that $M_{f,g}^{\perp(\Omega)} = \ker(f-1) + \ker(g-1)$. Hence $z + x + y = a + b$ where $a \in \ker(f-1)$ and $b \in \ker(g-1)$. We have $z = (g-1)(y) = (g-1)(y-b)$. It is sufficient to establish that $y - b \in \ker(fg-1)$. Since $g(b) = b$, $f(a) = a$, $f(x) = z + x$, one obtains $fg(y-b) = f(g(y) - b) = f(y + z - b) = f(a - x) = a - f(x) = a - z - x = y - b$. □

Proof of Proposition 37.8. Let $(f(t), g(t))$, $t \in [0,1]$, be a continuous path in $\mathrm{Sp}(V) \times \mathrm{Sp}(V)$, lying in a path component of (37.8a). Then $M_{f,g}$ depends continuously on t due to the constancy of its dimension, and so does the form C. Its signature does not vary if and only if $\dim(M_{f,g})^{\perp(C)}$ is constant. By Lemma 37.9 the latter is equivalent to the constancy of $\dim((g(t)-1)(\ker(f(t)g(t)-1))$. Since $\dim\ker(f(t)g(t)-1)$ is constant, the last condition is equivalent to the constancy of the dimension of

$$\ker(g(t)-1) \cap \ker(f(t)g(t)-1) = \ker(g(t)-1) \cap \ker(f(t)-1).$$

The assertion follows now from the equality

$$\dim(\ker(g-1) \cap \ker(f-1)) = 2n - p - q + s,$$

which holds for all f, g belonging to (37.8a). □

Kähler Structure

A *Kähler structure* on a symplectic vector space V is a map $J \in \mathrm{Sp}(V)$ with the property $J^2 = 1$.

Example 37.10. Given a symplectic vector base $\{e_i, 1 \leqq i \leqq 2n\}$ in V (see 1.15) and a vector $\varepsilon = \{\varepsilon_i, 1 \leqq i \leqq n\}$, $\varepsilon_i \in \{-1, +1\}$, define J by its values on e_i in the following manner: $Je_i = \varepsilon_i e_{n+i}$, $Je_{n+i} = -\varepsilon_i e_i$, $1 \leqq i \leqq n$.

Given a Kähler structure J, one defines a multiplication by a complex number $a + ib \in \mathbb{C}$, $a, b \in \mathbb{R}$, as $(a+ib)v = av + bJv$. Obviously, V becomes a complex vector space under such a multiplication and the old sum operation, and the dimension of V over the complex number field, $(\dim_{\mathbb{C}} V)$ equation whenever $\dim V = \dim_{\mathbb{R}} V \doteq 2n$.

The *bilinear form* of a Kähler structure J is

$$(37.11) \qquad B_J(u,v) = \Omega(Ju, v), \quad v \in V.$$

As one easily checks, B_J is symmetric, nonsingular, and invariant under J: $B(Ju, Jv) = B(u, v)$. The bilinear form of Example 37.10 has the matrix diag$\{\varepsilon_1, \varepsilon_2, \dots, \varepsilon_n\}$ in the base $\langle e_i, 1 \leqq i \leqq 2n \rangle$. This shows B need not be positively defined. A Kähler structure is called *positive* if its bilinear form is. Only positive Kähler structures will be considered further.

Fix a Lagrangian subspace W_0.

Proposition 37.12. *Given a positive Kähler structure J, define the bilinear form $b = B_J|_{W_0}$ on W_0 and the subspace $W = J(W_0)$. The form b is positively defined, and W is a Lagrangian complement to W_0. Conversely, given a positive bilinear form b on W_0 and a Lagrangian complement W to W_0, there is a uniquely defined Kähler structure J such that $b = B_{J|W_0}$ and $W = J(W_0)$.*

Proof. One easily checks the first assertion. Let us prove the converse. The pair (b, W) defines the pair of isomorphism: $\#(b) : W_0^* \longrightarrow W_0$ and $h : W \longrightarrow W_0^*$ (see the beginning of §1 and Proposition 1.11). The map $J : W_0 \oplus W \longrightarrow W_0 \oplus W$, if it exists, has the matrix expression

(37.12a)
$$\begin{pmatrix} 0 & \#(b) \circ h \\ -h^{-1} \circ \#(b)^{-1} & 0 \end{pmatrix}$$

as one can deduce from (37.11) and the formula (**) in the proof of Proposition 1.11. This proves the uniqueness of J. The existence follows from the fact map defined by the matrix (37.12a) has all the desired properties. □

We obtained the diversity of positive Kähler structures in V to be parameterized by a pair (b, W), b being a positive symmetric form on W_0, W being a transversal to W_0 a Lagrangian subspace; i.e. by a point of $\mathscr{P}(W_0) \times \mathscr{L}_{W_0}$. Here $\mathscr{P}(W_0)$ denotes the set of all positive symmetric bilinear forms on W_0. Both the spaces $\mathscr{P}(W_0)$ and $\mathscr{L}_{W_0}$ have natural topologies and smooth structures induced from the vector space of all forms on W_0 and the manifold $\mathscr{L}(V)$ correspondingly. Since $\mathscr{L}_{W_0}$ is diffeomorphic to an $n(n+1)/2$-dimensional ball (see 1.20) we have

Corollary 37.13. *The space of all positive Kähler structures $\mathscr{P}(W_0) \times \mathscr{L}_{W_0}$ on V is diffeomorphic to an $n(n+1)$-dimensional ball, $2n = \dim V$.*

We associate with a positive Kähler structure J on V the Hermitian inner product

(37.14)
$$\langle u, v \rangle = B_j(u, v) + \Omega(v, v)$$

which converts V, together with the above defined complex structure into an n-dimensional Hilbert space. The positive definiteness of (37.14) and the property $\langle u, v \rangle = \overline{\langle u, v \rangle}$ is obvious. The linearity of (37.14) with respect to the first variable is a consequence of a simple immediate calculation which uses the definition (37.11) and the equality $B_J(Ju, v) = -\Omega(u, v)$.

The *unitary group* $\mathscr{U}(V)$ consists of all $f \in GL(V)$ preserving the Hermitian inner product (37.14): $\langle fu, fu \rangle = \langle u, v \rangle \; \forall u, v \in V$. Such maps are called *unitary maps*. Note that $\mathscr{U}(V)$ depends on the choice of a positive Kähler structure, J.

Another additional structure will be useful. Fix a Lagrangian subspace $W_0 \subset V$. Then each $u \in V$ has a uniquely defined decomposition.

$$u = u' + Ju'', \quad u', u'' \in W_0.$$

Define the *conjugation map* $u \longmapsto \bar{u}$ by setting

$$\bar{u} = \bar{u}' - Ju''.$$

Evidently, the conjugation is an anti-linear map in the complex structure of V: $\overline{\alpha u + \beta u} = \bar{\alpha} \cdot \bar{u} + \bar{\beta} \cdot \bar{v}$, $\alpha, \beta \in \mathbb{C}$, $u, v \in V$, $\alpha \longmapsto \bar{\alpha}$ being the complex conjugation in $\mathbb{C}$. W_0 becomes the "real part" of V, $W = J(W_0)$ its "imaginary part". By a non-complicated calculation we get

$$\langle \bar{u}, \bar{v} \rangle = \overline{\langle u, v \rangle}. \tag{37.15}$$

If $A : V \longrightarrow V$ is a linear map, linearity being meant with respect to the complex linear structure of V, then we define the *complex conjugate* of A, the map $\bar{A} : V \longrightarrow V$, by the formula $\bar{A}v = \overline{A\bar{v}}$. It is evident that $\bar{A}$ is also linear in the same sense, and $\bar{\bar{A}} = A$. The following properties are simple consequences of the definitions:

$$\overline{AB} = \bar{A} \cdot \bar{B}, \qquad \overline{\mathrm{id}} = \mathrm{id}, \qquad \det, \bar{A} = \overline{\det A}, \tag{37.16}$$

$$\text{if } A \text{ is invertible, } \overline{A^{-1}} = (\bar{A})^{-1}, \tag{37.17}$$

$$\text{if } A \in \mathrm{Sp}(V), \text{ so is } \bar{A}, \tag{37.18}$$

$$\text{if } A \in \mathscr{U}(V), \text{ so is } \bar{A}. \tag{37.19}$$

Proposition 37.20. *Let $f, g \in \mathscr{U}(V)$. Then $\varphi(f, g)$ is even.*

Proof. Take the orthogonal with respect to the form B_j splitting $V = V_+ \oplus V_- \oplus V_0$ where V_i, $i \in \{+, -, 0\}$ are the subspaces where the form $C(u, u)$ defined by (37.3a) is respectively positive, negative, and zero-valued. Such splitting is uniquely defined (it is obtained from the spectral splitting of the symmetric operator corresponding to the form C in the space V supplied with the inner product B_J). The multiplication of u by a non-zero complex number c induces the multiplication of the form $C(u, u)$ by the positive number $|c|^2$ and preserves the orthogonality. Hence $cV_i = V_i$, i.e. the above splitting is the splitting in the complex linear structure too, the V_i are subspaces in the complex sense and have even real dimension. Note that $\varphi(f, g) = \dim V_+ - \dim V_-$, and so is even. □

On the Geometry of $\mathscr{L}(V)$

Here we consider again the manifold $\mathscr{L}(V)$ of Lagrangian subspaces of V. The symplectic group $\mathrm{Sp}(V)$ acts smoothly on $\mathscr{L}(V)$, the action being defined naturally as

$$(f, W) \longmapsto f(W), \quad f \in \mathrm{Sp}(V), \quad W \in \mathscr{L}(V).$$

Let us fix a positive Kähler structure J and restrict the said action onto the unitary subgroup $\hat{U}(V) \subset \mathrm{Sp}(V)$, $\hat{U}(V)$ corresponding to J.

Proposition 37.21. *The action of* $\hat{U}(V)$ *on* $\mathscr{L}(V)$ *is transitive. The stationary subgroup of* $W_0 \in \mathscr{L}(V)$, $\mathrm{St}(W_0) = \{f \in \mathscr{U}(V): f(W_0) = W_0\}$ *is isomorphic to* $\mathcal{O}(n)$, *the orthogonal group of* $\mathbb{R}^n$, *where* $2n = \dim V$.

Proof. It is well known that an orbit of a smooth action of a compact Lie group on a manifold is a smooth submanifold diffeomorphic to the homogeneous space of the group with respect to the stationary subgroup of an arbitrary point in the orbit [see e.g. Thom and Levin (1959) Proposition 2 of Sect. 2.1]. Let $W_0 \in \mathscr{L}(V)$. If $f \in \mathrm{St}(W_0)$ then $f|_{W_0}$ preserves the restriction on W_0 of the complex inner product (37.14), the latter becomes real on W_0 and so defines a Euclidean structure. Therefore, the map $j: \mathrm{St}(W_0) \ni f \longmapsto f|_{W_0}$ is a smooth homomorphism of $\mathrm{St}(W_0)$ to the group of orthogonal linear maps of W_0 onto itself. On the other hand, the map f commutes with the complex structure J, because $f \in \mathscr{U}(V)$, hence it leaves the splitting $W_0 \oplus JW_0$ invariant. Applying Proposition 1.12, we find j to be injective. Moreover, j is also onto. Indeed, each $h: W_0 \longrightarrow W_0$ can be uniquely extended, again by Proposition 1.12, to the symplectic splitting-preserving map $H: W_0 \oplus JW_0 \longrightarrow W_0 \oplus JW_0$. The map $H' = H \oplus (-JhJ): W_0 \oplus JW_0 \longrightarrow W_0 \oplus JW_0$ is another splitting-preserving extension of h Orthogonality of h proves H is symplectic: if $u \in W_0$ and $v \in JW_0$ then $\Omega(H'u, H'v) = \Omega(hu, -JhJv) = \Omega(Jhu, hJv) = b(hu, hJv) = b(u, Jv) = \Omega(Ju, Jv) = \Omega(u, v)$. So $H' = H$. We have H commutes with J and is unitary. Denote by $\mathcal{O}(W_0)$ the group of all orthogonal linear automorphisms of W_0. We obtained that $\mathrm{St}(W_0)$ and $\mathcal{O}(W_0)$ are isomorphic via j. In fact, this is an isomorphism of Lie groups. Since $\dim \mathscr{U}(V) = n^2$ and $\dim \mathcal{O}(W_0) = n(n-1)/2$ where $2n = \dim V$ (see Fuks and Rohklin (1984, Chapter 3, §2) we have: dimension of the orbit of $W_0 = n^2 - n(n-1)/2 = n(n+1)/2 = \dim \mathscr{L}(V)$. Therefore the orbit of W_0 is open. So it is a connected component of $\mathscr{L}(V)$ because it is also closed. Since $\mathscr{L}(V)$ is connected (Proposition 1.23), the orbit equals the whole space $\mathscr{L}(V)$. □

Fix $W_0 \in \mathscr{L}(V)$ and consider the conjugation operation associated with W_0 as it was defined in the previous subsection.

Proposition 37.22. *The stationary subgroup* $\mathrm{St}(W_0) \subset \mathscr{U}(V)$ *may be characterized as follows:* $f \in \mathrm{St}(W_0) \Leftrightarrow \bar{f} = f$.

Proof. Left to the reader. □

Denote by $\mathscr{W}(V)$ the subset of $\mathscr{U}(V)$ consisting of maps g satisfying the equation $g = \bar{g}^{-1}$, i.e. the *symmetric* maps. The latter designation is justified if one considers the matrix of g in the base $\{e_i, 1 \leqq i \leqq 2n\}$ with $\{e_i, 1 \leqq i \leqq n\} \subset W_0$ such that J is defined as in Example 37.10 with $\varepsilon_i = 1$. The said matrix is then symmetric.

Proposition 37.23. *The unitary map g belongs to $\mathscr{W}(V)$ if and only if it has a complete set of eigenvectors in W_0.*

Proof. Let $g_\lambda \in \mathscr{W}(V)$ and $gv = \lambda v, v \neq 0, \lambda \in \mathbb{C}$. It follows that $\bar{v}$ also is an eigenvector of g corresponding to the same eigenvalue λ. If $\bar{v} = av$, $0 \neq a \in \mathbb{C}$, take $b \in \mathbb{C}$ such that $\bar{b}b^{-1} = a$ and set $w = bv \in W_0$. If $\bar{v}$ and v are linearly independent (in the complex linear structure!), set $w_1 = (1/2)(v + \bar{v})$, $w_2 = (1/2i)(v - \bar{v}) \in W_0$. In such a manner we obtain the desired set of eigenvectors. The converse assertion is obvious. □

Proposition 37.24. *The relation*

(37.24a) $$\exists f \in \mathscr{U}(V): w = f(W_0) \text{ and } f(\bar{f})^{-1} = g$$

which connects $w \in \mathscr{L}(V)$ and $g \in \mathscr{W}(V)$ defines a diffeomorphism

$$\gamma: \mathscr{L}(V) \longrightarrow \mathscr{W}(V).$$

Proof. Let us first establish that (37.24a) defines a 1-to-1 correspondence. We are to prove for $f_i \in \mathscr{U}(V)$, $i = 1, 2$,

$$f_1(W_0) = f_2(W_0) \Leftrightarrow f_1(\bar{f}_1)^{-1} = f_2(\bar{f}_2)^{-1}.$$

The right-hand equation is equivalent to $f_2^{-1} f_1 = \bar{f}_2^{-1} \bar{f}_1$, or, in view of (37.22) and (37.16) $f_2^{-1} f_1 \in \mathrm{St}(W_0)$, i.e. to the left one. Due to (37.12), γ is defined on the whole space $\mathscr{L}(V)$. The inclusion $\gamma(\mathscr{L}(V)) \subset \mathscr{W}(V)$ is obvious. Let $g \in \mathscr{W}(V)$. Since a complex λ with $|\lambda| = 1$ can be represented as $\lambda = \beta\bar{\beta}^{-1}$, $|\beta| = 1$, we may construct $f \in \mathscr{W}(V)$ with $g = f\bar{f}^{-1}$ using the eigenbase in W_0 (see 37.23).

To prove the differential properties of γ, consider the map $w: f \longmapsto f\bar{f}^{-1}$. It is a smooth map of $\mathscr{U}(V)$ onto $\mathscr{W}(V)$ whose level sets are diffeomorphic images of $\mathrm{St}(W_0)$, its conjugacy classes. The kernel of $T_f w$ coincides with the tangent space to the class passing through f. From this it follows that γ is differentiable and $T_W \gamma$ is invertible. □

Proposition 37.25 *Let $g = \gamma(W)$, $W \in \mathscr{L}(V)$, then*

$$v \in W \Leftrightarrow v = g\bar{v}.$$

Proof. Immediate calculation. □

The following Lemma will be useful in the next subsection.

Lemma 37.26. *Let $k_0, k_1, k_2, \ldots, k_m \in \mathscr{L}(V)$. Then*

$$\dim \bigcap_{r=1}^{m} \ker(\gamma(k_0)^{-1}\gamma(k_r) - 1) = 2\dim(k_0 \cap k_1 \cap \cdots \cap k_m).$$

Proof. Denote $g_i = \gamma(k_i)$, $0 \leq i \leq m$. The subspace $k_1 \cap k_2 \cap \cdots \cap k_m$ may be defined by the equations $v = g_i\bar{v}$, $0 \leq i \leq m$, or, equivalently, by

(i) $$(g_0^{-1} g_i^{-1})v = 0, \quad 1 \leq i \leq m,$$

and

(ii) $$v = g_0 \bar{v}.$$

The equations (i) are linear in the complex linear structure. So they alone define the subspace

$$H = \bigcap_{k=1}^{m} \ker(\gamma(k_0)^{-1}\gamma(k_r) - 1)$$

in the sense of the complex structure. Equation (ii) defines $H_1 = k_0 \cap k_1 \cap \cdots \cap k_m$ as a subspace (in the real linear structure) of H. Take $h \in \mathscr{W}(V)$ satisfying the equation $g_0 = h^{-1}\bar{h}$. Such an h exists due to Proposition 37.24. Equation (ii) may be rewritten as $hv = \overline{hv}$. So $h(H_1) = h(H) \cap W_0$. Since h is unitary, it preserves complex structure, $h(H)$ is a subspace in a complex sense, and $h(H_1)$ consists of "real" vectors in $h(H)$. Hence $\dim_{\mathbb{C}} h(H) = \dim_{\mathbb{R}} h(H_1)$, and, since h does not change, the dimension, $\dim_{\mathbb{R}} H = 2 \dim_{\mathbb{R}} H_1$. □

Expression for Ternary Maslov's Index in Terms of *a* Cocycle φ

(37.27) $$\tau(k_1, k_2, k_3) = \tfrac{1}{2}\varphi(\gamma(k_1)^{-1}\gamma(k_2), \gamma(k_2)^{-1}\gamma(k_3)), \quad k_i \in \mathscr{L}(V),\ 1 \leqq i \leqq 2.$$

Proof. The right-hand side of (37.27) does not change if one deforms continuously the Kähler structure J and the subspace W. This follows from Lemma 37.26 and Proposition 37.8. As a matter of fact, it is independent of the choice of (J, W_0), for the pairs (J, W_0) form a connected set (see 37.13 and §1). So we may pick (J, W_0) as we like. Take the subspace K_2 for W_0. Let $m = \dim(K_1 \cap K_2)$, $2n = \dim V$. Choose a vector base $\{e_i, 1 \leqq i \leqq 2n\}$ in V so that $\{e_i, 1 \leqq i \leqq n\}$ is a base in K_2 and $\{e_1, e_2, \ldots, e_m, e_{n+m+1}, \ldots, e_{2n}\}$ is a base in K_1. Define a positive Kahler structure by setting $Je_i = e_{n+i}$, $Je_{n+i} = -e_i$. Denote $g_s = \gamma(K_s)$, $s = 1, 2, 3$. Our choice of (J, W_0) implies $f_2 = 1$. The vectors $\{e_i, 1 \leqq i \leqq n\}$ form a base in V as a complex vector space. The matrix of g_1 in this base has the form

(37.28) $$g_1 \sim \begin{pmatrix} I_m & 0 \\ 0 & -I_{n-m} \end{pmatrix},$$

where I_k denotes the $k \times k$ unit matrix. Consider the map $g_3 = f\bar{f}^{-1}$ with $f \in \mathscr{U}(V)$, $f(K_2) = K_1$. Let us represent f as $f = \xi + J\eta$, where $\xi = (1/2)(f + \bar{f})$ is the "real part", and $\eta = (1/2)J^{-1}(f - \bar{f})$ is the "imaginary part" of f. Both ξ and η leave K_2 invariant and are linear in the complex linear structure. We have

(37.29) $$g_3 = f\bar{f}^{-1} = (f - \bar{f})\bar{f}^{-1} + 1 = 2J\eta(\xi - J\eta)^{-1} + 1.$$

By definition, $\varphi(\gamma(K_1)\gamma(K_2), \gamma(K_2)\gamma(K_3)) = \varphi(g_1^{-1}, g_3)$ which equals the signature of the form

(37.30) $$C(u, v) = \Omega((g_1^{-1} - 1)^{-1}u + (g_3 - 1)^{-1}u + u, v)$$

defined on $M = \operatorname{im}(g_1^{-1} - 1) \cap (g_3 - 1)$. Take $u \in M$. There is a w such that $u = (g_3 - 1)w = (\text{use (37.29)})\ 2J\eta(\xi - J\eta)^{-1}(w)$, and, due to (37.28), $u = (g^{-1} - 1)(-u/2)$.

The argument in (37.29) reads $(g_1^{-1}-1)^{-1}u+(g_3-1)^{-1}u+u=-u/2+w+u= J\eta(\xi-J\eta)^{-1}(w)-(\xi-J\eta)(\xi-J\eta)^{-1}w+\xi(\xi-J\eta)^{-1}+w=\xi(\xi-J\eta)^{-1}(w)=\xi(z)+w$ where $z=(\xi-J\eta)^{-1}(w)$. It follows from (37.29) that

$$u=2J\eta(z). \tag{37.31}$$

After these transformations the expression for C becomes

$$C(u,v)=\Omega(\xi(z),v). \tag{37.32}$$

There is a splitting

$$M=(M\cap K_2)\oplus(M\cap J(K_2)). \tag{37.33}$$

To prove (37.33) note that (37.31) suggests the general form for $u\in\operatorname{im}(g_3-1)$. Substituting in (37.31) $z=(1/2)(y-Jx)$, $x,y\in K_2$, we obtain

$$u=\eta(x)+J\eta(y). \tag{37.34}$$

The fact that u belongs to M implies additional, due to (37.29), that the expansion of u in the chosen basis $\{e_i, 1\leqq i\leqq 2n\}$ (in the real linear structure) does not contain the terms with e_i, e_{i+n}, $1\leqq i\leqq m$. Both the terms $\eta(x)$ and $J\eta(y)$ obey this conditions, so $\eta(x)\in M\cap K_2$ and $J\eta(y)\in M\cap\mathscr{T}(K_2)$, and (37.34) is the decomposition of u in accordance with (37.33).

The splitting (37.33) is orthogonal with respect to the form C. Indeed, take $u=\eta(x)\in M\cap K_2$, $v=J\eta(y)\in M\cap J(K_2)$, $x,y\in K_2$, and consider the value of C by use of (37.32). We have $C(u,v)=\Omega(\xi(-(1/2)Jx),J\eta(y))=0$ since both arguments belong to $J(K_2)$, a Lagrangian subspace.

The subspace M and the form C are linear in the complex linear structure where J plays the role of an imaginary unit. The latter maps one summand of (37.33) onto another. From this, it follows that the restrictions of C onto these two summands are isomorphic via J (now as bilinear forms in the real linear structure). In other words, C is the orthogonal sum of two identical forms. Hence the signature of C equals twice that of the restriction on $M\cap J(K_2)$, for instance.

Consider the left-hand side of (37.27). The ternary index $\tau(k_1,k_2,k_3)$ equals the signature of the form $A(u,v)=\Omega(x',v)$, defined on $u,v\in(K_1+K_2)\cap K_3$ where $x'\in K_2$ is chosen such that $u-x'\in K_1$. To link this with the above considerations let us note that the second projection $p_2: K_2\oplus J(K_2)\longrightarrow J(K_2)$ has the property

$$p((K_1+K_2)\cap K_3)=M\cap J(K_2). \tag{37.35}$$

The right-hand side of (37.35) consists of all vectors of the form $J\eta(y)$, $y\in K_2$, which belong to the real linear hull of $\{e_{i+n}, m\leqq i\leqq n\}$, as was stated above. Since $K_3=(\xi+J\eta))K_2)$, $u\in(K_1+K_2)\cap K_3$ is equivalent to u having the form $u=\xi(x)+J\eta(x)$, $x\in K_2$ and having zero components in the e_i-expansion with $n+1\leqq i\leqq n+m$ (see the definition of $\{e_i\}$ in the beginning of the proof). The formula (37.35) follows from these two descriptions.

If $u,v\in(K_1+K_2)\cap K_3$ then

$$C(p(u),p(v))=C(J\eta(x),p(v))=(\text{use (37.31) and (37.32)}) \tag{37.36}$$
$$\tfrac{1}{2}\Omega(\xi(x),p(v))=\tfrac{1}{2}\Omega(\xi(x),v)=\tfrac{1}{2}A(u,v).$$

In (36.36) we took into account that $\xi(x)\in K_2$, so one may replace $\rho(v)$ by v. The vector $x'=\xi(x)$ satisfies $u-x'\in K_1$. Indeed $u-x'=J\eta(x)\in M\cap J(K_2)\subset K_1$. This justifies the last equality in (37.36). It follows from (37.36) that the signature of C restricted onto $M\cap J(K_2)$ equals the signature of A. The whose signature of C is twice that of the restricted one. □

The Universal Covering of the Unitary Group and Cochain Φ

In view of the result of the previous subsection, it is sufficient for our purposes to consider the cocycle φ only on the unitary subgroup $\mathscr{U}(V)\subset \mathrm{Sp}(V)$, the necessary structures are meant to be fixed. The restriction of φ onto $\mathscr{U}(V)$ will be denoted by the same letter. To motivate the following constructions, we observe that, according to group cohomology theory (see e.g. Kirillov (1976) §2), φ defines the class $[\varphi]\in H^2(\mathscr{U}(V),\mathbb{Z})$. If $[\varphi]$ was zero then φ would be the coboundary of a 1-chain. In fact $[\varphi]\neq 0$ (see Turaev (1985)), but the situation changes if we pass to the universal covering $p:\tilde{\mathscr{U}}(V)\longrightarrow\mathscr{U}(V)$. It will be useful for us to construct a "primitive" Φ for the lift of φ on $\tilde{\mathscr{U}}(V)$, that is a 1-cochain Φ which satisfies the equation

$$\varphi(f,g)=\Phi(\tilde f)+\Phi(\tilde g)-\Phi(\tilde f\tilde g), \tag{37.37}$$

where $f,g\in\mathscr{U}(V)$, $\tilde f,\tilde g\in\tilde{\mathscr{U}}(V)$ and $p(\tilde f)=f$, $p(\tilde g)=g$. At first, we shall prove that a "primitive" to φ exists in a local sense.

Lemma 37.38. *There is an open neighbourhood $\mathscr{V}$ of* 1 *in $\mathscr{U}(V)$ such that for any $f,g\in\mathscr{V}$*

$$\varphi(f,g)=\varphi(f,-1)+\varphi(g,-1)-\varphi(fg,-1). \tag{37.38a}$$

Proof. Write (37.4a) $h=-1$. If f and g are sufficiently close to 1, then $\mathrm{im}(g-1)=-\mathrm{im}(-fg-1)=V$, and (37.8) implies $\varphi(f,-g)$ does not depend on g in that neighbourhood, f being fixed. So $\varphi(f,-g)=\varphi(f,-1)$, and we have (37.38a). □

Let us proceed to construct explicitly the universal covering of $\mathscr{U}(V)$. Consider the direct product $\mathscr{U}(V)\oplus\mathbb{R}$ of Lie groups, with both group and smooth structures being involved. Consider the subset $\tilde{\mathscr{U}}(V)\subset\mathscr{U}(V)\oplus\mathbb{R}$ which consists of all pairs (f,α). $f\in\mathscr{U}(V)$, $\alpha\in\mathbb{R}$, such that $\det f=\exp(i\alpha)$. Evidently $\tilde{\mathscr{U}}(V)$ is a closed subgroup, hence a Lie subgroup of $\mathscr{U}(V)\oplus\mathbb{R}$. The restriction of the first projection of the product $\mathscr{U}(V)\oplus\mathbb{R}$ onto $\tilde{\mathscr{U}}(V)$, $p:\tilde{\mathscr{U}}(V)\longrightarrow\mathscr{U}(V)$, is a smooth covering and a group homomorphism. Let $\mathscr{S}\mathscr{U}(V)$ denote the subgroup of $\mathscr{U}(V)$ consisting of all maps $f\in\mathscr{U}(V)$ with $\det f=1$. Then the formula $(f,\alpha)\longrightarrow(fe^{i\alpha},n\alpha)$ defines a diffeomorphism between $\mathscr{S}\mathscr{U}(V)\times\mathbb{R}$ and $\tilde{\mathscr{U}}(V)$. Since $\mathscr{S}\mathscr{U}(V)$ and $\mathbb{R}$ are simply connected (see Fuks and Rokhlin (1984)) so is $\tilde{\mathscr{U}}(V)$. This proves $p:\tilde{\mathscr{U}}(V)\longrightarrow\mathscr{U}(V)$ is the universal covering.

Proposition 37.39. *There exists a uniquely defined function* $\Phi : \tilde{\mathscr{U}}(V) \longrightarrow \mathbb{Z}$ *such that* (37.37) *holds for all* $f, g \in \mathscr{U}(V)$ *and* $\tilde{f} \in p^{-1}(f)$, $\tilde{g} \in p^{-1}(g)$.

Proof. We may assume the neighbourhood $\mathscr{V}$ of Lemma 37.38 to be simply connected and so small that there is an open neighbourhood $\tilde{\mathscr{V}}$ of 1 in $\tilde{\mathscr{U}}(V)$ such that p maps $\tilde{\mathscr{V}}$ diffeomorphically onto $\mathscr{V}$. Then (37.37) holds for $\tilde{f}, \tilde{g} \in \tilde{\mathscr{V}}$ if one sets

(37.39a) $$\Phi(\tilde{f}) = \varphi(f, 1), \qquad p(\tilde{f}) = f.$$

We intend to extend Φ onto the whole space $\tilde{\mathscr{U}}(V)$ using (37.37). Call a continuous path $s:[0,1] \longrightarrow \tilde{\mathscr{U}}(V)$ a short path if for each $t_1, t_2 \in [0,1]$, $s(t_1)^{-1}s(t_2) \in \mathscr{V}$. Let s be a short path with Φ defined at $s(0) = \tilde{f}_0$. Then we define the *continuation of* Φ *along* s, the value at $s(1) = \tilde{f}_1$, as

(37.39b) $$\Phi(\tilde{f}_1) = \Phi(\tilde{f}_0) + \varphi(f_0^{-1} f_1, -1) - \varphi(f_0, f_0^{-1} f_1), \quad f_i = p(\tilde{f}_i),\ i = 0, 1.$$

This definition does not contain a contradiction if $\tilde{f}_0, \tilde{f}_1 \in \tilde{\mathscr{V}}$ as it then follows from (37.38a) (put $f = f_0, g = f_0^{-1} f_1$).

Any continuous path $s:[0,1] \longrightarrow \tilde{\mathscr{U}}(V)$ may be divided into a sequence of short paths by dividing the interval $[0,1]$ into small pieces and using the appropriate change of variable. If Φ is defined at the left end to be $s(0)$ then we define Φ at the right end to be $s(1)$ by applying the extensions along the short pieces of s successively. The following Lemma yields the consistency of such a definition.

Lemma 37.39c. *Let* f_i, $i = 1, 2, 3$, *be elementss of* $\tilde{\mathscr{U}}(V)$ *joined by short paths* s_i, $i = 1, 2, 3$, *as shown in* Fig. 56 (*the case when the* s_i *have intersections is not excluded*). *Let* Φ *is defined at* f_1. *Then the result of the successive extensions of* Φ *along* s_2 *and* s_3 *equals that along* s_1.

Proof. Denote by Φ_1 the result of the continuation along s_1, and by Φ_2 that of successive continuations along s_2 and s_3. The definition (37.39b) gives us

(37.39d) $$\Phi_1 = \Phi(\tilde{f}_1) + \varphi(f_1^{-1} f_2, -1) - \varphi(f_1, f_1^{-1} f_2),$$

(37.39e) $$\Phi_2 = \Phi(\tilde{f}_1) + \varphi(f_1^{-1} f_3, -1) - \varphi(f_1, f_1^{-1} f_3) + \varphi(f_3^{-1} f_2, -1) - \varphi(f_3, f_3^{-1} f_2).$$

Here $\tilde{f}_i = p(\tilde{f}_i)$, $i = 1, 2, 3$. To persuade ourselves the right-hand sides of (37.39d,e) coincide, we write out the equality (37.4a) substituting f_1 for f, $f_1^{-1} f_3$ for g, and $f_3^{-1} f_2$ for h:

$$\varphi(f_1, f_1^{-1} f_3) + \varphi(f_3, f_3^{-1} f_2) = \varphi(f_1, f_1^{-1} f_2) + \varphi(f_1^{-1} f_3, f_3^{-1} f_2)$$

and take into account (37.38a) with $f = f_1^{-1} f_3$, $g = f_3^{-1} f_2$:

So, the result of the continuation of Φ along a continuous path does not depend on the partition of the path into small pieces. Another consequence of Lemma 37.39c is that a small "triangle" deformation of a path (see Fig 57, where

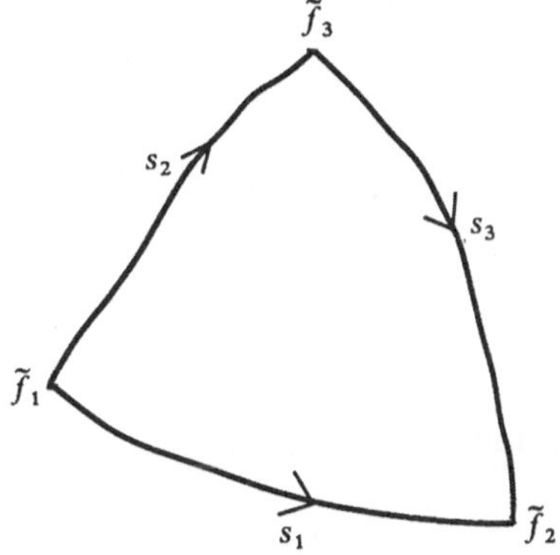

Fig. 56

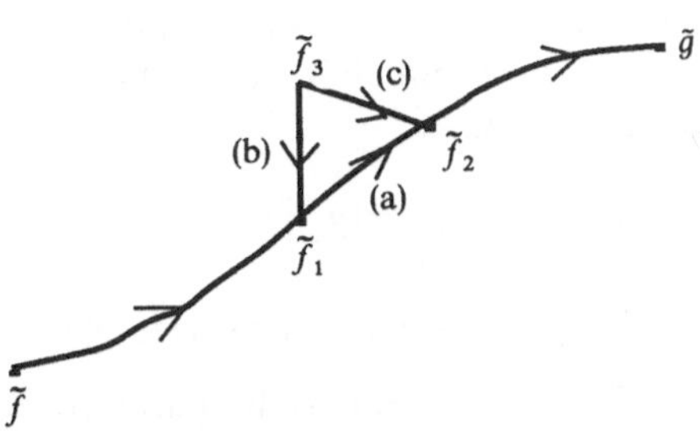

Fig. 57

a path joining $\tilde{f}$ with $\tilde{g}$ and containing a small segment (a) is replaced by that containing two small segments (b) and (c)) does not change the value $\Phi(\tilde{g})$ at the end of the path starting at $\tilde{f}, \Phi(\tilde{f})$ being already defined. It follows from this that the result of the extension from $\tilde{f}$ to $\tilde{g}$ does not depend on the choice of a path joing $\tilde{f}$ and $\tilde{g}$, for any two such paths may be successively deformed one to another via a sequence of small triangle deformations due to the simply connectednesss of $\tilde{\mathscr{U}}(V)$. Hence we have $\Phi(\tilde{f})$ well defined on $\tilde{\mathscr{U}}(V)$.

Let us prove the equation (37.37). If $\tilde{a}, \tilde{b} \in \tilde{\mathscr{U}}(V)$ then our method of extension via the formula (37.39a) implies

(37.39f)
$$\Phi(\tilde{b}) - \Phi(\tilde{a}) = \sum_{k=0}^{p-1} (\varphi(a_k^{-1} a_{k+1}, -1) - \varphi(a_k, a_k^{-1} a_{k+1})),$$

where $a_k = p(\tilde{a}_k)$, $0 \leqq k \leqq p$, $\tilde{a}_0 = \tilde{a}$, $\tilde{a}_p = \tilde{a}$, and $\tilde{a}_k$ lie on a path joining $\tilde{a}$ with $\tilde{b}$ so that the segments $[\tilde{a}_k, \tilde{a}_{k+1}]$ are short paths. Consider now two paths: the first joining 1 with $\tilde{g}$ and the second joining $\tilde{f}\tilde{g}$ with $\tilde{f}$as shown in Fig. 58. Divide the first path into short pieces denoting points of division $\tilde{g}_i$, $0 \leqq i \leqq p$, $\tilde{g}_0 = 1 \cdot \tilde{g}_p = \tilde{g}$. The points $\tilde{f}_i = \tilde{f}\tilde{g}_{p-1}$, $0 \leqq i \leqq p$, form a division of the second path. We choose

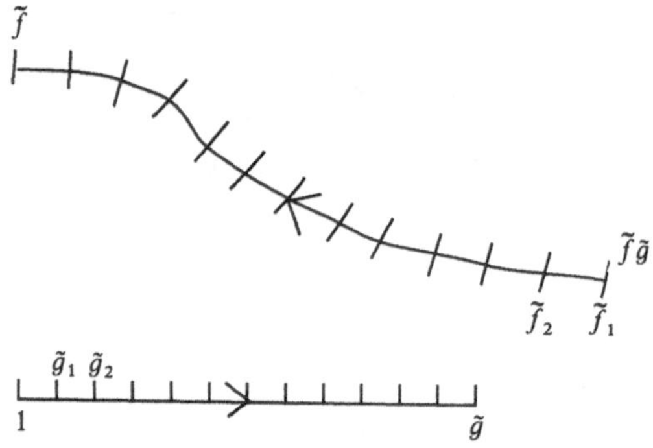

Fig. 58

the first division so fine that $[\tilde{f}_i, \tilde{f}_{i+1}]$ are short paths. Apply to these divisions the formula (37.39f), take into account that $\Phi(1) = 0$ due to (37.39a), and add up the two obtained equalities. We have

$$\Phi(\tilde{f}) + \Phi(\tilde{g}) - \Phi(\tilde{f}\tilde{g}) = \sum_{k=0}^{p-1} (\varphi(g_k^{-1}g_{k+1}, -1) - \varphi(g_k, g_k^{-1}g_{k+1}) + \varphi(g_{k+1}^{-1}g_k, -1) - \varphi(f g_{k+1}, g_{k+1}^{-1}g_k)). \quad (37.39g)$$

Here, as usual, $g_k = p(\tilde{g}_k)$. The expression in the parenthesis may be transformed as follows:

$$\begin{aligned}
&\varphi(g_k^{-1}g_{k+1}, -1) - \varphi(g_k, g_k^{-1}g_{k+1}) + \varphi(g_{k+1}^{-1}g_k, -1) - \varphi(f g_{k+1}, g_{k+1}^{-1}g_k) \\
&\quad = (\text{use (37.38a) with } f = g_k^{-1}g_{k+1},\ g = f^{-1})\varphi(g_k^{-1}g_{k+1}, g_{k+1}^{-1}g_k) \\
&\quad\quad - \varphi(g_k, g_k^{-1}g_{k+1}) - \varphi(f g_{k+1}, g_{k+1}^{-1}g_k) = (\text{use (37.1a) with } f = g_k, \\
&g = g_k^{-1}g_{k+1},\ h = g_{k+1}^{-1}g_k)\varphi(g_{k+1}, g_{k+1}^{-1}g_k) - \varphi(f g_{k+1}, g_{k+1}^{-1}g_k) \\
&\quad = (\text{use again (37.4a) with } f = f, g = g_{k+1}, h = g_{k+1}^{-1}g_k)\varphi(f, g_{k+1}) - \varphi(f, g_k).
\end{aligned}$$

Substituting this result into (37.39g) we obtain (37.37)

It remains for us to prove the uniqueness of $\Phi(\tilde{f})$ satisfying (37.37). If there were two such functions then their difference F would satisfy the equation $F(\tilde{f}) + F(\tilde{g}) = F(\tilde{f}\tilde{g})$, i.e. would define a homomorphism of $\tilde{\mathscr{U}}(V)$ into the additive group $\mathbb{Z}$. There is no such homomorphisms except the zero one, which one can easily be convinced of by considering the values $F(\tilde{f}^t)$ along any one-parameter subgroup $\tilde{f}^t$ of $\tilde{\mathscr{U}}(V)$. □

Information 37.40. One may prove the following explicit formula for $\Phi(\tilde{f})$ in terms of eigenvalues $\exp\{i\theta_k\}$, $1 \leq k \leq n$, of $f = p(\tilde{f})$ (see Turaev (1985))

$$\Phi(\tilde{f}) = \frac{2}{n}\left(\sum_{k=1}^{n} \theta_k - \alpha\right) - 2\sum_{k=1}^{n} \mu(\theta_k). \quad (37.40a)$$

Here $\tilde{f}=(f,\alpha)\in\tilde{\mathscr{U}}(V)$, $\mu:\mathbb{R}\longrightarrow\mathbb{Z}$ is the staircase-function defined by the equalities $\mu(2m\pi)=2m$, $\mu(]2m\pi,(2m+2)\pi[)=2m+1, m\in\mathbb{Z}$.

The Universal Covering of $\mathscr{L}(V)$ and the Binary Maslov's Index

To construct the universal covering of $\mathscr{L}(V)$ we consider the subset $\tilde{\mathscr{L}}(V)$ of $\mathscr{L}(V)\times\mathbb{R}$ defined by the condition

(37.41) $$(W,\theta)\in\tilde{\mathscr{L}}(V) \Leftrightarrow \det\gamma(W)=\exp(i\theta).$$

Since γ is smooth (Proposition 37.24) $\tilde{\mathscr{L}}(V)$ is a smooth submanifold. The group $\mathbb{Z}$ of all integers acts on $\tilde{\mathscr{L}}(V)$ as follows

(37.42) $$k(W,\theta)=(W,\theta+2\pi k),$$

and

$$\mathscr{L}(V)=\tilde{\mathscr{L}}(V)/\mathbb{Z}.$$

So, the restriction p of the first projection $\mathscr{L}(V)\times\mathbb{R}\longrightarrow\mathscr{L}(V)$ onto $\tilde{\mathscr{L}}(V)$ is a covering. Define the action of $\tilde{\mathscr{U}}(V)$ on $\tilde{\mathscr{L}}(V)$ by the formula

(37.43) $$(f,\alpha)(W,\theta)=(f(W),\theta+2\alpha).$$

The right-hand side of (37.43) belongs to $\tilde{\mathscr{L}}(V)$ indeed, by definition $\gamma(f(W))=f\gamma(W)\bar{f}^1$, so $\det\gamma(f(W))=\det\gamma(W)(\det f)^2=\exp i(\theta+2\alpha)$. The action (37.43) is transitive, since the action of $\mathscr{U}(V)$ on $\mathscr{L}(V)$ is. We have $\mathscr{L}(V)\cong\mathscr{U}(V)/\mathrm{St}(W_0,0)$. The stationary subgroup $\mathrm{St}(W_0,0)$ of $(W_0,0)$ consists of all points of $\tilde{\mathscr{L}}(V)$ of the form $(f,0)$ where $f\in\mathrm{St}(W_0)$ in $\mathscr{U}(V)$ and $\det f=1$ (see Proposition 37.22). So $\mathrm{St}(W_0,0)$ is isomorphic to the orthogonal group $\mathscr{SO}(n)$ (see Proposition 37.21 and take into account that $\det f=1$ if $(f,0)\in\tilde{\mathscr{L}}(V)$). Since $\mathscr{SO}(n)\cong\mathrm{St}(W_0,0)$ is connected (see Fuks and Rokhlin 1985, Chapter 3, §2) and $\tilde{\mathscr{U}}(V)$ is simply connected, the quotient space $\tilde{\mathscr{L}}(V)\cong\tilde{\mathscr{U}}(V)/\mathrm{St}(W_0,0)$ is simply connected. (Any closed path s in $\tilde{\mathscr{L}}(V)$ with the ends in $(W_0,0)$ can be lifted to a (not necessarily closed) path in $\tilde{\mathscr{U}}(V)$, the ends of the latter can be linked by a path in $\mathrm{St}(W_0,0)$. Take the sum of these two paths. The obtained closed path in $\tilde{\mathscr{U}}(V)$ can be deformed to a point. Take the projection onto $\tilde{\mathscr{L}}(V)$ of this deformation to obtain the deformation of s to a trivial path.)

We define the map $\Gamma:\tilde{\mathscr{L}}(V)\longrightarrow\tilde{\mathscr{U}}(V)$ by setting

(37.44) $$\Gamma(W,\theta)=(\gamma(W),\theta).$$

Evidently Γ is the covering maps of γ, i.e. $\Gamma\circ p=p\circ\gamma$, the letter p denoting both the universal projections.

The *binary Maslov's index* $\mathfrak{m}:\tilde{\mathscr{L}}(V)\times\tilde{\mathscr{L}}(V)\longrightarrow\mathbb{Z}$ is defined as

(37.45) $$\mathfrak{m}(\tilde{W}_1,\tilde{W}_2)=\tfrac{1}{2}\Phi(\Gamma(\tilde{W}_1)^{-1}\cdot\Gamma(\tilde{W}_2)).$$

Note that, due to Proposition 37.20, $\mathfrak{m}(\tilde{W}_1,\tilde{W}_2)$ is integer valued. Since $\Phi(\tilde{f}^{-1})=-\Phi(\tilde{f})$, the binary Maslov's index is skew-symmetric:

(37.46) $$\mathfrak{m}(\tilde{W}_1,\tilde{W}_2)=-\mathfrak{m}(\tilde{W}_2,\tilde{W}_1).$$

The following identity, the *Leray formula*, which links the binary Maslov's index with the ternary one, is the consequence of (37.27), (37.37), and (37.45):

$$\tau(K_1, K_2, K_3) = \mathfrak{m}(\tilde{K}_1, \tilde{K}_2) + \mathfrak{m}(\tilde{K}_2, \tilde{K}_3) + \mathfrak{m}(\tilde{K}_3, \tilde{K}_1), \tag{37.47}$$

where $p(\tilde{K}_i) = K_i$, $i = 1, 2, 3$.

§38. A Global Formula for Maslov's Operator

To sew together the local expressions of §35 in a patching manner, one needs to be convinced of their accordance at the intersections of the patchings. Comparing the expressions (35.17) and (35.22d), one observes that they differ, in their head terms, by multipliers $\exp(-(i/\hbar)L(z_0, \tilde{z}_0))$ and $\exp((i\pi/4) \operatorname{sign} M)$. Also, in the boundary case, two expressions for $\mathfrak{M}_{z_0}^{\text{oc,in}}$ and $\mathfrak{M}_{z_0}^{\text{loc,out}}$ differ on the boundary by the sign, or equivalently, by the multiplier $\exp(i\pi)$. The problem arises what to do with the said multipliers. We shall solve this problem by supplying $\mathfrak{M}_{z_0}^{\text{loc}}$'s with appropriate constant multipliers. Their construction is linked with certain cohomology classes of the Lagrangian submanifold $\mathscr{T}$.

The Liouville Class

Let $l: [0, 1] \longrightarrow \mathscr{T}$ be a smooth path. Consider the integral $L(l) = \int_l \Theta$ of the post-Liouville form over l. Since $d\Theta|_{\mathscr{T}\setminus\Pi} = -\Omega|_{\mathscr{T}\setminus\Pi} = 0$ and $\Theta|_{T\Pi}$ is uniquely defined because of (6.15), this integral is not changed when one deforms the path leaving the ends fixed. Hence $L(l)$ may be regarded as a function depending on $l(0)$ and $l(1)$, and defined up to a summand of the form $\sum_{k=1}^{N} m_k L_k$, m_k an integer, where $L_k = \int_{C_k} \Theta$, the integrals being over the basic cycles C_k of $\mathscr{T}$. The numbers L_k, $1 \leqq k \leqq N$, define the *Liouville class* $\mathscr{L} = \sum_{k=1}^{N} L_k [\mathfrak{e}_k] \in H^1(\mathscr{T}; \mathbb{R})$, the latter may be represented, in virtue of de Rham theorem, by a 1-form for instance, by $\sum_{k=1}^{N} L_k \mathfrak{e}_k$.

Maslov's Class

The following proposition enables us to express $\operatorname{sign} M$ in terms of the ternary Maslov's index $\tau(L, M, N)$ defined in §37. We denote by P the impulse subspace $P = L^{\varnothing}_{\{1,2,\ldots,n\}} \subset (\mathbb{R}^{2n}, \Omega_{st})$ (see 1.13).

Proposition 38.1. *Under the conditions and with the notations of* §35 *and* §37

$$\tau(L^{\mathfrak{a}}_{\bar{\mathfrak{a}}}, T_x\mathscr{T}', P) = \operatorname{sign}\left\{\frac{\partial q^{\bar{a}}_s}{\partial p_{\mathfrak{a}}}\right\}, \quad x = (q^{\bar{a}}, p_{\mathfrak{a}}) \in \mathscr{G}'.$$

Proof. Since $T_x\mathscr{T}'$ is transversal to $L^{\mathfrak{a}}_{\bar{\mathfrak{a}}}$, the domain of the form b in the definition of τ is the subspace P. The natural vector base in P is $\{\partial/\partial p_k, 1 \leqq k \leqq n\}$. Denote by

$v:\mathbb{R}^{2n} \longrightarrow L^{\mathfrak{a}}_{\bar{\mathfrak{a}}}$ the Lagrangian projection onto $L^{\mathfrak{a}}_{\bar{\mathfrak{a}}}$ along $T_x\mathcal{T}'$. The index, which we seek for, equals the signature of the matrix

$$\left\{-\Omega_{st}\left(v\frac{\partial}{\partial p_k}, \frac{\partial}{\partial p_l}\right), 1 \leqq k, l \leqq n\right\}. \tag{38.1a}$$

The equations of $\mathcal{T}'$ are given by (35.3), and we use the notations $p^s_i(q^{\bar{\mathfrak{a}}}, p_{\mathfrak{a}})$ and $q^i_s(q^{\bar{\mathfrak{a}}}, p_{\mathfrak{a}})$ for their right-hand sides. It is easy to verify that the vectors

$$f_i = \frac{\partial}{\partial q^i} + \sum_{k\in\mathfrak{a}} \frac{\partial q^k_s}{\partial q^i}\frac{\partial}{\partial q^k} + \sum_{l\in\bar{\mathfrak{a}}} \frac{\partial p^s_l}{\partial q^i}\frac{\partial}{\partial p_l},$$

$$f_j = \frac{\partial}{\partial p_j} + \sum_{k\in\mathfrak{a}} \frac{\partial q^k_s}{\partial p_j}\frac{\partial}{\partial q^k} + \sum_{l\in\bar{\mathfrak{a}}} \frac{\partial p^s_l}{\partial p^j}\frac{\partial}{\partial p_l},$$

$i\in\bar{\mathfrak{a}}$, $j\in\mathfrak{a}$, form a vector base in $T_x\mathcal{T}'$. We have $v(\partial/\partial p_i) = \partial/\partial p_i$, $i\in\bar{\mathfrak{a}}$ (these ones form a base in $L^{\mathfrak{a}}_{\bar{\mathfrak{a}}}\cap P$), and

$$v\left(\frac{\partial}{\partial p_j}\right) = -\sum_{k\in\mathfrak{a}} \frac{\partial q^k_s}{\partial p_j}\frac{\partial}{\partial q^k} - \sum_{l\in\bar{\mathfrak{a}}} \frac{\partial p^s_l}{\partial p_j}\frac{\partial}{\partial p_l}, \qquad j\in\mathfrak{a},$$

(for $v(f^j) = 0$). Substituting these values into (38.1a) and taking into account the formulae (2.17e), one obtains the matrix (38.1a) with the form

$\dfrac{\partial q^k_s}{\partial p_j}$	0
0	0

which proves the Proposition. □

Consider the *bundle of manifolds of Lagrangian subspaces* $\varkappa:\mathcal{L}(TZ)\longrightarrow Z$ over Z whose fiber over a point $z\in Z$ is $\mathcal{L}(T_zZ)$ (see §1). This is a smooth sub-bundle of the Grassmanian bundle $\mathcal{G}_n(TZ)\longrightarrow Z$ whose fibers are $G_n(T_zZ)$, the Grassmann manifolds of n-dimensional subspaces of T_zZ. Consider also

$$\varkappa_{\mathcal{T}} = i^!\varkappa: \mathcal{L}_{\mathcal{T}}(TZ) = \bigcup_{z\in\mathcal{T}} \mathcal{L}(TZ) \longrightarrow \mathcal{T},$$

the restriction of $\varkappa$ onto our Lagrangian submanifold $\mathcal{T}$, $i:\mathcal{T}\hookrightarrow Z$ being the inclusion. There are two remarkable sections of $\varkappa_{\mathcal{T}}$ which arise in the situation in question: the first one assigns to each $z\in\mathcal{T}$ the tangent space $L_z = T_z\mathcal{T}\in\mathcal{L}(T_zZ)$ of $\mathcal{T}$, the second one is $z\longmapsto P_z$ where $P_z = \ker T_z\pi' = \langle v\in T_zZ: T_z\pi'(v) = 0\rangle$, the *vertical sub-bundle* of the bundle $\pi': Z\longrightarrow Q$. Evidently, $T_z\beta(P_z) = P$, so $P_z\in\mathcal{L}(T_zZ)$. Proposition 38.1 and the formula (35.22f) yield the following expression for the matrix M of Proposition 35.22:

$$\operatorname{sign} M = \tau((T_z\tilde{\beta})^{-1}L^{\mathfrak{b}}_{\bar{\mathfrak{b}}}, L_z, P_z) - \tau((T_z\beta)^{-1}L^{\mathfrak{a}}_{\bar{\mathfrak{a}}}, L_z, P_z), \tag{38.2}$$

Each term in the right-hand side of (38.2) depends on $z\in U\cap\tilde{U}\cap\mathcal{T}$ separately, but their difference, due to the nonsingularity of M, is an integer number. To exclude the influence of the multiplier $\exp(i(\pi/4)M)$ one must attach to each local $\mathfrak{M}_{z_0}$ a constant multiplier. The exponents of the terms in the right-hand side of (38.2) multiplied by $i(\pi/4)$ are not suited for this purpose since they are *not constant*.

To gain the independance of z quantities, it is necessary to use the lifts of our Lagrangian subspaces of $T_z Z$ onto the universal covering $\tilde{\mathscr{L}}(T_z Z)$ of $\mathscr{L}(T_z Z)$. This is the crucial moment when Maslov's topological index arises. We use the notations shown in the following commutative diagram:

$$\begin{array}{ccc} \tilde{\mathscr{L}}(T_{\mathscr{T}} Z) & \xrightarrow{\tilde{\varkappa}_{\mathscr{T}}} & \\ \downarrow p & & \mathscr{T} \\ \mathscr{L}(T_{\mathscr{T}} Z) & \xrightarrow{\varkappa_{\mathscr{T}}} & \end{array} \tag{38.3}$$

Here $\tilde{\mathscr{L}}(T_{\mathscr{T}} Z)$ is the total space of a smooth bundle over $\mathscr{T}$ whose fibers are the universal covering spaces $\tilde{\mathscr{L}}(T_z Z)$ of $\mathscr{L}(T_z Z)$ and whose bundle projection is $\tilde{\varkappa}_{\mathscr{T}}$. The vertical arrow p is the covering projection. For each arrangement $\{(V,\alpha), U, \mathfrak{a}, \chi\}$ at a point $z_0 \in \mathscr{T}$ we pick a smooth section $U \cap \mathscr{T} \ni z \longmapsto \tilde{K}_z$ of the bundle $\tilde{\varkappa}_{\mathscr{T}} : \tilde{\mathscr{L}}(T_{\mathscr{T}} Z) \longrightarrow \mathscr{T}$ satisfying $p(\tilde{K}_z) = (T_z\beta)^{-1} L^{\mathfrak{a}}_{\bar{\mathfrak{a}}}$. Then the Leray formula (37.47) gives us

$$\tau((T_z\beta)^{-1} L^{\mathfrak{a}}_{\bar{\mathfrak{a}}}, L_z, P_z) = \mathfrak{m}(\tilde{K}_z, \tilde{L}_z) + \mathfrak{m}(\tilde{L}_z, \tilde{P}_z) + \mathfrak{m}(\tilde{P}_z, \tilde{K}_z) \tag{38.4}$$

where $\tilde{L}_z \in p^{-1}(L_z)$, $\tilde{P}_z \in p^{-1}(P_z)$ are arbitrary. If the latters are smooth and in concord for the arrangements $\{(V,\alpha), U, \mathfrak{a}, \chi\}$ and $\{(V',\alpha'), U', \mathfrak{b}; \chi'\}$ with $U \cap U' \cap \mathscr{T} \neq \emptyset$, then the substitution of (38.4) into (38.2) yields

$$\operatorname{sign} M = \mathfrak{m}(\tilde{K}'_z, \tilde{L}_z) + \mathfrak{m}(\tilde{P}_z, \tilde{K}'_z) - \mathfrak{m}(\tilde{K}_z, \tilde{L}_z) - \mathfrak{m}(\tilde{P}_z, \tilde{K}_z) \tag{38.5}$$

where $\tilde{K}'_z$ is the chosen section lifting, and we replaced the sign $\sim$ by the prime for marking the things associated with the second arrangement. The expression (38.5) has the desired property: it is divided into two parts corresponding to two domains: $U \cap \mathscr{T}$ and $U' \cap \mathscr{T}$, both parts being locally constant. The local constancy of $\mathfrak{m}(\tilde{K}_z, \tilde{L}_z)$ is the consequence of $L^{\mathfrak{a}}_{\bar{\mathfrak{a}}} \pitchfork T_z\beta(L_z)$, that of $\mathfrak{m}(\tilde{P}_z, \tilde{K}_z)$ is an obvious fact for both elements $\tilde{P}_z, \tilde{K}_z$ are the pre-images of the fixed coordinate Lagrangian subspaces in $(\mathbb{R}^{2n}, \Omega_{st})$. The success in removing the dependence on z, which the term $\mathfrak{m}(\tilde{L}_z, \tilde{P}_z)$ in (38.4) is responsible for, is repaid by passing to the covering space of $\mathscr{L}(T_{\mathscr{T}} Z)$.

For each continuous path $l:[0,1] \longrightarrow \mathscr{T}$ define a number $\mathfrak{m}(l)$ as follows. Consider the paths $L_{l(t)}$ and $P_{l(t)}$ in $\mathscr{L}(T_{\mathscr{T}} Z)$ and lift them onto $\tilde{\mathscr{L}}(T_{\mathscr{T}} Z)$ (see diagram (38.3)) to obtain continuous paths $\tilde{L}_{l(t)}$ and $\tilde{P}_{l(t)}$ with properties: $p(\tilde{L}_{l(t)}) = L_{l(t)}$ and $p(\tilde{P}_{l(t)}) = P_{l(t)}$. Then

$$\mathfrak{m}(l) = \mathfrak{m}(\tilde{L}_{l(1)}, \tilde{P}_{l(1)}) - \mathfrak{m}(\tilde{L}_{l(0)}, \tilde{P}_{l(0)}). \tag{38.6}$$

Lemma 38.7. *The right-hand side of* (38.6) *is independent of the choice of lifts* $\tilde{L}_{l(t)}$ *and* $\tilde{P}_{l(t)}$.

Proof. The lift $\tilde{L}_{l(t)}$ of the path $L_{l(t)}$ onto $\tilde{\mathscr{L}}(T_{\mathscr{T}} Z)$ is uniquely defined by $\tilde{L}_{l(0)}$. If $\tilde{L}'_{l(0)}$ is another lift of $L_{l(0)}$, then $\tilde{L}'_{l(0)} = k\tilde{L}_{l(0)}$ for some $k \in \mathbb{Z}$ (see (37.42)). Applying Leray's formula gives us

$$\mathfrak{m}(\tilde{L}'_{l(t)}, \tilde{P}_{l(t)}) - \mathfrak{m}(\tilde{L}_{l(t)}, \tilde{P}_{l(t)}) = \mathfrak{m}(\tilde{L}'_{l(t)}, \tilde{L}_{l(t)}) + \tau(L_{l(t)}, P_{l(t)}, L_{l(t)}). \tag{38.8}$$

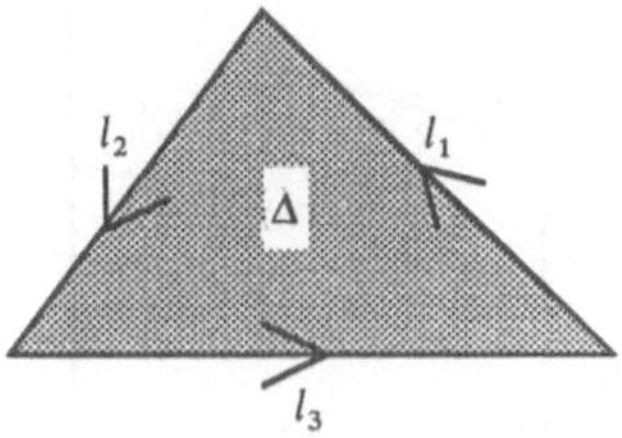

Fig. 59

Since $\tau(L_{l(t)}, P_{l(t)}, L_{l(t)}) = 0$ and $\mathfrak{m}(\tilde{L}'_{l(t)}, \tilde{L}_{l(t)}) = \mathfrak{m}(k\tilde{L}_{l(t)}, \tilde{L}_{l(t)}) = \frac{1}{2}\Phi(1, 2k\pi)$, the right-hand side of (38.8) is independent of t, and $\mathfrak{m}(l)$ given by (38.6) is not changed if one replaces $L_{l(t)}$ by $L'_{l(t)}$. The proof of the independence of the choice of a lift $\tilde{P}_{l(t)}$ goes analogously. □

The integer-valued function $\mathfrak{m}(l)$, being defined on 1-dimensional singular simplices (paths) in $\mathscr{T}$, may be uniquely extended onto the group $C_1(\mathscr{T} : \mathbb{Z})$ of 1-dimensional singular chains of $\mathscr{T}$. Since $\mathfrak{m}(l)$ is not changed if one deforms l leaving the ends fixed, $\mathfrak{m}(l_1) + \mathfrak{m}(l_2) + \mathfrak{m}(l_3) = 0$ if $l_1 + l_2 + l_3 = \partial\Delta$, where Δ is a 2-simplex (see Fig. 59). We have obtained that $\mathfrak{m}$ defines a 1-dimensional singular cocycle (see Dold (1972) or Spanier (1966) for the singular homology and cohomology theory), so it defines a class $\mathfrak{m} \in H^1(\mathscr{L}; \mathbb{Z})$, the *Maslov's class* of $\mathscr{T} \subset Z$. The expansion of $\mathfrak{m}$ onto the base $[e_k]$ is $\mathfrak{m} = \sum_{k=1}^{N} \mathfrak{m}_k[e_k]$, where $\mathfrak{m}_k = \mathfrak{m}(C_k)$, the value being independent of the choice of the beginning point is C_k.

The Boundary Class

Singular homology theory deals with singular simplices, chains etc. A singular simplex of dimension r in $\mathscr{T}$ is a continuous map $s : \Delta^r \longrightarrow \mathscr{T}$ where Δ^r is the standard r-dimensional ordered simplex. The singular chain complex of $\mathscr{T}$, $C*(\mathscr{T})$, is defined as a free abelian group generated by singular simplices and supplied with the appropriate boundary homomorphism. We shall distinguish a subcomplex $C_*^{\mathrm{reg}}(\mathscr{T}) \subset C_*(\mathscr{T})$ generated by *regular* simplices. The latters are defined inductively as follows: (1) 0-dimensional regular simplices are those singular ones (points) which are not in Π; (2) 1-dimensional regular simplices are the singular ones with properties (a) they are smooth, (b) their boundaries are 0-dimensional regular chains (linear combinations of 0-dimensional regular simplices), (c) they meet transversally to Π at each point of their intersection with Π; (3) if $r \geqq 2$, r-dimensional regular simplices are the singular ones whose boundaries are $(r-1)$-dimensional regular chains. Since any singular simplex may be continuously deformed to be a regular one, the inclusion $C_*^{\mathrm{reg}}(\mathscr{T}) \hookrightarrow C_*(\mathscr{T})$ is a homotopy equivalence, so one may build homology and cohomology theory dealing with regular chains only.

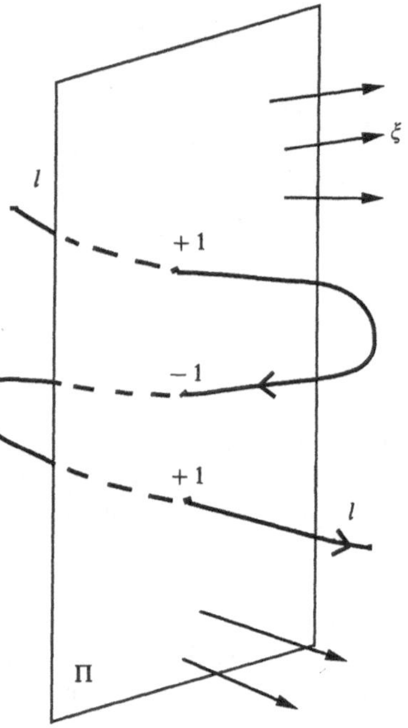

Fig. 60. Boundary index. Each point of intersection of l with Π gives contribution to $b(l)$ which is equal to ± 1 according to the direction of intersection

The *boundary index* of a regular 1-simplex $l\colon [\alpha, \beta] \longrightarrow \mathscr{T}$, $[\alpha, \beta] \subset \mathbb{R}$, is defined as

$$b(l) = \sum \pm 1, \tag{38.9}$$

where the summation runs over all points of intersection of l with Π, the sign $\pm$ being chosen in accordance with the direction of an intersection: one assigns $+1$ to a point of intersection if dl/dt and ξ look out the same side of Π at this point, and -1 in the opposite case.

The boundary index defines a 1-dimensional cochain which is a cocycle. Thus it defines the *boundary class* $\mathfrak{b} \in H^1(\mathscr{T}; \mathbb{Z})$.

In the case of $\mathscr{T}$ oriented, the boundary class corresponds via the Poincare duality $H_{n-1}(\mathscr{T}) \approx H^1(\mathscr{T}; \mathbb{Z})$ to the image in $H_{n-1}(\mathscr{T})$ of the fundamental class of $\Pi \cap \mathscr{T}$ under the map induced by the inclusion $\Pi \cap \mathscr{T} \hookrightarrow \mathscr{T}$, $\Pi \cap \mathscr{T}$ being oriented in accordance with ξ.

It is useful to extend the definition of the boundary index on those paths $l\colon [\alpha, \beta] \longrightarrow \mathscr{T}$, which have the ending point $l(1)$ in Π, the transversality condition being conserved, by the formula $\mathfrak{b}(l) = \lim_{\varepsilon \to 0} b(l|_{[\alpha, \beta - \varepsilon]})$.

The Patching Formula

Fix a finite or countable set $\{z_k\} \subset \mathscr{T}$, each point z_k being supplied with an arrangement $\{(\alpha_k, V_k), U_k, \mathfrak{a}_k, \psi_k\}$, so that the family $\{U_k\}$ is a *locally finite covering* of $\mathcal{T}$ (This means that each point $z \in \mathcal{T}$ has a neighbourhood U with $U \cap U_k \neq \emptyset$ for only a finite number of U_k's; T being paracompact gives the existence of such a family), and $\{\psi_k\}$ is a partition of unity on $\mathcal{T}$ subordinated to the covering $\{U_k \cap \mathcal{T}\}$ (see AI.35). Let us add to the arrangement of a point z_k two things. Fix a point $z^* \in \mathcal{T} \setminus \Pi$ and choose a path $l_k : [0, 1] \longrightarrow \mathcal{T}$ with $l_k(0) = z^*$ and $l_k(1) = z_k$, obeying the conditions (2) of the preceding subsection for a path to be a regular 1-simplex except probably the demand $l_k(1) \notin \Pi$.

To remove some discrepancies in patching the local expressions together, we need to introduce an additional multiplier which contains a quantity $\delta = (\delta_1, \delta_2, \ldots, \delta_N) \in [0, 2\pi[^N \subset \mathbb{R}^N \cong H^1(\mathcal{T}; \mathbb{R})$, and vector valued functions $\boldsymbol{\delta}_k : U_k \longrightarrow \mathbb{R}^N$ obeying the equations $d\boldsymbol{\delta}_k = \boldsymbol{e}$, $\boldsymbol{\delta}_k(z_k) = 0$. The existence and uniqueness of δ_k follows from the Poincaré lemma.

The patching formula for Maslov's operator $\mathfrak{M} : C_c^\infty(\tilde{\mathcal{T}}) \longrightarrow C^\infty(Q)$ reads

$$\mathfrak{M}v = \sum_k e^{i\vartheta_k} \mathfrak{M}_{z_k}^{\mathrm{loc}} e^{-i\langle \delta, \delta_k \rangle} \psi_k v, \tag{38.10}$$

where

$$\vartheta_k = \hbar^{-1} L(l_k) - \left\langle \delta, \int_{l_k} e \right\rangle + \frac{\pi}{4} [\tau((T_{z_k}\beta_k)^{-1} L_{\bar{\mathfrak{a}}_k}^{\mathfrak{a}_k}, L_{z_k}, P_{z_k}) - \mathfrak{m}(l_k) + 4\mathfrak{b}(l_k)]. \tag{38.11}$$

Since $\operatorname{supp} \vartheta$ is compact, the summation in (38.10) actually spreads over a finite number of k's. It follows immediately from (35.13) that, if $v \in C_{c,\rho}^\infty(\tilde{\mathcal{T}})$ then

$$\mathfrak{M}v|_{\partial Q} = 0. \tag{38.12}$$

As in the case of a local operator we shall use the subscript $\mathcal{T}$ as the symbol of a global Maslov's operator when we wish to indicate its dependence on the underlying Lagrangian manifold $\mathcal{T}$.

Quantum Conditions

Consider the two terms of (38.10) corresponding to points z_k and z_j. Let $U_k \cap U_j \cap \mathcal{T} \neq \emptyset$ and $\operatorname{supp} v \subset \rho^{-1}(U_k \cap U_j \cap \mathcal{T})$. Then the formulae (35.17) and (35.22d) show us that the leading terms of $e^{i\vartheta_k} \mathfrak{M}_{z_k}^{\mathrm{loc}} e^{-i\langle \delta, \delta_k \rangle} v$ and $e^{i\vartheta_j} \mathfrak{M}_{z_j}^{\mathrm{loc}} e^{-i\langle \delta, \delta_j \rangle} v$, after applying the pertinent Fourier transform, differ by the multiplier $\exp(i\vartheta_{kj})$ where

$$\begin{aligned} \vartheta_{kj} = {} & \hbar^{-1}[L(l_k) - L(l_j) + L(z_k, z_j)] - \left\langle \delta, \delta_k - \delta_j + \int_{l_k} e - \int_{l_j} e \right\rangle \\ & + \frac{\pi}{4}[\operatorname{sign} M + \tau((T_{z_k}\beta_k)^{-1} L_{\bar{\mathfrak{a}}_k}^{\mathfrak{a}_k}, L_{z_k}, P_{z_k}) \\ & - \mathfrak{m}(l_k) - \tau((T_{z_j}\beta_j)^{-1} L_{\bar{\mathfrak{a}}_j}^{\mathfrak{a}_j}, L_{z_j}, P_{z_j}) + \mathfrak{m}(l_j)] \\ & + \tfrac{1}{2}[\mathfrak{b}(l_k) + \varepsilon(\iota_k) - \mathfrak{b}(l_j) - \varepsilon(\iota_j)]. \end{aligned} \tag{38.13}$$

Here in the interior case $\varepsilon(\iota) = 0$, in the boundary case $\varepsilon(\iota) = 0$ if $\iota = \text{out}$, and $\varepsilon(\iota) = 1$ if $\iota = \text{in}$. Let l_{kj} be a path lying in $(U_k \cup U_j) \cap \mathcal{T}$ and joining z_k with z_j (see Fig. 61) so that the closed path $C = l_k + l_{kj} - l_j$ (orientation being taken into account), which

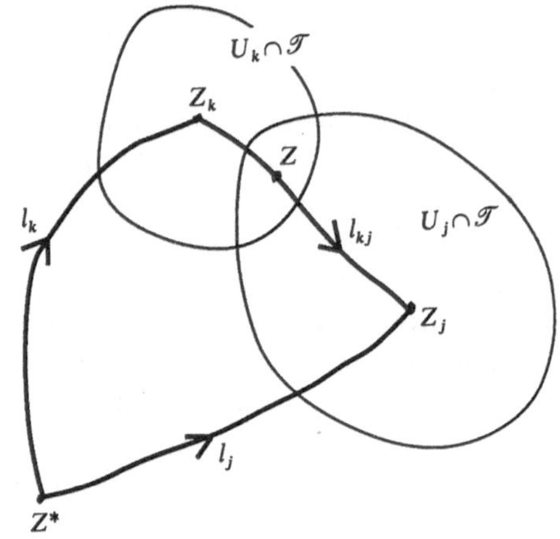

Fig. 61

starts and finishes at z_k, is regular. Such a path exists due to the conditions (i)–(iii) in the beginning of §35. It is clear that $L(z_k, z_j) = L(l_{kj})$,

$$L(l_k) - L(l_j) + L(l_{kj}) = L(C), \tag{38.14}$$

$$\delta_k - \delta_j + \int_{l_k} e - \int_{l_j} e = \int_c e. \tag{38.15}$$

$$\mathfrak{b}(l_k) + \varepsilon(\iota_k) - \mathfrak{b}(l_j) - \varepsilon(\iota_j) = \mathfrak{b}(C). \tag{38.16}$$

Note that (38.16) is valid in all possible cases. One must interpret $\mathfrak{b}(l_k) + \varepsilon(\iota_k)$ as $\mathfrak{b}(\tilde{l}_k)$ where $\tilde{l}_k$ is an extended path obtained by moving slightly the end z_k into $U_{k,l_k} \cap \mathscr{T}$ provided $z_k \in \Pi$. The same operations are to be made with l_j.

The sum of remaining terms in (38.13), multiplied by $4/\pi$, may be expressed, in virtue of (38.5) and (37.47), as follows. Take some $z \in U_k \cap U_j \cap \mathscr{T} \setminus \Pi$ and let $l(z)$ be the path joining z_k with z_j with is part of l_{kj}. We have

$$\begin{aligned} \operatorname{sign} M &+ \tau((T_{z_k}\beta_k)^{-1} L^{\mathfrak{a}_k}_{\bar{\mathfrak{a}}_k}, L_{z_k}, P_{z_k}) - \mathfrak{m}(l_k) - \tau((T_{z_j}\beta_j)^{-1} L^{\mathfrak{a}_j}_{\bar{\mathfrak{a}}_j}, L_{z_j}, P_{z_j}) + \mathfrak{m}(l_j) \\ &= \mathfrak{m}(\tilde{K}'_z, \tilde{L}_z) + \mathfrak{m}(\tilde{P}_z, \tilde{K}'_z) - \mathfrak{m}(\tilde{K}_z, \tilde{L}_z) - \mathfrak{m}(\tilde{P}_z, \tilde{K}_z) - \mathfrak{m}(l_k) + \mathfrak{m}(\tilde{K}_{z_k}, \tilde{L}_{z_k}) \\ &\quad + \mathfrak{m}(\tilde{L}_{z_k}, \tilde{P}_{z_k}) + \mathfrak{m}(\tilde{P}_{z_k}, \tilde{K}_{z_k}) - \mathfrak{m}(\tilde{K}'_{z_j}, \tilde{L}_{z_j}) - \mathfrak{m}(\tilde{L}_{z_j}, \tilde{P}_{z_j}) \\ &\quad - \mathfrak{m}(\tilde{P}_{z_j}, \tilde{K}'_{z_j}) + \mathfrak{m}(l). \end{aligned} \tag{38.17}$$

In (38.17) $\tilde{L}_z, \tilde{P}_{z_k}, \tilde{L}_{z_j}, \tilde{P}_z, \tilde{P}_{z_k}, \tilde{P}_{z_j}$ are obtained by lifting the families L_z, P_z along the paths $l_k + l(z), l_k, l_k + l_{kj}$ correspondingly onto $\tilde{\mathscr{L}}_{\mathscr{T}}(TZ)$ (see diagram (38.3)) and by taking the initial values so that $\mathfrak{m}(\tilde{L}_*, \tilde{P}_*) = 0$. The values $\tilde{K}_z, \tilde{K}_{z_k}$ are those of a chosen section of $\mathfrak{M}^{\text{loc}}_{z_k}$, the $\tilde{K}'_z, \tilde{K}'_{z_j}$ are of $\mathfrak{M}^{\text{loc}}_{z_k}$. The following pairs cancel:

$$\mathfrak{m}(\tilde{K}_z, \tilde{L}_z) = \mathfrak{m}(\tilde{K}_{z_k}, \tilde{L}_{z_k}), \qquad \mathfrak{m}(\tilde{P}_z, \tilde{K}_z) = \mathfrak{m}(\tilde{P}_{z_k}, \tilde{K}_{z_k}),$$

due to their constancy on $U_k \cap \mathscr{T}$, and

$$\mathfrak{m}(\tilde{K}'_z, \tilde{L}_z) = \mathfrak{m}(\tilde{K}'_{z_j}, \tilde{L}_{z_j}), \qquad \mathfrak{m}(\tilde{P}_z, \tilde{K}'_z) = \mathfrak{m}(\tilde{P}_{z_j}, \tilde{K}'_{z_j})$$

due to their constancy on $U_j \cap \mathcal{T}$. The remaining terms in the right-hand side of (38.17) are

$$\mathfrak{m}(l_k) + \mathfrak{m}(\tilde{K}_{z_k}, \tilde{L}_{z_k}) - \mathfrak{m}(\tilde{L}_{z_j}, \tilde{P}_{z_j}) + \mathfrak{m}(l_j) = -\mathfrak{m}(l_k - l_j + l_{kj}) = \mathfrak{m}(C).$$

Joining the results of our calculations together we may write the condition $\exp\{i\vartheta_{kj}\} = 1$ as

$$\frac{1}{2\pi\hbar} L(C) - \left\langle \frac{\delta}{2\pi}, \int_C e \right\rangle - \tfrac{1}{8}\mathfrak{m}(C) + \tfrac{1}{2}\mathfrak{b}(C) \in \mathbb{Z}. \tag{38.18}$$

We require (38.18) to be satisfied for all closed paths C lying on $\mathcal{T}$. This will be our *quantum condition.* It is sufficient to check (38.18) for basic cycles C_j, $1 \leqq j \leqq N$, on $\mathcal{T}$. Due to our notations it is equivalent to

$$\hbar^{-1} L_j - \delta_j - \frac{\pi}{2} m_j + \pi b_j = 2\pi k_j, \qquad 1 \leqq j \leqq N, \tag{38.19}$$

where k_j, the *quantum numbers* of $\mathcal{T}$ are integers. Both integer vector $k = (k_1, k_2, \ldots, k_N)$ and the deviation $\delta = (\delta_1, \delta_2, \ldots, \delta_N) \in [0, 1[^N$, are uniquely determined from the equations (38.19). They may be equivalently expressed as

$$k = \frac{1}{2\pi\hbar} L - \frac{\delta}{2\pi} - \frac{1}{2}\mathfrak{m} + \frac{1}{2}\mathfrak{b} \in H^1(\mathcal{T}; \mathbb{Z}).$$

From now on we shall assume the quantum condition to be fulfilled. Here we interpret the deviation δ as a cohomology class $\delta = \sum_{j=1}^N \delta_j [e_j]$, the *deviation class* of $\mathcal{T}$; $k = \sum_{j=1}^N k_j [e_j]$ is called *quantum number class* of $\mathcal{T}$.

Taking into account the other terms in (35.22d) yields the following

Proposition 38.20. *Given a pair of domains $U_k, U_j, U_k \cap U_j \cap \mathcal{T} \neq \emptyset$, there exists a sequence of differential operators $T_s^{(k,j)}: C_{U_k \cap U_j}(\mathcal{T}; \Pi) \longrightarrow C_{U_k \cap U_j}(\mathcal{T}; \Pi)$ $s = 1, 2, \ldots,$ the order of $T_s^{(k,j)}$ not exceeding $2s$, with smooth coefficients which are polynomials of δ, such that, given a positive integer N and $v \in C_{U_k \cap U_j}(\mathcal{T}; \Pi)$,*

$$e^{i\vartheta_k} \mathfrak{M}_{z_k}^{\text{loc}} e s^{-i\langle \delta, \delta_k \rangle} v = e^{i\vartheta_j} \mathfrak{M}_{z_j}^{\text{loc}} e^{-i\langle \delta, \delta_j \rangle} \left[v + \sum_{s=1}^{N} (i\hbar)^s T_s^{(k,j)} v \right] + \mathcal{O}(\hbar^{N+1}),$$

the estimate being in L_2-norm.

The Isometry Property of $\mathfrak{M}$

Proposition 38.21. $(\!(\mathfrak{M}u, \mathfrak{M}v)\!) = (\!(u, v)\!) + \mathcal{O}(\hbar)$.

Proof. $(\!(\mathfrak{M}u, \mathfrak{M}v)\!) = \sum_{k,j} (\!(e^{i\vartheta_k} \mathfrak{M}_{z_k}^{\text{loc}} e^{-i\langle \delta, \delta_k \rangle} \psi_k \sqrt{\psi_j} u, e^{i\vartheta_j} \mathfrak{M}_{z_j}^{\text{loc}} e^{-i\langle \delta, \delta_j \rangle} \sqrt{\psi_j} v)\!) =$ (using the equality of leading terms of different members) $\sum_{k,j} (\!(\mathfrak{M}_{z_j}^{\text{loc}} e^{-i\langle \delta, \delta_j \rangle} \psi_k \sqrt{\psi_j} u, \mathfrak{M}_{z_j}^{\text{loc}} e^{-i\langle \delta, \delta_j \rangle} \sqrt{\psi_j} v)\!) + \mathcal{O}(\hbar) =$ (apply (35.14)) $\sum_{k,j} (\!(\psi_k \sqrt{\psi_j} u, \sqrt{\psi_j} v)\!) + \mathcal{O}(\hbar) = (\!(u, v)\!) + \mathcal{O}(\hbar)$. □

The Commutation Rule

Proposition 38.22. *There exists a sequence of linear differential operators* $\mathscr{H}_s^{\square}$: $C_c^\infty(\tilde{\mathscr{F}}) \longrightarrow C_c^\infty(\tilde{\mathscr{F}})$ $s = 0, 1, 2, \ldots$, *where*

$$\mathscr{H}_0^{\square} v = Ev,$$

and

$$\mathscr{H}_1^{\square} v = -\tilde{\xi}|_{\tilde{\mathscr{F}}} v + \mathrm{i}\langle \delta, \rho^* e(\tilde{\xi}|_{\tilde{\mathscr{F}}})\rangle v, \tag{38.22a}$$

$\mathscr{H}_s^{\square}, s \geqq 2$, *is of order not exceeding 2s with coefficients in* $C^\infty(\tilde{\mathscr{T}})$ *which are polynomials in* δ, *such that, given a positive integer N and* $v \in C_c^\infty(\tilde{\mathscr{T}})$,

$$\mathscr{H}_\hbar \mathfrak{M} v = \mathfrak{M}\left(\sum_{s=0}^{N} (i\hbar)^s \mathscr{H}_s^{\square} v\right) + \mathcal{O}(\hbar^{N+1}), \tag{38.22b}$$

the estimate being in the L_2*-norm.*

Proof. Define the operators $B_s^{(k)}: C_{U_k}(\mathscr{T}; \Pi) \longrightarrow C_{U_k}(\mathscr{T}; \Pi)$ by setting $B_s^{(k)} v = \sum_j T_s^{(j,k)} \psi_j v$. This is a finite sum, and $B_s^{(k)}$ is a differential operator of the order not exceeding 2s, whose coefficients are in $C_{U_k}(\mathscr{T}; \Pi)$ and are polynomials of δ. The class of such operators $C_{U_k}(\mathscr{T}; \Pi) \longrightarrow C_{U_k}(\mathscr{T}; \Pi)$ will be denoted $\mathcal{O}_s^{(k)}$. There is a sequence $C_s^{(k)} \in \mathcal{O}_s^{(k)}$, $s = 1, 2, \ldots$, such that, for any positive integer N, $B^{(k)} C^{(k)} v = v + \mathcal{O}(\hbar^{N+1})$, where

$$B^{(k)} = 1 + \sum_{s=1}^{N} B_s^{(k)}, \qquad C^{(k)} = 1 + \sum_{s=1}^{N} C_s^{(k)}.$$

One easily obtains a recurrent equation for determining $C_s^{(k)}$. Now we get the assertion of the Proposition as follows:

$$\begin{aligned}
\mathscr{H}_\hbar \mathfrak{M} v &= \sum_k \mathscr{H}_\hbar e^{i\vartheta_k} \mathfrak{M}_{z_k}^{\mathrm{loc}} e^{-i\langle \delta, \delta_k\rangle} \psi_k v \\
&= (\text{apply } (36.9)) \sum_k e^{i\vartheta_k} \mathfrak{M}_{z_k}^{\mathrm{loc}} e^{-i\langle \delta, \delta_k\rangle} \left[E_0 \psi_k v + \sum_{s=1}^{N} (i\hbar)^s \mathscr{H}_{z_k,s}^{\square} \psi_k v \right] + \mathcal{O}(\hbar^{N+1}) \\
&= \mathfrak{M} E_0 v + \sum_k e^{i\vartheta_k} \mathfrak{M}_{z_k}^{\mathrm{loc}} e^{-i\langle \delta, \delta_k\rangle} B^{(k)} C^{(k)} \left[\sum_{s=1}^{N} (i\hbar)^s \mathscr{H}_{z_k,s}^{\square} \psi_k v \right] + \mathcal{O}(\hbar^{N+1}) \\
&= \mathfrak{M} E_0 v + \sum_{k,j} e^{i\vartheta} \mathfrak{M}_{z_k}^{\mathrm{loc}} e^{-i\langle \delta, \delta_k\rangle} \left[1 + \sum_{r=1}^{N} T_r^{(j,k)} \right] \psi_j C^{(k)} \left[\sum_{s=1}^{N} (i\hbar)^s \mathscr{H}_{z_k,s}^{\square} \psi_k v \right] \\
&= (\text{apply } (38.20))\ \mathfrak{M} E_0 v + \sum_{k,j} e^{i\vartheta_j} \mathfrak{M}_{z_j}^{\mathrm{loc}} e^{-i\langle \delta, \delta_j\rangle} \psi_j C^{(k)} \left[\sum_{s=1}^{N} (i\hbar)^s \mathscr{H}_{z_k,s}^{\square} \psi_k v \right] \\
&\quad + \mathcal{O}(\hbar^{N+1}) = \mathfrak{M} E_0 v + \mathfrak{M}\left(\sum_k C^{(k)} \left[\sum_{s=1}^{N} (i\hbar)^s \mathscr{H}_{z_k,s}^{\square} \psi_k v \right] \right) + \mathcal{O}(\hbar^{N+1}).
\end{aligned}$$

We obtained (38.22b) with $\mathscr{H}_0^{\square} = E$ and

$$\mathscr{H}_s^{\square} = \sum_k \left(\mathscr{H}_{z_k,s}^{\square} + \sum_{r=1}^{s-1} C_r^{(k)} \mathscr{H}_{z_k,s-r}^{\square} \right] \psi_k, \quad s \geqq 1.$$

Particular, the formula (38.22a) for $\mathscr{H}_1^{\square}$ follows from (36.9c) by virtue $\sum_k \psi_k = 1$. □

Notes to Chapter VI

The original work by Maslov (1972) contains all the proofs in the case $Q = \mathbb{R}^n$ but is very hard to read. A more readable exposition of Maslov canonical operator can found in maslov and Fedoryuk (1981), also in the case $Q = \mathbb{R}^n$. The theory of Maslov's index was developed by Arnol'd (1967). To expose this theory we followed Turaev (1985) (see also Lion and Vergne (1980)).

The proof of the formula (36.14) is contained in Lazutkin (1988).

Chapter VII. Quasimodes Attached to a KAM Set

Given a KAM set in the phase space of a generalized geodesic flow, on a manifold Q, $\dim Q = n$, we shall construct a family of quasimodes of arbitrary high order of discrepancy for the LBS-operator defined in Chapter V to which the generalized geodesic flow serves as a classical counterpart. The number of quasimodes appears to equal the phase-space volume occupied by the KAM set divided by $(2\pi\hbar)^n$, the elementary quantum-cell volume. The construction presented below is the most general one and involves no essential hypotheses.

§39. The Canonical Maslov's Operator Associated with a KAM Set

In this section we fix the notations and make some preparatory constructions to be able to use Maslov's operator for making quasimodes. A KAM set consists of a family of Lagrangian tori, each torus giving rise to Maslov's operator. A parameter I which labels tori, ranges over a Cantor like set $\mathscr{I}$, where all functions involved depend smoothly on $\mathscr{I}$ in the sense of Whitney. This suggests some modifications to be made in the general constructions described in Chapter VI. The latter involves some additional equipment (invariant measure, cohomology classes, etc.). Here we discuss them too. We start with a remain of the definition of a KAM set, the definition being improved slightly to be adopted to our purposes. All the improvements do not cause any loss of generality.

A KAM Set in the Phase Space of a Generalized Geodesic Flow

A KAM Set $\mathscr{K}$ in the phase space Z of a generalized Geodesic Flow (see §6) is defined by the triple $(\{g^t\}, \mathscr{I}, \Psi)$ where

(i) $g^t : A_D \longrightarrow A_D$, $t \in \mathbb{R}$, is a standard integrable symplectic flow in a symplectic annulus A_D (see Example 7.14) with a Hamiltonian $K : D \longrightarrow \mathbb{R}$ whose frequency map $\omega = \operatorname{grad} K : D \longrightarrow \mathbb{R}^n$ is an embedding;

(ii) $\mathscr{I}$ is a compact subset of D of the form $\mathscr{I} = \omega^{-1}(\mathscr{E})$ where $\mathscr{E} = \mathscr{E}_{\sigma,\gamma} \cap \Delta$, σ, γ are positive numbers, $\mathscr{E}_{\sigma,\gamma}$ is the Diophantum set in $\mathbb{R}^n$ defined in §9 and $\Delta \subset \omega(D)$ a compact convex polyhedron;

(iii) $\Psi : A_D \longrightarrow Z$ is a symplectic embedding of A_D into the phase space Z of a generalized geodesic flow which conjugates $\{g^t|_{\mathbb{T}^n \times \mathscr{I}}\}$ with a generalized geodesic flow restricted onto $\mathscr{K} = \Psi(\mathbb{T}^n \times \mathscr{I})$, the latter being an invariant subset of the generalized geodesic flow with complete dynamics.

It is useful to show all the objects involved in the following diagram (cf. diagram 34.1)

$$
\begin{array}{ccccccc}
 & & \tilde{A} & \subset & T^*\Omega\backslash\Sigma & \subset & T^*Q \\
 & & g^t & & \downarrow \rho & & \downarrow \pi \\
\mathbb{T}^n \times \mathscr{I} & \subset & \circlearrowleft A & \xrightarrow{\Psi} & Z & \xrightarrow{\pi'} & Q
\end{array}
\tag{39.1}
$$

Here $A_D = \rho^{-1}\Psi(A_D)$.

Further on in this section we shall make some additional assumptions. As it was mentioned they do not lead to loss of generality. The first one is the *Adaptness assumption.* Let H be the Hamiltonian function of the Generalized Geodesic Flow. Suppose that the Hamiltonian K of $\{g^t\}$ is adapted in the sense $K \circ p_{\tilde{2}}|_{\mathbb{T}^n \times J} = H \circ \Psi|_{\mathbb{T}^n \times J}$. The analogous equality is supposed to be valid for all derivatives of these functions. Note that in the more general situation discussed in §10 we posed the more complicated condition (HAC). Since our system possesses the Hamiltonian, only the adaptness condition is to be imposed.

The KAM set $K = \Psi(\mathbb{T}^n \times J)$ may be represented as the union of KAM tori $\mathscr{T}_I = \Psi(\mathbb{T}^n \times \{I\}), I \in \mathscr{I}$. The pre-image of the torus $\mathscr{T}_I$ under the gluing map ρ will be denoted $\tilde{\mathscr{T}}_I$. We are going to apply to the members of the family the construction of Chapter VI to obtain a family of Maslov's operators. That construction contains elements of two sorts: global ones such as Liouville, Maslov and boundary cohomology classes, the volume form σ and vector-valued 1-form $\boldsymbol{e}$, and a lot of local ones. Below we discuss all these elements sparing our attention to their dependence on the new parameter $I \in \mathscr{I}$.

Cohomology Classes

Since the inclusion $\mathbb{T}^n \times \{I\} \longrightarrow A_D$ is a homotopy equivalence one may identify the homology class of $\mathbb{T}^n \times \{I\}$ with those of A_D via the isomorphism induced by the inclusion. The space $H^1(A_D; \mathbb{R})$ was described in (2,5a). The classes $[d\varphi^j]$, $1 \leqq j \leqq n$, form the base in $H^1(A_D; \mathbb{R})$. The integer coefficients cohomology group $H^1(A_D; \mathbb{Z})$ is included naturally into $H^1(A_D; \mathbb{R})$ via the map which is the identity on $[d\varphi^j]$. The map $\sum_{i=1}^n v_i[d\varphi^i] \longmapsto (c_1, c_2, \ldots, c_n) \in \mathbb{R}^n$ identifies $H^1(A_D; \mathbb{R})$ with $\mathbb{R}^n$ and carries $H^1(A_D; \mathbb{Z})$ bijectively onto $\mathbb{Z}^n \subset \mathbb{R}^n$. Further we identify the pair $(H^1(A_D; \mathbb{R}), H^1(A_D; \mathbb{Z}))$ with $(\mathbb{R}^n, \mathbb{Z}^n)$. Also we identify the cohomology groups of $\mathscr{T}_1$ with those of $\mathbb{T}^n \times \{I\}$ via the isomorphism induced by Ψ. In virtue of the three identifications described, the cohomology groups $H^1(\mathscr{T}_1; \mathbb{R})$ and $H^1(\mathscr{T}; \mathbb{Z})$ coincide with $\mathbb{R}^n$ and $\mathbb{Z}^n$ respectively. The torus $\mathscr{T}_1$, regarded as a Lagrangian submanifold in Z, possesses three important cohomology classes: the *Liouville class* $\mathscr{L}(I) = (L_1(I), L_2(I), \ldots, L_n(I))$, *Maslov's class* $\mathfrak{m}(I) = (\mathfrak{m}_1(I), \mathfrak{m}_2(I), \ldots, \mathfrak{m}_n(I))$, and the *boundary class* $\mathfrak{b}(I) = (\mathfrak{b}_1(I), \mathfrak{b}_2(I), \ldots, \mathfrak{b}_n(I))$. The first one belongs to $H^1(\mathscr{T}_I;\mathbb{R})$, the latters two belong to $H^1(\mathscr{T}_1;\mathbb{Z})$, with all depending on $I\in\mathscr{I}$.

We impose the *cohomology classes conditions*:

$(SC1)$ $\mathscr{L}(I)=I$
$(SC2)$ *the classes* $\mathfrak{m}$ *and* $\mathfrak{b}$ *are independent of* I

Since $\mathfrak{m}_k(I)$ and $\mathfrak{b}_k(I)$ depend continuously on I and are integer numbers they are locally constant. One may divide D into simply connected pieces, each containing a part of $\mathscr{I}$ where the said functions are constant, and consider separately KAM sets corresponding to these pieces. So the cohomology classes condition, which concerns $\mathfrak{m}$ and $\mathfrak{b}$, may be accepted. As for the one concerning $\mathscr{L}$ let us consider the form $G=\psi^*\Theta-\sum_{k=1}^n I_k\,d\varphi^k$ and A_D. Since $dG=0$, this form may be represented as

$$G=\sum_{k=1}^{n} I_k^0\,d\phi^k+d\Phi,$$

where $I_k^0, 1\leqq k\leqq n$, are constants and Φ is a function. The change of variables $I_k\longmapsto I_k'=I_k+I_k^0$, or equivalently replacing D by a new domain D' by use of a shift, results in a new form G' which is a differential of a function. Then the formula $L_k=\int_{c_k}\Theta, 1\leqq k\leqq n$, yields $\mathscr{L}(I)=I$ for some chosen momentum variables.

The Volume Form σ^I and 1-form e.

There is a natural volume form on $\mathbb{T}^n\times\{I\}$ which is invariant under the flow g^t. This is $d\phi^1\wedge d\phi^2\wedge\cdots\wedge d\phi^n$. The form

$$\tilde{\sigma}^I=(\Psi^{-1}\circ\rho|_{\mathscr{T}_I})\,d\phi^2\wedge\cdots\wedge d\phi^2 \tag{39.2}$$

will serve as the volume form σ on $\tilde{\mathscr{T}}_I=\rho^{-1}(\mathscr{T}_I)$.

Define a vector-valued form e on $\Psi(A_D)$ by setting

$$e=(\Psi^{-1})^*(d\phi^1,d\phi^2,\ldots,d\phi^n). \tag{39.3}$$

Its restriction e^I *on the torus* $\mathscr{T}_I$ *will play the role of the form* e *of* §34.

Let us pass to the discussion of the local elements of the construction.

Arrangement of a Point

Given a point $(\phi_0,I^0)\in\mathbb{T}^n\times\mathscr{I}$, let us make some constructions. First choose a set of the form $D_0\times D_1\subset A_D$ where $D_0=\{\phi\in\mathbb{T}^n:|\phi-\phi_0|<r_1\}, D_1=\{I\in D:|I-I_0|<r_0\}$ are balls in $\mathbb{T}^n$ and D, the radii r_0 and r_1 being sufficiently small. Below we introduce some objects subjected to certain conditions, these being fulfilled due to the choice of the above radii. The mentioned objects are necessary for building the local Maslov's operator.

The domain U and the transversal $z_{0'}$. If $\Psi(\phi_0, I^0) \notin \Pi$ (interior case) then the set $U = \Psi(D_0 \times D_1)$ has no intersection with Π. In this case we define the transversal $z_0 : D_1 \longrightarrow Z$ by the formula $z_0(I) = \Psi(\phi_0, I)$. If $\Psi(\phi_0, I^0) \notin \Pi$ (a boundary case) then we suppose that $D_0 \times D_1 \cap \Psi^{-1}(\Pi)$ is the graph of a smooth function $\phi_0 : D_0 \longrightarrow D_1$. We define the transversal $z_0 : D_0 \longrightarrow Z$ in this case as $z_0(I) = \Psi(\phi_0(I), I)$. It follows that in the boundary case the image of the transversal lies in Π.
Attached Charts. There is a chart (V, α) in Q such that U is contained in $(\pi')^{-1}(V)$. In the interior case we require $V \cap \partial Q = \Phi$, and we consider the chart (W, β) in T^*Q where $\beta = T^{*-1}\alpha|_W$, $W = \rho^{-1}(U)$. In the boundary case the domain U is divided by Π into two subdomains U_{in} and U_{out}. The set $W = \rho^{-1}(U)$ consists of two connected components W_{in} and W_{out} such that $\partial W_{\text{in}} \subset C_{\text{in}}$, $\partial W_{\text{out}} \subset C_{\text{out}}$, and $\rho(W_\iota) = U_\iota$, $\iota \in \{\text{in}, \text{out}\}$. In this case we shall consider two attached charts (W_ι, β_ι), $\iota \in \{\text{in}, \text{out}\}$ in T^*Q with $\beta_\iota = T^{*-1}\alpha|_{W_\iota}$.
Sheets and the set $\mathfrak{a}$. In the interior case the intersection $\tilde{\mathscr{T}}_I \cap W$, $I \in D_1 \cap \mathscr{I}$, will be called the sheet of the torus $\mathscr{T}_I$. It is diffeomorphic to an n-ball, in fact it is the image of D_0 via $\Psi(., I)$. In the boundary case we have two sheets $\tilde{T}_I \cap W_\iota$, $\iota \in \{\text{in}, \text{out}\}$, each diffeomorphic to half an n-ball. Each sheet meets transversally the pre-images of the transversal z_0 under the map ρ.

Consider the images of the sheet under the attached charts: $\mathscr{T}'_I = \beta(\tilde{\mathscr{T}}_I \cap W)$ in the interior case and $\mathscr{T}'_{I,\iota} = \beta(\tilde{\mathscr{T}}_I \cap W_\iota)$, $\iota \in \{\text{in}, \text{out}\}$, in the boundary case. There is a subset $a \subset \{1, 2, \ldots, n\}$ such that the standard Lagrangian projection $\pi^{\bar{\mathfrak{a}}}_{\mathfrak{a}}$ maps each $\mathscr{T}'_I$ (or $\mathscr{T}'_{I,\iota}$) diffeomorphically onto its image G_I, $(0 + G_{I,\iota})$ in $L^{\bar{\mathfrak{a}}}_{\mathfrak{a}}$, all the derivatives of the restriction of $\pi^{\bar{\mathfrak{a}}}_{\mathfrak{a}}$ onto $\mathscr{T}'_I$ (or $\mathscr{T}'_{I,\iota}$) being bounded uniformly with respect to the variable $I \in D_1$.
The map $\gamma_{I(\iota)}$. The letter γ_I (or $\gamma_{I,\iota}$) will denote the inverse of $(\pi^{\bar{\mathfrak{a}}}_{\mathfrak{a}}|_{\mathscr{T}'_I}) \circ \beta$ (or correspondingly, of $(\pi^{\bar{\mathfrak{a}}}_{\mathfrak{a}}|_{\mathscr{T}'_{I,\iota}}) \circ \beta_\iota$).
The generating function $S_{I(\iota)}$. In the interior case, since $\mathscr{T}'_I$ has a nondegenerated projection onto G_I, it possesses the degenerating function $S_I : G_I \longrightarrow \mathbb{R}$ depending on the variables $(q^{\bar{\mathfrak{a}}}, p_{\mathfrak{a}})$ which run G_I, and on $I \in D_1 \cap \mathscr{I}$ as a parameter. The equations of $\mathscr{T}'_I$ read

$$
(39.4) \qquad \begin{cases} p_i = p^s_i(I, q^{\bar{\mathfrak{a}}}, p_{\mathfrak{a}}) \overset{\text{def}}{=} \dfrac{\partial S_I(a^{\bar{\mathfrak{a}}}, p_{\mathfrak{a}})}{\partial q^i}, & i \in \mathfrak{a}, \\[2ex] q^i = q^i_s(I, q^{\bar{\mathfrak{a}}}, p_{\mathfrak{a}}) \overset{\text{def}}{=} -\dfrac{\partial S_I(q^{\bar{\mathfrak{a}}}, p_{\mathfrak{a}})}{\partial p_j}, & j \in \mathfrak{a}. \end{cases}
$$

We subjected S_I to be norming condition

$$
(39.5) \qquad S_I(a^{\bar{\mathfrak{a}}}, p_{\mathfrak{a}}) + \sum_{j \in \mathfrak{a}} q^j_s(I, q^{\bar{\mathfrak{a}}}, p_{\mathfrak{a}}) p_j|_{\rho \circ \gamma_I(q^{\bar{\mathfrak{a}}}, p_{\mathfrak{a}}) = z_0(I)} = 0.
$$

In the boundary case the same definitions of the generating function $S_{I,\iota}$ and the same formulae (39.4) and (39.5) are true but, one must attach the indices $\iota \in \{\text{in}, \text{out}\}$ in the appropriate places. The formulae (39.4) and (39.5) are the analogues of (35.3) and (35.4). The obvious analogues of (35.5) and (35.6) are also true. Note that

$S_{I(\iota)}(q^{\bar{\mathrm{a}}}, p_{\mathrm{a}})$ depends smoothly on the variable I too in the sense of Whitney. In fact, it is a smooth function of the variables $(I, q^{\bar{\mathrm{a}}}, p_{\mathrm{a}})$.
The switching function. This is a smooth function $\chi: \mathbb{R}^n \longrightarrow \mathbb{R}$ satisfying $\operatorname{supp}\chi \subset \alpha(V)$ and $\chi = 1$ on $\alpha \circ \pi'(U)$.

All the constructions are quite similar to those of §35, the family of Lagrangian manifolds $\mathscr{T}_I$ being involved instead of one $\mathscr{T}$. Fixing $I \in \mathscr{I} \cap D_1$, the collection $\langle \{V, \alpha), U, \mathrm{a}, \chi\rangle$ becomes the arrangement at a point $z_0(I)$ in the sense of §35. So, the local Maslov's operator can be attached to a sheet $\tilde{\mathscr{T}}_I \cap W_{(\iota)}$.

Maslov's Operator

Going on our construction, we cover the KAM set $\mathscr{K} = \Psi(\mathbb{T}^n \times \mathscr{I})$ by a finite collection U_r, $1 \leqq r \leqq R$ of domains of the form described in the previous subsection, all the above conditions being supposed to be fulfilled with all the arrangements attached: (V_r, α_r), a_r, etc., to each U_r. We assume without loss of generality that there would be *enough interior arrangements* (eia). This means *each torus $\mathscr{T}_I, I \in \mathscr{I}$, has a point $z \in \mathscr{I}_I \backslash \Pi$ such that there is some $r \in \{1, 2, \ldots, R)$ with the property $z \in U_r$, $V_r \cap \partial Q = \varnothing$, and $\psi_r = 1$ in some neighbourhood of z.*

It is preferable to regard I as an additional variable ranging over a Cantor-like set $\mathscr{I}$ rather than a parameter. With this in mind, we introduce a slightly modified family of Maslov's operators: $\mathfrak{M}^{(I)}: C^\infty(\tilde{\mathscr{K}}) \longrightarrow C^\infty(Q)$, $\tilde{\mathscr{K}} = \rho^{-1}(\mathscr{K})$, $I \in \mathscr{I}$, by setting

$$\mathfrak{M}^{(I)} = \mathfrak{M}_I \circ r_I \tag{39.6}$$

where $r_I: C^\infty(\tilde{\mathscr{K}}) \longrightarrow C^\infty(\tilde{\mathscr{T}}_I)$ is the restriction map: $r_I v = v|_{\tilde{\mathscr{T}}_I}$.

As a consequence of (38.12) we have

$$\text{If } v \in C^\infty_\rho(\tilde{\mathscr{K}}) \text{ then } \mathfrak{M}^{(I)} v|_{\partial Q} = 0 \quad \forall I \in \mathscr{I}. \tag{39.7}$$

The Commutation Rule

The commutation rule stated in Proposition 38.22 may be now rewritten as follows.

Proposition 39.8. *There exists a sequence of linear differential operators.* $\tilde{\mathscr{H}}^{\square}_s$: $C^\infty_c(\tilde{\mathscr{K}}) \longrightarrow C^\infty_c(\tilde{\mathscr{K}})\, s = 0, 1, 2, \ldots,$ *such that, given a positive integer N and $v \in C^\infty_c(\tilde{\mathscr{K}})$,*

$$\mathscr{H}_\hbar \mathfrak{M} v = \mathfrak{M}^{(I)}\left(\sum_{s=0}^{N} (i\hbar)^s \tilde{\mathscr{H}}^{\square}_s v\right) + \mathcal{O}(\hbar^{N+1}), \tag{39.8a}$$

the estimate $\mathcal{O}(\hbar^{N+1})$ being in the L_2-norm and being uniform with respect to $I \in \mathscr{I}$. The first $\tilde{\mathscr{H}}^{\square}_s$'s are given by the formulae:

$$\tilde{\mathscr{H}}^{\square}_0 v = \tilde{H}|_{\tilde{\mathscr{K}}} v, \tag{39.8b}$$

$$\tilde{\mathscr{H}}^{\square}_1 v = \tilde{\xi}|_{\tilde{\mathscr{K}}} v + i\langle \delta, \rho^* e(\tilde{\xi}|_{\tilde{\mathscr{K}}})\rangle v, \tag{39.8c}$$

where $\tilde{H}$ is the pre-Hamiltonian and $\mathscr{H}_0^{\square}$ is the multiplication by this function; $\tilde{\xi}$ is the pre-Hamiltonian vector field (see §6). The subsequent operators.$\mathscr{H}_s^{\square}$, $s \geqq 2$, are differential operators which contain the derivations in the tangent directions to $\mathscr{T}_I$ of order not exceeding 2s. Their coefficients are polynomials of the variable δ with coefficients in $C^\infty(\mathscr{K})$.

§40. Quantum Conditions and the Set Λ

The formula for $\mathfrak{M}^{(I)}$ contain a quantity $\beta \in \mathbb{R}^n$ which is to be determined from quantum condition (38.19). In our case the number N of (38.19) equals n. Another quantity which arises is the quantum class $k = (k_1, k_2, \ldots, k_n) \in H^1(\mathscr{S}_I; \mathbb{Z})$. The question we shall consider here is about the domain Λ where k ranges if I ranges in $\mathscr{I}$. We required earlier $\delta \in [0, 2\pi[^n$; here we impose on δ a stronger constraint, namely $|\delta| < C\hbar^{\alpha-1}$ where $C > 0$ and $\alpha \in]1, (\sigma/n) - 1[$ are some fixed constants. This implies the following constraint on the quantum class:

$$|I - 2\pi\hbar[k + (1/8)\mathfrak{m} - (1/2)\mathfrak{b}]| < C\hbar^\alpha \tag{40.1}$$

which follows from (38.19) and the equality (SC1). Here $\mathfrak{m}$ and $\mathfrak{b}$ are Maslov's and boundary classes respectively. Recall that they are supposed to be constants.

Given $I \in \mathscr{I}$, there is at most one $k \in \mathbb{Z}^n$ satisfying (40.1) provided $\hbar$ is sufficiently small. Denote by Λ the subset of $\mathbb{Z}^N$ consisting of all k such that there exists $I \in \mathscr{I}$ satisfying (40.10). Such an I is not unique. For each $k \in \Lambda$ we choose some $I_k \in \mathscr{I}$ obeying (40.1). So we fix a function $\Lambda \longrightarrow \mathscr{I}$ which assigns a KAM torus to each admissible quantum class. In our further constructions the quasimodes will be labelled by the quantum class k which ranges over Λ. It is important for us to know an estimate for, $|\Lambda|$ the number, the points of the set Λ.

Proposition 40.2. *There is a constant $C_1 > 0$, depending on C, α, and on the KAM set $\mathscr{K}$ such that*

$$\left| |\Lambda| - \frac{\operatorname{Vol}\mathscr{K}}{(2\pi\hbar)^n} \right| < C_1 \hbar^{-n+1-\alpha n/(\sigma-n)}. \tag{40.2a}$$

Proof. Since the conjugating map, Ψ, preserves the symplectic volume, $\operatorname{Vol}\mathscr{K} = \operatorname{Vol}(\mathbb{T}^n \times \mathscr{I}) = \mathscr{M}es\,\mathscr{I}$, we will compare $\mathscr{M}es\,\mathscr{I}$ with the number of points of the lattice $2\pi\hbar\mathbb{Z}^n$ in the set $[\mathscr{I}]^\varepsilon = \{I: \operatorname{dist}(I, \mathscr{I}) < \varepsilon\}$ where $\varepsilon = C\hbar^\alpha$. One may assume $\mathfrak{m} = 0$ and $\mathfrak{b} = 0$ in (40.1). Otherwise replace $\mathscr{I}$ by $\tilde{\mathscr{I}} = \mathscr{I} - 2\pi\hbar[(1/8)\mathfrak{m} - (1/2)\mathfrak{b}]$, D by $\tilde{D} = D - 2\pi\hbar[(1/8)\mathfrak{m} - (1/2)\mathfrak{b}]$, $\omega(I)$ by $\tilde{\omega}(I) = \omega(I + 2\pi\hbar[(1/8)\mathfrak{m} - (1/2)\mathfrak{b}]$, and Ψ by $\tilde{\Psi}$ where $\tilde{\Psi}(\phi, I) = \Psi(\phi, I + 2\pi\hbar[(1/8)\mathfrak{m} - (1/2)\mathfrak{b}]$.

Let S_m, $m \in \mathbb{Z}^n$, denote the strip $\{\omega \in \mathbb{R}^n: |\langle m, \omega \rangle| \gamma |m|^{-\sigma}\}$ and let $\tilde{S}_m = (\operatorname{grad} K)^{-1}(S_m)$. Given $I \notin \mathscr{I}$ consider the ball $B_\varepsilon(I)$ with center I and radius ε, such that $B_\varepsilon(I) \cap \mathscr{I} = \varnothing$, and set $m_{I,\varepsilon} = \min\{|m|: S_m \cap B_\varepsilon(I) \neq \varnothing\}$. There is a const > 0, such that

$$m_{I,\varepsilon} \leqq m_\varepsilon = \operatorname{const} \varepsilon^{-1/(\sigma-n)}. \tag{40.2b}$$

Indeed, $B_\varepsilon(I)$ is covered by the union of $\tilde{S}_M$'s with $|m| \geqq M_{I,\varepsilon}$. The measure of the intersection $B_\varepsilon(I) \cap \tilde{S}_m$ is less than const $|m|^{-\sigma}\varepsilon^{n-1}$. Estimating the volume of $B_\varepsilon(I)$ from above by that of the union of all $B_\varepsilon(I) \cap \tilde{S}_m$ yields

$$\varepsilon^n \leqq \text{const}\,\varepsilon \sum_{|m| \geqq m_{I,\varepsilon}} |m|^{-\sigma} \leqq \text{const}\,\varepsilon^{n-1} m_{I,\varepsilon}^{-\sigma+n}$$

from which (40.2b) follows.

Consider the ε-expanded strips $[\tilde{S}_m]^\varepsilon = \{I\colon \operatorname{dist}(I, \tilde{S}_m) < \varepsilon\}$, and let $D \setminus \bigcup_{|m| \leqq m_\varepsilon} [\tilde{S}_m]^\varepsilon = \bigcup_\alpha D_\alpha^{-\varepsilon}$ where $D_\alpha^{-\beta}$ are connected components. Also let $D \setminus \bigcup_{|m| \leqq m_\varepsilon} \tilde{S}_m = \bigcup_j D_j$ be the decomposition into connected components D_j. If $I \notin [\mathscr{I}]^\varepsilon$, then I is the center of a ball $B_\varepsilon(I)$ which has no intersection with $\mathscr{I}$, so there exists m with $|m| = m_{I,\varepsilon}$ such that $B_\varepsilon(I) \cap \tilde{S}_m \neq \varnothing$. Hence $I \in [\tilde{S}_m]^\varepsilon$. It follows that $D \setminus [\mathscr{I}]^\varepsilon$, or

(40.2c) $$\bigcup_\alpha D_\alpha^{-\varepsilon} \subset [\mathscr{I}]^\varepsilon \subset \bigcup_j [D_j]^\varepsilon.$$

The right-hand inclusion in (40.2c) is a consequence of $\mathscr{I} \subset D \setminus \bigcup_{|m| \leqq m_\varepsilon} \tilde{S}_m$. The number of points of the lattice $2\pi\mathbb{Z}^n$ is estimated from both sides by those of the right- and left-hand sets in (20.2c). Denote these numbers by N_r and N_l respectively. Let us prove the estimates

(40.2d) $$\left| \frac{\mathscr{M}es(D \bigcup_{|m| \leqq m_r} S_m)}{(2\pi\hbar)^n} - N_{r(l)} \right| \leqq \text{const}\,\hbar^{-n+1-\alpha n/(\sigma-n)}.$$

The left (right)-hand side of (40.2c) is the sum of domains obtained by removing enlarged (diminished) ε strips from D. Consider the paving of $\mathbb{R}^n$ by the regular cubes with centers in the points of $2\pi\hbar\mathbb{Z}^n$ and with length of edge $2\pi\hbar$. We say a cube belongs to the 1st kind if it is contained wholly in $\bigcup_\alpha D_\alpha^{-\varepsilon}$ (or $\bigcup_j [D_j]^\varepsilon$), it belongs to the IInd kind if it has nonempty intersection with $\bigcup_\alpha D_\alpha^{-\varepsilon}$ (or $\bigcup_j [D_j]^\varepsilon$) and is not of the 1st kind. Let N_I and N_II denote the numbers of cubes of the 1st, and the IInd kinds respectively. Evidently $|N_{r(l)} - N_\text{I}| \leqq N_\text{II}$ and

$$(2\pi\hbar)^n N_\text{I} \leqq \mathscr{M}es(\textstyle\bigcup_\alpha D_\alpha^{-\varepsilon}) \leqq (2\pi\hbar)^n (N_\text{I} + N_\text{II})$$

(or the analogous estimate for $\mathscr{M}es(\bigcup_j [D_j]^\varepsilon)$). N_II may be estimated from above by the number of strips $\tilde{S}_m$ multiplied by the const $\hbar^{-n+1}$, the estimate from above of the number of II-cubes having nonempty intersection with $[\tilde{S}_m]^\varepsilon$ (or $[\tilde{S}_m]_\varepsilon$, the diminished strip). The number of stripes does not exceed $m_\varepsilon^n = \text{const}\,\varepsilon^{-n/(\sigma-n)}$. So we have obtained the estimate (40.2d) with $D \setminus \bigcup_{|m| \leqq m_\varepsilon} [\tilde{S}_m]^{\pm\varepsilon}$ instead of $D \setminus \bigcup_{|m| \leqq m_\varepsilon} \tilde{S}_m$. The difference does not exceed the number of strips multiplied by $\varepsilon\hbar^n \leqq \hbar^{-n+1}$ which is less than the right-hand side of (40.2d). Lastly, $\mathscr{M}es(D \setminus \bigcup_{|m| \leqq m_\varepsilon} \tilde{S}_m)$ differs from $\mathscr{M}es\,\mathscr{I}$ by the sum of $\mathscr{M}es\,\tilde{S}_m$ with $|m| > m_\varepsilon$. The latter quantity is less than

$$\sum_{|m| > m^\varepsilon} \text{const}\,|m|^{-\sigma} \sim \text{const} \sum_{k < m_\varepsilon} k^{-\sigma+n-1} \sim \text{const}\,(m_\varepsilon)^{n-\sigma} = \text{const}\,\varepsilon = \text{const}\,2C\hbar^\alpha.$$

Since $\alpha > 1$, $\mathscr{M}es(D \setminus \bigcup_{|m| \leqq m_\varepsilon} \tilde{S}_m)$ may be replaced by $\mathscr{M}es\,\mathscr{I}$ in (40.2d) without changing the order of error. The desired estimate (40.2a) follows from this and (40.2c). □

§41. Construction of Quasimodes

The explicit formulae for quasimodes of $\mathscr{H}_\hbar$ attached to $\mathscr{K}$ will be given here in terms of the modified global Maslov's operator defined in §38. These quasimodes will be labeled by the multi-index $k=(k_1,k_2,\dots,k_n)$ which runs over the set Λ constructed in the previous section. Fix a positive integer N, the proclaimed quasimodes $\{(u_k,E_k),\ k\in\Lambda\}$ are of the form

(41.1) $$u_k = c_k \mathfrak{M}^{(I_k)}\left(\sum_{s=0}^{N}(i\hbar)^s v_s(\delta_k)\right),$$

(41.2) $$E_k = \sum_{s=0}^{N}(i\hbar)^s E_s(\delta_k),$$

where $E_s(\delta)$, $0\leqq s\leqq N$, are polynomials of variable δ with coefficients belonging to $C^\infty(\vartheta)$ (in the sense of Whitney), $v_s(\delta)$ are also polynomials on δ with coefficients belonging to $C^\infty_\rho(\tilde{\mathscr{K}})$, c_k is a norming constant which is determined by requiring

(41.3) $$\|u_k\| = 1.$$

We set

(41.4) $$v_0 = 1$$

Proposition 38.21 yields

(41.5) $$c_k = 1+\mathcal{O}(\hbar),$$

the estimate being uniform on $k\in\Lambda$. The values of I_k were fixed in the previous section, δ_k are determined by the formula

(41.6) $$\delta_k = k-(2\pi)^{-1}I_k+\tfrac{1}{8}\mathfrak{m}-\tfrac{1}{2}\mathfrak{b},$$

and obey the inequality (40.1).

It follows from 39.7 that the u_k satisfy the Dirichlet boundary condition. Hence u_k belong to the domain of definition of the LBS operator $\mathscr{H}_{\hbar,0}$.

Proposition 41.7. *There exist a sequence $E_s(\delta), s=0,1,2,\dots,$ of polynomials with coefficients belonging to $C^\infty(\mathscr{I})$ and a sequence of polynomials $v_s(\delta)$, $s=0,1,2,\dots,$ with coefficients belonging to $C^\infty_\rho(\tilde{\mathscr{K}})$, such that, given a positive integer N, the quasimodes defined by* (41.1, 2, 3, 4, 6) *obey the inequality*

(41.7a) $$\|\mathscr{H}_\hbar u_k - E_k\| \leqq \text{const}\,\hbar^{N+1}.$$

(41.7a) *holds uniformly on $k\in\Lambda$, the positive const depends on $\mathscr{K}$, N and on the choice of the constants C and α in the quantum condition. The formulae for the first E_s are:*

(41.7b) $$E_0(I) = K(I),$$

(41.7c) $$E_1(I,\delta) = i\langle\delta,\omega(I)\rangle,$$

where K is the Hamiltonian of a model system $\{g^t\}$ on $\mathscr{K}$, and $\omega(I) = \operatorname{grad} K(I)$ is its frequency vector.

Corollary 41.8. *Let E_k, $k \in \Lambda$, be given by,* (41.2) *with $E_s(I, \delta)$ from Proposition* 41.7. *The distance from E_k to be spectrum of $\mathscr{H}_\hbar$ does not exceed const $\hbar^{N+1}$. Particularly, since the spectrum lies on the real axis, we have*

$$|\operatorname{Im} E_k| \leqq \text{const}\, \hbar^{N+1}.$$

In the next section we shall prove the approximate orthogonality of quasimodes.

Proposition 41.9. *Let E_k and v_k be chosen as in Proposition* 41.7. *Then uniformly on $k, l \in \Lambda, k \neq l$,*

$$|(\!(u_k, u_l)\!)| \leqq \text{const}\, \hbar^{[(N+1)/2]-1}, \tag{41.9a}$$

where $[\cdot]$ denotes the integer part and const depends on the same set of quantities as in Proposition 41.7.

Summarizing the statements of these propositions and taking into account Proposition 40.2 we reach our final statement:

Theorem 41.10. *Let $\mathscr{K}$ be a KAM set in the phase space of a generalized geodesic flow in a Riemannian manifold (Q, g) supplied with a potential V, the conditions* (30.15) *being fulfilled. Fix, an integer $N > 2n + 2$. Let the sequences $\{E_s\}$ and $\{v_s\}$, $0 \leqq s \leqq N$, be chosen as in Proposition* 41.7. *Then the formulae* (41.1, 2, 4) *provide us with the family of quasimodes with discrepancy of order $\hbar^{N+1}$ and with deviation from orthogonality of order $\hbar^{[(N+1)/2]-1}$.*

Let the spectrum of $\mathscr{H}_\hbar$ in the interval $[\min K(I), \max K(I)]$ be discrete for all $\hbar \in]0, \hbar_0[$, $\hbar_0 > 0$. Then for each $k \in \Lambda$ and $\hbar \in]0, \hbar_0[$ there is an eigenvalue $E_k^*(\hbar)$ of $\mathscr{H}_\hbar$ such that $E_k(\hbar)$ given by (41.2) *with E_s given by Proposition* 41.7 *obeys*

$$|E_k(\hbar) - E_k^*(\hbar) \leqq \text{const}\, \hbar^{N+1}, \tag{41.10a}$$

const depending only on the KAM set, N, C and α.

Denote by $\mathscr{E}_\hbar(A)$ the spectral projection of $\mathscr{H}_\hbar$ corresponding to the subset $A \subset \mathbb{R}$. Let

$$\Delta = [E_k^*(\hbar) - \hbar^p, E_k^*(\hbar) + \hbar^p], \quad k \in \Lambda, \quad p > 0. \tag{41.10b}$$

Then

$$\| \mathscr{E}_\hbar(\Delta) u_k - u_k \| \leqq \text{const}\, \hbar^{N+1-p}. \tag{41.10c}$$

Denote by N^* the total dimension of the subspace $\mathscr{E}_\hbar(\bigcup_\Delta \Delta) \mathscr{L}\{u_k, k \in \Lambda\}$ where Δ ranges over all intervals of the form (41.10b), *and $\mathscr{L}\{u_k, k \in \Lambda\}$ is the linear hull of quasimodes $\{u_k, k \in \Lambda\}$. Let $p < N + 1 - n$, then*

$$\left| N^* - \frac{\operatorname{Vol} \mathscr{K}}{(2\pi\hbar)^n} \right| \leqq \text{const}\, \hbar^{n-1+\alpha}. \tag{41.10d}$$

In the estimates (41.10c, d) const's *depend only on the KAM set, N, C, α and p.*

Proof of Proposition 41.7.

Using the commutation formula (39.8a) yields

$$(\mathscr{H}_\hbar - E_k)u_k = c_k \mathfrak{M}^{(I_k)}\left(\sum_{s=0}^{N} (i\hbar)^s \sum_{l+r=s} (\tilde{\mathscr{H}}_l^{\square} - E_l)v_r + \mathcal{O}(\hbar^{N+1}) \right).$$

Let us require the term of order $\hbar^s$, $0 \leqq s \leqq N$, to be zero. This will be reached if we subject v_s and E_s to the equations

$$\sum_{l+r=s} (\tilde{\mathscr{H}}_l^{[\,]} - E_l)v_r = 0, \quad 0 \leqq s \leqq N. \tag{41.11}$$

Since $v_0 = 1$, the first equation (41.11), (39.8c), and the Hamiltonian Adaptness Condition give (41.7b). The equation (41.11) with $s = 1$ results in (41.7c). Indeed, the term containing v_1 cancels, and the remaining term, due to (39.8c), gives

$$E_1(I, \delta) = i\langle \delta, \rho^* e(\tilde{\xi}|_{\mathscr{T}_I})\rangle. \tag{41.12}$$

In fact the right-hand side of (41.12) does not depend on the point of $\mathscr{T}_I$. Since the conjugating map Ψ conjugates $\{g^t|_{\mathbb{T}_n \times \mathscr{I}}\}$ and the generalized geodesic flow restricted on $\Psi(\mathbb{T}^n \times \mathscr{I})$, it follows that

$$(\Psi^{-1} \circ \rho)_* \tilde{\xi}\Big|_{\mathbb{T}^n \times \mathscr{I}} = \sum_{k=1}^{n} \omega^k \frac{\partial}{\partial \phi^k}\Big|_{\mathbb{T}^n \times \mathscr{I}}, \tag{41.13}$$

where by definition $(\Psi^{-1} \circ \rho)_* \tilde{\xi} = T(\Psi^{-1} \circ \rho) \circ \tilde{\xi} \circ \Psi$, and the multi-valuedness of the inverse of the gluing projection, ρ^{-1}, does not spoil the map, since the gluing was made by $\tilde{\xi}$. As a consequence of (41.43) and the definition (39.3) of e, we obtain for $I \in \mathscr{I}$:

$$\begin{aligned}(\rho^{-1} \circ \Phi)^* \rho^* e(\tilde{\xi}|_{\mathscr{T}_I}) &= (d\phi^1, d\phi^2, \ldots, d\phi^n)|_{\mathbb{T}^n \times \{I\}} \left(\sum_{k=1}^{n} \omega^k \frac{\partial}{\partial \phi^k} \right) \\ &= (\omega^1(I), \omega^2(I), \ldots, \omega^n(I)).\end{aligned}$$

Substituting this into (41.12) gives (41.7c).

Consider the equation (41.11) with $s \geqq 2$. Let the elements $v_{s'}$ and $E_{s'+1}$, $0 \leqq s' < s-1$, be already defined. In view of (41.7b), (39.8c), (41.4), and (41.12), the equation in question may be rewritten as

$$\tilde{\xi} v_{s-1} + E_s = f_s, \tag{41.14}$$

where f_s depends on already known elements. In order to solve (41.14) we set

$$E_s(I, \delta) = \int f_s(\delta)|_{\mathscr{T}_I} \sigma^{(1)}. \tag{41.45}$$

Then we find v_{s-1} by applying to (41.14) the following proposition.

Proposition 41.16. *The equation* $\tilde{\xi}v = f, f \in C^\infty(\tilde{\mathscr{K}})$, *has a solution* $v \in C_\rho^\infty(\tilde{K})$ *for all* $I \in \mathscr{I}$ *provided*

$$\int f|_{\mathscr{T}_I} \sigma^{(1)} = 0. \tag{41.16a}$$

Proof. Extend f to a smooth function which belongs to $C_c(\rho^{-1} \circ \Psi(A_D))$, keeping the same notation for the extended function. Let us consider firstly the simple case when $\Psi(A_D) \cap \Pi = \varnothing$. Then $\Psi^{-1} \circ \rho$ maps $\rho^{-1} \circ \Psi(A_D)$ diffeomorphically onto A_D. In virtue of (41.13) and (39.2), the problem can be restated as: find $v' \in C^\infty(\mathbb{T}^n \times \mathscr{I})$ satisfying

$$\sum_{i=1}^{n} \omega^i \frac{\partial}{\partial \phi^i} v' = f'|_{\mathbb{T}^n \times \mathscr{I}}, \tag{41.16b}$$

where $f' = (\rho^{-1} \circ \Psi)^* f$, provided

$$\int f'|_{\mathbb{T}^n \times \mathscr{I}} d\phi^1 \wedge d\phi^2 \wedge \cdots \wedge d\phi^n = 0 \quad \forall I \in \mathscr{I}. \tag{41.16c}$$

If such a v' is found then $v = (\Psi^{-1} \circ \rho)^* v'$ is a solution to (41.16a). Choose a smooth function $\mu : \mathbb{R} \longrightarrow \mathbb{C}$ such that (i) $\mu(x) \quad x \, x$ if $|x| \geqq \gamma$, and (ii) $|\mu(x)| \geqq \gamma/2$ for all $x \in R$, where γ enters the definition of the Diophantum set $\mathscr{E}_{\sigma,\gamma}$. Define the function $\hat{v}$ by the formula

$$\hat{v}(\phi, I) = \sum_{k \in \mathbb{Z}^n \setminus \{0\}} \frac{|k|^\sigma f_k(I)}{2\pi i \mu(|k|^\sigma \langle \omega(I), k \rangle)} e^{i2\pi \langle k, \phi \rangle}, \tag{41.16d}$$

where

$$f_k(I) = \int f'|_{\mathbb{T}^n \times \mathscr{I}} e^{-i2\pi \langle k, \phi \rangle} d\phi^1 \wedge d\phi^2 \wedge \cdots \wedge d\phi^n, \tag{41.16e}$$

and $\mathbb{T}^n \times \{I\}$ is oriented by $d\phi^1 \wedge d\phi^2 \wedge \cdots \wedge d\phi^n$. Since the Fourier coefficients (41.16e) tend to zero faster than any degree of $|k|^{-1}$ when $|k| \longrightarrow \infty$, we have $\hat{v} \in C_c^\infty(A_D)$. The definitions of the Diophantum set $\mathscr{E}_{\sigma,\gamma}$, the set $\mathscr{I}$, and the above definition of μ yield $v' = \hat{v}|_{\mathbb{T}^n \times \mathscr{I}}$ to be a solution to (41.16b).

Let us proceed to a more complicated situation when $\Pi \cap \Psi(A_D) \neq \varnothing$. The immediate application of the map induced by $\Psi^{-1} \circ \rho$ is useless for f has no smooth pre-image defined on A_D. It is necessary beforehand to subtract from f a part which gives rise to singularities.

Recall that $f^{-1}(\Pi) = \partial(T^*Q \setminus \Sigma)$ was divided in §35 into two sets $C_{\text{in}} \cup C_{\text{out}}$ according to the direction of the field $\tilde{\xi}$ at points of those sets. Each C_ι, $\iota \in \{\text{in}, \text{out}\}$, is a hypersurface of T^*Q. Choose a subset $V \subset C_{\text{in}}$ which is a neighbourhood of $\operatorname{supp} f \cap C_{\text{in}}$ in C_{in} ans has compact closure. There is an $\varepsilon > 0$ such that for any $x \in V$ there is an integral curve $x_t, t \in [0, 4\varepsilon]$, of the field $\tilde{\xi}$, starting at x: $x_0 = x$, and the map $\lambda : (x, t) \longmapsto x_t$ is an embedding of $V \times [0, 4\varepsilon]$ in $T^*Q \setminus \Sigma$. Denote by W the image of that embedding. The image of $\tilde{\xi}|_W$ under the map induced by λ^{-1} is d/dt. Choose two smooth functions $\chi_k : [0, 4\varepsilon] \longrightarrow \mathbb{R}$, $k = 1, 2$, such that:

$\chi_1(t)=1$ if $t \leqq \varepsilon/2$ and $\chi_1(t)=0$ if $t \geqq \varepsilon$, $\chi_2(t)=0$ if $t \leqq 2\varepsilon$ or $t \geqq 3\varepsilon$, and $\int \chi_2(t)\,dt=1$. Define $f_k \in C^\infty_W(\rho^{-1}\circ\Psi(A_D))$ by the formulae

$$f_1\circ\lambda(x,t)=\chi_1(t)f\circ\lambda(x,t), \qquad f_2\circ\lambda(x,t)=\chi_2(t)\int_0^\varepsilon f_1\circ\lambda(x,t')\,dt'$$

and set $f_{\text{in}}=f_1-f_2$. Then the equation $\tilde{\xi}v=f_{\text{in}}$ has the solution $v_{\text{in}}\in C^\infty_{0,\rho,W}$ defined as $v_{\text{in}}\circ\lambda(x,t)=\int_0^t f_{\text{in}}\circ\lambda(x,t')\,dt'$. Let $v'_{\text{in}}\in C^\infty(A_D)$ be a function such that $(\Psi^{-1}\circ\rho)^*v'_{\text{in}}=v_{\text{in}}$. In fact v'_{in} is C^∞ excepting the points of the image of C_{in}. We have, if $I\in\mathscr{I}$,

$$\text{(41.16f)} \qquad \int f_{\text{in}}|_{\mathscr{T}_I}\sigma^{(1)}=\int \tilde{\xi}v_{\text{in}}|_{\mathscr{T}_I}\sigma^{(1)}=\int\left(\sum_{i=1}^n \omega^i\frac{\partial}{\partial\phi^i}v'_{\text{in}}\,d\phi'\wedge d\phi^2\wedge\cdots\wedge d\phi^n=0\right).$$

The function f_{out} and the solution v_{out} of the equation $\tilde{\xi}v=f_{\text{out}}$ are constructed in the same way, by substituting C_{in} by C_{out} and $[0,4\varepsilon]$ by $[-4\varepsilon,0]$. The function $\tilde{f}=f-f_{\text{in}}-f_{\text{out}}\in C_c(\rho^{-1}\circ\Psi(A_D))$ vanishes in some open neighbourhood of $C_{\text{in}}\cup C_{\text{out}}$ and obeys, like f, the orthogonality condition (41.16a) (take into accout (41.16f) and the analogous equation for f_{out}). So there is $f'\in C^\infty(A_D)$ such that $\tilde{f}=(\Psi^{-1}\circ\rho)^*f'$, f' satisfying (41.16c). The equation $\tilde{\xi}\tilde{v}=\tilde{f}|_{\tilde{\mathscr{K}}}$ is carried by $\Psi^{-1}\circ\rho$ into an equation of the form (41.16b) giving a new unknown function $v'\in C^\infty(\mathbb{T}^n\times\mathscr{I})$ which is connected with $\tilde{v}$ by the formula $\tilde{v}=(\Psi^{-1}\circ\rho)^*v'$. (41.16b) can now be solved in the same way as in the preceding case when $\Pi\cap\Psi(A_D)=\varnothing$. Returning to the initial setting, we have

$$v=(\Psi^{-1}\circ\rho)^*v'+(v_{\text{in}}+v_{\text{out}})|_{\tilde{\mathscr{K}}}\in C^\infty_\rho(\tilde{\mathscr{K}})$$

is a solution to $\tilde{\xi}v=f$. □

§42. Orthogonality

This whole section is devoted to the proof of Proposition 41.9.

Let $k,l\in\Lambda$, $k\neq l$, and let (u_k,E_k) and (u_l,E_l) be quasimodes given by the formulae (41.1), (41.2) cooresponding with the quantum classes k and l, the v_s's and E_s's being chosen as in Proposition 41.7. We distinguish three cases.

Case I. $|E_k-E_l|\geqq\hbar^\mu$, $\mu=[N/2]+2$. The simple calculation, usually used when proving the orthogonality of eigenvectors of a self-adjoint operator, goes, in this case, as follows:

$$\begin{aligned}(\!(u_k,u_l)\!)(E_k-E_l)&=(\text{use Corollary 41.8})\ (\!(E_ku_k,u_l)\!)-(\!(u_k,E_lu_l)\!)+\mathcal{O}(\hbar^{N+1})\\&=(\!(\mathscr{H}_\hbar u_k,u_l)\!)-(\!(u_k,\mathscr{H}_\hbar u_l)\!)+\mathcal{O}(\hbar^{N+1})=\mathcal{O}(\hbar^{N+1}).\end{aligned}$$

Dividing by E_k-E_l yields (41.9a) for those values of k,l.

Case II. $|E_k-E_l|<\hbar^\mu$, $|I_k-I_l|\geqq\hbar^\nu$ where ν is a fixed number in the interval $]0,1/3[$. The main idea is to estimate the integral $(\!(u_k,u_l)\!)$ by use of the stationary phase method. In fact the desired approximate orthogonality holds on the level of the local canonical operator. Let us define, given a set $s=\{D_0\times D_1\subset A_D, z_0\colon D_1\longrightarrow Z, (V,\alpha), \mathfrak{a}, \chi\}$ like in §39, the family of modified local canonical operators $\mathfrak{M}_s^{\mathrm{loc}(1)}\colon C_W^\infty(\mathscr{K})\longrightarrow C^\infty(Q)$, with $I\in\mathscr{I}\cap D_1$, $w=(\Psi^{-1}\circ\rho)^{-1}\times(D_0\times D_1)$ by setting $\mathfrak{M}_s^{\mathrm{loc}(1)}=\mathfrak{M}^{\mathrm{loc}}_{z_0(I),\mathscr{T}_I}\circ r_I$, $r_I\colon C^\infty(\mathscr{K})\longrightarrow C^\infty(\widetilde{\mathscr{T}}_I)$ being, as above, the restriction map: $r_I v=v|_{\mathscr{T}_I}$. Let $\hat{s}=\{\hat{D}_0\times\hat{D}_1\subset A_D, \hat{z}_0\colon\hat{D}_1\longrightarrow Z, (\hat{V},\hat{\alpha}), \hat{\mathfrak{a}}, \hat{\chi}\}$ be another set for building the modified local operator. Consider first the situation when one of the arrangements, say $\hat{s}$, belongs to the interior case. It is sufficient to estimate from above, given $v\in C_W^\infty(\widetilde{\mathscr{K}})$ and $\hat{v}\in C_W^\infty(\widetilde{\mathscr{K}})$, the integral

$$(42.1)\qquad (\!(\mathfrak{M}_s^{\mathrm{loc}(I_k)}v, \mathfrak{M}_s^{\mathrm{loc}(I_l)}\hat{v})\!)$$
$$=\int_{\alpha(V\cap\hat{V})} dq\,\chi F_{\mathfrak{a}}^{-1}\sqrt{\lambda}e^{(i/\hbar)S}\gamma^* v(\hat{\alpha}\circ\alpha^{-1})^*\hat{\chi}F_{\mathfrak{b}}\sqrt{\hat{\lambda}}e^{-(i/\hbar)\hat{S}}\hat{\gamma}^*\hat{v}.$$

We use systematically the sign ^ to indicate the same things attributed to the second operator (one exception is for the notation $\mathfrak{b}$ instead of $\hat{\mathfrak{a}}$). The expression on the right-hand side of (42.1) is a typical stationary phase integral. The crucial role is played by the phase function which is equal, in our case, to

$$(42.2)\qquad \Phi(q,p_{\mathfrak{a}},\hat{p}_{\mathfrak{b}})=\sum_{j\in\mathfrak{a}}p_jq^j+S(q^{\bar{\mathfrak{a}}},p_{\mathfrak{a}})-\sum_{j\in\mathfrak{b}}\hat{p}_j\hat{q}^j(q)-\hat{S}(\hat{q}^{\bar{\mathfrak{b}}}(q),\hat{p}_{\mathfrak{b}}).$$

Here $\hat{q}^j(q)$ denotes the functions of the coordinate expression of the map $\hat{\alpha}\circ\alpha^{-1}$, and we adopt the abbreviated notations S and $\hat{S}$ for the corresponding generating functions, the subscripts I_k and I_l being omitted.

Let us write out the explicit formulae for the components of grad Φ:

$$(42.3)\qquad \frac{\partial\Phi}{\partial q^j}=p_j-\sum_{j\in\mathfrak{b}}\hat{p}_j\frac{\partial\hat{q}^j(q)}{\partial q^j}-\sum_{i\in\bar{\mathfrak{b}}}p_i^{\hat{S}}(\hat{q}^{\bar{\mathfrak{b}}}(q),\hat{p}_{\mathfrak{b}})\frac{\partial\hat{q}^i(q)}{\partial q^j},\quad j\in\mathfrak{a},$$

$$(42.4)\qquad \frac{\partial\Phi}{\partial q^i}=p_i^S(q^{\bar{\mathfrak{a}}},p_{\mathfrak{a}})-\sum_{j\in\mathfrak{b}}\hat{p}_j\frac{\partial\hat{q}^j(q)}{\partial q^i}-\sum_{r\in\bar{\mathfrak{b}}}\hat{p}_r^{\hat{S}}(\hat{q}^{\bar{\mathfrak{b}}}(q),\hat{p}_{\mathfrak{b}})\frac{\partial\hat{q}^r(q)}{\partial q^j},\quad i\in\bar{\mathfrak{a}},$$

$$(42.5)\qquad \frac{\partial\Phi}{\partial p_j}=q^j-q_S^j(q^{\bar{\mathfrak{a}}},p_{\mathfrak{a}}),\quad j\in\mathfrak{a},$$

$$(42.6)\qquad \frac{\partial\Phi}{\partial\hat{p}_j}=-\hat{q}^j(q)+q_{\hat{S}}^j(\hat{q}\mathfrak{b}(q),\hat{p}_{\mathfrak{b}}),\quad j\in\mathfrak{b}.$$

Here we also omitted the variable I in the notations for q_S and $q_{\hat{S}}$ (see (39.4)), the former taking the values I_k and I_l correspondingly.

Consider the compact subset

$$K=\operatorname{supp}\hat{\chi}\times p_{\bar{\mathfrak{b}}}^{\hat{S}}(\operatorname{supp}\hat{\gamma}^*\hat{v})\times pr_{\hat{p}_{\mathfrak{b}}}(\operatorname{supp}\hat{\gamma}^*\hat{v})\subset\mathbb{R}^{2n}.$$

Denote by L the Lipschitz constant of the map $f=T^{*-1}\alpha\circ\hat{\alpha}^{-1}$ restricted on K.

Lemma 42.7. *There is a positive consant c, depending only on the KAM set and on L such that at points of the support of the integrand of* (42.1)

$$|\operatorname{grad}_{q,p_{\mathfrak{a}},\hat{p}_{\mathfrak{b}}}\Phi| \geqq c\hbar^{\nu}.$$

Proof. Denote $\mathscr{T}' = T^{*-1}\alpha(\tilde{\mathscr{T}}_{I_k})$, $\mathscr{T}'' = T^{*-1}\alpha(\tilde{\mathscr{T}}_{I_l})$, and $\hat{\mathscr{T}}' = T^{*-1}\hat{\alpha}(\tilde{\mathscr{T}}_{I_l})$. Since $|I_k - I_l| \geqq \hbar^{\nu}$, there exists $d > 0$ such that $\operatorname{dist}(\mathscr{T}', \mathscr{T}'') \geqq d\hbar^{\nu}$, d depending only on the KAM set. Let $(q, p_{\mathfrak{a}}, \hat{p}_{\mathfrak{b}})$ be a point where we wish to estimate grad Φ. Denote $\hat{q} = \alpha \circ \hat{\alpha}^{-1}(q)$,

$$x = (q, p_{\bar{\mathfrak{a}}}^{S}(q^{\bar{\mathfrak{a}}}, p_{\mathfrak{a}}), p_{\mathfrak{a}}) \in \mathbb{R}^{2n}, \qquad \hat{x} = (\hat{q}, p_{\bar{\mathfrak{a}}}^{\hat{S}}(\hat{q}^{\bar{\mathfrak{b}}}, \hat{p}_{\mathfrak{b}}), \hat{p}_{\mathfrak{b}}) \in \mathbb{R}^{2n},$$
$$x' = (q^{\bar{\mathfrak{a}}}, q_{S}^{\mathfrak{a}}(q^{\bar{\mathfrak{a}}}, p_{\mathfrak{a}}), p_{\bar{\mathfrak{a}}}^{S}(q^{\bar{\mathfrak{a}}}, p_{\mathfrak{a}}), p_{\mathfrak{a}}) \in \mathscr{T}' \subset \mathbb{R}^{2n},$$
$$\hat{x}' = (\hat{q}^{\bar{\mathfrak{b}}}, q_{\hat{S}}^{\mathfrak{b}}(\hat{q}^{\bar{\mathfrak{b}}}, \hat{p}_{\mathfrak{b}}), p_{\bar{\mathfrak{b}}}^{\hat{S}}(\hat{q}^{\mathfrak{b}}, \hat{p}_{\mathfrak{b}}), \hat{p}_{\mathfrak{b}}) \in \hat{\mathscr{T}}' \subset \mathbb{R}^{2n}.$$

One observes that $\hat{x}, \hat{x}' \in K$ if $(q, p_{\mathfrak{a}}, \hat{p}_{\mathfrak{b}})$ belongs to the support of the integrand. So $|f(\hat{x}) - f(\hat{x}')| \leqq L|\hat{x} - \hat{x}'|$. Suppose that

$$|\operatorname{grad}_{q,p_{\mathfrak{a}},\hat{p}_{\mathfrak{b}}}\Phi| < c\hbar^{\nu}, \tag{42.7a}$$

where c is some positive number. Then, comparing x and x' and taking into account (42.5), yields $|\hat{x} - \hat{x}'| < c\hbar^{\nu}$. Analogously, regarding (42.6), we obtain $|x - x'| < c\hbar^{\nu}$. In turn, the equations (42.3) and (42.4) imply

$$|x - f(\hat{x})| < c\hbar^{\nu}. \tag{42.7b}$$

Recall that $f = T^{*-1}\alpha \circ \hat{\alpha}^{-1}$ and $\hat{q} = \alpha \circ \hat{\alpha}^{-1}(q)$. Since $x' \in \mathscr{T}'$ and $f(x'') \in \mathscr{T}''$, we have $d\hbar^{\nu} \leqq |x' - f(\hat{x}')| \leqq |x' - x| + |x - f(\hat{x})| + |f(\hat{x}) - f(\hat{x}')| < (2 + L)^{-1}$. So, (42.7a) fails for that value of c. □

To obtain the desired estimation for (42.1) it remains for us to apply the trick of integration by parts described in the proof of AII.5. Each step of that trick gives the multiplier $\hbar^{1-3\nu}$ to an estimate. A finite number of steps is sufficient.

Let us pass to the situation when both arrangements s and $\hat{s}$ belong to the boundary case. The domain of integration in (42.1) contains now the boundary $q^1 = 0$, and the application of the above integration-by-parts trick gives rise to the boundary terms. Denote by Φ_1 the restriction of Φ given by (42.2) onto the hyperplane $q^1 = 0$. This is a function defined on a subset in $\mathbb{R}^{n+|\mathfrak{a}|+|\mathfrak{b}|-1}$, the variables being denoted by $\{q^k, 2 \leqq k \leqq n, p_j, j \in \mathfrak{a}, \hat{p}_l, l \in \mathfrak{b}\}$. The said boundary terms are also of the form of the oscillatory integral but with Φ_1 as phase function. The same procedure gives the same estimation for these boundary integrals if we prove

Lemma 42.8. *There is a positive constant c_1, depending only on the KAM set and on L, such that $|\operatorname{grad}\Phi_1| \geqq c_1\hbar^{\nu}$ at points of the support of the integrand in* (42.1) *intersected with the hyperplane $q^1 = 0$.*

Proof. Suppose that

$$|\operatorname{grad}\Phi_1| < c_1\hbar^{\nu}, \tag{42.8a}$$

where c_1 is some positive number and follow the same scheme as in the proof of Lemma 42.7, keeping the same notation. Of course, $q^1 = \hat{q}^1 = 0$ under those considerations. The scheme fails in proving (42.7b) since (42.8a) gives no information about $(\partial/\partial q^1)\Phi_1$. Instead we obtain

(b) $$\sum_{k=2}^{n} |p_k - p_k^*|^2 \leqq c_1^2 \hbar^{2\nu}$$

and

(c) $$|x - f(\hat{x})| = |p - p^*| \leqq c_1 \hbar^\nu + |p_1 - p_1^*|,$$

where we used for brevity the following notations for the vectors in $\mathbb{R}^n$: $p = (p_{\bar{a}}^S(q^{\bar{a}}, p_a), p_a)$, $\hat{p} = (p_{\bar{b}}^S(\hat{q}^{\bar{b}}, \hat{p}_b), \hat{p}_b)$, $(q, p^*) = f(\hat{q}, \hat{p})$. To obtain an estimate for $|p_1 - p_1^*|$, consider the form $T(p') = \frac{1}{2}\sum g^{ik}(q)p'_i p'_k$. The condition $|E_k - E_l| < \hbar^\mu$ implies

(d) $$|T(p) - T(p^*)| < c_2 \hbar,$$

where $c_2 > 0$ depends on the KAM set (see formulae (41.2) and (41.7b)). If $\tilde{p}(t) = p + t\,\mathrm{ort}_1$, $t \in \mathbb{R}$, then

(e) $$T(\tilde{p}(t)) - T(p) = tv_1 + 2^{-1}t^2 g^{11},$$

where v_1 is the first component of the vector $z^{\#(g)}$ and $z = (T^{*-1}\alpha)^{-1}x$, the latter coinciding with $T_z \pi \xi$ (Proposition 6.6). Since ξ is transversal to the boundary of $T^*Q \backslash \Sigma$, there is a positive constant c_3 such that $|v_1| \geqq c_3$. Also $g^{11} \geqq c_4 > 0$. Set $t = p_1^* - p_1$. Then (b) implies

(f) $$|\tilde{p}(t) - p^*| \leqq c_1 \hbar^\nu.$$

Comparing (d) with (e) and taking into account (f), one deduces $|tv_1 + 2^{-1}t^2 g^{11}| \leqq \mathrm{const}\,\hbar^\nu$, or $|t + v_1/g^{11}| \leqq \mathrm{const}\,\hbar^\nu$. The latter inequality may be rewritten, in view of the chosen value of t, after multiplication by g^{11} as $|g^{11}p_1^* + \sum_{k=2}^n g^{k1}p_k| \leqq \mathrm{const}\,\hbar^\nu$, or, again taking into account (b), as $|v_1^*| \leqq \mathrm{const}\,\hbar^\nu$, where v_1^* is the first component of the vector $\hat{z}^{\#(g)} = T_{\hat{z}}\xi$, $\hat{z} = (T^{*-1}\hat{\alpha})^{-1}\hat{x}$ this contradicts the transversality of ξ with respect to the boundary. Hence, the first inequality $|t| = |p_1 - p_1^*| \leqq \mathrm{const}\,\hbar^\nu$ holds, which, together with (c), enables us to finish the proof in the same way as that of Lemma 42.7. □

The last case is

Case III. : $|E_k - E_l| < \hbar^\mu$, $|I_k - I_l| < \hbar^\nu$.

In fact, due to the quantum condition (40.1) and inequality $k \neq l$, the momenta obey the two-sided estimation:

(42.9) $$\mathrm{const}\,\hbar < |I_k - I_l| < \hbar^\nu.$$

Let us formulate the following Proposition which gives us the solution to the problem stated in case III.

Proposition 42.10. *Given a point $I^* \in \mathscr{I}$ and a point ϑ^* of the unit sphere $S^{n-1} \subset \mathbb{R}^n$, there are: a neighbourhood $W_1 \subset D$ of the point I^*, a neighbourhood W_2 of ϑ^* in*

S^{n-1}, *and a positive number c depending only on the KAM set, such that, if* $I_k, I_l \in W_1$, $(I_k - I_l)/|I_k - I_l| \in W_2$, *and k, l belong to case III, then*

$$|\langle u_k, u_l \rangle| \leqq c\hbar^{[(N+1)/2]-1}. \tag{42.10a}$$

Given $I^* \in \mathscr{I}$, cover S^{n-1} by a finite number of neighbourhoods of type W_2. Let W be the intersection of the corresponding W_1's. Then, if $I_k, I_l \in W$ and belong to case III, the estimate (42.10a) holds for an appropriate c. Since $\mathscr{I}$ may be covered by a finite number of such neighbourhoods W, the desired estimate (41.9a) also holds. We now persuade ourselves of validity of the Proposition 42.10.

Proof of Proposition 42.10. The main idea is to consider a perturbed operator $\mathscr{H}_\hbar^{\text{pert}} = \mathscr{H}_\hbar + \mathscr{B}$ where $\mathscr{B}$ is a suitably chosen symmetric linear operator. Here we regard the operators as acting from $C_0^\infty(Q)$ to $C^\infty(Q)$, the symmetry is meant with respect to the inner product (39.9). Let the perturbation be so small that the perturbed operator possesses "perturbed" quasimodes $(u_k^{\text{pert}}, E_k^{\text{pert}})$, $k \in \Lambda$, such that

$$\|u_k^{\text{pert}} - u_k\| \leqq \text{const}\, \hbar^{M-1}, \tag{b}$$

whilst E_k^{pert} and E_l^{pert} are "normally" separated:

$$|E_k^{\text{pert}} - E_l^{\text{pert}}| \geqq \text{const}\, \hbar^{M+1} \tag{c}$$

provided k, l obey the conditions of Proposition 42.10, and the neighbourhoods W_1 and W_2 are chosen in an appropriate manner. In (b) and (c) $M = [(N/2)] + 1$, with the positive constants in the right-hand sides of these estimates depending only on the KAM set and on I^*. The considerations in Case I are applicable to the perturbed quasimodes and give the estimate $|(\!(u_k^{\text{pert}}, u_l^{\text{pert}})\!)| \leqq \text{const}\, \hbar^{N-M}$. The latter, together with (b), yields (42.10a). It remains for us to construct such a perturbation.

Take a point $z \in \mathscr{T}_{I^*} \setminus \Pi$ satisfying (eia) (see §39 subsection "Maslov's Operator"). Let $\phi \in \mathbb{T}^n$ be such that $\Psi(\phi^*, I^*) = z$, and let $(V, \alpha) = (V^r, \alpha^r)$ be the chart in Q given by (eia) for our I^* and z. To construct a perturbation we shall use the operator $\mathfrak{O}_{a,b}$ attached to (V, α) (see §36). Take $W_3 = B_{2\varepsilon}(I^*) \subset D$, $W_4 = B_{2\varepsilon}(\phi^*) \subset \mathbb{T}^n$, the balls of radii 2ε and centers I^* and ϕ^* correspondingly. Let the number ε be so small that $\Psi(W_3 \times W_4) \subset U^r$ where U^r is the second element of the arrangement. Further on we shall decrease ε if necessary. Choose a constant $d > 0$ such that $\langle \vartheta^*, I \rangle + d > 0$ if $I \in W_3$. Choose functions $\chi_3 \in C_c^\infty(D)$, $\chi_4 \in C^\infty(\mathbb{T}^n)$ such that $\operatorname{supp} \chi_3 \subset W_3$, $\chi_3 = 1$ on $B_\varepsilon(I^*)$, $\operatorname{supp} \chi_4 \subset W_4$, and $\int \chi^2\, d\phi^1 \wedge d\phi^2 \wedge \cdots \wedge d\phi^n = 1$.

Define $a, b \in C_{\text{symb}}^\infty(\mathbb{R}^{2n})$ by the formulae

$$\begin{aligned} &(T^{*-1}\alpha \circ \rho^{-1} \circ \Psi)^* \sqrt[4]{g}\, a(I, \phi) = R\sqrt{\langle \vartheta^*, I \rangle + d}\, \chi_3(I)\chi_4(\phi), \\ &b(q, p) = a(q, p)\sqrt{g(q)}, \end{aligned} \tag{d}$$

where $g(q)$ is the determinant of the matrix g_{ik} of the Riemannian metric in the chart (V, α), and R is a constant which we shall choose later.

Our perturbation will be of the form $\mathscr{B} = \hbar^M \mathfrak{D}_{a,b}$. The condition (d) guarantees the symmetry of $\mathfrak{D}_{a,b}$: $\langle \mathfrak{D}_{a,b} u, v\rangle = \langle u, \mathfrak{D}_{a,b} v\rangle$ $\forall u, v \in C_V^\infty(Q)$, as one obtains by straightforward calculation.

To obtain quasimodes for the perturbed operator $\mathscr{H}_\hbar^{\text{pert}}$, we use the same procedure as that described in §41. The set Λ will be the same as for the non-perturbed operator. As will be the coefficients $v_{s-1}(\phi, I, \delta)$, $E_s(I, \delta)$, $s < M$, for the change will affect only the terms of order $\hbar^M$ and higher. This implies the estimate (b) if we take into account (38.21).

The first perturbation in the equation of the type (41.14) will be with $s = M$. Let us write it out explicitly:

$$\xi v_{M-1}^{\text{pert}} + E_M^{\text{pert}} = \sum_{l=2}^{M-1} (\tilde{\mathscr{H}}_l^\square - E_l) v_{M-l} + \tilde{\mathscr{H}}_M^\square \cdot I + (-i)^M (\mathfrak{D}_{a,b})_0^\square,$$

using the commutation rule 36.4. We find $(\mathfrak{D}_{a,b})_0^\square = \beta_r^*(a \cdot b)$, and the change in E_M, which arose due to the perturbation, appears to be

$$E_M^{\text{pert}} - E_M = (-i)^M \int \beta_r^*(a \cdot b)|_{\mathscr{T}_I} \sigma^I.$$

To obtain the estimation (c), it is sufficient to estimate from below the following expression:

(e)
$$\int \beta_r^*(a \cdot b)|_{\mathscr{T}_{I_k}} \sigma^{I_k} - \int \beta_r^*(a \cdot b)|_{\mathscr{T}_{I_l}} \sigma^{I_l}$$
$$= \int_{\mathbb{T}^n} [A(\phi, I_k) - A(\phi, I_l)]\, d\phi^1 \wedge d\phi^2 \wedge \cdots \wedge d\phi^n,$$

where $A(\phi, I) = R^2(\langle \vartheta^*, I\rangle + d)\chi_3^2(I)\chi_4^2(\phi)$. Take $W_1 = B_\varepsilon(I^*)$. Then $\chi_3(I_k) = \chi_3(I_l) = 1$ if $I_k, I_l \in W_1$, and the right-hand side of (e) is equal to $R^2\langle \vartheta^*, I_k - I_l\rangle$. Take W_2, a neighbourhood of ϑ^* in S^{n-1}, so that $\langle \vartheta^*, \vartheta\rangle \geqq 1/2$ if $\vartheta \in W_2$. Then

(f)
$$|\langle \vartheta^*, I_k - I_l\rangle| \geqq |I_k - I_l|\,|\langle \vartheta^*, (I_k - I_l)/|I_k - I_l|\rangle| \geqq |I_k - I_l|/2$$

if $(I_k - I_l)/|I_k - I_l| \in W_2$. We have

(g)
$$E_k^{\text{pert}} - E_l^{\text{pert}} = (E_k^{\text{pert}} - E_k) - (E_l^{\text{pert}} - E_l) + (E_k - E_l)$$
$$= \hbar^M R^2 \langle \vartheta^*, I_k - I_l\rangle + P + \mathcal{O}(\hbar^{M+1}).$$

Here we denoted by P the contribution of the elements of (41.2) with $s > M$. One can easily deduce

(h)
$$|P| \leqq \text{const}\, \hbar^{M+1} R^2 |I_k - I_l|,$$

where const is independent of R. The estimates (f), (g), (h), and (42.9) give us

$$|E_k^{\text{pert}} - E_l^{\text{pert}}| \geqq (1/2)\hbar^M R^2 |I_k - I_l| - \text{const}\, \hbar^{M+1} R^2 |I_k - I_l|$$
$$- \text{const}\, \hbar^{M+1} \geqq c\hbar^{M+1}, \quad c > 0,$$

provided R is chosen sufficiently large, and $\hbar$ is sufficiently small. This proves (c) and, hence, Proposition 42.10. □

Notes to Chapter VII

Einstein (1917) was the first to formulate quantum conditions in a very geometrical form and they were later slightly developed in Keller and Rubinov (1960), and Maslov (1972) by introducing indices.

A set of quasimodes of positive density was constructed in Lazutkin (1973) for the case of a planar convex domain and in a more general case by Colin de Verdière (1977).

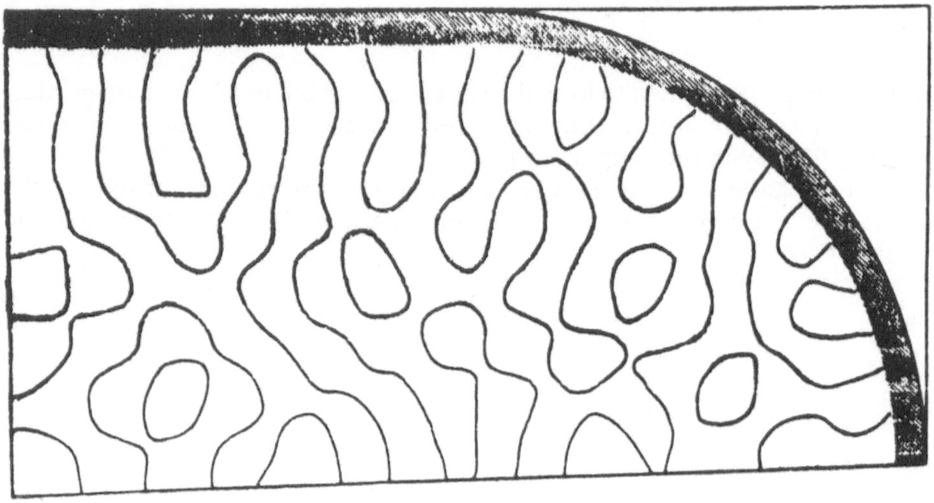

Fig. 62. Ergodic eigenfunction. Nodal curves $[\psi(x,y)=0]$ for the eigenfunction of the equation $(\Delta+k^2)\psi=0$ ($\psi=0$ on the boundary) with eigenvalue $k=50.158$ in the stadium composed by the square and two semicircles with radii 0.665 (only one quadrant being drawn) (by S.W. McDonald and A.N. Kaufman (1979) with permission)

Addendum

On the Asymptotic Properties of Eigenfunctions in the Regions of Chaotic Motion

By A. I. Shnirelman

Introduction

In the asymptotic theory of high-frequency eigenfunctions of elliptic operators the most attention was paid to the quasiclassic eigenfunctions. They are associated with the simplest invariant sets of the phase flow, namely stable closed orbits and invariant tori. Much less is known about other eigenfunctions. If, for example, the phase flow is ergodic on the constant energy surface (with respect to the Liouville measure); then none such sets exist, and the methods described above fail. But there remains the problem, what are the asymptotic properties of eigenfunctions with large numbers. This is the subject of the first section of the present Addendum. We treat the sequence of eigenfunctions as something like a random function, and estimate some quadratic functionals displaying the "mass distribution" of an eigenfunction. We prove that if the phase flow is ergodic, then almost all of the eigenfunctions are "uniformly distributed" (rigorous definition is given below).

In the second section we deal with the KAM situation. But our purpose is to investigate the eigenfunctions associated with the gaps between the invariant tori. We present the construction of such quasimodes. This construction is not completely explicit; it uses the "true" eigenfunctions as a raw material. We can only prove, that these quasimodes exist, and that they are numerous enough.

A nontrivial implication of our results is that in a case of a 2-dimensional torus and an operator on it, the principal part of which is close enough to the standard Laplacian, the whole sequence of eigenvalues is asymptotically multiple. This means that the splitting between the nearest eigenvalues decreases faster than any power of their number. This fact is a little strange, because there is no explicit symmetry in the problem.

The main subject of the study is the "mass" distribution of eigenfunctions. We shall obviously explain what we mean by that. Let $u_k(x)$ be an eigenfunction of a second order elliptic operator $\hat{L}$ on a manifold M with eigenvalue λ_k^2; then $u(x,t) = u_k(x)e^{i\lambda_k t}$ is a solution of the wave equation $u_{tt} = \hat{L}u$. Let us consider an instrument measuring the angle and position distribution of the wave energy flow. An example of such an instrument, for the electromagnetic waves, is our eye together with the corresponding part of our brain. It transforms the electromagnetic wave into our "visible picture" of the world. The energy flow depends on the position of the one looking and the direction of their sight. So, the "visible picture" is situated in S^*M, i.e. the manifold of unit cotangent vectors to M (we don't consider the colour, i.e. the frequency distribution, for the eigenfunctions are strictly monochromatic).

If $u_k(x)$ is a quasiclassical eigenfunction, i.e. it is locally of the form $u_k(x)=\sum_{p=1}^{P} a_p(x)e^{i\lambda_k(x)\varphi_p(x)}$, and the eye is at the point $x\in M$, then it will "see" P bright "stars" on the dark "sky" S^*M; their intensities are $|a_p(x)|^2$, and their directions are grad $\varphi_p(x)$. If we change our position on M, then the "stars" change their positions on the "sky" too. If we approach the caustic and cross it, then we'll see two to the "stars" approaching each other, flashing up brightly and then disappearing.

The situation is quite different in the opposite case of ergodic phase flow. We shall prove in Section AD1 that for most of the eigenfunctions $u_k(x)$ the "eye" will not see separate stars, but it will "see" the "sky" uniformly bright whatever position and direction we choose!

The visible "brightness of the sky" for the eigenfunction $u_k(x)$ is in fact some non-negative measure $\mu_k(dz)$ on S^*M. This measure is the main object of our study; it absorbs the whole of the rough structure of an eigenfunction. The similar notion called a "Wiegner function" was introduced by E. Wiegner (see the reference in [1]) many years ago. These are the oscillatory functions $E(x,\xi)$ in the phase space T^*M asymptotically (for $k\longrightarrow\infty$) coinciding with our measure $\mu_k(dz)$ "in the mean" (i.e. like functionals on the space of smooth functions). Our results, however, make sense only for the measures $\mu_k(dz)$. In particular, the statement that the quasimodes constructed in Section AD.2 are concentrated in some regions of the phase space means that corresponding measures are concentrated there.

Our main tool is the theory of pseudodifferential operators (p.d.o. theory). It absorbs most of the necessary asymptotical techniques, and therefore our proofs are simple enough. We use the most classical version of the p.d.o. theory [2], and almost nothing of the Fourier integral operators theory, except the Egorov theorem [3].

This work was done and the results announced about 1975 [4, 5]. Since that time the proofs of the results of Section AD.1 were published by several authors [6, 7, 8]; they are all based on similar ideas to our original one presented here. The proof of the results of Section AD.2 are published here for the first time.

I am very grateful to professor V. F. Lazutkin for his kind proposal to participate in this book and for his infinite patience.

AD.1. Ergodic Properties of Eigenfunctions

All of our results are valid for an arbitrary positive self-adjoint elliptic operator on a compact manifold, but we shall restrict ourselves to the case of a Laplace–Beltrami operator Δ on a compact Riemanian manifold M without boundary.

Let $u_1, u_2, \ldots$ be the sequence of eigenfunctions of $(-\Delta)$, orthonormal in $L^2(M)$, and let $\lambda_1^2 \leqq \lambda_2^2 \leqq \ldots$ be the sequence of corresponding eigenvalues (it is convenient to denote by λ_j the eigenfrequencies).

Let $M' \subset M$ be an open set; our primary question is, what may be said about the quantities

$$\int_{M'} |u_k(x)|^2 \, dx$$

when $k \longrightarrow \infty$? Note that $\int_M |u_k(x)|^2 \, dx \equiv 1$. In other words, what may be said about the "mass distribution" of the eigenfunction $u_k(x)$ for large k? If, for example, $M = \mathbb{T}^n = \mathbb{R}^n/\mathbb{L}$ is an n-dimensional torus ($\mathbb{L}$ is some lattice in $\mathbb{R}^n$), then $u_k(x) = \sqrt{2}|M|^{-1/2} \sin(b_k, x)$, where b_k is a covector of the dual lattice $\mathbb{L}^*$. Hence, $\int_{M'} |u_k(x)|^2 \, dx \approx |M'|/|M|$ (here and below $|M'| =$ volume (M') for each domain $M' \subseteq M$). This means that the eigenfunctions $u_k(x)$ are asymptotically uniformly distributed on $\mathbb{T}^n$. Another example illustrates different behaviour of eigenfunctions. It is a subsequence $\{u_k\}$ of eigenfunctions concentrated near the closed stable geodesic line γ on M [15]. In this case, $\int_{M'} |u_{k_j}(x)|^2 \, dx \longrightarrow \int_{M'} \delta_\gamma(x) \, dx$ $(j \longrightarrow \infty)$, where $\delta_\gamma(x) \, dx$ denotes a measure distributed uniformly along γ. We see that in this case the distribution is extremely remote from the uniform one.

In the case of non-integrable phase flow, the eigenfunctions cann't so far be found either precisely or asymptotically. But there is a class of cases, for which we can find the asymptotical distribution of the eigenfunctions.

Let $\langle , \rangle_x$ be a scalar product in $T_x^* M$ (M is a Riemanian manifold); let $\Omega = S^*M = \{(x, \xi) \in T^*M \mid \langle \xi, \xi \rangle = 1\}$ be a fibre bundle of unit covectors on M. Let $g(x, \xi) = (\partial H/\partial \xi, -\partial H/\partial x)$ be a Hamiltonian vector field on T^*M for $H(x, \xi) = \langle \xi, \xi \rangle_x^{1/2} = |\xi|$. Vector field g is tangent to Ω, for $gH \equiv 0$, and therefore it induces a tangent vector field $g(z)$ on Ω, $z \in \Omega$. Let $G_t : \Omega \longrightarrow \Omega$, $t \in \mathbb{R}^1$, be a one-parameter transformation group (flow) generated on Ω by the vector field $g(z)$. It conserves the volume element $dz = (dx \wedge d\xi)/dH$ on Ω (here $dx \wedge d\xi$ is the Liouville volume element in T^*M, and dz is a Leray–Gelfand $(2n-1)$-form, i.e. such a form that $dz \wedge dH = dx \wedge d\xi$).

The main result of this section is the following

Theorem AD1.1. *Suppose that the flow G_t is ergodic on Ω with respect to the measure dz. Then, there exists a subsequence $\{u_{k_j}\}_{j=1}^\infty$ of density 1, such that for every open set $M' \subseteq M$ with a smooth boundary*

$$\int_{M'} |u_{k_j}(x)|^2 \, dx \longrightarrow |M'|/|M| \quad (j \longrightarrow \infty).$$

Here the expression "*subsequence of density* 1" means that $\lim_{j \to \infty} (k_j/j) = 1$.

Our theorem means that if the phase flow is ergodic, then "almost all" of the eigenfunctions are asymptotically uniformly distributed on M. (The inverse theorem is not true; see the above example $M = \mathbb{T}^n$).

We shall prove a more general assertion concerning the uniform distribution of u_k on Ω. Its formulation and proof are based on the theory of pseudo-differential operators (p.d.o. theory; see [2, 3]).

Let $\hat{A}$ be a p.d.o. on M of order 0 with principal symbol $A(x, \xi)$, $A(x, t\xi) = A(x, \xi)$ for all $t > 0$ (we shall further denote the p.d.o. by putting a cap ($\hat{}$) over its symbol as it is usually done in quantum theory). The symbol is completely determined by its restriction $A(z)$ onto Ω, for it is homogeneous in ξ of order 0. Let us consider the sequence $(\!(\hat{A}u_k, u_k)\!)$, $k = 1, 2, \ldots$, where $(\!(\ \ )\!)$ denotes the scalar product in $L^2(M)$. It is bounded, and

$$\limsup_{k\to\infty} |(\!(\hat{A}u_k, u_k)\!)| \leqq \sup_{z\in\Omega} |A(z)|;$$

see [3].

If $A^{(1)}, A^{(2)}$ are two operators with the same principal symbol $A(z)$ (and in general different subordinate terms), then $\lim_{k\to\infty} |(\!((A^{(1)} - A^{(2)})u_k, u_k)\!)| = 0$. Hence, the asymptotic behaviour of the sequence $(\!(\hat{A}u_k, u_k)\!)$ depends only on the principal symbol $A(z)$.

Given a symbol $A(z)$, we may construct a p.d.o. $\hat{A}$ (see [3]). This construction requires some auxiliary elements, such as partition of unity on M, local coordinate systems etc. If we choose these elements in a different manner, we obtain a p.d.o. A' such that $(A' - A)$ is a p.d.o. of order (-1); this difference does not affect the asymptotic behaviour of the sequence $(\!(\hat{A}u_k, u_k)\!)$. Let us fix all the elements of the construction. For each symbol $A(z)$ we get a p.d.o. $\hat{A} = \mathrm{Op}(A)$.

The correspondence $\mathrm{Op}: A(z) \longrightarrow \hat{A}$ is linear; hence, the quantity $(\!(\hat{A}u_k, u_k)\!) = (\!(\mathrm{Op}Au_k, u_k)\!)$ linearly depends on the function $A(z)$, i.e. $(\!(Au_k, u_k)\!)$ is a linear functional on the space of smooth symbols $A(z)$.

We may therefore set

$$(\!(\hat{A}u_k, u_k)\!) = \int_\Omega U_k(z) A(z)\, dz,$$

where $U_k(z)$ is some function (in general, distribution) on Ω. We shall investigate the asymptotic behaviour of the functions $U_k(z)$ for $k \longrightarrow \infty$.

Lemma AD.1.2. *There exists a sequence of positive measures $\mu_k(dz)$ on Ω, $\mu_k(\Omega) = 1$, such that for each $A(z) \in C^\infty(\Omega)$,*

$$\text{(AD.1.2')} \qquad \left| \int_\Omega U_k(z) A(z)\, dz - \int_\Omega A(z) \mu_k(dz) \right| \longrightarrow 0 \quad (k \longrightarrow \infty)$$

uniformly for $A(z) \in B$, where B is an arbitrary bounded set in $C^\infty(\Omega)$.

Proof. The Gårding inequality (see [3]) states that for each p.d.o. $\hat{A}$ of order 0 with real symbol $A(z)$, $z \in \Omega$, for all $u(x) \in H_0$,

$$\mathrm{Re}(\!(\hat{A}u, u)\!) \geqq \inf_{z\in\Omega} A(z) \|u\|_0^2 - C\|u\|_{-1/2}^2,$$

$$|\mathrm{Im}(\!(\hat{A}u, u)\!)| \leqq C\|u\|_{-1/2}^2,$$

where C is bounded provided $A(z) \in B$ for each bounded set $B \subset C^\infty(\Omega)$. (Recall that H_α is a Sobolev space, $\|u\|_\alpha^2 = \|u\|_{H_\alpha}^2 = \|\Delta^{\alpha/2}u\|_{L^2}^2$, where Δ is the Laplace operator on M; $H_0 = L^2$.) If, in particular, $u = u_k$ is the k-th eigenfunction of $(-\Delta)$, then

$$\operatorname{Re}\langle\!\langle \hat{A}u_k, u_k \rangle\!\rangle \geqq \inf_{z\in\Omega} A(z) - C\lambda_k^{-1},$$

$$|\operatorname{Im}\langle\!\langle \hat{A}u_k, u_k \rangle\!\rangle| \leqq C\lambda_k^{-1};$$

i.e.,

$$\operatorname{Re}\int_\Omega U_k(z)A(z)\,dz \geqq \inf_{z\in\Omega} A(z) - C\lambda_k^{-1},$$

$$\left|\operatorname{Im}\int_\Omega U_k(z)\,dz\right| \leqq C\lambda_k^{-1}.$$

If $A(z) \equiv 1$, then $\hat{A} = \operatorname{Op} A = \mathit{Identity}$, and $\langle\!\langle \hat{A}u_k, u_k \rangle\!\rangle \equiv 1$. Hence,

$$\int_\Omega U_k(z)\,dz \equiv 1.$$

Let $K_j(z_1, z_2) \in C^\infty(\Omega \times \Omega)$, $j = 1, 2, \ldots$, be such kernels, that

(i) $K_j(z_1, z_2) \geqq 0$;

(ii) $\int_\Omega K_j(z_1, z_2)\,dz_1 \equiv 1$;

(iii) $\int_\Omega K_j(z_1, z_2)\,dz_2 \equiv 1$;

(iv) $K_j(z_1, z_2) \longrightarrow \delta(z_1 - z_2)$ in $D'(\Omega \times \Omega)$, if $j \longrightarrow \infty$.

Let

$$U_k^{(j)}(z) = \int_\Omega K_j(z, w)U_k(w)\,dw;$$

then $U_k^{(j)} \longrightarrow U_k(z)$ $(j \longrightarrow \infty)$ in $D'(\Omega)$. For each fixed j,

$$\int_\Omega U_k^{(j)}(z)A(z)\,dz = \int_\Omega U_k(w)A^{(j)}(w)\,dw,$$

where

$$A^{(j)}(w) \equiv \int_\Omega K_j(z, w)\,dz.$$

If $A(z)$ is in a bounded set $B \subset D'(\Omega)$, then $A^{(j)}$ is in some bounded set B_j in $C^\infty(\Omega)$, and hence

$$\operatorname{Re}\int_\Omega U_k^{(j)}(z)A(z)\,dz \geqq \inf_{z\in\Omega} A^{(j)}(z) - C_j\lambda_k^{-1}. \tag{AD.1.3}$$

$$\left|\operatorname{Im}\int_\Omega U_k^{(j)}(z)A(z)dz\right| \leqq C_j\lambda_k^{-1},$$

$$\int_\Omega U_k^{(j)}(z)\,dz \equiv 1.$$

(iv) also implies, that $A^{(j)} \longrightarrow A$ $(j \longrightarrow \infty)$ in $D'(\Omega)$. Taking in (AD.1.3) $A(z) = \delta(z-w)$, we see, that

$$\operatorname{Re} U_k^{(j)}(z) \geqq C_j\lambda_k^{-1}, \qquad |\operatorname{Im} U_k^{(j)}(z)| \leqq C_j\lambda_k^{-1}$$

for all $w \in \Omega$.

We may choose a sequence $\{j_k\}_{k=1}^\infty$ in such a manner that $j_k \longrightarrow \infty$ and $C_{j_k}\lambda_k^{-1} \longrightarrow 0$, if $k \longrightarrow \infty$. Now, let us set

$$m_k(z) = \operatorname{Re} U_k^{(j_k)}(z) + C_{j_k}\lambda_k^{-1};$$

$$\mu_k(dz) = m_k(z)\,dz \Big/ \int_\Omega m_k(z)\,dz.$$

This sequence of measures possesses all the properties claimed in Lemma AD.1.2. □

It is evident that the sequence $\{\mu_k\}$ of measures satisfying (AD.1.2′) is asymptotically unique, i.e. if $\{\mu_k'\}$ is another such sequence, then

$$\int_\Omega A(z)\mu_k(dz) - \int_\Omega A(z)\mu_k'(dz) \longrightarrow 0 \quad (k \longrightarrow \infty)$$

for each $A(z) \in C^\infty(\Omega)$.

It is the measures $\mu_k(dz)$ that express the "energy" distribution of the eigenfunction $u_k(x)$ both in position and direction (see the introduction to the present Addendum).

Each p.d.o. $\hat{A}$ describes some "measuring instrument", and the quantity $(\!(\hat{A}u_k, u_k)\!)$ may be regarded as a result of the measurement. The analysis of real instruments, for example, the optic ones, shows that, in fact, for each instrument such an operator may be found (possibly, it may belong to a more wide class of operators than the classical p.d.o). This question is close to the measuring problem in quantum theory, but is, of course, much simpler.

Now we can formulate the main result of this section.

Theorem AD.1.4. *Suppose, that the phase flow G_t is ergodic on Ω with respect to the measure dz. Then there exists a subsequence $\{\mu_{k_j}\}$ of density* 1, *such that for each function $A(z) \in C^\infty(\Omega)$,*

(AD.1.4′) $$\int_\Omega A(z)\mu_{k_j}(dz) \longrightarrow \frac{1}{|\Omega|}\int_\Omega A(z)dz, \quad j \longrightarrow \infty.$$

This theorem immediately implies Theorem AD.1.1.

Proof of Theorem AD.1.1. Let $M' \subset M$ be an open set with a smooth boundary $\partial M'$, and $\varepsilon > 0$. Let $a_1(x)$, $a_2(x)$ be such real functions on M, that (i) $0 \leqq a_1(x) \leqq a_2(x) \leqq 1$ for all $x \in M$; (ii) $a_1(x) \equiv 0$, $x \in M \backslash M'$; (iii) $a_2(x) \equiv 1$, $x \in M'$; (iv) $(1/|M|) \int_M (a_2(x) - a_1(x))dx < \varepsilon$. Such functions exist for all $\varepsilon > 0$. Let $\hat{a}_i$, $i = 1, 2$, be the operators of multiplication by $a_i(x)$: $\hat{a}_i v(x) = a_i(x) \cdot v(x)$. These operators are p.d.o. of order 0 with the symbols $a_i(z) = a_i(x, \omega) \equiv a_i(x)$, $\omega = \xi/|\xi|$. Let $\{\mu_{k_j}\}$ be the subsequence of density 1, whose existence is stated in Theorem AD.1.4, and let $\{u_{k_j}\}$ be the corresponding subsequence of eigenfunctions. Then

$$(\!(\hat{a}_i u_{k_j}, u_{k_j})\!) \sim \int_\Omega a_i(z) \mu_{k_j}(dz) \longrightarrow \frac{1}{|\Omega|} \int_\Omega a_i(z)\, dz = \frac{1}{|M|} \int_M a_i(x)\, dx, \quad (j \longrightarrow \infty).$$

On the other hand,

$$(\!(\hat{a}_i u_{k_j}, u_{k_j})\!) = \int_M a_i(x) |u_{k_j}(x)|^2\, dx.$$

This implies that

$$|M'|/|M| - \varepsilon \leqq \liminf_{j \to \infty} \frac{1}{|M|} \int_{M'} |u_{k_j}(x)|^2\, dx$$

$$\leqq \limsup_{j \to \infty} \frac{1}{|M|} \int_{M'} |u_{k_j}(x)|^2\, dx \leqq |M'|/|M| + \varepsilon.$$

This is true for all $\varepsilon > 0$, and hence

$$\lim_{j \to \infty} \frac{1}{|M|} \int_{M'} |u_{k_j}(x)|^2\, dx = |M'|/|M|;$$

Thus, Theorem AD.1.1 is proved. □

In the rest of this section we shall prove Theorem AD.1.2. Our next lemma concerns the uniform distribution of the measures μ_k "in the mean".

Lemma AD.1.5. *For each $A(z) \in C^\infty(\Omega)$,*

$$\frac{1}{N} \sum_{k=1}^{N} \int_\Omega A(z) \mu_k\, dz) \longrightarrow \frac{1}{|\Omega|} \int_\Omega A(z)\, dz \quad (N \longrightarrow \infty)$$

uniformly, if A is in a bounded set $B \subset C^\infty(\Omega)$.

Proof: Reproducing the well known proof of the asymptotical formula for the number of eigenvalues $\leqq \lambda$ [9.10], let us set

$$P_k(s) = e^{-\lambda_k^2 s} \Big/ \sum_{j=\infty}^{\infty} e^{-\lambda_j^2 s}.$$

We shall first prove that

$$\sum_{k=1}^{\infty} P_k(s) \int_\Omega A(z) \mu_k(dz) \longrightarrow \frac{1}{|\Omega|} \int_\Omega A(z)\, dz,$$

if $s \longrightarrow 0$. Consider the following initial-value problem for the heat equation on M:

$$\frac{\partial u(x,t)}{\partial t} = \Delta_x u(x,t),$$

$$u(x,t)|_{t=0} = \delta(x-x').$$

Let $G(x,x',t)$ be the solution of this problem. Then $G(x,x',t) = \sum_k u_k(x)\bar{u}_k(x')e^{-\lambda_k^2 t}$.
Let $\hat{A}$ be a p.d.o. of order 0 with the symbol $A(z)$. Then

$$\text{(AD.1.6)} \qquad \int_M \hat{A}_{x'} G(x,x',t)|_{x=x'}\, dx = \sum_k \langle\!\langle \hat{A}u_k, u_k \rangle\!\rangle e^{-\lambda_k^2 t}$$

$$= \left(\sum_k e^{-\lambda_k^2 t}\right) \sum_k P_k(t) \langle\!\langle \hat{A}u_k, u_k \rangle\!\rangle.$$

If $t \searrow 0$, then $P_k(t) \longrightarrow 0$ for each individual k, and

$$\text{(AD.1.7)} \qquad \sum_k P_k(t) \langle\!\langle \hat{A}u_k, u_k \rangle\!\rangle \sim \sum_k P_k(t) \int_\Omega A(z)\mu_k(dz).$$

On the other hand, if $t \searrow 0$, then

$$G(x,x',t) \sim (4\pi t)^{-n/2} \exp\left(-\frac{|x-x'|}{4t}\right),$$

where $|x-x'|$ is the Riemanian distance between x and x', and $G(x,x',t)$ is exponentially small, if $|x-x'| > const$ and $t \longrightarrow 0$.

We may asymptotically compute $\int_M \hat{A}_{x'} G(x,x',t)|_{x=x'} dx$ for $t \longrightarrow 0$, using the local coordinate system with the origin at the point x:

$$\hat{A}_{x'} G(x,x',t)| \sim (2\pi)^{-n} \int e^{i\langle (x'-x),\xi\rangle} A(x',\xi) \cdot \mathcal{F}_{x' \to \xi} G(x,\xi,t)\, d\xi |_{x'=x}$$

$$\sim (2\pi)^{-n} \int A(x,\xi) e^{-|\xi|^2 t}\, d\xi = (2\pi t)^{-n/2} \cdot \frac{1}{|S^{n-1}|} \int_{S^{n-1}} A(x,\omega)\, d\omega,$$

where S^{n-1} is a unit $(n-1)$-sphere in $T_x^* M$, $d\omega$ is its angle element, $|S^{n-1}| = \int_{S^{n-1}} d\omega$, and $\mathcal{F}_{x' \to \xi}$ denotes the Fourier transform in the variable x' (ξ is its dual variable). Hence,

$$\text{(AD.1.8)} \qquad \int_M \hat{A}_{x'} G(x,x',t)|_{x=x'}\, dx \sim (2\pi t)^{-n/2} \frac{1}{|S^{n-1}|} \int_M dx \int_{S_x^*} A(x,\omega)\, d\omega$$

$$= (2\pi t)^{-n/2} \frac{1}{|S^{n-1}|} \int_\Omega A(z)\, dz.$$

If we take $\hat{A} = \mathit{Identity}$, i.e. $A(z) \equiv 1$, we see, that

$$\text{(AD.1.9)} \qquad \sum_k e^{-\lambda_k^2 t} \sim (2\pi t)^{-n/2} |M|.$$

If we compare (AD.1.6)–(AD.1.9), we see that

$$|M|(2\pi t)^{-n/2}\sum_k P_k(t)\int_\Omega A(z)\mu_k(dz) \sim (2\pi t)^{-n/2}\frac{1}{|S^{n-1}|}\int_\Omega A(z)\,dz;$$

(AD.1.10)
$$\sum_k P_k(t)\int_\Omega A(z)\mu_k(dz) \longrightarrow \frac{1}{|\Omega|}\int_\Omega A(z)\,dz \quad (t \longrightarrow 0).$$

Now, let us set

$$a_k = \int_\Omega A(z)\mu_k(dz); \qquad a = \frac{1}{|\Omega|}\int_\Omega A(z)dz; \qquad w(\sigma) = \sum_{\lambda_k^2<\sigma} a_k.$$

Then (AD.1.10) means that

$$\int_0^\infty e^{-\sigma t}\,dw(\sigma) = a\sum_k e^{-\lambda_k^2 t}(1+o(1))$$
$$= a(2\pi t)^{-n/2}|M|(1+o(1)) \quad (t \longrightarrow 0).$$

By the Karamata tauberian theorem [3],

$$w(\sigma) = a|M|\sigma^{n/2}/(n/2)\Gamma(n/2)(2\pi)^{n/2}(1+o(1)) = aN(\sigma)(1+o(1)) \quad (\sigma \longrightarrow \infty),$$

where $N(\sigma) = \sum_{\lambda_k^2<\sigma} 1$. Hence,

$$\frac{1}{N(\sigma)}\sum_{k=1}^{N(\sigma)}\int_\Omega A(z)\mu_k(dz) \longrightarrow \frac{1}{|\Omega|}\int_\Omega A(z)\,dz \quad (\sigma \longrightarrow \infty);$$

this is equivalent to the statement of the lemma. □

Our next lemma concerns the behaviour of the measures μ_k with respect to the phase flow $G_t : \Omega \longrightarrow \Omega$.

Lemma AD.1.11. *For each $A(z) \in C^\infty(\Omega)$, $t \in \mathbb{R}$,*

$$\int_\Omega (A(z) - A(G_t^{-1}(z)))\mu_k(dz) \longrightarrow 0 \quad (k \longrightarrow \infty);$$

convergence is uniform if $|t| \leq t_0$ and $A \subset B$, for arbitrary $t_0 > 0$ and arbitrary bounded set $B \subset C^\infty(\Omega)$.

In other words, the measures μ_k are asymptotically ($k \longrightarrow \infty$) invariant under the Hamiltonian phase flow.

Proof. Let $\hat{\Lambda} = +\sqrt{-\Delta}$; it is a positive p.d.o. of order 1 with the principal symbol $\Lambda(x,\xi) = |\xi|$. Let us consider the following initial-value problem:

$$\frac{\partial u}{\partial t} = i\hat{\Lambda}u,$$
$$u|_{t=0} = u_0.$$

It is a first order hyperbolic p.d. problem; its solution is $u_t = e^{it\hat{\Lambda}}u_0$. The Egorov theorem [3] states, that if $\hat{A}$ is a p.d.o. of arbitrary order with principal symbol $A(x, \xi)$, then $A_t = e^{-it\hat{\Lambda}}\hat{A}e^{it\hat{\Lambda}}$ is also a p.d.o., and its principal symbol $A_t(x, \xi) = A_0(G_t^{-1}(x, \xi))$. (All the estimates are uniform, if t and A are in bounded sets of $\mathbb{R}$ and $C^\infty(\Omega)$.) Here G_t is a Hamiltonian phase flow in T^*M generated by the Hamiltonian function $H = \Lambda(x, \xi) = |\xi|$. Note that H is homogeneous in ξ of order 1. This implies that G_t transforms the rays $\{(x, t\xi),\ t > 0\}$ into rays of the same kind; on these rays G_t is homogeneous in ξ of order 1 and conserves $\Lambda(x, \xi) = |\xi|$.

Hence, G_t is correctly defined on $\Omega = \{(x, \xi) \mid |\xi| = 1\}$ and conserves the volume element $dz = (dx \wedge d\xi / d|\xi|)$. If $\hat{A}$ is a p.d.o. of order 0 with the symbol $A(z)$, $z \in \Omega$, then $\hat{A}_t = e^{-it\hat{\Lambda}}\hat{A}e^{it\hat{\Lambda}}$ has a symbol $A_t(z) = A(G_t^{-1}(z))$.

Now,

$$\begin{aligned}\int_\Omega A(G_t^{-1}(z))\mu_k(dz) &= \int_\Omega A_t(z)\mu_k(dz) = (\!(\hat{A}_t u_k, u_k)\!) + o(1) \\ &= (\!(e^{-it\hat{\Lambda}}\hat{A}e^{it\hat{\Lambda}}u_k, u_k)\!) + o(1) = (\!(\hat{A}e^{it\hat{\Lambda}}u_k, e^{it\hat{\Lambda}}u_k)\!) + o(1) \\ &= (\!(\hat{A}e^{it\lambda_k}u_k, e^{it\lambda_k}u_k)\!) + o(1) = (\!(\hat{A}u_k, u_k)\!) + o(1) \\ &= \int_\Omega A(z)\mu_k(dz) + o(1) \quad (k \longrightarrow \infty)\end{aligned}$$

uniformly on the bounded sets $B \subset C^\infty$ and $|t| < t_0$. The lemma is proved. □

Proof of Theorem AD.1.4. Let $A(z)$ be an arbitrary fixed smooth function on Ω. We shall first prove that there exist a subsequence $\{\mu_{k_j}\}$ of density 1 such that

$$\int_\Omega A(z)\mu_{k_j}(dz) \longrightarrow \frac{1}{|\Omega|}\int_\Omega A(z)\,dz \quad (j \longrightarrow \infty). \tag{AD.1.12}$$

The Birkhoff theorem [11] together with the ergodicity of G_t implies that for almost all $z \in \Omega$,

$$\lim_{T \to \infty} \frac{1}{T}\int_0^T A(G_t^{-1}(z))\,dt = \frac{1}{|\Omega|}\int_\Omega A(z)\,dz = \langle A \rangle. \tag{AD.1.13}$$

For $T > 0$, $\varepsilon > 0$ we shall define a domain

$$\Omega_{T,\varepsilon} = \left\{ z \in \Omega \,\middle|\, \left| \frac{1}{T}\int_0^T A(G_t^{-1}(z))\,dt - \langle A \rangle \right| < \varepsilon \right\}.$$

(AD.1.13) implies that for each $\varepsilon > 0$, $\delta > 0$ there exists $T > 0$ such that

$$|\Omega_{T,\varepsilon}| > \left(1 - \frac{\delta}{2}\right)|\Omega|.$$

We shall suppose that the domain $\Omega_{T,\varepsilon}$ has a smooth boundary; otherwise we may find a subdomain $\tilde{\Omega}_{T,\varepsilon} \subset \Omega_{T,\varepsilon}$ with a smooth boundary such that $|\tilde{\Omega}_{T,\varepsilon} \backslash \Omega_{T,\varepsilon}| < (\delta/2)|\Omega|$. By the Lemma AD.1.11, there exists k_0 such that for all

$k \geqq k_0$, $0 \leqq t \leqq T$,

$$\left| \int_\Omega A(G_t^{-1}(z))\mu_k(dz) - \int_\Omega A(z)\mu_k(dz) \right| < \frac{\delta}{2};$$

hence,

$$\left| \int_\Omega \frac{1}{T} \int_0^T A(G_t^{-1}(z))\, dt \mu_k(dz) - \int_\Omega A(z)\mu_k(dz) \right| < \frac{\delta}{2}.$$

Let

$$B(z) = \frac{1}{T} \int_0^T A(G_t^{-1}(z))\, dt - \langle A \rangle.$$

Then

$$\begin{aligned} \int_\Omega A(z)\mu_k(dz) - \langle A \rangle &= \int_\Omega A(z)\mu_k(dz) - \int_\Omega \langle A \rangle \mu_k(dz) \\ &= \int_\Omega B(z)\mu_k(dz) + R_k, \end{aligned}$$

where $|R_k| < \delta/2$ for all $k > k_0$.

Let us estimate

$$\int_\Omega B(z)\mu_k(dz) = \int_{\Omega_{T,\varepsilon}} \cdots + \int_{\Omega \backslash \Omega_{T,\varepsilon}} \cdots$$

By Lemma AD.1.5, we may find n_0 such that for all $N > n_0$,

$$\left| \frac{1}{N} \sum_{k=k_0+1}^{k_0+N} \int_{\Omega \backslash \Omega_{T,\varepsilon}} \mu_k(dz) - |\Omega \backslash \Omega_{T,\varepsilon}|/|\Omega| \right| < \frac{\delta}{2};$$

hence,

(AD.1.14)
$$\frac{1}{N} \sum_{k=k_0+1}^{k_0+N} \int_{\Omega \backslash \Omega_{T,\varepsilon}} \mu_k(dz) < \delta.$$

Each term in the left part of (AD.1.14) is non-negative. Hence,

$$\#\left\{ k \,|\, k_0 + 1 \leqq k \leqq k_0 + N, \int_{\Omega \backslash \Omega_{T,\varepsilon}} \mu_k(dz) < \delta^{1/2} \right\} > N(1 - \delta^{1/2}).$$

(Here $\#K$ means the number of elements of the finite set K.) For such k,

$$\left| \int_{\Omega_{T,\varepsilon}} B(z)\mu_k(dz) \right| < \varepsilon|\Omega|$$

$$\left| \int_{\Omega \backslash \Omega_{T,\varepsilon}} B(z)\mu_k(dz) \right| \leqq \sup_{z \in \Omega} |B(z)|\, |\Omega \backslash \Omega_{T,\varepsilon}| \leqq 2 \sup_{z \in \Omega} |A(z)| \delta^{1/2},$$

and

$$\left|\int_\Omega A(z)\mu_k(dz) - \langle A\rangle\right| < \varepsilon|\Omega| + \frac{\delta}{2} + 2\sup_{z\in\Omega}|A(z)|\delta^{1/2}.$$

Let

$$K(a,N) = \left\{k \leqq N \,\middle|\, \left|\int_\Omega A(z)\mu_k(dz) - \langle A\rangle\right| < a\right\}. \tag{AD.1.15}$$

We have already proved that for each $a>0$, $b>0$

$$\liminf_{N\to\infty}\frac{1}{N}\#K(a,N) > 1-b;$$

This implies that

$$\lim_{N\to\infty}\frac{1}{N}\#K(a,N) = 1. \tag{AD.1.16}$$

For each sequence $a_j \searrow 0$, let us define the set $K \subset \mathbb{Z}^+$,

$$K = \bigcup_{j=1}^{\infty}(K(a_j, 2^j)\cap\{k\,|\,2^{j-1}\leqq k\leqq 2^j\}).$$

If we take a sequence $\{a_j\}$ decreasing sufficiently slowly, we obtain by (AD.1.16) a sequence $K\subset\mathbb{Z}^+$ of density 1. By (AD.1.15),

$$\lim_{\substack{k\to\infty\\ k\in K}}\left|\int_\Omega A(z)\mu_k(dz) - \langle A\rangle\right| = 0;$$

this means, that $K=\{k_j\}$ is the required subsequence of density 1, such that (AD.1.12) is true.

Subsequence K depends on the function $A(z)$, $K=K(a)$. Let $\{A_j(z)\}_{j=1}^{\infty}$ be a sequence of smooth functions on Ω dense in $C^\infty(\Omega)$, and $p_1 < p_2 < \cdots$, $p_j\in\mathbb{Z}^+$. Let us define the sequence

$$\bar{K} = \bigcap_{j=1}^{\infty}(K(A_j)\cup\{k\,|\,1\leqq k\leqq p_j\}).$$

If the sequence $\{p_j\}$ increases sufficiently rapidly, then the sequence $\bar{K}$ has density 1. For each A_j,

$$\lim_{\substack{k\to\infty\\ k\in\bar{K}}}\left|\int_\Omega A_j(z)\mu_k(dz) - \langle A\rangle\right| = 0,$$

because $(k\in K, k>p_j)\Rightarrow(k\in K(A_j))$. But $\{A_j\}$ is a dense set in $C^\infty(\Omega)$; hence, for each $A(z)\in C^\infty$,

$$\lim_{\substack{k\to\infty\\ k\in\bar{K}}}\left|\int_\Omega A(z)\mu_k(dz) - \langle A\rangle\right| = 0.$$

Thus $\bar{K}$ is the desired sequence of density 1, and Theorem AD.1.5 is completely proved. □

If we could find numerically some high-number eigenfunctions in the case when the conditions of Theorem AD.1.1 are satisfied, it would be interesting to compare their properties with the statement of this theorem. In fact, it was done for the Laplace operator with Dirichlet boundary conditions in the flat domain, in which the billiard flow is ergodic [12] (Fig. 62). Theorem AD.1.1 seems to be true in this case, although its proof must require much more technical efforts. The picture obtained in this work displays the nodal lines of the high-frequency eigenfunction in the "stadium" domain, in which the billiard flow is really ergodic [16]. The eigenfunction in this domain seems to be close to homogeneous and isotropic, in accordance with Theorem AD.1.4. It will be very interesting to do the same computation for a compact surface of negative curvature without boundary (computations may be done in the fundamental domain of a discrete group of motions of the Lobachevsky plane).

AD.2. Quasimodes Associated with the Gaps Between Invariant Tori

The quasimodes constructed in the preceding chapters of this book are of special kind. They are associated with invariant tori of the phase flow. In this section we shall describe other types of quasimode, which are concentrated in the gaps between invariant tori. Of course, this makes sense only in the 2-dimensional case, for only in this case do the invariant tori divide locally the constant energy surface in the phase space.

An immediate consequence of the existence of such quasimodes is the fact that the spectrum is asymptotically multiple. Namely, for each $N>0$, the distance between the eigenvalue λ_k^2 and its closest neighbour (λ_{k-1}^2 or λ_{k+1}^2) decreases faster than k^{-N}. The "quasiclassic" quasimodes constructed in the preceding chapters, are also asymptotically multiple, but they form only a subsequence of the spectrum having positive density <1. Our result is stronger: the whole sequence of eigenfunctions is asymptotically multiple. (We guess that in fact, in the generic case, this sequence is asymptotically *double*.)

Let M be a 2-dimensional torus with the coordinates $x=(x_1,x_2)$ on it, $0\leqq x_i\leqq 2\pi$. Let $\hat{D}$ be a positive symmetric second order operator on M with smooth coefficients. Suppose, that its principal part is close to the negative of the standard Laplace operator (e.g., $\hat{D}$ may be a negative of the Laplace–Beltrami operator for the metric close to the standard one; another example is the Schrödinger operator with arbitrary smooth potential). Let $\hat{\Lambda}=+\sqrt{\hat{D}}$ and let $\Lambda_1(x,\xi)$ be its principal symbol. $\Lambda_1(x,\xi)$ is homogeneous in ξ of order 1 and close to $|\xi|$.

Let us consider the phase flow $G_t\colon T^*M\longrightarrow T^*M$ generated by the Hamiltonian vector field $\dot{x}=\partial\Lambda_1/\partial\xi$, $\dot{\xi}=-\partial\Lambda_1/\partial x$. This flow transforms the rays

$\{(x, t\xi) | t > 0\}$ into rays of the same kind and preserves $\Lambda_1(x, \xi)$. Hence, G_t is correctly defined on the space of the rays; we identify it with $\Omega = \{(x, \xi) | \Lambda_1(x, \xi) = 1\}$.

If $\Lambda_1(x, \xi)$ is sufficiently close to $|\xi|$, then the KAM theorem (see above) states the existence of invariant tori T_ω of the flow G_t on Ω. ω is the rotation number of the flow restricted on the invariant torus, and such tori exist for all the rotation numbers ω which are slowly approximable by rational numbers. The tori T_ω are close to the invariant tori of the unperturbed system having the form $\xi_2/\xi_1 = \omega$, and form a set of positive measure in Ω. In what follows, $\tilde{T}_\omega$ denotes the following conic set: $\tilde{T}_\omega = \bigcup_{t>0} \{(x, t\xi) | (x, \xi) \in T_\omega\} \subset T^*M$.

Let $T_1 = T_{\omega_1}$, $T_2 = T_{\omega_2}$ be two invariant tori, dividing Ω in two domains; denote them Ω_1 and Ω_2. Let $\tilde{\Omega}_i = \bigcup_{t>0} \{(x, t\xi) | (x, \xi) \in \Omega_i\}$. We shall consruct below in section AD.3 an operator $\hat{Q}$, which may be called a "quasi-projector" associated with Ω_1. Its properties are summarized in the following lemma.

Lemma AD.2.1. *There exists an operator $\hat{Q}$, bounded in H_s for all s, such that*

(i) *$\hat{Q}$ "almost commutes" with $\hat{\Lambda}$, i.e. for all s, N there exists $C_{s,N} > 0$ such that*

$$\| [\hat{Q}, \hat{\Lambda}] \|_{s+N} \leqq C_{s,N} \| u \|_s;$$

(ii) *Let $A(x, \xi)$, $B(x, \xi)$ be homogeneous in ξ of order* 0, *and $A(z) \in C_0^\infty(\Omega_1)$, $B(z) \in C_0^\infty(\Omega_2)$, $z \in \Omega = \{(x, \xi) | \Lambda_1(x, \xi) = 1\}$. Let $\hat{A}, \hat{B}$ be corresponding p.d.o. of order* 0. *Then for each $s > 0$ there exists $C_s > 0$* such *that for all $u(x) \in L^2(M)$,*

$$\| \hat{A}\hat{Q}u - \hat{A}u \|_0 \leqq C_s \| u \|_{-s},$$

$$\| \hat{B}\hat{Q}u \|_0 \leqq C_s \| u \|_{-s};$$

(iii) *There exist $\gamma > 0$, $C > 0$, such that for each p.d.o. $\hat{A}$ of order* 0

$$\| [\hat{A}, \hat{Q}]u \|_0 \leqq C \| u \|_{-\gamma};$$

if $A(x, \xi)$ is the (complete) symbol of $\hat{A}$, and $A(x, \xi) = 0$ in a conical neighbourhood of $\partial\Omega_1 = T_1 \cup T_2$, then

$$\| [\hat{A}, Q]u \|_0 \leqq C_s \| u \|_{-s}$$

for all $s > 0$.

The same is true for $\hat{Q}^*$.

(Here $\| \ \|_s$ is the Sobolev norm of order s, $\| u \|_s = \| \Delta^{s/2} u \|_{L^2}$).

Roughly speaking, $\hat{Q}$ is a p.d.o. of order 0 having the symbol $Q(x, \xi) = 1$ in Ω_1, $Q(x, \xi) = 0$ in Ω_2. Its complete construction, described in section AD.3, is rather complicated and requires several steps. In this section we use this operator as a main tool.

Let $u_1, u_2, \dots$ be the eigenfunctions of $\hat{\Lambda}$ with eigenvalues $\lambda_1 \leqq \lambda_2 \leqq \cdots$ Let $\{\mu_k(dz)\}$ be the sequence of the measures on Ω associated with the sequence $\{u_k\}$ of eigenfunctions (see Lemma AD.1.2). We shall begin with a similar construction for the sequence $\{\hat{Q}u_k\}$.

Lemma AD.2.2. *There exists a sequence $\{\nu_k(dz)\}$ of positive measures on Ω, such that*

(i) *For each p.d.o.* $\hat{A}$ *of order* 0 *with the symbol* $A(z)$,

$$|(\!(\hat{A}\hat{Q}u_k, \hat{Q}u_k)\!) - \int_\Omega A(z)v_k(dz)| \longrightarrow 0 \quad (k \longrightarrow \infty);$$

(ii) *There exists* $C > 0$, *such that for each symbol* $A(z) \in C^\infty(\Omega)$, *and each* $\varepsilon > 0$,

$$\int_\Omega |A(z)|^2 v_k(dz) \leqq C \int_\Omega |A(z)|^2 \mu_k(dz) + \varepsilon$$

for all sufficiently large k;

(iii) *If* $A(z) \equiv 0$ *near* $\partial\Omega_1$, *then*

$$\left|\int_\Omega A(z)v_k(dz) - \int_\Omega A(z)\chi_{\Omega_1}(dz)\right| \longrightarrow 0 \quad (k \longrightarrow \infty),$$

where $\chi_{\Omega_1}(z)$ *is the characteristic function of* Ω_1.

Proof. (i) for each p.d.o. $\hat{A}$ of order 0 with the real-valued symbol $A(z)$, the Gårding inequality is true:

$$\mathrm{Re}(\!(\hat{A}\hat{Q}u_k, \hat{Q}u_k)\!) \geqq \inf_{z\in\Omega} A(z)\|\hat{Q}u_k\|_0^2 - C_A\|\hat{Q}u_k\|_{-1/2}^2,$$

$$|\mathrm{Im}(\!(\hat{A}\hat{Q}u_k, \hat{Q}u_k)\!)| \geqq C_A\|\hat{Q}u_k\|_{-1/2}^2.$$

Now we may go on as in the proof of Lemma AD.1.2; but first it is necessary to prove that $\|Qu_k\|_{-1/2}^2 \longrightarrow 0$. In fact,

$$\begin{aligned}\|\hat{Q}u_k\|_{-1/2}^2 &= (\!(\hat{\Lambda}^{-1}\hat{Q}u_k, \hat{Q}u_k)\!) = (\!(\hat{Q}\hat{\Lambda}^{-1}u_k, \hat{Q}u_k)\!) + (\!([\hat{\Lambda}^{-1}, \hat{Q}]u_k, \hat{Q}u_k)\!)\\ &= \lambda_k^{-1}(\!(\hat{Q}u_k, \hat{Q}u_k)\!) + (\!(\hat{\Lambda}^{-1}[\hat{Q}, \hat{\Lambda}]\hat{\Lambda}^{-1}u_k, \hat{Q}u_k)\!)\\ &= \lambda_k^{-1}(\!(\hat{Q}u_k, \hat{Q}u_k)\!) + \lambda_k^{-1}(\!([\hat{Q}, \hat{\Lambda}]u_k, \hat{\Lambda}^{-1}\hat{Q}u_k)\!) \leqq C\lambda_k^{-1},\end{aligned}$$

and (i) is proved (we use the boundedness of $\hat{Q}$ in L^2 and Lemma AD.2.1).

(ii)
$$\int_\Omega |A(z)|^2 v_k(dz) \sim (\!(\hat{A}^*\hat{A}\hat{Q}u_k, \hat{A}\hat{Q}u_k)\!) = (\!(\hat{A}\hat{Q}u_k, \hat{A}\hat{Q}u_k)\!)$$

$$\sim (\!(\hat{Q}\hat{A}u_k, \hat{Q}\hat{A}u_k)\!) \leqq C(\!(\hat{A}u_k, \hat{A}u_k)\!) = C(\!(\hat{A}^*\hat{A}u_k, u_k)\!) \sim C\int_\Omega |A(z)|^2 \mu_k(dz).$$

(iii) Follows immediately from Lemma AD.2.1(ii). □

Let u_k be the eigenfunction of $\hat{\Lambda}$; let us set $v_k = \hat{Q}u_k$, $w_k = (I - \hat{Q})u_k$. Lemma AD.2.1(i) implies, that for each $N > 0$, $\|(\hat{\Lambda} - \lambda_k)v_k\|_0$ and $\|(\hat{\Lambda} - \lambda_k)w_k\|_0$ decrease faster than $\lambda_k^{-N}(k \longrightarrow \infty)$. In fact,

$$\begin{aligned}\|(\hat{\Lambda} - \lambda_k)v_k\|_0 &= \|(\hat{\Lambda} - \lambda_k)\hat{Q}u_k\|_0 = \|[\hat{\Lambda}, \hat{Q}]u_k + \hat{Q}\hat{\Lambda}u_k - \hat{Q}\lambda_k u_k\|_0\\ &= \|[\hat{\Lambda}, \hat{Q}]u_k + \hat{Q}(\hat{\Lambda} - \lambda_k)u_k\|_0 = \|[\hat{\Lambda}, \hat{Q}]u_k\|_0 \leqq C_N\lambda_k^{-N},\end{aligned}$$

and similarly for w_k. Hence, the orthonormal combinations of v_k and w_k,

$$v'_k = \frac{v_k}{\|v_k\|_0}, \qquad w'_k = \frac{w_k - (v'_k, w_k)v'_k}{\|w_k - (v'_k, w_k)v'_k)\|_0}$$

are quasimodes (i.e. $\|(\hat{\Lambda}-\lambda_k)v'_k\|_0$, $\|(\hat{\Lambda}-\lambda_k)w'_k\|_0$ decrease faster than any power of λ_k), if the denominators are not too small.

Lemma AD.2.3. *Let $A(z)\in C_0^\infty(\Omega_1)$, $B(z)\in C_0^\infty(\Omega_2)$; let $\hat{A}$ and $\hat{B}$ be p.d.o. of order 0 with the symbols $A(z)$ and $B(z)$. Let u_k be an eigenfunction of $\hat{\Lambda}$ with eigenvalue λ_k, $k>k_0$. Suppose that*

(i) $$\|\hat{A}\|<C,\ \|\hat{B}\|<C \text{ in } L^2(M);$$

(ii) $$\left|\int_\Omega A(z)\mu_k(dz)\right|>C^{-1},\ \left|\int_\Omega B(z)\mu_k(dz)\right|>C^{-1};$$

(iii) $$\|\hat{Q}\|<C \text{ in } L^2(M).$$

Then

(i) $C^{-1}<\|v_k\|_0<C, C^{-1}<\|w_k\|_0<C$;

(ii) *Let α be the angle between v_k and w_k in $L^2(M)$. Then* $\sin\alpha>C^{-2}$.

(iii) *Let $v'_k=a_{11}v_k+a_{12}w_k$, $w'_k=a_{21}v_k+a_{22}w_k$ be any orthonormal linear combination of v_k and w_k. Then $|a_{ij}|<C^3$ for all i, j*

Proof. (i) By Lemma AD.2.1(ii), and Lemma AD.2.2,

$$(\!(\hat{A}v_k, v_k)\!)\sim\int_\Omega A(z)v_k(dz)\sim\int_\Omega A(z)\mu_k(dz)>C^{-1},$$

$$(\!(\hat{B}w_k, w_k)\!)\sim\int_\Omega B(z)v_k(dz)\sim\int_\Omega B(z)\mu_k(dz)>C^{-1},$$

for sufficiently large k. But $|(\!(\hat{A}v_k, v_k)\!)|\leqq\|\hat{A}\|\,\|v_k\|^2<C\|v_k\|^2$, and hence $\|v_k\|^2>C^{-2}$, $\|v_k\|>C^{-1}$. The same is true for w_k. Condition (ii) implies that $\|v_k\|<C$, $\|w_k\|<C$.

(ii) Let us estimate the angle between v_k and w_k in $L^2(M)$. Let $w_k=av_k+w'_k$, $a\in\mathbb{C}$. Then

$$|(\!(\hat{B}w_k, w_k)\!)|=|a^2(\!(\hat{B}v_k, v_k)\!)+a(\!(\hat{B}v_k, w'_k)\!)+a(\!(w'_k, \hat{B}^*v_k)\!)+(\!(\hat{B}w'_k, w'_k)\!)|>C^{-1}.$$

The first three terms tend to 0 when $k\longrightarrow\infty$, and hence for large k $|(\!(\hat{B}w'_k, w'_k)\!)|>C^{-1}$. But $\|\hat{B}\|<C$; hence, $\|w'_k\|>C^{-1}$. This implies (ii).

(iii) Follows directly from (i) and (ii).

Corollary AD.2.4. *Suppose that the conditions of Lemma AD.2.3 are valid for an infinite subsequence $\{u_{k_j}\}$. Then we may find a sequence of pairs of orthonormal functions $\{v'_{k_j}, w'_{k_j}\}$ such that for each $N>0$,*

$$\|(\hat{\Lambda}-\lambda_{k_j})v'_{k_j}\|_0<C_N\lambda_{k_j}^{-N},\qquad \|(\hat{\Lambda}-\lambda_{k_j})w'_{k_j}\|_0<C_N\lambda_{k_j}^{-N}.$$

This implies that there exist at least two eigenvalues of $\hat{\Lambda}$, λ'_{k_j} and λ''_{k_j}, such that $|\lambda'_{k_j}-\lambda''_{k_j}|<C_N\lambda_{k_j}^{-N}$ see above, Chapter 5, Props. 32.1, 32.4). Thus we have found a subsequence $\{\lambda_{k_j}\}$ of asymptotically multiple (at least double) eigenvalues. In fact, if M is a 2-dimensional torus, and $\Lambda_1(x,\xi)$ is sufficiently close to $|\xi|$, the whole of the spectrum $\{\lambda_k\}$ of $\hat{\Lambda}$ is asymptotically multiple!

Multiplicity of the spectrum is usually connected with some symmetry present in the problem [13]. In our case an approximate symmetry is true for the measures $\mu_k(dz)$.

Lemma AD.2.5. *Let* $J:\Omega \longrightarrow \Omega$ *be an involution* $(x,\omega) \longrightarrow (x,-\omega)$. *Then the measures* $\mu_k(dz)$ *are asymptotically invariant with respect to* J. *This means that for each odd symbol* $A(z)$, *i.e. such a symbol that* $A(Jz) = -A(z)$, $\int_\Omega A(z)\mu_k(dz) \longrightarrow 0$ $(k \longrightarrow \infty)$.

Proof. For each imaginary odd symbol $A(z)$, a p.d.o. $\hat{A} = \mathrm{Op}(A)$ has a real distributional kernel $A(x,y)$. It is odd, i.e. $A(x,y) = -A(y,x) +$ subordinate terms. The eigenfunctions u_k are real; consequently,

$$(\!(\hat{A}u_k, u_k)\!) = -(\!(u_k, \hat{A}u_k)\!) + O(\lambda_k^{-1}) = -\overline{(\!(\hat{A}u_k, u_k)\!)} + O(\lambda_k^{-1}) = -(\!(\hat{A}u_k, u_k)\!) + O(\lambda_k^{-1}).$$

Hence,

$$\int_\Omega A(z)\mu_k(dz) = O(\lambda_k^{-1}). \qquad \square$$

Now we arrive at the main result of this section. Let M be a 2-dimensional torus. Suppose, that the following condition is valid:

Condition AD.2.6. $\Lambda_1(x,\xi)$ *is sufficiently close to* $|\xi|$, so *that the KAM theorem guarantees the existence of at least 4 invariant tori* T_1, $T_2 = JT_1$, T_3, $T_4 = JT_3$, *which are close to the invariant tori of the unperturbed system.* (*Here* J *is the involution defined in* lemma AD.2.5.)

Theorem AD.2.7. *If* M *is a 2-dimensional torus, and Condition* AD.2.1 *is valid, then the spectrum of* $\hat{\Lambda}$ *is asymptotically multiple, i.e. for each* $N > 0$ *there exists* $C_N > 0$ *such that*

$$\min(\lambda_k - \lambda_{k-1}, \lambda_{k+1} - \lambda_k) < C_N \lambda_k^{-N}.$$

Proof. The tori T_1 and T_2 divide Ω into two domains, denote them Ω_3 and Ω_4; the tori T_3 and T_4 also divide Ω into two domains, denote them Ω_1 and Ω_2. Their numeration is chosen in such a way that $T_i \subset \Omega_i$ for $i = 1,2,3,4$. Let T_i^ε be ε-neighbourhoods of T_i, and assume that ε is so small that $T_i^\varepsilon \cap T_j^\varepsilon = \varnothing$ if $i \neq j$.

Let us divide the sequence $\{u_k\}$ of eigenfunctions into two sequences, $\{u_{p_i}\}$ and $\{u_{q_j}\}$. We put u_k into the first subsequence if $\mu_k(T_1^\varepsilon) > 1/5$ and $\mu_k(T_2^\varepsilon) > 1/5$. Otherwise we put u_k into the second subsequence.

Now, we choose real non-negative functions $A_i(z) \in C_0^\infty(\Omega_i)$, $i = 1,2,3,4$, such that $A_1(z) \equiv 1$ on T_1^ε. $A_2(z) \equiv 1$ on T_2^ε, $A_3(z) \equiv 1$ on $\Omega_3 \backslash (T_1^\varepsilon \cup T_2^\varepsilon)$, $A_4(z) \equiv 1$ on $\Omega_4 \backslash (T_1^\varepsilon \cup T_2^\varepsilon)$. Suppose that the symmetry condition is also satisfied: $A_1(z) = A_2(Jz)$, $A_3(I) = A_4(Jz)$. Let $\hat{A}_i$ be the p.d.o. of order 0 with the symbols $A_i(z)$. Lemma AD.2.5. implies that

$$\left|\int_\Omega (A_1(z)\mu_k(dz) - A_2(z)\mu_k(dz))\right| \longrightarrow 0 \quad (k \longrightarrow \infty),$$

$$\left|\int_\Omega (A_3(z)\mu_k(dz) - A_4(z)\mu_k(dz))\right| \longrightarrow 0 \quad (k \longrightarrow \infty).$$

If the first subsequence $\{u_{p_i})$ is infinite, then the conditions of Lemma AD2.3. are satisfied for the tori T_3, T_4 and domains Ω_1, Ω_2, because $\int_\Omega A_1(z)\mu_{p_i}(dz) > 1/6$, $\int_\Omega A_2(z)\mu_{p_i}(dz) > 1/6$ for all i large enough. Similarly, if the second subsequence $\{u_{q_j}\}$ is infinite, then $\int_\Omega A_3(z)\mu_{q_j}(dz) > 1/6$, $\int_\Omega A_4(z)\mu_{q_j}(dz) > 1/6$, (because $\mu_{q_j}(\Omega) = 1$), and we may apply Lemma AD.2.3 for the tori T_1, T_2 and the domains Ω_3 and Ω_4. In both cases we may apply Corollary AD.2.4. This completes the proof. □

Let T_1, T_2 be arbitrary invariant tori dividing Ω into two domains, Ω_1 and Ω_2. Let $\hat{Q}$ be the operator described in Lemma AD.2.1, associated with the domain Ω_1. Lemma AD.1.5 implies that the mean value of $\int_{\Omega_1} \mu_k(dz)$ is $|\Omega_1|/|\Omega|$; hence, for all N large enough

$$\frac{1}{N}\#\left\{k \geqq N \left| \int_{\Omega_1} \mu_k(dz) > \frac{|\Omega_1|}{2|\Omega|}\right.\right\} > \frac{|\Omega_1|}{2|\Omega|}.$$

Thus, the subsequence $\{u_{k_j}\}$ of eigenfunctions such that $\|\hat{Q}u_{k_j}\| > |\Omega_1|/2|\Omega|$, has a positive (lower) density. The sequence $v_j = \hat{Q}u_{k_j}/\|\hat{Q}u_{k_j}\|$ may be called a *quasimode* concentrated in the gap Ω_1 between the invariant tori T_1 and T_2. In fact, $\|v_j\| = 1$, and by Lemma AD.2.1, $\|\hat{\Lambda}v_j - \lambda_{k_j}v_j\| \leqq C_N\lambda_{k_j}^{-N}$ for each N.

What is the nature of these "quasi-eigenfunctions"? They may be attached to other invariant tori lying in Ω_1 (i.e. corresponding measures μ_k are concentrated near these tori). They may also be associated with closed trajectories of the phase flow G_t in Ω_1, or with the "thin" invariant tori, encircling these trajectories, or with other invariant sets of G_t in Ω_1. But these measures may occur "diffuse", like the ones described above in section AD.1. This depends a great deal on the structure of the phase flow G_t in the gaps between the invariant tori, which is yet poorly known.

In the next section we shall prove Lemma AD.2.1.

Notes. In the preceding sections of this book (see Chapter VII) the subsequence of the spectrum was constructed, which is associated with the Kolmogorov set in the phase space (i.e. the union of the invariant tori of the phase flow). Our results mean that for every gap between the invariant tori there also exists a subsequence of the spectrum which is associated with this gap.

The gaps between invariant tori are the most "tame" examples of invariant sets of the phase flow G_t having positive measure. For such sets we have constructed the quasimodes, i.e. the sequences of quasi-proper functions of positive density, such that corresponding measures μ_k are concentrated on the invariant sets. On the another hand, if there is no nonrivial invariant sets, i.e. if G_t is ergodic on Ω, then almost all of the measures μ_k are asympttically uniformly distributed. Now, two natural questions may be asked.

1. If Ω' is an arbitrary invariant set of positive measure, does there exist a quasimode, concentrated on Ω'? What conditions on Ω' are necessary or sufficient for the existence of such a quasimode?

2. If G_t is not ergodic on Ω, is it true that "almost all" eigenfunctions are asymptotically uniformly distributed on "almost all" of the ergodic components? What conditions are necessary or sufficient for such uniform distribution?

We know nothing about these general problems.

AD.3. Construction of the Operator $\hat{Q}$

This construction requires several steps. Let T_1, T_2 be two invariant tori of the phase flow G_t on Ω, ω_1 and ω_2 being the rotation numbers on T_1 and T_2. These tori divide Ω into two domains, Ω_1 and Ω_2. Assume that the natural projection $(x, \omega) \to x$ is a one-to-one mapping of T_i onto M (this is true if the Hamiltonian function $\Lambda_1(x, \xi)$ is sufficiently close to $|\xi|$).

1. Let us begin with a partition of unity on Ω,

$$1 \equiv \varphi_1(z) + \varphi_2(z) + \psi_1(z) + \psi_2(z),$$

where

$$\begin{aligned} &\varphi_i \equiv 1 \quad \text{in } T_i^{\varepsilon/2}, \qquad \varphi_i \equiv 0 \quad \text{outside } T_i^{\varepsilon}; \\ &\psi_i \equiv 1 \quad \text{in } \Omega_i \backslash (T_1^{\varepsilon} \cup T_2^{\varepsilon}), \\ &\psi_i \equiv 0 \quad \text{outside } \Omega_i \backslash (T_1^{\varepsilon/2} \cup T_2^{\varepsilon/2}); \end{aligned}$$

here T_i^{δ} is the δ-neighbourhood of T_i. Let $\hat{\varphi}_1, \hat{\varphi}_2, \hat{\psi}_1, \hat{\psi}_2$ be corresponding p.d.o.,s of order 0 with the symbls $\varphi_1, \varphi_2, \psi_1, \psi_2$ respectively, and with $\hat{\varphi}_1 + \hat{\varphi}_2 + \hat{\psi}_1 + \hat{\psi}_2 =$ Identity.

Set $\hat{Q}u = \hat{\psi}_1 u + \hat{Q}\hat{\varphi}_1 u + \hat{Q}\hat{\varphi}_2 u$. In what follows we shall denote $\hat{\varphi}_1 u = v$ and construct $\hat{Q}v$; $\hat{Q}\varphi_2 u$ is constructed similarly.

2. We are going to construct operators $\hat{Q}_N$, such that $[\hat{Q}_N, \hat{\Lambda}]$: $H_s \longrightarrow H_{s+N}$ is continuous for all s. Then we shall "glue them together" in the following manner. Choose a function $\Psi(\lambda) \in C^{\infty}[0, +\infty)$, $\Psi(\lambda) \geqq 0$, $\Psi(\lambda) \equiv 1$ for $0 \leqq \lambda \leqq 1/2$, $\Psi(\lambda) \equiv 0$ for $\lambda \geqq 1$. If $u_1, u_2, \ldots$ are the eigenfunctions of $\hat{\Lambda}$ with eigenvalues $\lambda_1, \lambda_2, \ldots$, and $u = \sum_k a_k u_k$, then we set

$$\hat{Q}u = \sum_N \sum_k (\Psi(\alpha_N \lambda_k) - \Psi(\alpha_{N-1} \lambda_k)) a_k \hat{Q}_n u_k, \tag{AD.3.1}$$

where $\alpha \longrightarrow 0$ sufficiently rapidly as $N \longrightarrow \infty$.

3. Let $\Lambda(x, \xi)$ be a complete symbol of $\hat{\Lambda}$. We start our construction from the particular case of Λ depending on ξ only. In this case $\tilde{T}_1$ has the equation $\xi_2/\xi_1 = \text{const}$ (recall that $\tilde{T}_1$ is a conic set in T^*M with the base T_1). It is convenient to introduce a new coordinate system (x, η_1, η_2) in T^*M, such that T^*M has the equation $\eta_2 = 0$; let us assume, that $\eta_1 > 0$, $\eta_2 > 0$ in $\Omega_1 \cap$(conic neighbourhood of $\tilde{T}_1$). Consider the symbol

$$q(x, \eta_1, \eta_2) = \Phi(\eta_2/\eta_1^{1-\gamma}) \chi_1(x, \eta/|\eta|).$$

Here $\chi_1(x,\omega)=\chi_1(z)$ is a smooth function with support in T_1^ε, and such that $\chi_1(z)\varphi_1(z)\equiv\varphi_1(z)$; $\Phi(\zeta)\in C^\infty(\mathbb{R}^1)$, $\Phi'(\zeta)\geqq 0$, $\Phi(\zeta)\equiv 0$ for $\zeta<-1$, $\Phi(\zeta)\equiv 1$ for $\zeta>1$; $0<\gamma<1/2$. Let $q(x,\xi)$ be the same function in (x,ξ)-coordinates.

The symbol $q(x,\xi)$ is not homogeneous in ξ; but it is in the Hörmander class $S^0_{1-\gamma,0}$ [2], i,e. $|\partial_\xi^\alpha\partial_x^\beta q(x,\xi)|\leqq C_{\alpha\beta}(1+|\xi|)^{(\gamma-1)|\alpha|}$ for all multi-indices α. Let $\hat{q}$ be the p.d.o. with the symbol $q(x,\xi)$; it is bounded in H_s for all s, and $[\hat{q},\hat{\Lambda}]=0$.

For each p.d.o. $\hat{A}$ of order 0 with the complete symbol $A(x,\xi)$, the commutator $[\hat{q},\hat{A}]=\hat{B}$ is a p.d.o. with the complete symbol [2]

$$B(x,\xi)\sim\sum_\alpha\frac{(-i)^{|\alpha|}}{\alpha!}(\partial_x^\alpha q\partial_\xi^\alpha A-\partial_\xi^\alpha q\partial_x^\alpha A);$$

hence, $B(x,\xi)\in S^{\gamma-1}_{1-\gamma,0}$, and $B(x,\xi)\in s^{-\infty}_{1-\gamma,0}$ if $A(x,\xi)\equiv 0$ in the conic neighbourhood of $\tilde{T}_1$. Consequently, $[\hat{q},\hat{A}]$: $H_s\longrightarrow H_{s-\gamma+1}$ is continuous, and it is continuous $H_s\longrightarrow H_\infty$ if $A(x,\xi)\equiv 0$ in the neighbourhood of $\tilde{T}_1$. Thus, $\hat{q}$ possesses all the properties claimed in Lemma AD.2.1, and we may set $\hat{Q}=\hat{q}$.

4. Now, let us consider the general case. We are going to construct an operator $\hat{V}_N$ such that the operator $\hat{L}=\hat{V}_N^{-1}\hat{\Lambda}\hat{V}_N$ has a symbol sufficiently close to constant in x near $\tilde{T}_1$. Namely, $\hat{L}$ is a p.d.o. of order 1 with a complete symbol $L(x,\xi)=L^{(0)}(\xi)+L^{(1)}(x,\xi)$, such that

$$|\partial_x^\alpha L^{(1)}(x,\xi)|\leqq C_\alpha(|\eta_2/\eta_1|^{N_1}\eta_1+|\eta|^{-N_2}),\tag{AD.3.2}$$

where N_1 and N_2 are arbitrary fixed numbers (here (η_1,η_2) are the coordinates in T^*M used together with the coordinates (ξ_1,ξ_2) dual to (x_1,x_2)).

Then, by the commutation formula, the operator $\hat{F}=[\hat{q},\hat{L}]=[\hat{q},\hat{L}^{(1)}]$ has a symbol

$$F(x,\xi)\sim\sum_{|\alpha|\geqq 1}\frac{(-i)^{|\alpha|}}{\alpha!}(\partial_\xi^\alpha q\cdot\partial_x^\alpha L^{(1)}-\partial_\xi^\alpha L^{(1)}\cdot\partial_x^\alpha q).\tag{AD.3.3}$$

Now,

$$|\partial_\xi^\alpha q|\leqq C_\alpha(1+\eta_1)^{(\gamma-1)|\alpha|},$$

and

$$\partial_\xi^\alpha q\equiv 0,\quad\text{if }|\eta_2|>C\eta_1^{1-\gamma};$$
$$\partial_x^\alpha q\equiv 0.$$

Hence,

$$F(x,\xi)\sim\sum_{|\alpha|\geqq 1}\frac{(-i)^{|\alpha|}}{\alpha!}\partial_\xi^\alpha q\cdot\partial_x^\alpha L^{(1)},$$

and

$$\begin{aligned}|\partial_\xi^\alpha q\cdot\partial_x^\alpha L^{(1)}|&\leqq C_\alpha(1+\eta_1)^{(\gamma-1)\cdot|\alpha|}(\eta_1^{-\gamma N_1+1}+\eta_1^{-N_2})\\&\leqq C_\alpha[(1+\eta_1)^{(\gamma-1)\cdot|\alpha|-\gamma N_1+1}+(1+\eta_1)^{(\gamma-1)|\alpha|-N_2}].\end{aligned}$$

We see, that the orders of the terms of the asymptotic expansion of the symbol $F(x,\xi)$ of $\hat{F}=[\hat{q},\hat{L}]$ tend to $-\infty$ and do not exceed $\max(-\gamma N_1+1,-N_2)$. If we choose N_1 and N_2 sufficiently large, we may obtain arbitrary accuracy of the commutation of $\hat{q}$ and $\hat{L}$ (i.e. $[\hat{q}, \hat{L}]$ may be made a p.d.o. with the symbol decreasing as rapidly as $|\xi|^{-N}$ for each N).

If the operator $\hat{V}_N$ is constructed, we may set

$$\hat{Q}_N = \hat{V}_N \hat{q} \hat{V}_N^{-1};$$

then,

$$[\hat{\Lambda}, \hat{Q}_N] = \hat{V}_N(\hat{V}_N^{-1}\hat{\Lambda}\hat{V}_N, \hat{q}]\hat{V}_N^{-1} = \hat{V}_N(\hat{L}, \hat{q})\hat{V}_N^{-1},$$

and we see that $[\hat{\Lambda}, \hat{Q}_n]$: $H_s \longrightarrow H_{s+N}$ is continuous. Other necessary properties of $\hat{Q}_N$ follow from the fact that the symbol of $\hat{Q}_N$ is equal to 0 or to 1 outside any arbitrary conical neighbourhood of $\tilde{T}_1$ for large $|\xi|$.

It remains to construct the operators $\hat{V}_N$. In its turn, this requires several steps.

5. Suppose that the invariant torus T_1 lying on the surface Ω: $\Lambda_1(x, \xi) = 1$ is mapped one-to-one onto M under the natural projection $\pi: (x, \xi) \longrightarrow x$. Then the phase flow on Ω is transferred into the flow $\pi_* G_t$ on M with the same rotation number ω and invariant 2-form $\pi^{*-1} dz$. Since ω is slowly approximable by rational numbers, there exists by the Kolmogorov theorem [14] a smooth change of coordinates transferring the flow $\pi_* G_t$ into the rectilinear flow with the constant velocity field $\dot{x} = \mathfrak{b} = (b_1, b_2) = \text{const}$ (the rotation number $\omega = b_2/b_1$). In these new coordinates T_1 has the equation $\xi = \xi(x) = \xi_0 + ds$, where s is some smooth univalent function on M (because T_1 is a lagrangian manifold in T^*M). But on the invariant torus T_1, $(\dot{x}, \xi(x)) = \Lambda_1 = 1$, for Λ_1 is homogeneous in ξ of order 1; hence, $(\mathfrak{b}, ds) = \text{const} = 0$. This implies that $s = \text{const}$ (for ω is irrational), and T_1 has the equation $\xi(x) = \xi_0$.

Now, we make more concrete the choice of coordinates (η_1, η_2) in T^*M. There is a constant velocity vector $\mathfrak{b}$ on M, and a constant covector $\xi_0 (\mathfrak{b} = \partial\Lambda_1(x, \xi)/\partial\xi|_{\xi=\xi_0})$. Let us choose a constant vector $\mathfrak{a}$ orthogonal to ξ_0, and denote $\eta_1 = \eta_1(\xi) = (\mathfrak{b}, \xi)$, $\eta_2 = \eta_2(\xi) = (\mathfrak{a}, \xi)$, $\xi \in T^*M, x \in M$.

6. Let $s(x, \xi)$ be a real function, homogeneous in ξ of order 1, and let $\hat{s}$ be the corresponding p.d.o. of order 1. We define a homogeneous flow $h_t: T^*M \longrightarrow T^*M$ generated by the Hamiltonian equations $\dot{x} = \partial s/\partial\xi$, $\dot{\xi} = -\partial s/\partial x$. Let $\hat{h}_t$ be a one-parameter group of operators in $H_s(M)$ generated by the hyperbolic first order equation

$$\frac{\partial u}{\partial t} = i\hat{s}u.$$

If $\hat{A}$ is a p.d.o. with the principal symbol $A(x, \xi)$, then $\hat{h}_t^{-1}\hat{A}\hat{h}_t$ is also a p.d.o. with a principal symbol $A(h_t^{-1}(x, \xi))$ (Egorov theorem).

Our next step is to find such a function $s(x, \xi)$, that the principal symbol of the operator $\hat{L} = \hat{h}_t^{-1}\hat{\Lambda}\hat{h}_t$ is $L_1(x, \xi) = L_1^{(0)}(\xi) + L_1^{(1)}(x, \xi)$, where $L_1^{(1)}(x, \xi)$ satisfies, in the conical neighbourhood of $\tilde{T}_1$, the inequalities

(AD.3.4) $$|\partial_\xi^\alpha \partial_x^\beta L_1^{(1)}(x, \xi)| \leqq C_{\alpha\beta} |\eta_2/\eta_1|^{N_1 - |\alpha|} \eta_1^{1-|\alpha|}, \quad |\alpha| \leqq N_1.$$

The principal symbol $\Lambda_1(x, \xi)$ possesses the following obvious properties:

(i) $$\Lambda_1(x, t\xi) = t\Lambda_1(x, \xi), \quad t > 0;$$

(ii) $(\partial\Lambda_1/\partial x)|_{\eta_2=0} = 0;$

(iii) $(\partial\Lambda_1/\partial\eta_1)|_{\eta_2=0} = 1;$

(iv) $(\partial\Lambda_1/\partial\eta_2)|_{\eta_2=0} = 0 \quad (x\in M, \eta_1 = 0).$

This implies that $\Lambda_1(x,\eta_1,\eta_2)=\eta_1+\Lambda_1^{(1)}(x,\eta)$, where $\Lambda_1^{(1)}|_{\eta_2=0}=0$, $(\partial\Lambda_1^{(1)}/\partial\eta_2|_{\eta_2=0}=0$. Hence,

$\Lambda_1^{(1)}(x,\eta_1,\eta_2)=(\eta_2/\eta_1)^2\eta_1 f^{(2)}(x)+$ terms of 3-d and higher orders in η_2/η_1.

Let us find $s(x,\xi)$, with $s(x,t\xi)=ts(x,\xi)$, such that $\{s,\Lambda_1\} = -\Lambda_1^{(1)}(x,\xi)+\Lambda_1'(\xi)+$ terms of 3-d and higher orders in η_2/η_1 (here $\{\cdots\}$ denotes the Poisson brackets). First, we set

$$\Lambda_1'(\xi)=(2\pi)^{-1}\int_M \Lambda_1(x,\xi)\,dx.$$

Let

$$f^{(2)}(x)=\sum_k e^{i\langle k,x\rangle} f_k^{(2)}$$

be a Fourier expansion of $f^{(2)}(x)$; then we set

$$s(x,\xi)=\sum_{k\neq 0} e^{i\langle k,x\rangle} s_k(\xi),$$

where

$$s_k(\xi)=-\frac{1}{i\langle k,\mathrm{b}\rangle}(\eta_2/\eta_1)^2\eta_1 f_k^{(2)}.$$

(We assumed a priori that $\omega=b_2/b_1$ is slowly approximated by the rational numbers, hence $s(x,\xi)\in C^\infty$.)

Let us now consider the Hamiltonian flow h_t in T^*M generated by $s(x,\xi)$. h_t is homogeneous in ξ of order 1, and $\tilde{T}$ is fixed.

The flow h_t transforms Λ in the following manner:

$$\Lambda(h_1^{-1}(x,\xi))-\Lambda(x,\xi)=\frac{d}{dt}\Lambda(h_t^{-1}(x,\xi)|_{t=0}(1+O(\eta_2/\eta_1))$$
$$=\{\Lambda,s\}(1+O(\eta_2/\eta_1))=-\{s,\Lambda\}(1+O(\eta_2/\eta_1)).$$

Consider the p.d.o. $\hat{L}=\hat{h}_1^{-1}\hat{\Lambda}\hat{h}_1$; its principal symbol $L_1(x,\xi)=\Lambda(h_1^{-1}(x,\xi))=L_1^{(0)}(\xi)+L_1^{(1)}(x,\xi)$ is such that

$$|\partial_\xi^\alpha\partial_x^\beta L_1^{(1)}(x,\xi)|\leqq C_{\alpha\beta}(\eta_2/\eta_1)^{3-|\alpha|}\eta_1^{1-|\alpha|},\quad |\alpha|\leqq 3.$$

Thus we have eliminated the term in $\Lambda_1^{(1)}$ having second order in (η_2/η_1).

After having repeated a similar construction (N_1-3) times, we transform $\hat{\Lambda}$ into a p.d.o. $\hat{\tilde{L}}=\hat{P}_1^{-1}\hat{\Lambda}\hat{P}_1$ with principal symbol $\tilde{L}(x,\xi)=\tilde{L}_1^{(0)}(\xi)+\tilde{L}_1^{(1)}(x,\xi)$ such that

$$|\partial_\xi^\alpha\partial_x^\beta\tilde{L}(x,\xi)|\leqq C_{\alpha\beta}(\eta_2/\eta_1)^{N_1-|\alpha|}\eta_1^{1-|\alpha|},\quad |\alpha|\leqq N_1.$$

This coincides with (AD.3.4).

7. We have already transformed our initial operator $\hat{\Lambda}$ into a p.d.o. $\hat{L}=\hat{P}_1^{-1}\hat{\Lambda}\hat{P}_1$ such that its principal symbol $L_1(x,\xi)=L_1^{(0)}(\xi)+L_1^{(1)}(x,\xi)$ satisfies (AD.3.4) (we write $\hat{L}$ instead of $\hat{\hat{L}}$). But $\hat{L}$ has subordinate terms (having orders $\leqq 0$), and we are going to transform $\hat{L}$ so that (AD.3.2) were valid. Let the complete symbol of $\hat{L}$ be $L_1^{(0)}(\xi)+L_1^{(1)}+L_0(x,\xi)$, where $L_1^{(0)}$, $L_1^{(1)}$ are homogeneous of order 1 in ξ, $L_1^{(1)}$ satisfies (AD.3.4), and L_0 is a nonhomogeneous symbol of order 0: $L_0(x,\xi)=L_0^{(0)}+L_0^{(-1)}+\cdots$, ord $L_0^{(-j)}=-j$.

Let $\hat{P}$ be an invertible p.d.o. of order 0 with principal symbol $P(x,\xi)$. Consider the operator $\hat{P}^{-1}\hat{L}\hat{P}-\hat{L}=\hat{P}^{-1}[\hat{L},\hat{P}]$. By the commutation formula (AD.3.3), its principal symbol is

$$i\frac{\{L,P\}}{P}=i\{L,\ln P\},$$

where $\{\ \ \}$ denotes the Poisson brackets.

Let us choose $\hat{P}$ in order to eliminate the term $L_0^{(0)}(x,\xi)$ in the symbol of $\hat{P}^{-1}\hat{L}\hat{P}$ on $\tilde{T}_1$. Let

$$L_0^{(0)}(x,\xi)=L_{0,0}^{0}(\xi)+\sum_{k\neq 0}L_{0,k}^{(0)}(\xi)e^{i\langle k,x\rangle}$$

be the Fourier expansion in x of $L_0^{(0)}$. We add the constant (in x) term $L_{0,0}^{(0)}(\xi)$ to $L_1^{(0)}(\xi)$ and set

$$\ln P(x,\xi)=\sum_{k\neq 0}p_k(\xi)e^{i\langle k,x\rangle},$$

where

$$p_k(\xi)=\frac{1}{i\langle k,\mathfrak{b}\rangle}L_{0,k}^{(0)}(\xi).$$

$\ln P(x,\xi)$ is homogeneous in ξ of order 0, and hence $P(x,\xi)$ is a symbol of a Fredholm p.d.o. of order 0. Let $\hat{P}$ be a p.d.o. with the symbol $P(x,\xi)$; we assume that $\hat{P}$ is invertible, otherwise we may improve it by adding some smooth finite-dimensional operator. The transformed operator $\hat{P}^{-1}\hat{L}\hat{P}$ has a symbol

$$L^{(1)}(x,\xi)=L_1^{(1,0)}(\xi)+L_1^{(1)}(x,\xi)+L_0^{(1)}(x,\xi)+\cdots,$$

where

$$|\partial_\xi^\alpha\partial_x^\beta L_0^{(1)}(x,\xi)|\leqq C_{\alpha\beta}\left[\left|\frac{\eta_2}{\eta_1}\right|^{1-|\alpha|}\eta_1^{-|\alpha|}\right],\quad |\alpha|\leqq 1.$$

(Note that this transformation does not change the principal symbol of $\hat{L}$.)

Now we repeat the similar construction N_0 times ($N_0<N_1$), and obtain a p.d.o. $\hat{L}^{(N_0)}=\hat{P}_0^{-1}\hat{L}\hat{P}_0$ with the symbol

$$L^{(N_0)}(x,\xi)=L_1^{(0)}(\xi)+L_0^{(N_0,0)}(\xi)+L_1^{(1)}(x,\xi)+L_0^{(N_0,1)}(x,\xi)+L_{-1}^{(N_0)}(x,\xi)+\cdots,$$

where the term $L_0^{(N_0,1)}$ of the 0-th order satisfies the inequalities

$$|\partial_\xi^\alpha\partial_x^\beta L_0^{(N_0,1)}(x,\xi)|\leqq C_{\alpha\beta}\left[\left|\frac{\eta_2}{\eta_1}\right|^{N_0-|\alpha|}\eta_1^{-|\alpha|}\right],\quad |a|\leqq N_0.$$

8. Now, we are going to eliminate on $\tilde{T}_1$ the term $L_{-1}(x,\xi)$ of order (-1). Let us take a p.d.o. $\hat{P}$ with the symbol $P(x,\xi)=1+P_{-1}(x,\xi)+\cdots$, $\operatorname{ord} P_{-j}=-j$. The commutation formula (AD.3.3) implies that the complete symbol of the p.d.o. $\hat{F}=\hat{P}^{-1}\hat{L}\hat{P}$,

$$F(x,\xi)=L(x,\xi)-i\{L_1,P_{-1}\}+\cdots.$$

We are going to find $P_{-1}(x,\xi)$, such that if for a given N_{-1}, $L=L_1(\xi)+L_0(\xi)+L_{-1}^{(N_{-1})}(x,\xi)+\cdots$ is a complete symbol of $\hat{P}^{-1}\hat{L}\hat{P}$, then

$$|\partial_\xi^\alpha\partial_x^\beta L_{-1}^{(N_{-1})}|\leqq C_{\alpha\beta}\left[\left|\frac{\eta_2}{\eta_1}\right|^{N_{-1}-|\alpha|}\eta_1^{-1-|\alpha|}\right],\quad |\alpha|\leqq N_{-1};$$

the higher order terms L_1 and L_0 remain unchanged. To construct $P_{-1}(x,\xi)$, we solve N_{-1} times the homological equation

$$\left(\mathfrak{b},\frac{\partial P_{-1}^{(N)}}{\partial x}\right)=iL_{-1}^{(N-1)},\qquad L_{-1}^{(N)}=L_{-1}^{(N-1)}-i\{L_1,P_{-1}^{(N)}\},\quad N=1,\dots,N_{-1}.$$

Then we set $P_{-1}=P_{-1}^{(1)}+\cdots+P_{-1}^{(N_1)}$.

9. This construction may be repeated to eliminate on $\tilde{T}_1$ the terms $L_{-2},\dots,L_{-K}$ of orders $(-2),\dots,(-K)$. We arrive at a p.d.o. $\hat{L}$ with the symbol

$$L^{(0)}(\xi)+L_1^{(N_1)}(x,\xi)+L_0^{(N_0)}(x,\xi)+\cdots+L_{-K}^{(N_{-K})}(x,\xi)+\cdots,$$

such that

$$|\partial_\xi^\alpha\partial_x^\beta L_{-j}^{(N_{-j})}|\leqq C_{\alpha\beta}\left[\left|\frac{\eta_2}{\eta_1}\right|^{N_{-j}-|\alpha|}\eta_1^{-j-|\alpha|}\right],\quad -1\leqq j\leqq K,\ |\alpha|\leqq N_j.$$

If $N_1,N_0,N_{-1},\dots,N_{-K}$ and K are large enough, then (AD.3.2) is true.

Let $\hat{P}_1,\hat{P}_0,\hat{P}_{-1},\dots,\hat{P}_{-K}$ be the transformation operators eliminating the x-dependent terms of orders $1,0,-1,\dots,-K$ of the symbol of $\hat{L}$ in the infinitesimal neighbourhood of $\tilde{T}_1$, and $\hat{V}=\hat{P}_{-K}\cdots\hat{P}_0\hat{P}_1$. Then $\hat{L}=\hat{V}^{-1}\hat{\Lambda}\hat{V}$ satisfies (AD.3.2), and $\hat{Q}_N=\hat{V}\hat{q}\hat{V}^{-1}$ is constructed. Now we obtain the operator $\hat{Q}$ by the "glueing" formula (AD.3.1). Thus, Lemma AD.2.1 is proved. □

References

1. Feynman, R. P. Statistical mechanics. California Institute of Technology. Reading, Mass., 1972
2. Hörmander, L.: Pseudo-differential operators. Comm. Pure Appl. Math. *18*, number 3, 501–517 (1965)
3. Taylor, M. E.: Pseudo-differential operators. Princeton University Press, Princeton, New Jersey, 1981
4. Shnirelman, A. I.: Ergodic properties of eigenfunctions. Uspehi Math. nauk, *29*, number 6, 181–182 (1974) [Russian]
5. Shnirelman, A. I.: On the asymptotic multiplicity of the spectrum of the Laplace operator. Uspehi Math. nauk, *30*, number 4, 265–266 (1975) [Russian]
6. Collin de Verdiere, Y.: Ergodicite et fonctions propres du Laplasien. In: Equations aux derivees partielles, Seminaire Bony-Sjöstrand-Meyer, expose 13, pp. XIII-1 – XIII-7. Ecole Politechnique, paris, 1984–85

7. Zelditch, S.: Eigenfunctions on compact Riemann surfeces of genus $g \geqq 2$. preprint, N.Y. (1984)
8. Helffer, B., Martinez, A., Robert, D.: Ergodicte et limite semiclassique. Comm. Math. Phys., *109*, number 2, 313-326 (1987)
9. Bailey, P. B., Brownell, S. H.: Removal of the log factor in the asymptotic estimates of poligonal membrane eigenvalues. J. Math. Anal. Appl., *4*, number 2, 212–239 (1962)
10. Rosenblum, G. V., Solomjak, M. Z., Shubin, M. A.: Spectral theory of differential operators. In: Partial differential equations-7, Advances of Science and Engineering, Modern Problems of Mathematics, Fundamental directions, *64*. Moscow: VINITI, pp. 96–97 (1989) [Russian]
11. Kornfeld, I. P., Sinai, Ja. G., Fomin, S. V.: Ergodic Theory. Moscow: Nauka, 1980 [Russian]
12. Mc. Donald, S. W., Kaufman, A. N.: Spectrum and Eigenfunctions for a Hamiltonian with stochastic trajectories. Phys. Rev. Lett., *42*, number 18, 1189–1191 (1979)
13. Arnold, V. I.: Modes and quasi-modes. Funct. Anal. and Appl., *6*, number 2, 12–20 (1972) [Russian]
14. Kolmogorov, A. N.: On the dynamic systems on the torus having integral invariant. Dokl. Akad. Nauk SSSR, N.S., *93*, 763–766 (1953) [Russian]
15. Babich, V. M., Buldyrev, V. S.: Asymptotic methods in the difraction problems of the short waves. Moscow: Nauka, pp. 228–269 (1972) [Russian]
16. Bunimovich, L. A.: On he ergodic properties of nowhere dispersing billiards. Comm. Math. Phys., *65*, pp. 295–312 (1979)

Appendix I. Manifolds

The goal of this Appendix is to provide the reader with necessary definitions, notations and statements concerning manifolds. For basic definitions connected with the concept of topological space the reader is referred to Bourbaki (1965), Kelley (1955) or to Fuks and Rokhlin (1984).

Bundles

AI.1. A *bundle* is a triple (E, p, X) where E and X are topological spaces called correspondingly the *total space* and the *base* of a bundle, and $p: E \longrightarrow X$ is a continuous map called the bundle *projection.* We also say that E is a bundle *over* X with projection p, and write $p: E \longrightarrow X$, or simply E for denoting the bundle (the same letter for a bundle and its total space). The base X of a bundle E will be sometimes denoted by $\operatorname{bs} E$. For each $x \in \operatorname{bs} E$ the set $E_x = p^{-1}\{x\}$ is called the *fibre* over a point x. The *product bundle* with base X and fibre F is a bundle of a form $(X \times F, p_1, X)$ where p_1 is the first projection of a product: $p_1(x, f) = x$, $(x, f) \in X \times F$.

AI.2. A *morphism* from a bundle (E, p, X) to a bundle (E', p', X') is a pair (F, f) of continuous maps $F: E \longrightarrow E'$ and $f: X \longrightarrow X'$ such that the following diagram commutes:

$$\begin{array}{ccc} E & \xrightarrow{\;F\;} & E' \\ \downarrow{\scriptstyle p} & & \downarrow{\scriptstyle p'} \\ X & \xrightarrow{\;f\;} & X' \end{array}$$

i.e. $p' \circ F = f \circ p$. It follows that the map F carries fibres of E into fibres of E': $F(E_x) \subset E_{f(x)}$ for each $x \in X$. We shall often refer to the first member F of a pair (F, f) as a morphism. The map f is called the *base map* of a morphism F and is denoted $\operatorname{bs} F$. A morphism $F: E \longrightarrow E'$ is called a *morphism over* X if $\operatorname{bs} E = \operatorname{bs} E' = X$ and $\operatorname{bs} F = 1_X$. A morphism over X is called an *equivalency* if it maps the total spaces homeomorphically one onto another.

The identity map 1_E of the total space E is an equivalency. The superposition of morphisms (morphisms over X, equivalencies) is again a morphism (morphism over X, equivalency).

A bundle is called *trivial* if it is equivalent to a product bundle.

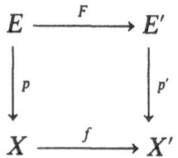

AI.3. Let E be a bundle with projection p, and $A \subset \operatorname{bs} E$ an arbitrary subset. If we supply $p^{-1}(A) \subset E$ and $A \subset \operatorname{bs} E$ with the induced topologies, then the triple $(p^{-1}(A), p|p^{-1}(A), A)$ becomes a bundle which is called the *restriction of a bundle* E onto A and is denoted by $E|_A$. The inclusion maps $j: p^{-1}(A) \hookrightarrow E$ and $i: A \hookrightarrow \operatorname{bs} E$ form a morphism of a bundle $E|_A$ into E.

AI.4. Let E be a bundle with projection p, and $f: X \longrightarrow \operatorname{bs} E$ a continuous map. The *pullback* of E along f is a bundle $f^! E$ over X whose total space is a subspace of XE consisting of all pairs (x, e) such that $f(x) = p(e)$, and whose projection is $(x, e) \longmapsto x$, the restriction of the first projection of the product. The restriction of the second projection, $(x, e) \longmapsto e$ forms the *adjoint* morphism $\operatorname{ad} f\colon f^! E \longrightarrow E$ with $\operatorname{bs} \operatorname{ad} f = f$.

Any morphism $F: E \longrightarrow E'$ of bundles can be decomposed into the superposition $F = \operatorname{corr} F \circ \operatorname{ad} \operatorname{bs} F$, where the *correction map* $\operatorname{corr} F: E \longrightarrow (\operatorname{bs} F)^! E'$ is a morphism over $\operatorname{bs} E$ defined by the formula $\operatorname{corr} F(e) = (p(e), F(e)), e \in E, p$ being the projection of E.

The restriction $E|_A$ of a bundle E on a subset $A \hookrightarrow \operatorname{bs} E$ may be canonically identified with $i^! E$, where $i: A \hookrightarrow \operatorname{bs} E$ is the inclusion, by means of the map $(x, e) \longmapsto e$, $e \in E_x$.

AI.5. Let E be a bundle and $E' \subset E$ be an arbitrary subset of its total space. Then E' is a bundle supplied with the induced topology, with the restriction of the bundle projection of E onto E' as the projection, and with $\operatorname{bs} E$ as the base. The obtained bundle is called a *subbundle* of E. The inclusion map $j: E' \hookrightarrow E$ is a bundle morphism over $\operatorname{bs} E$.

AI.6. Let E be a bundle with the projection p, and $\sim$ be an equivalence relation which is stronger than that defined by the bundle projection (i.e. if $e \sim e'$ then $p(e) = p(e))$, $e, e' \in E$. Then one may consider the *quotient bundle* $E/\!\sim$ over the same base, the total space of $E/\!\sim$ consisting of equivalence classes, the projection being induced by p. The natural map $E \longrightarrow E/\!\sim$ is a bundle morphism.

AI.7. The *product* of a finite collection of bundles (E_i, p_i, X_i), $1 \leqq i \leqq n$, is the bundle with total space $E_1 \times E_2 \times \cdots \times E_n$, with the base $X_1 \times X_2 \times \cdots \times X_n$, and with the projection $p_1 \times p_2 \times \cdots \times p_n$, the latter means $p_1 \times p_2 \times \cdots \times p_n(e_1, e_2, \ldots, e_n) = (p_1(e_1), p_2(e_2), \ldots, p_n(e_n)) \in X_1 \times X_2 \times \cdots \times X_n$ for $(e_1, e_2, \ldots, e_n) \in E_1 \times E_2 \times \cdots \times E_n$. The pair (Π_j, π_j), where $\Pi_j: E_1 \times E_2 \times \cdots \times E_n \longrightarrow E_j$ is the jth projection, and so is $\pi_j: X_1 \times X_2 \times \cdots \times X_n \longrightarrow X_j$, is a bundle morphism of the product-bundle in E_j for each $j \in \{1, 2, \ldots, n\}$.

AI.8. The *Whitney sum* of a finite collection of bundles E_i, $1 \leqq i \leqq n$, over the same base X is the bundle $\Delta^!(E_1 \times E_2 \times \cdots \times E_n)$, where $\Delta: X \longrightarrow \underbrace{X \times X \times \cdots \times X}_{n\text{-times}}$ is the diagonal map $x \longmapsto (x, x, \ldots, x)$.

AI.9. A *section* of a bundle (E, p, X) is a map $s: X \longrightarrow E$ with property $p \circ s = 1_X$. A section assigns a point in E_x to each point $x \in X$. The set of all sections of a bundle E is denoted by $\Gamma(E)$. The value of a section s at a point $x \in X$ will be denoted by s_x, or $s|_x$. The notation for the restriction of s onto a subset $A \subset X$ is $s|_A$. The latter is a member of $\Gamma(E|_A)$. Let $f: Y \longrightarrow X$ be a continuous map. A *lift* of f is a map $g: Y \longrightarrow E$ such that the following diagram commutes

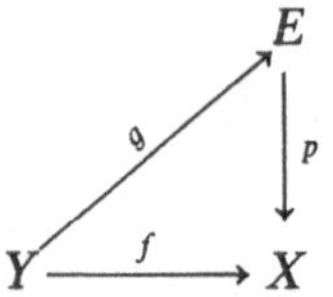

The set of all lifts of f will be denoted by $L(E,f)$. The superposition operation $s \longmapsto s \circ f$ defines the map $f^0: \Gamma(E) \longrightarrow L(E,f)$. If g is a lift of f then the map $I_f g: x \longmapsto (x, g(x))$ is a section of $f^! E$, and we have the map $I_f: L(E,f) \longrightarrow (f^! E)$. The superposition $f^! = I_f \circ f^0$ carries sections of E into sections of $f^! E$. We call this map $f^!: \Gamma(E) \longrightarrow \Gamma(f^! E)$ a *pull-back of sections*.

Vector Bundles

AI.10. A (real) *vector bundle* is a bundle E whose fibres E_x are endowed with structures of real finite dimensional vector space of the same dimension, say k, so that the following condition of local triviality is fulfilled: each point $x \in \operatorname{bs} E$ has a neighbourhood U such that the restriction $E|_U$ is equivalent to the product bundle $U \times \mathbb{R}^k$ via an equivalency T whose restriction to each fibre $T_x: E_x \longrightarrow \{x\} \times \mathbb{R}^k$ is linear. (The vector space structure of $\{x\} \times \mathbb{R}^k$ is defined via the usual identification $\{x\} \times \mathbb{R}^k \cong \mathbb{R}^k$). Such an equivalency T is called a *trivialisation* of a vector bundle E.

If the dimension of fibres of a vector bundle E equals k, we also say that E is a *k-dimensional vector bundle.*

Any vector bundle E possesses the *zero-section* which assigns to each point $x \in \operatorname{bs} E$ the zero vector of E_x. The zero-section maps $\operatorname{bs} E$ homeomorphically onto its image in E. We may identify the latter with $\operatorname{bs} E$, so we shall assume further that $\operatorname{bs} E \subset E$, and the zero-section is the inclusion.

The *product-bundle* $X \times \mathbb{R}^k$ is a k-dimensional vector bundle.

The *zero-bundle* 0 over the topological space X is a zero-dimensional vector bundle $(X, 1_X, X)$.

AI.11. A *vector bundle morphism* between vector bundles is a bundle morphism which is linear on fibres. Analogous definitions are for a *vector bundle morphism over X* and for a *vector bundle equivalency.* The identity map of the total space of a vector bundle is a vector bundle equivalency. The superposition of vector bundle morphisms (morphisms over X, equivalencies) is again a vector bundle morphism (morphism over X, equivalency).

For an arbitrary vector bundle E there are unique vector bundle morphisms $0 \longrightarrow E$ and $E \longrightarrow 0$ of the zero-bundle 0 over the $\operatorname{bs} E$, the morphisms also being over $\operatorname{bs} E$.

AI.12. Restrictions, pullbacks, products and Whitney sums (see AI.3, 4, 7, 8) of vector bundles have the canonical structures of vector bundles, the involved maps (inclusions, projections) being vector bundle morphisms.

AI.13. A subbundle E' of a k-dimensional vector bundle E is called an *s-dimensional vector subbundle*, $0 \leqq s \leqq k$, if (i) all fibres E'_x of E' are vector subspaces of E_x, the fibres of E are of dimension s, and (ii) each point $x \in \operatorname{bs} E = \operatorname{bs} E'$ has a neighbourhood U and a trivialisation $T: p^{-1}(U) \longrightarrow U \times \mathbb{R}^k$ of E such that $T(E'_x) = \{x\} \times \mathbb{R}^s$, $\mathbb{R}^s$ being included in $\mathbb{R}^k$ in the standard way: $(x_1, x_2, \ldots, x_s) \longmapsto (x_1, x_2, \ldots, x_s, 0, \ldots, 0)$. A vector subbundle of a vector bundle has the canonical structure of a vector bundle. The vector structures on fibres are defined by (i), the maps $T|p^{-1}(U) \cap E'$ may be taken for trivialisations, where T is as in (ii). The inclusion map $j: E' \hookrightarrow E$ is a vector bundle morphism.

AI.14. Let E be a k-dimensional vector bundle with projection p and let E' be its s-dimensional vector subbundle. The relation $p(e_1) = p(e_2)$ and $e_1 - e_2 \in E'_x$ where $x = p(e_1)$ is an equivalence relation in E which is stronger then that defined by p. The corresponding quotient vector bundle E/E' has the canonical structure of a vector bundle whose fibres are quotient vector spaces E_x/E'_x. The natural projection $p: E \longrightarrow E/E'$ is a vector bundle morphism. There is an exact short diagram of vector bundles and vector bundle morphisms over $\operatorname{bs} E = \operatorname{bs} E' = \operatorname{bs} E/E'$

(AI.15) $$0 \longrightarrow E' \xrightarrow{j} E \xrightarrow{p} E/E' \longrightarrow 0$$

The exactness means that at any point x of the base and in any member of the sequence the images and the kernels of corresponding linear maps coincide.

AI.16. We say that a vector bundle E is a *sum* of a finite collection E_i, $1 \leqq i \leqq n$, of its subbundles and write $E = E_1 \oplus E_2 \oplus \cdots \oplus E_n$ if the equality $E_x = E_{1,x} \oplus E_{2,x} \oplus \cdots \oplus E_{n,x}$ holds for all $x \in \operatorname{bs} E$. The *Whitney sum* E of a finite collection E_i, $1 \leqq i \leqq n$, of vector bundles over the same space X turns out to be a sum of E_i's as subbundles if one identifies each E_i with a subbundle of E by the map

$$E_{i,x} \ni e \longmapsto (x, 0, 0, \ldots, e, \ldots, 0) \in E_x \; X E_1 E_2 \cdots E_n.$$
$$\nwarrow$$
$$i\text{th place}$$

This justifies the notation $E_1 \oplus E_2 \oplus \cdots \oplus E_n$ for the Whitney sum of vector bundles E_i, $1 \leqq i \leqq n$.

AI.17. The set $\Gamma(E)$ of all sections of the vector bundle is supplied in a natural way with the structure of a real vector space (the sum and multiplication by a number are defined in a pointwise manner). The same is true for the set $L(E, f)$ of all lifts along a map f into $\operatorname{bs} E$. The maps $f^0: \Gamma(E) \longrightarrow L(E, f)$, $I_f: L(E, f) \longrightarrow \Gamma(f^! E)$ and $f^! = I_f \circ f^0$ (see definitions in AI.9) are linear in these structures.

AI.18. Let V be a finite dimensional real vector space, and p a positive integer. Denote by $F^p(V)$ the vector space of all *p-linear forms* on V, i.e. maps

$\underbrace{V \times V \times \cdots \times V}_{p\text{-times}} \longrightarrow \mathbb{R}$ which are linear in each of their p variables, and denote by $S^p(V)$ and $\Lambda^p(V)$ the subspaces of $F^p(V)$ consisting of symmetric and skew-symmetric forms correspondingly. A form is called *symmetric* (or *skew-symmetric*) if its value does not change (is multiplied by -1) under the permutation of the values of two of its different variables. By definition, $F^1(V) = S^1(V) = \Lambda^1(V) = V^*$, the adjoint space. Let us define $F^0(V) = S^0(V) = \Lambda^0(V) = \mathbb{R}$. If $\dim V = n$, then $\dim F^p(V) = n^p$, $\dim S^p(V) = \binom{n+p-1}{p}$, $\dim \Lambda^p(V) = \binom{n}{p}$. Let $T: V \longrightarrow V'$ be a linear isomorphism between two vector spaces V and V'. It induces the linear isomorphism T^p: $F^p(V) \longrightarrow F^p(V')$ by the formula

(AI.18a)
$$(T^p\psi)(v_1, v_2, \ldots, v_p) = \psi(T^{-1}v_1, T^{-1}v_2, \ldots, T^{-1}v_p),$$
$$\psi \in F^p(V),\ v_i \in V,\ 1 \leqq i \leqq p.$$

The isomorphism T^p preserves the subspaces of symmetric and skew-symmetric forms.

AI.19. Let E be a vector bundle over X and p a nonnegative integer. Define the *bundle of p-forms* on E, $F^p(E)$, as a vector bundle over X whose total space is the union $F^p(E) = \bigcup_{x \in X} F^p(E_x)$ and whose projection assigns the point x to a form $\psi \in F^p(E_x)$. Trivialisations of $F^p(E)$ are obtained from those of E (see AI.10) by applying the inducing operation (AI.18a) to the linear isomorphisms $T_x: E_x \longrightarrow \mathbb{R}^k$ which a trivialisation T induces on fibres, and by identifying $F^p(\mathbb{R}^k)$ with $\mathbb{R}^{k^p}$. The topology on $F^p(E)$ is defined via said trivialisations.

The symmetric and skew-symmetric p-forms form subbundles $S^p(E)$ and $\Lambda^p(E)$ of $F^p(E)$ correspondingly. We have $F^1(E) = S^1(E) = \Lambda^1(E) = E^*$, the *adjoint vector* bundle, and $F^0(E) = S^0(E) = \Lambda^0(E) = X \times \mathbb{R}$, the one-dimensional product vector bundle. A morphism $f: E \longrightarrow E'$ between vector bundles induces the morphism $f^p: F^p(E) \longrightarrow F^p(E')$ in a fibrewise manner by the formula (AI.18a). f^p preserves the subbundles of symmetric and skew-symmetric forms.

AI.20. A *Euclidean structure* on a vector bundle E is a continuous section G of $S^2(E)$ such that each its value G_x is a positively defined bilinear symmetric form on E_x, i.e. $G_x(v, v) > 0$ if $0 \neq v \in E_x$. Such a structure exists if $\operatorname{bs} E$ is paracompact.

Topological Manifolds

AI.21. Two topological spaces X and Y are said to be *locally isomorphic* in points $x \in X$ and $y \in Y$ if there exists neighbourhoods U of x in X, V of y in Y, and a homeomorphism $\varphi: U \longrightarrow V$ such that $\varphi(x) = y$. A triple (U, V, φ) as above is called a *local homeomorphism.* We say that a topological space X is *modelled* by a topological space X_0 if for each point $x \in X$ there exists a point $y \in X_0$ such that X and X_0 are locally isomorphic in points x and y.

AI.22. Let us denote by $\mathbb{R}^n_+$, $n \geqq 1$ an integer, the subspace of $\mathbb{R}^n$ consisting of all $x = (x_1, x_2, \ldots, x_n)$ $\mathbb{R}^n$ with $x_1 \geqq 0$. The boundary $\partial\mathbb{R}^n_+$ of $\mathbb{R}^n_+$ is the set of all points of the form $x = (0, x_2, \ldots, x_n)$. Define $\mathbb{R}^0$ as the single-point space $\{*\}$ and put $\mathbb{R}^0_+ = \partial\mathbb{R}^0 = \phi$, the void set. Denote by $\mathbb{R}^n \cup \mathbb{R}^n_+$ the topological sum of $\mathbb{R}^n$ and $\mathbb{R}^n_+$. Recall that the *topological sum* $\coprod_{k\in K} X_k$ of a collection X_k, $k \in K$, of topological spaces is their disjoint union endowed with the topology whose open sets are precisely those of the form $\bigcup_{k\in K} i_k(U_k)$, $i_k : X_k \longrightarrow \coprod_{k\in K} X_k$ being the inclusion maps, U_k open in X_k, $k \in K$.

AI.23. A topological space X is called a *topological manifold* if it (i) is Hausdorff, (ii) has a denumerable base, and (iii) is modelled by $\mathbb{R}^n \amalg \mathbb{R}^n_+$ for some integer $n \geqq 0$.

It is an immediate consequence of Brower's domain invariance theorem that the integer n in (iii) above is uniquely defined. This number n is called the *dimension* of X and denoted by $\dim X$.

Another consequence of the mentioned theorem is that of *invariancy of boundary*, that is if a point $x \in X$ is carried by a local homeomorphism $X \supset U \xrightarrow{\varphi} V \subset \mathbb{R}^n \amalg \mathbb{R}^n_+$ into a point $\varphi(x)$ belonging to $i_2(\partial\mathbb{R}^n_+)$, $i_2 : \mathbb{R}^n_+ \longrightarrow \mathbb{R}^n \amalg \mathbb{R}^n_+$ being the inclusion, then this property of a point is independent of the choice of local homeomorphism. A point of X possessing the said property is called a *boundary* point. The *boundary* ∂X of a topological manifold is the set of all its boundary points. The boundary of a 0-dimensional manifold is empty by definition. A zero-dimensional manifold itself is a finite or denumerable set endowed with discrete topology.

AI.24. The spaces $\mathbb{R}^n$ and $\mathbb{R}^n_+$ are topological manifolds of dimension n. Any non-empty open subset of a topological manifold itself is a topological manifold of the same dimension. The topological sum $\coprod_{k\in K} X_k$ of finite or countable collection X_k, $k \in K$, of topological manifolds of the same dimension again is a topological manifold, and $\partial \coprod_{k\in K} X_k = \coprod_{k\in K} \partial X_k$, $\dim \coprod_{k\in K} X_k = \dim X_k$. The product $\prod_{k\in K} X_k$ of finite collection X_k, $k \in K$, of topological manifolds again is a topological manifold, and $\partial \prod_{k\in K} X_k = \bigcup_{j\in K} \prod_{k\in K} X_{kj}$ where $X_{kj} = X_k$ if $k \neq j$, and $X_{jj} = \partial X_j$. We have $\dim \prod_{k\in K} X_k = \sum_{k\in K} \dim X_k$.

AI.25. If the boundary ∂X of a topological manifold X is non-empty then it is itself a topological manifold (with an empty boundary) with the induced topology, and $\dim \partial X = \dim X - 1$.

Smooth Structures

AI.26. Let X be an n-dimensional topological manifold. A *chart* in X is a pair (U, α) where U is a non-empty open subset of X, and $\alpha : U \longrightarrow \mathbb{R}^n$, or $\alpha : U \longrightarrow \mathbb{R}^n_+$, maps the set U homeomorphically onto its image, the latter being *open* in $\mathbb{R}^n$, or $\mathbb{R}^n_+$ correspondingly. The set U is called the *domain of a* chart and α the *coordinate map* of a chart. Whereby we consider two sorts of charts: ones with images in $\mathbb{R}^n$

and others with those in $\mathbb{R}^n_+$. The formers will be called $\mathbb{R}^n$-*charts*, the latters $\mathbb{R}^n_+$-*charts* if one needs to emphasize the kind of chart. For convenience let us adopt the notation $\mathbb{R}^n_{(+)}$ which we shall use instead of "$\mathbb{R}^n$ or $\mathbb{R}^n_+$". The *coordinate functions* of a chart $\alpha: U \longrightarrow \mathbb{R}^n_{(+)}$ are $pr_i \circ \alpha: U \longrightarrow \mathbb{R}$ (or $\mathbb{R}_+$ if $i = 1$ and $(+) = +$) where pr_i is the i'th projection map of the product $\mathbb{R}^n = \mathbb{R} \times \mathbb{R} \times \cdots \times \mathbb{R}$ (n-times) or $\mathbb{R}^n_+ = \mathbb{R}_+ \times \underbrace{\mathbb{R} \times \cdots \times \mathbb{R}}_{(n-1)\text{-times}}$. Usually we shall use the notation x_i for the ith coordinate function but sometimes other notation will be preferable.

AI.27. Two charts (U, α) and (V, β) in X are said to be *smoothly compatible* if both the maps $\alpha \circ \beta^{-1}$ and $\beta \circ \alpha^{-1}$ are smooth. Note that they are inverse one to another and are homeomorphic mappings between the sets $\alpha(U \cap V)$ and $\beta(U \cap V)$. The latters are open in $\mathbb{R}^n$ or $\mathbb{R}^n_+$ correspondingly, so the term "smooth" is sensible, it means C^∞ elsewhere in this book. The "void" map which occurs if $U \cap V = \varnothing$ is smooth by definition. An *atlas* in X is a collection of charts whose domains cover X. An atlas is called *smooth* if any two of its charts are smoothly compatible. Two smooth atlases are said to be *equivalent* if their union is again a smooth atlas. A *smooth structure* on X is a class of equivalent smooth atlases. A *smooth manifold* is a topological manifold with a chosen smooth structure. Further we shall always omit the word "smooth" before "manifold". The term "manifold" means "smooth manifold". A chart in a manifold is called *smooth* if it belongs to some smooth atlas of its smooth structure. All smooth charts form the *maximal smooth atlas.* To define a smooth structure on a topological manifold it is sufficient to pick one smooth atlas. A chart will be smooth if and only if it is smoothly compatible with all the charts of a chosen atlas. Any n-dimensional manifold has a smooth atlas consisting only of $\mathbb{R}^n_+$-charts, an $\mathbb{R}^n_+$-*atlas.* Any n-dimensional manifold without boundary (=with no boundary) has a smooth atlas consisting only of $\mathbb{R}^n$-charts, an $\mathbb{R}^n$-*atlas.* A zero-dimensional topological manifold has unique smooth structure. Its smooth atlas consists of the collection of pairs $(\{x\}, m_x)$, x running over the manifold, m_x being the maps $m_x: \{x\} \longrightarrow \mathbb{R}^0 = \{*\}$.

AI.28. If the boundary ∂X of a manifold X is non-empty then it possesses the naturally defined smooth structure. Smooth $\mathbb{R}^{n-1}$-charts of this structure, where $n = \dim X$, are those of the form $(\partial X \cap U, \pi \circ \alpha|_{\partial X \cap U})$. Here (U, α) runs over smooth $\mathbb{R}^n_+$-charts of X with $U \cap \partial X \neq \varnothing$, $\pi: \mathbb{R}^n_+ \longrightarrow \mathbb{R}^{n-1}$ is the projection $\mathbb{R}^n_+ \ni (x_1, x_2, \ldots, x_n) \longmapsto (x_2, \ldots, x_n) \in \mathbb{R}^{n-1}$.

AI.29. Let X and Y be two manifolds of dimensions m and n respectively, and $f: X \longrightarrow Y$ be a map. The *local representative* of f in charts (U, α) in X and (V, β) in Y is the map $\beta \circ f \circ \alpha^{-1}$. It is defined on the open set $\alpha(U \cap f^{-1}(V)) \subset \mathbb{R}^m_{(+)}$ and takes its values in $\mathbb{R}^n_{(+)}$. A map f is said to be *smooth* if all its local representatives in smooth charts are smooth. A smooth map is obviously continuous. The identity map is smooth, the composition of smooth maps again is a smooth map. A map $f: X \longrightarrow Y$ between two smooth manifolds is called a *diffeomorphism* if it is a bijection and both f and f^{-1} are smooth. If there is a diffeomorphism between two manifolds, they are said to be *diffeomorphic*.

A map $f: U \longrightarrow X$ is called a *local diffeomorphism*, U and X being manifolds, if f is a diffeomorphism of U onto some non-empty open subset of X.

The inclusion map $\partial X \hookrightarrow X$ of the boundary is smooth.

AI.30. Any non-empty open subset A of a manifold X inherits a smooth structure, the induced one. Smooth charts of the induced structure are those of the form $(U \cap A, \alpha|_A)$, (U, α) running over smooth charts of X. The inclusion map $i: A \hookrightarrow X$ is smooth.

AI.31. The product $\prod_{k \in K} X_k$ of the finite collection X_k, $k \in K$, of manifolds, at most one of X_k having non-empty boundary, has the naturally defined smooth structure. For a smooth atlas in $\prod_{k \in K} X_k$ one may take the collection of charts of the form $(\prod_{k \in K} U_k, \prod_{k \in K} \alpha_k)$ where (U_k, α_k), $k \in K$, runs over the $\mathbb{R}^{n_k}$-atlas of X_k, $n_k = \dim X_k$, if $\partial X_k = \varnothing$, and $\mathbb{R}^{n_k}_+$-atlas otherwise, the latter case occurring for at most one value of k. The map $\prod_{k \in K} \alpha_k$ assigns to a point $\{x_k\} \in \prod_k U_k$ the point $\{\alpha_k(x_k)\} \in \prod_k Y_k$, where $Y_k = \mathbb{R}^{n_k}_{(+)}$. The space $\prod_k Y_k$ may be identified with $\mathbb{R}^{\sum n_k}_{(+)}$ in the standard way. The projection maps $\pi_r: \prod_{k \in K} X_k \longrightarrow X_r$, $r \in K$, are smooth ones.

AI.32. The topological sum $\coprod_{k \in K} X_k$ of a finite or countable collection $X_k, k \in K$, of manifolds of the same dimension also has the naturally defined smooth structure in which the inclusion maps $i_r: X_r \longrightarrow \coprod_{k \in K} X_k$, $r \in K$, are smooth. The charts of the form $(i_k(U), \alpha \circ [i_k|_{i_k(U)}]^{-1})$, where k runs over K, (α, U) runs over some smooth atlas of X_k, make up a smooth atlas of this structure.

AI.33. $\mathbb{R}^n$ and $\mathbb{R}^n_+$ possess the *standard smooth structures*. In both cases we define them by taking the *standard atlases* consisting of only one map, namely the identity.

AI.34. Two smooth charts (U, α) and (V, β) in a manifold are called *orientably compatible* if the determinant of the matrix of the first derivatives of $\alpha \circ \beta^{-1}$ is elsewhere positive. A smooth atlas is *orientable* if any two of its charts are orientably compatible. A manifold is called *orientable* if its smooth structure has an orientable atlas. Two orientable smooth atlases are orientably equivalent if their union again is an orientable atlas. The *orientation* of an orientable manifold is the class of orientably equivalent atlases. It contains the maximal orientable atlas, the union of all orientable atlases of the orientation.

AI.35. Let X be a manifold. A smooth real valued function on X is a smooth map $f: X \longrightarrow \mathbb{R}$. The support supp f of a function f is the closure in X of the set $\{x \in X: f(x) \neq 0\}$. Let $\{U_k, k \in K\}$ be an arbitrary covering of X by open sets. A *partition of unity subordinated to a covering* is a family $\{\eta_i, i \in I\}$ of smooth functions such that (i) the values of η_i lie in $[0, 1]$, (ii) supp η_i is compact and is contained in some U_k, (iii) the collection of sets $\{\text{supp}\, \eta_i, i \in I\}$ is a locally-finite covering of X, i.e. each point $x \in X$ belongs to some supp η_i and has a neighbourhood which intersects only a finite number of supp η_i's, (iv) $\sum_{i \in I} \eta_i(x) = 1$ for all $x \in X$. The sum in (iv) makes sense as it actually spreads over only a finite number of nonzero summands.

Given an open covering, there always exists a partition of unity subordinate to it.

Smooth Vector Bundles

AI.36. Let E be a vector bundle over a topological manifold X. Then AI.10 shows that E, the total space, is a topological manifold whose dimension equals $n+k$, $n = \dim X$, k being the dimension of fibres of E.

A *smooth vector bundle* is a vector bundle over a manifold whose total space is supplied with a smooth structure defined by a smooth atlas consisting of smooth trivialisations. Note that the total space of the product-bundle $U \times \mathbb{R}^k$ which occurs in the definition of a trivialisation (see AI.10), U being open in a manifold, has the smooth structure defined in AI.30–31. The projection of a smooth vector bundle is a smooth map.

The boundary of the total space of a smooth vector bundle coincides with the total space of the restriction of the bundle onto the boundary of the base. The restriction also being a smooth vector bundle.

The zero bundle 0 over a manifold X is a smooth vector bundle.

The product vector bundle $X \times \mathbb{R}^k$ where X is a manifold is a smooth vector bundle.

We may consider smooth vector bundle morphisms between smooth vector bundles, smooth vector bundle morphisms over a manifold and smooth equivalencies.

Smooth sections of a smooth vector bundle E and smooth lifts of a smooth map f of some smooth manifold into bsE form correspondingly the vector subspaces $\Gamma^\infty(E)$ of $\Gamma(E)$ and $L^\infty(E,f)$ of $L(E,f)$.

AI.37. The product of a finite collection of smooth vector bundles is again a smooth vector bundle provided at most one member of the collection has non-empty boundary. The sum of a finite or countable collection of smooth vector bundles of the same dimension of bases and the same dimension of fibres is again a smooth vector bundle. In both constructions the total space of the resulting bundle is endowed with the smooth structure of the product or the sum of smooth manifolds respectively.

AI.38. A *vector subbundle* E' of a smooth vector bundle E is called *smooth* if trivialisations in AI.13 may be chosen to be smooth. Such trivialisations convert E' itself into a smooth vector bundle.

AI.39. The bundle of p-forms $F^p(E)$ on a smooth vector bundle E has the natural structure of a smooth vector bundle. Its smooth trivialisations are obtained from those of E by applying the induction operation described in AI.19.

The vector bundles $S^p(E)$ and $\Lambda^p(E)$ of symmetric and skew-symmetric forms in a smooth vector bundle E are smooth vector subbundles of $F^p(E)$.

AI.40. Any smooth vector bundle E admits a smooth Euclidean structure, i.e. a smooth section $g\colon \operatorname{bs} E \longrightarrow S^2(E)$ such that g_x is positively defined on E_x for all $x \in \operatorname{bs} E$.

AI.41. The quotient bundle E/E' of a smooth vector bundle E with respect to its smooth vector subbundle E' has the canonically defined smooth structure of the smooth vector bundle, the canonical projection $p: E \longrightarrow E/E'$ being smooth.

AI.42. The pullback $f^!E$ of a smooth vector bundle E along a smooth map $f: X \longrightarrow \operatorname{bs} E$ again has the canonical structure of a smooth vector bundle. Note that, if both X and E have non-empty boundary, the space $X \times E$ is not a smooth manifold but nevertheless $f^!E \subset X \times E$ has the canonically defined smooth structure "induced" by those of X and E. The adjoint map $\operatorname{ad} f: f^!E \longrightarrow E$ is a smooth vector bundle morphism. The maps f^0, I_f and $f^!$ defined in AI.9, 17 carry smooth sections into smooth ones.

The Whitney sum of smooth vector bundles over the same base again is a smooth vector bundle being supplied with the canonically defined smooth structure. The same reservation as in the case of the pullback must be made when defining the smooth structure.

The Tangent Bundle

AI.43. A *local function* at a point x of a manifold X is a smooth real-valued function defined in some open neighbourhood of x. The set of all local functions in the point $x \in X$ is denoted by $\mathscr{F}_x$, or $\mathscr{F}_x(X)$. Three operations are defined on $\mathscr{F}_x$ in a pointwise manner: the sum $\varphi_1, \varphi_2 \longmapsto \varphi_1 + \varphi_2$ and the product $\varphi_1, \varphi_2 \longmapsto \varphi_1 \cdot \varphi_2$ of two local functions φ_1, φ_2, and the product $c, \varphi \longmapsto c\varphi$ of a real number c and a local function φ, $c\varphi$ being defined on the same open set in X as φ, the domain of $\varphi_1 + \varphi_2$ and $\varphi_1 \cdot \varphi_2$ being the intersection of those of φ_1 and φ_2. A *tangent vector* in the point x is a map $v: \mathscr{F}_x \longrightarrow \mathbb{R}$ such that

(i) $v(c_1\varphi_1 + c_2\varphi_2) = c_1 v(\varphi_1) + c_2 v(\varphi_2)$ for all $c_1, c_2 \in \mathbb{R}$ and $\varphi_1, \varphi_2 \in \mathscr{F}_x$ (linearity),

(ii) $v(\varphi_1 \cdot \varphi_2) = \varphi_1(x) v(\varphi_2) + \varphi_2(x) v(\varphi_1)$ for all φ_1 and $\varphi_2 \in \mathscr{F}_x$ (differential property).

The multiplication by the number and the sum, being applied to tangent vectors regarded as real valued functions on $\mathscr{F}_x$, yield again tangent vectors. So, the set T_xX of all tangent vectors at the point x inherits the structure of a real vector space, it is called the *tangent space* of X as the *point* x. If $\dim X = 0$, then T_xX consists of a single point, the zero tangent vector: $0(\varphi) = 0 \in \mathbb{R}$ for all $\varphi \in \mathscr{F}_x = \mathbb{R}$. Let $\dim X = n \geqq 1$, and let (U, α) be a smooth chart in X with $x \in U$. Then the set of maps $\partial/\partial x_i|_x: \mathscr{F}_x \longrightarrow \mathbb{R}$, $1 \leqq i \leqq n$, defined by

$$\left.\frac{\partial}{\partial x_i}\right|_x (\varphi) = \frac{\partial}{\partial x_i}(\varphi \circ \alpha^{-1})(x_1, x_2, \ldots, x_n)|_{(x_1, x_2, \ldots, x_n) = 0(x)},$$

$(x_1, x_2, \ldots, x_n)$ being the coordinates in $\mathbb{R}^n$, $\partial/\partial x_i$ in the right-hand side of the formula being the usual partial derivatives of a real-valued function in Euclidean space with respect to the variable x_i, form the vector base in T_xX. Thereby, $\dim T_xX = \dim X$.

Any smooth map $f: X \longrightarrow Y$ between two manifolds induces a linear map, the *tangent map in the point* $x \in X$, $T_x f: T_x X \longrightarrow T_{f(x)} Y$ by the formula

$$((T_x f)(v))(\varphi) = v(\varphi \circ f) \quad \text{for all } \varphi \in \mathscr{F}_{f(x)}(Y),\ v \in T_x X.$$

AI.44. The *tangent bundle* of a manifold X is the bundle (TX, p_X, X) whose total space is the union $TX = \bigcup_{x \in X} T_x X$ of all tangent spaces at points of X, and whose projection p_X assigns the point x to a vector $v \in T_x X$. If $\dim X = 0$ then $TX = X$, the topology being discrete. Let $\dim X = n \geqq 1$. In this case any smooth chart $\alpha: U \longrightarrow \mathbb{R}^n_{(+)}$ generates the map

(AI.44a) $$T\alpha: T_u X = \bigcup_{x \in U} T_x X \longrightarrow \mathbb{R}^n_{(+)} \times \mathbb{R}^n$$

by the formula $T\alpha(v) = (\alpha \circ p_X(v), (v_1, v_2, \ldots, v_n))$, the numbers v_i, $1 \leqq i \leqq n$, being the coefficients in the expansion

$$v = \sum_{i=1}^{n} v_i \left.\frac{\partial}{\partial x_i}\right|_x, \quad x = p_X(v).$$

The topology in TX is defined by the requirement for all the maps (AI.44a) to be continuous and by the property to be the weakest among topologies satisfying the first requirement. Together with the vector space structure in the $T_x X$'s this topology converts TX into a vector bundle. The maps (AI.44a) are charts, they form a smooth atlas in TX provided (U, α) runs over all smooth charts in X. Thus TX becomes a manifold. The pair $(T_U X, (\alpha^{-1} \times 1_{\mathbb{R}^n}) \circ T\alpha)$ is a smooth trivialisation, so TX is a smooth vector bundle. The smooth chart $(T_U X, T\alpha)$ in TX is called the *chart associated with a chart* (U, α). The restriction of TX onto a non-empty open subset $U \subset X$ is denoted by $T_U X$. Any smooth map $f: X \longrightarrow Y$ between manifolds X and Y generates a smooth morphism $Tf: TX \longrightarrow TY$, the *tangent map* of f, which coincides with $T_x f$ on the fibre $T_x X$ for each $x \in X$. Natural properties of the tangent map are: $T1_x = 1_{TX}$ and $T(f \circ g) = Tf \circ Tg$ for smooth maps $g: X \longrightarrow Y$ and $f: Y \longrightarrow Z$.

AI.45. Sections of TX are called *vector fields* on X. Smooth vector fields form a subspace of $\Gamma(TX)$ which we denote by $\mathscr{X}$, or $\mathscr{X}(X)$. If $\dim X = n \geqq 1$, any smooth chart (U, α) in X *generates* n smooth vector fields $\partial/\partial x_i$, $1 \leqq i \leqq n$, on U. The latters are sections of $T_U X$, their values at a point $x \in U$ are $\partial/\partial x_i|_x$ correspondingly. Any vector field ξ on a manifold X has the following local representation in a chart (U, α):

(AI.45a) $$\xi|_U = \sum_{i=1}^{n} \xi_i \frac{\partial}{\partial x_i}$$

where ξ_i are real-valued functions. The latters are smooth provided ξ is.

If ξ is a vector field on a manifold X which is nonzero at a point x_0, then there is a chart (U, α), $U \ni x_0$, such that in this chart $\xi|_U = \partial/\partial x_1$ (*local straightening of a vector field*). Otherwise, if $\xi x_0 = 0$, the problem of describing ξ in a neighbourhood of a point x_0 becomes quite difficult.

Let $\varphi: U \longrightarrow \mathbb{R}$ be a smooth function, $U \subset X$ a non-empty open subset. We define the *application* of a vector field $\xi \in \mathscr{X}(X)$ to a function φ as $\xi\varphi \in C^\infty(U)$ where $\xi\varphi(x) = \xi_x(\varphi)$, $x \in U$, φ being regarded in the right-hand side of the last expression as a member of $\mathscr{F}_x$ (see AI.43). The map $\mathscr{X}(X) \times C^\infty(U) \longrightarrow C^\infty(U)$ defined by the formula $(\xi, \varphi) \longmapsto \xi\varphi$ is bilinear.

The space $\mathscr{X}(X)$ of all smooth vector fields on a manifold X is a Lie algebra. The *Lie bracket* $[\xi_1, \xi_2]$ of two vector fields ξ_1, ξ_2 is the field whose value at a point x is the vector $[\xi_1, \xi_2]_x : \mathscr{F}_x \longrightarrow \mathbb{R}$ defined as

$$[\xi_1, \xi_2]_x(\varphi) = \xi_{2x}(\xi_1\varphi) - \xi_{1x}(\xi_2\varphi) \tag{AI.45b}$$

where the expressions in parenthesis mean the applications of ξ_i to a function φ. One can easily verify that the right-hand side of (AI.45b) satisfies the conditions (i) and (ii) of AI.43. The Lie bracket is bilinear and skew-symmetric. The *Lie identity*

$$[[\xi_1, \xi_2], \xi_3] + [[\xi_2, \xi_3], \xi_1] + [[\xi_3, \xi_1], \xi_2] = 0 \tag{AI.45c}$$

holds for any vector fields ξ_i, $i = 1, 2, 3$.

AI.46. The tangent bundles of $\mathbb{R}^n$ and $\mathbb{R}^n_+$ are naturally isomorphic to the product vector bundles $(\mathbb{R}^n \times \mathbb{R}^n, p_1, \mathbb{R}^n)$ and $(\mathbb{R}^n_+ \times \mathbb{R}^n, p_1, \mathbb{R}^n_+)$ respectively, p_1 being the first projection, via the morphisms (AI.44a) with $(U, \alpha) = (\mathbb{R}^n, 1_{\mathbb{R}^n})$ and $(\mathbb{R}^n_+, 1_{\mathbb{R}^n_+})$. The said charts generate the *standard vector fields* $\partial/\partial x_i$, $1 \leqq i \leqq n$, on $\mathbb{R}^n$ and $\mathbb{R}^n_+$ correspondingly. In the case $n = 1$ we often use the letter t for denoting the variable and write d/dt for the standard vector field.

AI.47. Let X_k, $k \in K$, be a finite collection of manifolds, with at most one of the X_k's having non-empty boundary. Consider the product $X = \prod_{k\in K} X_k$ (see definition AI.31) and let $p_j : X \longrightarrow X_j$ be the jth projection map. Then the tangents $T_x p_j : T_x X \longrightarrow T_{p_j(x)} X_j$, $j \in K$, induce the natural isomorphism

$$T_x X \cong \sum_{k\in X} \oplus T_{p_k(x)} X_k, \quad x \in X,$$

so that the tangent bundle TX of the product is naturally and smoothly equivalent to the Whitney sum $\sum_{k\in K} \oplus p_k^! TX_k$ via these pointwise isomorphisms. We may regard $p^! TX_k$ as a subbundle of TX, the subbundle of vectors tangent to X_k. Define linear maps $p_k^! : \mathscr{X}(X_k) \longrightarrow \mathscr{X}(X)$ by taking the superposition of that defined in AI.9 (the pullback of sections with respect to p_k) with the obvious inclusion $\mathscr{X}(p_k^! X_k) \hookrightarrow \mathscr{X}(X)$.

AI.48. Let $p : E \longrightarrow X$ be a smooth vector bundle. Then the commutative diagram

$$\begin{array}{ccc} TE & \xrightarrow{\;Tp\;} & TX \\ \Big\downarrow{\scriptstyle p_{TE}} & & \Big\downarrow{\scriptstyle p_X} \\ E & \xrightarrow{\;\;p\;\;} & X \end{array}$$

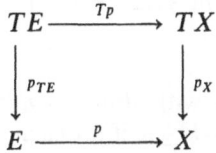

may be extended to the following commutative diagram

(AI.48a)

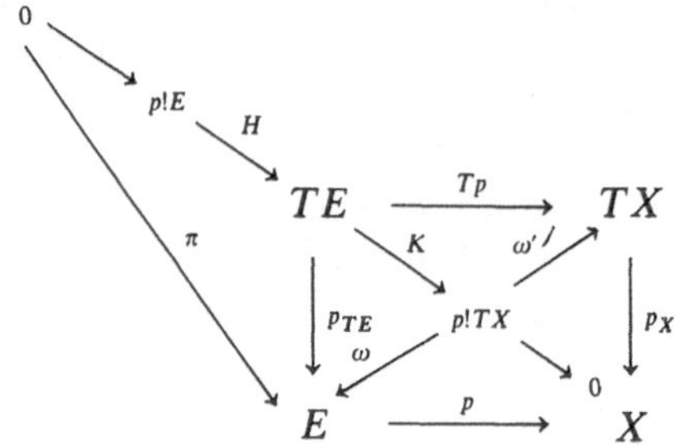

The diagonal sequence

(AI.48b)
$$0 \longrightarrow p^! E \xrightarrow{H} TE \xrightarrow{K} p^! TX \longrightarrow 0$$

is a short exact sequence of smooth bundles *over* E. The maps π and ω in (AI.48a) are the projections of the pullbacks $p^! E$ and $p^! TX$, $\omega' = \operatorname{ad} p$. The morphism H over E is defined by the formula $H(e, e') = w$, where e, $e' \in E_x$, the same fibre of E, and $w \in T_e E$ acts on a local function $\varphi \in \mathscr{F}_e(E)$ as the derivative in the direction e':

$$w(\varphi) = \frac{d}{dt}(e + te')|_{t=0}.$$

The map H naturally embeds the bundle $p^! E$ into TE as a smooth vector subbundle, the subbundle of vertical tangent vectors. The latters are precisely those which belong to $\ker T_e p$, $e \in E$. The formula for K is $K(w) = (p_{TE}(w), T_p(w)) \in p^! TX$.

Submanifolds

AI.49. Consider a smooth map $f : X \longrightarrow Y$ between manifolds X and Y and its tangent map in a point $x \in X$:

(+)
$$T_x f : T_x X \longrightarrow T_{f(x)} Y.$$

The map f is said to be
 (i) *immersive* in the point x if (+) is injective,
 (ii) *submersive* in the point x if (+) is surjective.

A smooth map is an *immersion* (*submersion*) if it is immersive (submersive) in each point of its domain. A smooth map $f : X \longrightarrow Y$ between the manifolds X and Y is called an *embedding* if it is an immersion and maps the space X homeomorphically onto its image.

AI.50. The inclusion $i : \mathbb{R}^n_+ \hookrightarrow \mathbb{R}^n_+$, $1 \leqq k \leqq n$, defined by $i(x_1, x_2, \ldots, x_k) = (x_1, x_2, \ldots, x_k, 0, \ldots, 0)$, is an embedding. The same is true for $j : \mathbb{R}^n_+ \hookrightarrow \mathbb{R}^n$, these superposition of i and the inclusion $\mathbb{R}^n_+ \hookrightarrow \mathbb{R}^n$.

AI.51. Let X be a manifold with $\dim X = n \geqq 1$. A subset A of X is called a *k-dimensional submanifold*, $1 \leqq k \leqq n$, if for each point $x \in A$ there are: a neighbourhood U of x in X, an *embedding* $g: U \longrightarrow \mathbb{R}^n$, and an *open* set V in $\mathbb{R}^k_+$ (the latter being embedded into $\mathbb{R}^n$ via the map j defined in AI.50) such that $A \cap U = g^{-1}(V)$.

A *zero-dimensional submanifold* A of an arbitrary manifold X is, by definition, a subset of X whose inherited topology is discrete.

The *codimension* of a k-dimensional submanifold A of an n-dimensional manifold X equals $n-k$ and is denoted by codim A.

A k-dimensional submanifold is itself a k-dimensional topological manifold regarded as a subspace. It also has the naturally defined smooth structure, the *induced*, or *inherited* one. If $k \geqq 1$, the pairs $(A \cap U, g|A \cap U)$, where U and g are from the above definition, form a smooth atlas of the induced structure. Further, any submanifold will be meant as a (smooth) manifold, the smooth structure being the induced one.

The boundary ∂X of a manifold is a submanifold provided it is non-empty. In this case codim $\partial X = 1$.

The inclusion map $i: A \hookrightarrow X$ of a submanifold of a manifold X is an embedding. Its tangent $Ti: TA \longrightarrow TX$ is also an embedding, and we identify TA with its image under Ti. Thus, for each $x \in A$ the tangent space $T_x A$ is a subspace of $T_x X$.

A subset of a manifold is a submanifold if and only if it is the image of some embedding.

AI.52. Let A be a submanifold of a manifold X, $i_A: A \hookrightarrow X$ be an inclusion, and $p: E \longrightarrow X$ be a smooth vector bundle. Then $E|_A = i_A^! E$ is a smooth vector bundle over A (we identify the restricted bundle $\bigcup_{x \in A} E_A$ with the pullback bundle $\bigcup_{x \in A} \{x\} \times E_x$ in the obvious way). In particular, $i_A^! TX = T_A X$ is a smooth vector bundle over A. The tangent map $Ti_A: TA \longrightarrow TX$ may be decomposed into the superposition

$$TA \xrightarrow{\text{corr } Ti_A} T_A X \xrightarrow{\text{ad } i_A} TX$$

The correction map corr Ti_A (see AI.4) embeds TA in $T_A X$ as a smooth vector subbundle. The quotient vector bundle $T_A X / TA$ over A is called the *normal bundle* of a submanifold A.

AI.53. Let X and Y be manifolds and $f: X \longrightarrow Y$ a smooth map. A point $x \in X$ is called *regular* if f is submersive in x (see AI.49). Otherwise x is called *critical*, and $f(x) \in Y$ is called a critical value. A point $y \in f(X)$ is a *regular value* if it is not a critical value.

A submanifold $A \subset X$ is called *tame* if $\partial A = A \cap \partial X$ and for each $x \in A \cap \partial X$ there is a smooth chart $\alpha: U \longrightarrow \mathbb{R}^n_+$ of X, $n = \dim X$, such that $x \in U$ and $A \cap U = \alpha^{-1}(\mathbb{R}^k_+)$, $k = \dim A$, $\mathbb{R}^n_+$ being embedded in $\mathbb{R}^n_+$ via the map i of AI.50. For a zero-dimensional submanifold A this definition reads as $A \cap \partial X = \varnothing$.

If $y \in Y \setminus \partial Y$ is a regular value both of f and $f|\partial X$ then $f^{-1}(\{y\})$ is a tame submanifold of X, and codim $f^{-1}(\{y\}) = \dim Y$.

A smooth map $f: X \longrightarrow Y$ is said to be *transversal* to a submanifold $B \subset Y$ if for each point $x \in X$ such that $f(x) \in B$ the subspaces $T_x f(T_x X)$ and $T_{f(x)} B$ span $T_{f(x)} Y$.

Let Y be a manifold and B be its tame submanifold. If $f: X \longrightarrow Y$ is a smooth map transversal to B and $f(X) \cap \partial Y = \varnothing$, then $f^{-1}(B)$ is a tame submanifold of X and codim $f^{-1}(B) =$ codim B (see Hirsch (1976)).

AI.54. (Tubular Neighbourhood Theorem) *Let A be a tame submanifold of a manifold X. Then A has an open neighbourhood in X which is diffeomorphic to the normal bundle $T_A X/TA$* (see AI.52) *via a diffeomorphism leaving A fixed.* (Recall that in accordance with our convention in AI.10 the space A is embedded into any vector bundle over A as the image of the zero section.) Such a neighbourhood together with the said diffeomorphism is called a *tubular neighbourhood* of A in X (see Hirsch (1976) Theorem 6.3).

AI.55. (Collar Theorem) *Let X be a manifold with non-empty boundary ∂X. Then ∂X has an open neighbourhood in X which is diffeomorphic* to $\partial X \times \mathbb{R}_+$ *via a diffeomorphism which leaves ∂X fixed, the latter being included in $\partial X \times \mathbb{R}_+$ as $\partial X \times \{0\}$* (see Hirsch (1976) Theorem 6.1).

Such a neighbourhood together with the said diffeomorphism, say g, is called a *collar* of X. If ∂X has many connected components, and C is a subset of ∂X consisting of the whole components then we may consider the *partial collar* $(g^{-1}(C \times \mathbb{R}_+), g|g^{-1}(C \times \mathbb{R}_+))$.

Exterior Algebra of a Manifold

AI.56. Let X be a manifold of dimension $n \geqq 1$, and p be an integer, $0 \leqq p \leqq n$. The *bundle of exterior p-forms* on X is, by definition, the bundle of skew-symmetric forms $\Lambda^p TX$ of the tangent bundle (see (AI.19, 39, for brevity we omit the parenthesis in the notation). We mark two particular cases of this bundle. The bundle $\Lambda^1 TX$ coincides with the bundle of linear forms on the tangent space, the *cotangent bundle* of a manifold X, and is denoted also by T^*X, or by (T^*X, π_X^*, X). The bundle $\Lambda^0 TX$ coincides with the product bundle $X \times \mathbb{R}$.

The space of all smooth sections of $\Lambda^p TX$ will be denoted by $\mathscr{E}^p$, or $\mathscr{E}^p(X)$, members of $\mathscr{E}^p(X)$ are called *exterior p-forms* on a manifold X. In particular, $\mathscr{E}^0(X)$ coincides with the space $C^\infty(X)$ of real valued smooth functions on X. Define $\mathscr{E}^p(X) = \{0\}$, the zero-dimensional vector space if $p < 0$ or $p > n$.

AI.57. The *exterior product* of exterior forms $\alpha \in \mathscr{E}^p$ and $\beta \in \mathscr{E}^q$ is defined in a pointwise manner (see 1.24) and again is an exterior form which belongs to $\mathscr{E}^{p+q}$.

The *inner product* of a vector field $\xi \in \mathscr{X}$ and an exterior form $\alpha \in \mathscr{E}^p$, $p \geqq 1$, is an exterior form $\xi \lrcorner \alpha \in \mathscr{E}^{p-1}$ defined by the formula

$$(\xi \lrcorner \alpha)_x(v_1, v_2, \ldots, v_{p-1}) = \alpha(\xi_x, v_1, v_2, \ldots, v_{p-1}) \quad \forall v_1, v_2, \ldots, v_{p-1} \in T_x X, \ \forall x \in X.$$

Both the products are bilinear with respect to the vector space structures on $\mathscr{E}^p$, $\mathscr{E}^q$, and $\mathscr{X}$.

AI.58. The *differential* $d: \mathscr{E}^p \longrightarrow \mathscr{E}^{p+1}$ is a linear map which is uniquely defined by the properties

(i) $(d\varphi)_x = T_x\varphi : T_xX \longrightarrow T_{\varphi(x)}\mathbb{R} = \mathbb{R}$ if $\varphi \in \mathscr{E}^0 = C^\infty(X)$,

(ii) $d \circ d = 0$,

(iii) $d(\alpha \wedge \beta) = (d\alpha) \wedge \beta + (-1)^p \alpha \wedge d\beta$ if $\alpha \in \mathscr{E}^p$, $\beta \in \mathscr{E}^q$, $p, q \geqq 0$.

The following property may be taken also as the equivalent definition. If $\alpha \in \mathscr{E}^p$, $p \geqq 0$, $\xi_i \in \mathscr{X}$, $1 \leqq i \leqq p+1$, then

$$\text{(iv)}\quad d\alpha(\xi_1, \xi_2, \ldots, \xi_{p+1}) = \sum_{i=1}^{p+1} (-1)^{i+1} \xi_i \alpha(\xi_1, \xi_2, \ldots, \hat{\xi}_i, \ldots, \xi_{p+1}) + \sum_{i<j} (-1)^{i+j} \alpha([\xi_i, \xi_j], \xi_1, \ldots, \hat{\xi}_i, \ldots, \hat{\xi}_j, \ldots, \xi_{p+1}).$$

Here $\hat{}$ denotes the absence of the corresponding variable, $\xi\varphi$ denotes the application of a vector field ξ to a function φ, $[\xi_i, \xi_j]$ denotes the Lie bracket of vector fields ξ_i and ξ_j (see AI.45).

AI.59. Any smooth map $f : X \longrightarrow Y$ between manifold X and Y induces the bilinear map $f^* : \mathscr{E}^p(Y) \longrightarrow \mathscr{E}^p(X)$ which acts by the formula

$$f^*\alpha(v_1, v_2, \ldots, v_p) = \alpha(T_xfv_1, T_xfv_2, \ldots, T_xfv_p),$$

if $\alpha \in \mathscr{E}^p(Y)$, $v_1, v_2, \ldots, v_p \in T_xX$, $x \in X$, $1 \leqq p \leqq \dim Y$, and $f^*\varphi = \varphi \circ f$ if $\varphi \in \mathscr{E}^0(Y)$.

If U is a non-empty open subset of X then the restriction $\alpha|_U$ of an exterior form α coincides with $i^*\alpha$ where $i : U \hookrightarrow X$ is the inclusion.

The map f^* preserves exterior products and differentials:

$$f^*(\alpha \wedge \beta) = (f^*\alpha) \wedge (f^*\beta),$$
$$f^*d\alpha = d(f^*\alpha).$$

In particular $(\alpha \wedge \beta)|_U = \alpha|_U \wedge \beta|_U$ and $(d\alpha)|_U = d(\alpha|_U)$.

We have also the contravariancy property $(f \circ g)^* = g^* \circ f^*$.

AI.60. Let $\alpha : U \longrightarrow \mathbb{R}^n_{(+)}$ be a smooth chart in a manifold X, and $pr_i : \mathbb{R}^n \longrightarrow \mathbb{R}$ be the ith projection, $1 \leqq i \leqq n$. Then the composition $x^i = pr_i \circ \alpha$ is a smooth function on U, called the *ith coordinate*. The differential dx^i of these functions are members of $\mathscr{E}^1(U) = \Gamma^\infty(T^*_UX)$. Their values at the point $x \in X$ form the vector base in T^*_xX, the dual to the basis $\{\partial/\partial x^i\}$ in T_xX. The formulae

$$\text{(60a)}\qquad dx^i\left(\frac{\partial}{\partial x^j}\right) = \delta^i_j \quad \text{(the Kronecker symbol)}$$

follow immediately from the definition.

Any cotangent vector $p \in T^*_xX$ has the expansion $p = \sum_{i=1}^n p_i \, dx^i|_x$. Thus we have the *associated chart* $(T^*_UX, \tilde{\alpha})$ in the cotangent bundle T^*X where

$$\tilde{\alpha}(p) = (x^1, x^2, \ldots, x^n, p_1, p_2, \ldots, p_n) \in \mathbb{R}^n_{(+)} \times \mathbb{R}^n,$$
$$(x^1, x^2, \ldots, x^n) = \alpha(x), \quad x = \pi^*_X(p).$$

The values of the differentials $\{dx^i, dp_i, 1 \leqq i \leqq n\}$ of these coordinates in the point $p \in T^*X$ form the associated vector base in $T^*_p T^*X$, and so do the values of vector fields

$\{\partial/\partial x^i, \partial/\partial p_i, 1 \leqq i \leqq n\}$ with respect to $T_p T^*X$.

The exterior products $dx^{i_1} \wedge dx^{i_2} \wedge \cdots \wedge dx^{i_p}$, $1 < i_1 < i_2 < \cdots < i_p \leqq n$ form a vector base in $\Lambda^p_x TX$, $p \geqq 1$, being valued at the point $x \in U$. Any exterior p-form β has a local representation in a chart (U, α):

(60b) $$\beta|_U = \sum_{i_1 i_2 \cdots i_p} \beta_{i_1 i_2 \cdots i_p} dx^{i_1} \wedge dx^{i_2} \wedge \cdots \wedge dx^{i_p},$$

where $\beta_{i_1 i_2 \cdots i_p}$ are smooth real-valued functions in U. The local representations of the differential are

(60c) $$d\varphi|_U = \sum_{i=1}^{n} \frac{\partial \varphi}{\partial x^i} dx^i, \quad \varphi \in \mathscr{E}^0 = C^\infty(X),$$

(60d) $$d\beta|_U = \sum_{i_1 i_2 \cdots i_p} (d\beta_{i_1 i_2 \cdots i_p}) \wedge dx^{i_1} \wedge dx^{i_2} \wedge \cdots \wedge dx^{i_p},$$

$\beta \in \mathscr{E}^p$, $p \geqq 1$ having the local representation (60b). In (60c) the function $\partial\varphi/\partial x^i$ may be regarded as the result of the application of the vector field $\partial/\partial x^i$ to φ (see AI.45) or, equivalently, $\partial\varphi/\partial x^i = d\varphi(\partial/\partial x^i)$. The 1-form $d\beta_{i_1 i_2 \cdots i_p}$ in (60d) is given by (60c).

AI.61. The *de Rham complex* of a manifold X is the sequence $\mathscr{E}^p(X)$, $p \in \mathbb{Z}$ supplied with the differential operation $d\colon \mathscr{E}^p(X)$, $\mathscr{E}^{p+1}(X)$ defined in AI.58. The *de Rham cohomology groups* $H^p_{\partial R}(X)$, $p \in \mathbb{Z}$, are those of the de Rham complex. Recall that

$$H^p_{\partial R}(X) = \operatorname{im} d|_{\mathscr{E}^{p-1}} / \ker d|_{\mathscr{E}^p}$$

de Rham's theorem asserts that there is a natural isomorphism between de Rham cohomology groups and singular cohomology groups with real coefficients: $H^p_{\partial R}(X) \cong H^p(X; \mathbb{R})$. It follows from algebraic topology that the latters are homotopical invariants (see Dold (1972)).

AI.62. (*Orientation forms*). Let X be an n-dimensional manifold. A form $\Omega \in \mathscr{E}^n(X)$ is called a *volume form* if it vanishes nowhere, i.e. $\Omega_x \neq 0$ $\forall x \in X$. A manifold X is orientable if and only if it has a volume form. Any volume form defines the orientation (see AI.34). A chart (U, α) belongs to the maximal orientable atlas of this orientation if the local representative of Ω in (U, α) is $\Omega|_U = a\, dx^1 \wedge dx^2 \wedge \cdots \wedge dx^n$ where a is a *positive* function. Conversely, given the orientation of X, one can construct, by means of partition of unity, the volume form which generates the orientation in the said sense.

Families of Maps

AI.63. A one parameter family $f_t\colon X \longrightarrow Y$ of smooth maps between manifolds X and Y, t running over an open interval Δ in $\mathbb{R}$, is said to be *smooth* if the map $F\colon X \times \Delta \longrightarrow Y$ defined by $F(x, t) = f_t(x)$, $x \in X$, $t \in \Delta$, is smooth. The words "one-parameter" and "smooth" will be omitted in future. Unless otherwise stated, "family" means "one-parameter smooth family". The map F which occurred in the above definition will be referred to as the *defining map* of a family.

A *time-independent family* $f_t = f : X \longrightarrow Y$ has the defining map $(x, t) \longmapsto f(x)$. A *family of inclusions* $i_t : X \longrightarrow X \times \Delta$, $t \in \Delta$, defined by $i_t(x) = (x, t)$, has the identity $1_{X \times \Delta}$ as the defining map. An arbitrary family $f_t : X \longrightarrow Y$, $t \in \Delta$, can be written by means of its defining map F and the family $i_t : X \longrightarrow X \times \Delta$ of inclusions as $f_t = F \circ i_t$.

The superposition $g_t \circ f_t : X \longrightarrow Z$, $t \in \Delta \cap \Gamma$, of families $f_t : X \longrightarrow Y$, $t \in \Delta$, and $g_t : Y \longrightarrow Z$, $t \in \Gamma$, is again a family with $(x, t) \longmapsto G(F(x, t), t)$ as the defining map, G and F being those of g_t and f_t correspondingly.

AI.64. The *derivative* df_t/dt of a family f_t, $t \in \Delta$, is defined by the formula

$$\frac{df_t}{dt} = TF \circ \frac{\partial}{\partial t} \circ i_t.$$

Here F is the defining map of f_t, i_t ($t \in \Delta$) is the family of inclusions and $\partial/\partial t$ is the "vertical" vector field on $X \times \Delta$ defined as $\partial/\partial t = p_2^! d/dt$, where $p_2 : X \times \Delta \longrightarrow \Delta$ is the second projection, the map $p_2^! : \mathscr{X}(\Delta) \longrightarrow \mathscr{X}(X \times \Delta)$ is defined in AI.47 and the vector field $d/dt \in \mathscr{X}(\Delta)$ is the restriction on Δ of the standard vector field d/dt on $\mathbb{R}$ defined in AI.46.

The map df_t/dt is a lift of f_t in the tangent bundle:

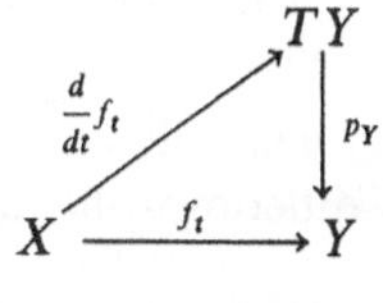

The family df_t/dt, $t \in \Delta$, has $TF \circ \partial/\partial t$ as the defining map.

The following formula is valid for the derivative of the composition of families:

(AI.64a) $$\frac{d}{dt}(g_t \circ f_t) = \left(\frac{d}{dt} g_t\right) \circ f_t + (Tg_t) \circ \frac{d}{dt} f_t.$$

AI.65. Let us consider the particular case of a family of maps, namely, the case of a family of sections $s_t : X \longrightarrow E$, $t \in \Delta$, of a smooth vector bundle $p : E \longrightarrow X$. Fixing $x \in X$, we may calculate the ordinary derivative $d/dt\, s_t(x) \in E_x$ regarding $s_t(x)$, $t \in \Delta$, as a vector valued function. Thus we obtain the family of sections $d/dt\, s_t : X \longrightarrow E$, $t \in \Delta$. This family is linked to the family $d/dt\, s_t : X \longrightarrow TE$ by the formula

$$\frac{ds_t}{dt} = \left(s_t, \frac{ds_t}{dt}\right) : X \longrightarrow p^! E \subset E \times E$$

if one identifies the bundle $p^! E$ with a subbundle of TE via the map H described in AI.48.

AI.66. Let $f_t : X \rightarrow Y$, $t \in \Delta$, be a family of maps and α_t, $t \in \Gamma$, be a family of exterior p-forms on Y, $\alpha_t \in \mathscr{E}^p(Y)$. The *differentiating formula* for the family of sections $f_t^* \alpha_t$,

$t \in \Delta \cap \Gamma$, is

(AI.66a) $$\frac{d}{dt} f_t^* \alpha_t = f_t \frac{d}{dt} \alpha_t + f_t^* \lrcorner\, d\alpha_t + d(f_t^* \lrcorner\, \alpha_t).$$

Here we use the notation $f_t^* \lrcorner\, \beta$, $\beta \in \mathscr{E}^p(Y)$ for the element of $\mathscr{E}^{p-1}(X)$ defined by the formulas:

(AI.66b) $$(f_t^* \lrcorner\, \beta)(v_1, v_2, \ldots, v_{p-1}) = \beta\left(\frac{d}{dt} f_t, T_x f_t v_1, T_x f_t v_2, \ldots, T_x f_t v_{p-1}\right),$$

$$v_i \in T_x X,\ 1 \leqq i \leqq p-1,\ x \in X \text{ if } p \geqq 1,$$

and

(AI.66c) $$f_t^* \lrcorner\, \varphi = \varphi \circ f_t \quad \text{if } \varphi \in \mathscr{E}^0(Y) = c^\infty(Y).$$

Ordinary Differential Equations and Flows

AI.67. Consider a family $\xi_t : X \longrightarrow TX$, $t \in \Gamma$, of vector fields on a manifolds X. A family $f_t : U \longrightarrow X$, $t \in \Delta$, U being another manifold, is an *integral* of ξ_t, $\Delta \subset \Gamma$ and for all $t \in \Delta$

$$\frac{d}{dt} f_t = \xi_t \circ f_t.$$

If U is a single point $\{*\}$, an integral is called an *integral curve*. Another interesting case is that if each f_t is a local diffeomorphism (see AI.29). Such an integral is called a *general integral.*

The theorems of the theory of ordinary differential equations (Hartman (1964) chapters II and V) assert that *given a family* $\xi_t : X \longrightarrow TX$, $t \in \Gamma$, *of vector fields* (1) *the manifold* $X \times \Gamma$ *can be covered by open sets of the form* $V \times \Delta$, *so that for each such that* V *there is a general integral* $f_t : U \longrightarrow X$, $t \in \Delta$, *with* $f_t(U) \supset V$, $\forall_t \in \Delta$ (existence of the solution and smooth dependence on initial data); (2) *if two integral curves* $s_t^{(i)} : \{*\} \longrightarrow X$, $t \in \Delta^{(i)}$, $i = 1, 2$, coincide at one *point* $t_0 \in \Delta^{(1)} \cap \Delta^{(2)}$ *then they coincide on* $\Delta^{(1)} \cap \Delta^{(2)}$ (uniqueness of the solution).

AI.68. Let $f_t : U \to X$, $t \in \Delta$, be an integral of a vector field $\xi_t : X \longrightarrow TX$, $t \in \Gamma \supset \Delta$. By virtue of AI.67 (a), the differentiation formula (AI.66a) now reads

$$\frac{d}{dt} f_t^* \alpha_t = f_t^* \frac{d}{dt} \alpha_t + f_t^* (\xi_t \lrcorner\, d\alpha_t) + d f_t^* (\xi_t \lrcorner\, \alpha_t)$$

for any family α_t, $t \in \Delta$, of p-forms on X, $p \geqq 1$.

AI.69. A family of local diffeomorphisms $f_t : U \longrightarrow X$, $t \in \Delta$, is called a *local flow* if for all $a, b, c \in \Delta$ such that $c + b - a \in \Delta$ the identity

(+) $$f_c \circ f_a^{-1} = f_{c+b-a} \circ f_b^{-1}$$

holds on $f_a(U) \cap f_b(U)$.

A family $f_t: U \longrightarrow X$, $t \in \Delta$, is a *normed family* if $0 \in \Delta$, $U \subset X$, and f_0 is the inclusion of U into X. Any family of local diffeomorphisms $f_t: U \longrightarrow X$, $t \in \Delta =]\alpha, \beta[$ may be turned into a normed family $g_t: U' \longrightarrow X$, $t \in \Delta'$, if one chooses $t \in \Delta$ and lets $g_t = f_{t_0+t} \circ f_{t_0}^{-1}$, $U' = f_{t_0}(U)$, and $\Delta' = \Delta - t_0 =]\alpha - t_0, \beta - t_0[$.

If f_t is an integral of a family of vector fields ξ_t, $t \in \Gamma \supset \Delta$, then g_t, constructed as above, is an integral of ξ_{t_0+t}, $t \in \Gamma' = \Gamma - t_0$.

For a normed family of local diffeomorphisms the condition $(+)$ is equivalent to

$(++)$ for any $a, C \in \Delta$ such that $a + b \in \Delta$ the identity $f_a \circ f_b = f_{a+b}$ holds on $U \cap f_b^{-1}(U)$.

Proposition AI.70. *A family of vector fields is time-independent if and only if all its general integrals are local flows.*

Proof. It is sufficient to consider the case of a normed general integral. If $f_t: U \longrightarrow X$, $t \in \Delta$, is a normed general integral of time-independent vector field $\xi_t = \xi$, $t \in \Delta$, and a, b, $a + b \in \Delta$, then for each $x \in U \cap f_b^{-1}(U)$ the curves $s_t = f_{t+b}(x)$ and $r_t = f_t(f_b(x))$, t running over some open interval containing 0 and a, are integral curves of the vector field ξ passing through the same point $f_b(x)$ at $t = 0$. By the uniqueness theorem they coincide at $t = a$. This proves $f_a \circ f_b = f_{a+b}$. Conversely, let this equality hold for each normed general integral of a family of vector fields ξ_t, $t \in \Gamma \ni 0$ for said values of a, b, and x. Let f_t be a normed general integral defined in some open neighbourhood of a point x, and let $\varphi \in \mathscr{F}_x$. Then, by definition,

$$\begin{aligned}
\xi_t \varphi &= \left[\frac{\partial}{\partial \tau}(\varphi \circ f_{t+\tau})\right]_{\tau=0} \circ f_t^{-1} \\
&= \left[\frac{\partial}{\partial \tau}(\varphi \circ f_\tau) \circ f_t\right]_{\tau=0} \circ f_t^{-1} \\
&= \left[\frac{\partial}{\partial \tau}(\varphi \circ f_\tau)\right]_{\tau=0} \circ f_t \circ f_t^{-1} = \xi_0 \varphi.
\end{aligned}$$

This is true for sufficiently small t, and thereby for all $t \in \Gamma$. Thus $\xi_t = \xi_0$. □

If ξ is a vector field, then all general integrals of the time-independent family $\xi_t = \xi$, $t \in \mathbb{R}$, are local flows. We call them *local flows generated by a field* ξ, or *local flows of a field* ξ.

AI.71. A vector field ξ on a manifold X is called *complete* if its time-independent family $\xi_t = \xi$, $t \in \mathbb{R}$, has a normed integral. $f_t: X \longrightarrow X$ defined at all real values of t, each map f_t being a diffeomorphism of X onto itself. Such an integral is unique, it is called the *flow generated* by ξ. We shall write

$$f_t = \exp t\xi.$$

A vector field on a compact manifold is complete if any only if it is tangent to the boundary. The tangency of a vector field $\xi \in \mathscr{X}(X)$ to ∂X means that $\xi|_{\partial X}$ takes its values in $T\partial X$, the latter being embedded in $T_{\partial X}X$ as in AI.52.

Appendix II. Derivatives of Superposition

The goal of this Appendix is to give some estimate for the derivatives of the superposition of smooth maps in Euclidean spaces. At first we define appropriate norms to be used here and in other chapters of this book.

r-(semi) norms

For a smooth real or complex valued function u defined on an open subset $A \subset \mathbb{R}^m$ and for a multi-index $\rho = (\rho_1, \rho_2, \ldots, \rho_m) \in \mathbb{Z}_+^m$ we set

(AII.1) $$D^\rho u(x) = \frac{\partial^{\rho_1}}{\partial x^1} \circ \frac{\partial^{\rho_2}}{\partial x^2} \circ \cdots \circ \frac{\partial^{\rho_m}}{\partial x^m} u(x).$$

The r-seminorm of u, $r \in \mathbb{Z}_+$, is defined as

(AII.2) $$|u|_r = \sup_{x \in A, |\rho| = r} |D^\rho u(x)|.$$

Here $|\rho| = \rho_1 + \rho_2 + \cdots + \rho_m$. The supremum in (AII.2) is taken over all multi-indices $\rho \in \mathbb{Z}_+^m$ with $|\rho| = r$ and over all $x \in A$. The r-norm is

(AII.3) $$\| u \|_r = \max_{0 \leqq r' \leqq r} |u|_{r'}.$$

A vector valued function $u = (u^1, u^2, \ldots, u^1)$: $A \longrightarrow \mathbb{R}^1$ (or $\mathbb{C}^1$) also possesses the r-seminorm and the r-norm both defined as maxima of those of its components u^i.

The Formula for the Derivative of Superposition

Consider the superposition of smooth maps $A \xrightarrow{g} B \xrightarrow{W} \mathbb{R}^l$, where $A, B \subset \mathbb{R}^m$ are open domains. Let us write out explicitly the formula for the derivative D^ρ of $W \circ g$, $\rho \in \mathbb{Z}_+^m$, $|\rho| \geqq 1$:

(AII.4) $$D^\rho W \circ g = \sum_{1 \leqq |s| \leqq |\rho|} (D^s W) \circ g \sum_{\lambda \in \Lambda(\rho, s)} C_{s\lambda} \prod_k (D^k g)^{\lambda}(k).$$

The variable $k = (k_1, k_2, \ldots, k_m)$ ranges over the set $\mathbb{Z}_+^m \setminus \{0\}$, $\Lambda(\rho, s)$ is the set of all maps $\lambda: \mathbb{Z}_+^m \setminus \{0\} \longrightarrow \mathbb{Z}_+^m$ obeying the two equalities

(AII.5) $$\sum_k k |\lambda(k)| = \rho \qquad \sum_k \lambda(k) = s.$$

A typical multiplier in (AII.4) is

(AII.6) $$(D^k g)^{\lambda(k)} = (D^k g^1)^{\lambda_1(k)}(D^k g^2)^{\lambda_2}(k)\cdots(D^k g^m)^{\lambda_m(k)},$$

where g^i, $1 \leqq i \leqq m$, stand for the components of g.

One can easily deduce that $\Lambda(\rho, s)$ is finite and any $\lambda \in \Lambda(\rho, s)$ vanishes for all but a finite number of k's. So the sums and the products in (AII.4) spread over a finite number of summands and multipliers, $C_{s\lambda}$ are nonnegative integer numbers depending only on m, ρ, s, and λ.

We single out the special element $\lambda^* \in \Lambda(\rho, s)$ which is defined by the relations:

(AII.7) $$\lambda^*(\mathrm{ort}_i) = \rho_i \mathrm{ort}_i, \quad 1 \leqq i \leqq m, \ \lambda^*(k) = 0 \text{ for other values of } k.$$

One computes immediately $C_{\rho,\lambda^*} = 1$, and the corresponding term in (AII.4) is

(AII.8) $$(D^\rho W)\circ g \prod_{i=1}^{m} \left(\frac{\partial g^i}{\partial x^i}\right)^{\rho_i}.$$

The reason for distinguishing this term follows.

Further we shall be interested mostly in the case when g is of the form $\mathrm{id} + u$, u being a smooth vector valued function, and our goal will be to get estimates containing the (semi) norms of u. In this case $\partial g^i/\partial x^i$ in (AII.8) must be substituted by $1 + \partial u^i/\partial x^i$ while all other derivatives of g^i coincide with those of u^i.

Estimates of r-norms

Let us adopt the notations:

(AII.9) $$\mu(\lambda) = \sum_{i=1}^{m} (\lambda(\mathrm{ort}_i))_i,$$

$$|\lambda(k)|' = \begin{cases} |\lambda(k)| = \lambda(k)_1 + \lambda(k)_2 + \cdots + \lambda(k)_m & \text{if } |k| > 1 \\ |\lambda(\mathrm{ort}_i)| - (\lambda(\mathrm{ort}_i))_i & \text{if } k = \mathrm{ort}_i. \end{cases}$$

Then the typical product in (AII.4) admits the following estimate, where $g = \mathrm{id} + u$;

(AI.10) $$\left|\prod_k (D^k g)^{\lambda(k)}\right| \leqq (1 + |u|_1)^{\mu(\lambda)} \prod_k |u|_{|k|}^{|\lambda(k)|'}$$

Note that $\mu(\lambda) < |s|$, if $\lambda \in \Lambda(\rho, s)$ provided $\lambda \neq \lambda^*$ as it follows from the relations (AII.5). In the case $\lambda = \lambda^*$ an immediate calculation give $\mu(\lambda^*) = |\rho|$ and $|\lambda^*(k)|' = 0$ for all k.

The opposite case is that with $|s| = 1$, or $s = \mathrm{ort}_i$ for some $i \in \{1, 2, \ldots, m\}$. In that case $\Lambda(\rho, s)$ consists of a single element $\lambda_{\rho i}$ defined by equalities: $\lambda_{\rho i}(\rho) = \mathrm{ort}_i$ and $\lambda_{\rho i}(k) = 0$ if $k \neq \rho$. The corresponding term is

(AII.11) $$(D^{\mathrm{ort}_i} W)\circ g \cdot D^\rho u^i$$

excepting the case $\rho = \mathrm{ort}_i$, when $\lambda_{\rho i} = \lambda^*$. We call these terms, (AII.8) and (AII.11), the marked terms. The following two estimates are obvious:

(AII.12) $$|\text{marked terms}| \leqq |W|_r(1+\text{const}(|u|_1+|u|_1^r))+|W|_1|u|_r,$$

(AII.13) $$|\text{marked terms} - D^\rho W| \leqq |W|_{r+1}|u|_0 + \text{const}|W|_r(|u|_1+|u|_1^r) + |W|_1|u|_r.$$

Here $r=|\rho|$, and const depends only on m and r. It is easy to obtain an estimate for the remaining, nonmarked terms in the following way. Take into account (AII.9), (AII.10), and the relations (AII.5). We have

(AII.14) $$\begin{aligned}|C_{s\lambda}(D^s W)\circ g \prod_k (D^k u)^{\lambda(k)}| &\leqq |C_{s\lambda}|\, \|W\|_s(1+|u|_1)^{\mu(\lambda)}\|u\|_{r-1}^{|s|-\mu(\lambda)}\\ &\leqq |C_{s\lambda}|\,\|W\|_r(1+\|u\|_{r-1})^{|s|-1}\|u\|_{r-1}.\end{aligned}$$

To obtain this, we replaced all but one of the multipliers $\|u\|_{r-1}$ by $(1+\|u\|_{r-1})$. Summarizing (AII.12), (AII.14) or (AII.13), (AII.14), the conclusion may be formulated as follows:

AII.15. Lemma. *There is a positive constant C depending only on r, m, and l, such that*

(a) $$\|W\circ(\text{id}+u)\|_r \leqq \|W\|_r(1+C(\|u\|_r+\|u\|_{r-1}^r)),$$

(b) $$\|W\circ(\text{id}+u)-W\|_r \leqq C\|W\|_{r+1}(\|u\|_r+\|u\|_{r-1}^r).$$

The estimate (A.II.15b) may be generalized as follows.

Lemma A.II.16. *Consider two local diffeomorphisms $g_i: A \longrightarrow B$, $i=1,2$, $A, B \subset \mathbb{R}^m$ open, such that $\|g_i\|_r + \|g_i^{-1}\|_r \leqq C_1$. There is a positive constant C_2 depending only on r, m, and C_1 such that*

$$\|W\circ g_1 - W\circ g_2\|_r \leqq C_2\|W\|_{r+1}\|g_1-g_2\|_r.$$

Proof. Set $g_2 = g_1 + u$ and write

$$W\circ g_1 - W\circ(g_1+u) = (W - W\circ(\text{id}+ug_1^{-1}))\circ g_1.$$

Combining the estimates of the preceding Lemma, we have

$$\begin{aligned}\|W\circ g_1 - W\circ g_2\|_r &\leqq \text{const}\,\|W - W\circ(\text{id}+u\circ g_1^{-1})\|_r\\ &\leqq \text{const}\,\|W\|_{r+1}(\|u\circ g_1^{-1}\|_r + \|u\circ g_1^{-1}\|_{r-1}^r)\\ &\leqq \text{const}\,\|W\|_{r+1}\|u\circ g_1^{-1}\|_r \leqq \text{const}\,\|W\|_{r+1}\|u\|_r.\end{aligned}$$

Here const's in different places of the formula denote different constants depending only on m, r, and C_1. □

The Weierstrass Property

The estimates of derivatives of the superposition may be essentially improved if the function possesses some additional properties. We say a real valued function u

defined on $\mathbb{R}^m$ has the *Weierstrass property* if it is C^∞ and for each multi-index $\rho\in\mathbb{Z}_+^m$ there is a point in $\mathbb{R}^m$ where $|D^\rho u|$ reaches its maximum. The usual examples of functions with the Weierstrass property are those of periodic smooth functions and those of smooth functions with compact supports. A complex valued (vector valued) function is Weierstrass if its real and imaginary parts (each vector component) are Weierstrass. A function defined on an open subset of $\mathbb{R}^m$ is Weierstrass if it acquires this property being extended onto the whole space $\mathbb{R}^m$ by the zero function. Consider firstly the most simple case $u:\mathbb{R}\longrightarrow\mathbb{R}$. Let x_0 be a point where $|u'(x)|$ takes its maximum. Let, for instance, $u'(x_0)>0$. The domain bounded by the graph of u' and the x-axis contains an isosceles triangle with height $a=u'(x_0)$ and half base b. Take the maximal triangle with such a property. Evidently $a/b\leqq\max_x|u''(x)|$ and the area ab of the triangle does not exceed $\int_\alpha^\beta u'(x)\,dx=u(\beta)-u(\alpha)\leqq 2\max|u(x)|$ (see Fig. 63). It follows that $\max|u'(x)|\leqq\sqrt{2\max|u|\max|u''|}$. This calculation may be easily generalized to the case of a function of many variables as well as to a complex-valued function.

We find the seminorms (A.II.2) of a Weierstrass function u obey the inequality

$$|u|_{r+1}\leqq\sqrt{2|u|_r|u|_{r+2}}.$$

Using induction one can state without any difficulty the following more general interpolation inequality

$$|u|_r\leqq\text{const}\,(u_{r-a})^{b/(a+b)}(u_{r+b})^{a/(a+b)}, \tag{AII.17}$$

where $r,a,b,r-a\in\mathbb{Z}_+$, and const depends only on r, a,b, and m. The inequality obtained enables us to estimate the typical product in (A.II.10) as

$$\left|\prod_k(D^kg)^{\lambda k}\right|\text{const}\,(1+|u|_1)^{\mu(\lambda)}|u|_1^{l-1\mu(\lambda)}|u|_{r-l+1}, \tag{AII.18}$$

where $r=|\rho|$, $l=|s|$. When proving (AIII.18) we applied (AII.17) to each term in the product in the right-hand side of (AII.10) and took into account the relations (AII.5). This method works if $|s|=l<r|\rho|$. However, (AII.18) is true even in the

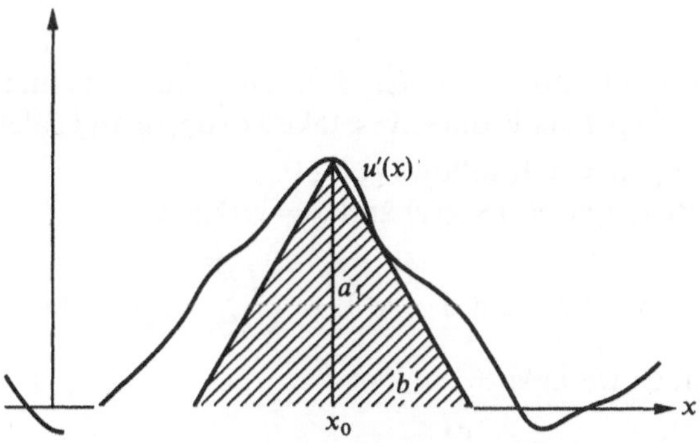

Fig. 63

case $r = l$. In the latter case we have $k = 1$ if $\lambda(k) \neq 0$ in the product over k, due to (AII.5), and (AII.18) coincides with (AII.10).

The differentiation formula (AII.4), and the formulae (AII.12), (AII.13), and (AII.18) yield the following estimate.

AII.19. *There exists a positive constant C depending only on r, m, l,* such that

(a) $$|W\circ(\mathrm{id}+u)|_r \leqq |W|_r + C\sum_{l=1}^{r} |W|_1(1+|u|_1)^{l-1}|u|_{r-l+1},$$

(b) $$|W\circ(\mathrm{id}+u) - W|_r \leqq |W|_{r+1}|u|_0 + C\sum_{l=1}^{r} |W|_1(1+|u|_1)^{l-1}|u|_{r-l+1},$$

provided u possesses the Weierstrass property.

The Inverse Diffeomorphism

Here we establish some sufficient conditions for a smooth map to be a diffeomorphism. The following Lemma about matrices will be useful.

Lemma AII.20. *Let A be an* $n \times n$ *matrix with coefficients* a_{ik} *satisfying the inequality* $|a_{ik}| < 1/n$, $1 \leqq i, k \leqq n$. *Then* $\det(1 + A) \neq 0$.

Here 1 denotes the unit matrix.

Proof. Suppose that $\det(1 + A) = 0$. Then there is a vector $v \neq 0$ such that $v = -Av$. Let v_{i_0} be the component of v with maximal absolute value. We have

$$|v_{i_0}| \leqq \sum_{i=1}^{n} |a_{i_0 i}||v_i| < \max_i |v_i| = |v_{i_0}|,$$

a contradictory inequality.

Lemma AII.21. *Let* $u: D \longrightarrow \mathbb{R}^m$ *be a smooth map,* $D \subset \mathbb{R}^m$, *such that* $\operatorname{supp} u \subset [D]_a$ *and* $\|u\|_1 \leqq \varepsilon$. *If* $\varepsilon < \min\{m^{-1}, am^{-1/2}\}$ *then the map* $g = \mathrm{id} + u$ *is a diffeomorphism of D onto itself.*

Proof. (1) g is an immersion since $|\partial u^i/\partial x^k| < m^{-1}$ (use Lemma AII.20)

(2) $g(D) \subset D$. To persuade ourselves, take $x \in \operatorname{supp} u$ and calculate $\|g(x) - x\| = \sqrt{\sum_{i=1}^{m}(u^i(x))^2} \leqq \varepsilon\sqrt{m} < a$. If follows $g(x) \in D$.

(3) g is injective. Let $g(x) = g(y)$ and $x \neq y$. Then

$$\|x - y\| = \|u(y) - u(x)\| = \sqrt{\sum_{i=1}^{m} u^i(y) - u^i(x)|^2}.$$

For scalar functions we have

$$u^i(y) - u^i(x) = \int_0^1 \langle \operatorname{grad} u^i(x + (y - x)t), y - x \rangle \, dt.$$

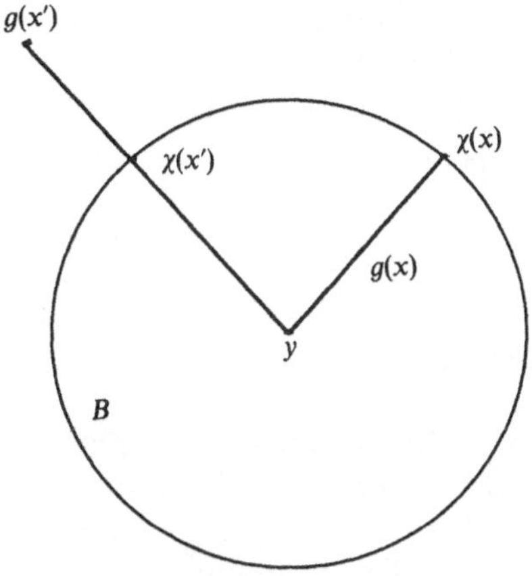

Fig. 64

Hence

$$|u^i(y)-u^i(x)| \leqq \varepsilon\sqrt{m}\,\|y-x\| < m^{-1/2}\|y-x\|.$$

Inserting this in the above inequality for $\|x-y\|$ we obtain $\|x-y\| < \|x-y\|$. Contradiction.

(4) g is onto D. Take $y \in D$. We seek for an x such that $g(x)=y$. One may assume $y \in \operatorname{supp} u$, otherwise $g(y)=y$. Take the closed ball B with y as a center and of radius a. Since $y \in \operatorname{supp} u$, $B \subset D$. Suppose there is no $x \in B$ such that $g(x)=y$. Consider the map $\chi: B \longrightarrow \partial B$ which assigns to a point x the intersection with ∂B of the ray sterting at y and passing through $g(x)$ (see Fig. 64). The map $\chi|_{\partial B}$ is homotopic to the identity. To obtain the homotopy one can consider the family $g_t = \mathrm{id} + tu$, $t \in [0,1]$, and define $\chi_t|_{\partial B}$ by use of above construction. So $\chi|_{\partial B}$ induces the identity map of the homotopy group $\pi_{m-1}(\partial B) = \mathbb{Z}$. On the other hand $\chi|_{\partial B}$ may be factored into the superposition of the inclusion $\partial B \hookrightarrow B$ and $\chi: B \longrightarrow \partial B$. So can be the induced map of the $(M-1)$th homotopy group. This contradicts the triviality of $\pi_{m-1}(B)$ (cf. Alexandroff and Hopf (1935)). □

Lemma AII.22. *Let $g: B \longrightarrow D$ be a diffeomorphism, $B \cdot D \subset \mathbb{R}^m$, whose inverse $k = g^{-1}$ has bounded first derivatives: $|\partial k^i/\partial x^j \leqq R$, $1 \leqq i, j \leqq m$. Let a smooth function $v: B \longrightarrow \mathbb{R}^m$ satisfy the conditions $\|v\|_1 \leqq \varepsilon$, $\operatorname{supp} v \subset [B]_a$. If $\varepsilon < \min\{m^{-1}, R^{-1}m^{-2}, aR^{-1}m^{-3/2}\}$, then $g+v$ maps B diffeomorphically onto D.*

Proof. Denote $h = g + v$ and consider the map $h_1 = h \circ g^{-1} = \mathrm{id} + v \circ g^{-1}: D \longrightarrow \mathbb{R}^m$. Let us prove that $u = v \circ g^{-1}$ obeys the conditions of Lemma AII.21 with ε and a replaced by $\varepsilon_1 = \max\{\varepsilon, \varepsilon Rm\}$ and $a_1 = R^{-1}m^{-1}a$ correspondingly. Differentiating the superposition yields

$$\frac{\partial u^i}{\partial x^l} = \left|\sum_{j=1}^{m} \frac{\partial v^i}{\partial y^j} \cdot \frac{\partial k^j}{\partial x^l}\right| \leqq \varepsilon Rm.$$

To obtain $\operatorname{supp} u \subset [D]_{a_1}$ let us prove $g([B]_a) \subset [D]_{a_1}$. The $y \in g([B]_a)$ and let $y = g(x)$, $x \in [B]_a$. Take a small positive number δ and let $y^* \in g(\partial[B]_\delta) = \partial g([B]_\delta)$ be a nearest point to y, and let $y^* = g(x^*)$, $x^* \in \partial[B]_\delta$. We have $\| x - x^* \| \geqq a - \delta$, and

$$\| y - y^* \| = \| g(x) - g(x^*) \| \geqq R^{-1} m^{-1} \| x - x^* \| \geqq R^{-1} m^{-1} (a - \delta).$$

Since $\delta > 0$ is arbitrary, $\| y - y^* \| \geqq a_1$. The closed ball of radius a_1 with a center at y lies wholly in D. Hence $y \in [D]_{a_1}$.

Applying Lemma AII.21 shows $h_1 = \mathrm{id} + u$ to be a diffeomorphism of D onto itself, which implies the Lemma. □

Remark AII.23. If in the preceding Lemmata we replace maps by families of maps depending smoothly on a parameter, the estimates being satisfied uniformly with respect to the parameter, then the resulting families of inverse maps depend smoothly on the parameter too. This is an immediate consequence of the implicit function theorem.

The next two Lemmata deal with functions and diffeomorphisms defined in $\mathbb{R}^n \times D$. The variables will be denoted (x, y), $x \in \mathbb{R}^n$, $y \in D$. The spaces $C^\infty_{\mathrm{per},\,a}(\mathbb{R}^n \times D)$ and $\mathrm{Diff}_{\mathrm{per},\,a}(\mathbb{R}^n \times D)$ are defined.

If $u \in C^\infty_{\mathrm{per},\,a}(\mathbb{R}^n \times D)$, then it possesses the Weierstrass property.

Lemma AII.24. *Let $u = (u_i, 1 \leqq i \leqq n)$ be a vector valued function with components belonging to $C^\infty_{\mathrm{per},\,a}(\mathbb{R}^n \times D)$. Consider the map $r:(x,y) \longrightarrow (x + u(x, y), y)$. If $\| u \|_1 < \min\{n^{-1}, an^{-1/2}\}$, then $r \in \mathrm{Diff}_{\mathrm{per},\,a}(\mathbb{R}^n \times D)$.*

Proof. Apply Lemma AII.21 to the map $g_y : x \longrightarrow x + u(x, y)$, regarding y as a parameter. We obtain a family of inverse diffeomorphisms $g_y^{-1} : \mathbb{R}^n \longrightarrow \mathbb{R}^n$, which depend smoothly on $y \in D$ (use Remark AII.23). Since u is periodic, g_y commutes with integer shifts and so does g_y^{-1}. The map $h:(x, y) \longrightarrow (g_y^{-1}(x), y)$ is the inverse to g. □

Lemma AII.25. *Let $v = (v_i, 1 \leqq i \leqq n)$ be a vector valued function with components balonging to $C^\infty_{\mathrm{per},a}(\mathbb{R}^n \times D)$. Consider the* map $l:(x, y) \longrightarrow (x, y + v(x, y))$. *If $\| v \|_1 < \min\{n^{-1}, an^{-1/2}\}$, then $l \in \mathrm{Diff}_{\mathrm{per},\,a}(\mathbb{R}^n \times D)$.*

Proof. Similar to that of Lemma AII.24. □

The following assertions contain useful information about r-(semi)norms of the inverse diffeomorphism.

Lemma AII.26. *There is a positive constant d and a map $C : \mathbb{Z}_+ \longrightarrow \mathbb{R}_+ \backslash \{0\}$ depending only on m and a such that, under the conditions and with the notations of Lemma* AII.21, *if $|u|_1 < d$, and u and $g^{-1} - \mathrm{id}$ possess the Weierstrass property, then*

(a) $$|g^{-1} - \mathrm{id}|_r \leqq C(r) |u|_r, \quad r \in \mathbb{Z}_+.$$

Proof. Denote $g^{-1}(x) = x + v(x)$. Then the function v obeys the equation

(b) $$v(x) = -u(x + v(x)).$$

This immediately, gives the estimate (a) for $r=0$ with $C(0)=1$. Applying AII.19 to (b) yields

(c) $$|v|_r \leqq |u|_r + C_1(m)|u|_1|v|_r + C_2(m,r)\sum_{p=2}^{r} |u|_p(1+|v|_1)^{p-1}|v|_{r-p+1}.$$

Note that the constant $C_1(m)$ in (c) may be chosen independent of r as one can check by writing out explicitly the term containing the highest derivative of v. We choose $d<\min\{1/2C_1(m), m^{-1}, am^{-1/2}\}$. Then (c) considered for $r=1$ yields $|v|_1 \leqq 2|u|_1$. The estimate (a) for higher values of r will be obtained with use of (c) by induction. Suppose (a) is true for $r<s$. Then, by (c):

(d) $$|v|_s \leqq 2|u|_s + \text{const}\sum_{p=2}^{s} |u|_p|u|_{s-p+1}.$$

Applying (AII.17) to a typical term in (d), we obtain

$$|u|_p \leqq \text{const}|u|_1^{(s-p)/(s-1)}|u|_s^{(p-1)/(s-1)}, \qquad |u|_{s-p+1} \leqq \text{const}|u|_1^{(p-1)/(s-1)}|u|_s^{(s-p)/(s-1)},$$

and $|u|_p|u|_{s-p+1} \leqq \text{const}\ |u|_1|u|_s$. Inserting this into (d) and taking into account $|u|_1 \leqq \text{const}$ yields (a) for $r=s$. □

Proposition AII.27. *There is a map* $C:\mathbb{Z}_+ \times (\mathbb{R}_+\backslash\{0\}) \longrightarrow \mathbb{R}_+\backslash\{0\}$ *such that the following holds. Under the conditions of Lemma* AII.22, *let* v *possesses the Weierstrass property. Then, for* $r\in\mathbb{Z}_+$,

(a) $$\|(g+v)^{-1}\|_r \leqq C(r, \|g^{-1}\|_r)\cdot\|v\|_r,$$

(b) $$\|(g+v)^{-1} - g^{-1}\|_r \leqq C(r, \|g^{-1}\|_{r+1})\cdot\|v\|_r.$$

Proof. Denoting $k=g^{-1}$ and $g\circ(g+v)^{-1} = \text{id} + w$, we have

(c) $$(\text{id} + v\circ k)^{-1} = \text{id} + w,$$

(d) $$(g+v)^{-1} - g^{-1} = k\circ(\text{id}+w) - k.$$

It follows from (c), AII.26 and AII.15(a) that

(e) $$|w|_r \leqq \text{const}\,|v|_r$$

where const depends only on m, r, and $\|k\|_r$. Applying AII.19(b) to the right-hand side of (d) yields, in view of (e)

$$\|(g+v)^{-1} - g^{-1}\|_r \leqq \text{const}\sum_{l=1}^{r+1} \|k\|_1\,\|w\|_{r-l+1}$$

$$\leqq C(r, \|g^{-1}\|_{r+1})\,\|v\|_r.$$

One obtains the inequality (a) in an analogous way be applying AII.19(a) to (d) and dropping the last terms $-g^{-1} = -k$ in both sides. □

Appendix III. The Stationary Phase Method

This method provides us, under certain regularity conditions, with the asymptotic formula for the values of an integral

(AII.1) $$I(A, \Phi, \hbar, x) = \int_D A(x, y)\, e^{(i/\hbar)\Phi(x,y)}\, dy,$$

the real parameter h tending to zero, $x \in \mathbb{R}^m$, $y \in \mathbb{R}^n$, $dy = dy^1\, dy^2 \cdots dy^n$. Both the functions A and Φ, the *amplitude* and the *phase* of the integral (AIII.1), are supposed to be smooth and defined on a nonempty, open set $D \subset \mathbb{R}^{m+n}$. The integration in (AII.1) spreads over the x-slice

(AIII.2) $$D^{(x)} = \{y \in \mathbb{R}^n\colon (x,y) \in D \subset \mathbb{R}^m \times \mathbb{R}^n\}.$$

We suppose that

(i) suup A *is a compact subset of* D,

(ii) Φ *is real valued.*

Let f be a complex-valued function defined on $Z \times\,]0, 1[$, Z being an arbitrary set. Given real N, we write $f(z, h) = \mathcal{O}(\hbar^N)$ if there is $C > 0$ such that

$$|f(z, h)| < C\hbar^N \quad \text{for all} \quad (z, h) \in Z \times\,]0, 1[.$$

Also we write $f(z, h) = \mathcal{O}(\hbar^\infty)$ if $f(z, \hbar) = \mathcal{O}(\hbar^N)$ for all $N > 0$.

It follows from the above conditions (i) and (ii) that

(AIII.3) $$I(A, \Phi, \hbar, x) = \mathcal{O}(1).$$

A point $(x, y) \in D$ is called a *critical point of* Φ *with respect to the y-variable* or, briefly, a *critical point*, if

(AIII.4) $$\operatorname{grad}_y \Phi(x, y) = 0.$$

Here

$$\operatorname{grad}_y \Phi(x, y) = \left(\frac{\partial\Phi(x, y)}{\partial y^1}, \frac{\partial\Phi(x, y)}{\partial y^2}, \ldots, \frac{\partial\Phi(x, y)}{\partial y^n}\right).$$

Critical points of Φ will be also called the *stationary-phase points* of the integral (AIII.1)

Lemma AIII.5. *If* supp A *contains no stationary-phase points of* (AII.1) *then*

(AIII.5a) $$I(A, \Phi, \hbar, x) = \mathcal{O}(\hbar^\infty).$$

Proof. The formula for integration by parts yields

$$(\text{AIII.5b}) \qquad I(A, \Phi, \hbar, x) = -i\hbar \int A(x, y) \| \operatorname{grad}_y \Phi \|^{-2} \sum_{k=1}^{n} \frac{\partial \Phi}{\partial y^k} \frac{\partial}{\partial y^k} e^{(i/\hbar)\Phi} \, dy$$

$$= i\hbar I(\operatorname{grad}_y [A \| \operatorname{grad}_y \Phi \|^{-2}], \Phi, \hbar, x).$$

$\operatorname{grad}_y \Phi$ does not vanish on $\operatorname{supp} A$, so the right-hand side of (AIII.5b) is well defined. Repeating (AIII.5b) many times and taking into account (AIII.3), one obtains (AIII.5a). □

We are intending to find the $\hbar \longrightarrow 0$ asymptotics of (AII.1) up to the terms of order $\mathcal{O}(\hbar^\infty)$. Lemma AII.5 shows us that only the stationary-phase points give an essential contribution to the integral (AIII.1). In general, the set of all critical points may be complicated, and so is the posed problem. Nevertheless the latter has an exhaustive solution in the case of nondegenerate stationary points. A critical point (x, y) of Φ is called *nondegenerate* if the $n \times n$-matrix

$$\left\{ \frac{\partial^2 \Phi(x, y)}{\partial y^i \partial y^k}, \quad 1 \leqq i, k \leqq n \right\}$$

is invertible. This is equivalent to the condition:

$$(\text{AIII.6}) \qquad \det \left\{ \frac{\partial^2 \Phi(x, y)}{\partial y^i \partial y^k} \right\} \neq 0.$$

Henceforward we add the condition

(iii) *all critical points of* Φ, *which are contained in* $\operatorname{supp} A$, *are nondegenerate.*

Each point $(x_0, y_0) \in \operatorname{supp} A$ has an open neighbourhood $U \subset D$ such that
(α) if (x_0, y_0) is not critical, then neither are all the points of U;
(β) if (x_0, y_0) is critical, then the condition (AIII.6) is fulfilled for all $(x, y) \in U$.

In the case (α) the subset of critical points in U is an m-dimensional submanifold (apply AI.53 to the equation (AIII.4)).

Another consequence of (iii) is that the tangent space to the said manifold at the point (x_0, y_0) is transversal to $T_{(x_0, y_0)}\{x_0\} \times \mathbb{R}^n$, i.e. it has a nondegenerate projection onto $T_{(x_0, y_0)}\mathbb{R}^m \times \{y_0\}$. It follows in the case (β) that the neighbourhood U can be chosen so that

(γ) it has the form $U = V \times W$, $V \subset \mathbb{R}^m$, $W \subset \mathbb{R}^n$, and for each $x \in V$ equation (AII.4) has a unique solution $y = y(x) \in W$, the latter being a smooth function of x.

Cover $\operatorname{supp} A$ with a finite set $\{U_k, 1 \leqq k \leqq s\}$ of open neighbourhoods such that either (α), or (β) and (γ), are fulfilled (denote $U_k = V_k \times W_k$ in the latter case), add to this covering the set $U_0 = D \backslash \operatorname{supp} A$ to obtain a covering of D, and take a partition of unity $\{\chi_k, 0 \leqq k \leqq s\}$ in D subordinate to the covering $\{U_k, 0 \leqq k \leqq s\}$. Then

$$I(A, \Phi, \hbar, x) = \sum_{k=0}^{s} I(A\chi_k, \Phi, \hbar, x),$$

and

$$I(A\chi_k, \Phi, \hbar, x) = \mathcal{O}(\hbar^\infty)$$

if either we have the case (α) or the case (β), (γ) with $x \notin V_k$. The remaining terms correspond to the case (β), (γ) with $x \in V_k$. Thus our problem is reduced to that of finding the asymptotics of (AII.1) when $\hbar \longrightarrow 0$ under the additional condition:

(iv) *the domian D has the form $D = V \times W$, $V \subset \mathbb{R}^m$, $W \subset \mathbb{R}^n$, and equation* (AII.4) *defines a submanifold in $V \times W$ which is the graph of a smooth function $y: V \longrightarrow W$.*

To formulate the answer we need two definitions. The *signature* of a square symmetric matrix M with real elements is defined as

sign M = the number of positive eigenvalues − the number of negative eigenvalues, each eigenvalue being counted as many times as its multiplicity.

The symbol $f \sim \sum_{k=0}^{\infty} f_k \hbar^k$ means that for each nonnegative integer N

$$f - \sum_{k=0}^{N} f_k \hbar^k = \mathcal{O}(h^{N+1}).$$

Theorem AIII.7. *Under the conditions* (i)–(iv), *the integral* (AIII.1) *has the following asymptotics when $\hbar \longrightarrow Q$:*

$$\text{(AIII.7a)} \qquad I(A, \Phi, \hbar, x) \sim \frac{(\pi h)^{n/2}}{\sqrt{\left|\det\left\{\dfrac{\partial^2 \Phi_{(x,y)}}{\partial y^i \partial y^k}\right\}\right|}}\Bigg|_{y=y(x)} e^{i(\pi/4)\operatorname{sign}\{(\partial^2\Phi)/(\partial y^i \partial y^k)\}|_{y=y(x)}}$$

$$\times\, e^{(i/\hbar)\Phi(x,y(x))} \sum_{p=0}^{\infty} (i\hbar)^p R_p(A, \Phi)(x)$$

where

$$\text{(AIII.7b)} \qquad R_0 = 1,$$

$$\text{(AIII.7c)} \qquad R_p(A, \Phi)(x) = \sum_{|k| \leqq 2p} R_{pk}(\Phi)(x) D_y^k A(x, y)|_{y=y(x)},$$

$$R_{pk}: V \longrightarrow \mathbb{R}, \quad p \geqq 1,\ k = (k^1, k^2, \ldots, k^n) \in \mathbb{Z}_+^n,$$

are smooth functions depending only on Φ.

The rest of this Appendix is devoted to the proof of Theorem AIII.7.

The first step is to introduce appropriate coordinates (x, η) so that the phase function Φ is quadratic in the η-variable. Morse's Lemma provides us with such coordinates locally, in a small neighbourhood of a given point of the critical set.

Lemma AIII.8 (M. Morse). *Let $\Phi: V \times W \longrightarrow \mathbb{R}$, $V \subset \mathbb{R}^m$, $W \subset \mathbb{R}^n$, be a smooth function such that*

(a) *the critical set $C_\Phi = \{(x, y) \in V \times W\colon \operatorname{grad}_y \Phi(x, y) = 0\}$ is a submanifold, moreover, it has the equation $y = y(x)$ where $y: V \longrightarrow W$ is a smooth map;*

(b) *the matrix of second derivatives of* $\Phi(x,y)$ *with respect to the y-variables is nondegenerate at the points of* C_Φ, *i.e.*

$$\det\left\{\frac{\partial^2\Phi_{(x,y)}}{\partial y^i \partial y^k}\right\}\Bigg|_{(x,y(x))} \neq 0.$$

Then for each $x_0 \in V$ *there are: a sequence* $\varepsilon_k = \pm 1$, $1 \leqq k \leqq n$, *a neighbourhood* $V' \subset V$ *of* x_0 *in* $\mathbb{R}^m$, *a neighbourhood* W' *of the origin in* $\mathbb{R}^n$, *and a local diffeomorphism* $\Psi: V' \times W' \longrightarrow V \times W$ *of the form* $(x,\eta) \longmapsto (x, g(x,\eta))$ *such that*

(c) Ψ *carries the x-coordinate subspace into the critical set:* $\Psi(V' \times \{0\}) = C_\Phi \cap (V \times W)$, *or, equivalently,* $g(x,0) = (x, y(x))$, *and for each* $x \in V'$, $\eta = (\eta_1, \eta_2, \ldots, \eta_n) \in W'$ *the following equalities hold:*

(d)
$$\Phi_0 \Psi(x,\eta) = \Phi(x, y(x)) + \sum_{k=1}^{N} \varepsilon_k \eta_k^2,$$

(e)
$$\det\left\{\frac{\partial g_i(x,\eta)}{\partial \eta k}\right\}\Bigg|_{\eta=0} = \left|\det\left\{\frac{\partial^2\Phi_{(x,y)}}{\partial y^i \partial y^k}\right\}\right|^{-1/2}\Bigg|_{y=y(x)},$$

(f)
$$\operatorname{sign}\left\{\frac{\partial^2\Phi_{(x,y)}}{\partial y^i \partial y^k}\right\}\Bigg|_{y=y(x)} = \sum_{k=1}^{n} \varepsilon_k.$$

The situation above is shown in Fig. 65.

Proof of Lemma AIII.8. There are neighbourhoods V_1 of x_0 in V and W_1 of the origin in $\mathbb{R}^n$ such that the formula $\Psi_1: (x,\eta) \longrightarrow (x, \eta + y(x))$ defines a local diffeomorphism of $V_1 \times W_1$ into $V \times W$ and the equality $\Psi_1(V_1 \times \{0\}) = C_\Phi \cap$

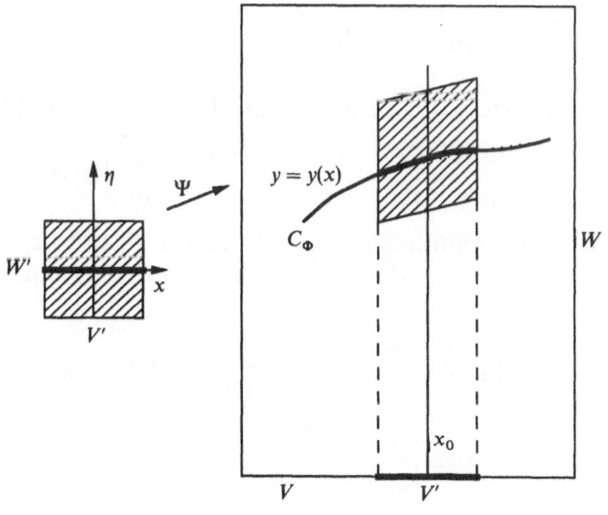

Fig. 65

$(V_1 \times W)$ holds. Thus the problem reduces to that with $W \ni 0$ and $y(x)=0$. Henceforth we assume these conditions to be fulfilled.

AIII.8′. Auxiliary Lemma. Let F be a smooth real-valued function defined on $V \times W$, $W \ni 0$. Then there are smooth real-valued functions F_i on $V \times W$, $1 \leqq i \leqq n = \dim W$, such that

$$F(x, y) = F(x, 0) + \sum_{i=1}^{n} y^i F_i(x, y)$$

and

$$F_i(x, 0) = \frac{\partial F}{\partial y^i}(x, 0).$$

Proof. $$F(x, y) = F(x, 0) + \int_0^1 \frac{d}{dt} F(x, ty)\, dt = F(x, 0) + \sum_{i=1}^{n} y^i \int_0^1 \frac{\partial F}{\partial y^i}(x, ty)\, dt. \quad \square$$

Applying twice the Auxiliary Lemma to $\Phi(x, y)$ we obtain smooth functions Φ_{ij} such that

$$\Phi(x, y) = \Phi(x, 0) + \sum_{i,j=1}^{n} y^i y^j \Phi_{ij}(x, y) \tag{g}$$

and

$$\Phi_{ij}(x, 0) = \frac{\partial^2 \Phi}{\partial y^i \partial y^j}(x, 0).$$

Furthermore we use induction: let there be a local diffeomorphism $\Psi_2 : V_2 \times W_2 \longrightarrow V \times W$, $V_2 \ni x_0$, $W_2 \ni 0$, of the form $\Psi_2(x, \eta) = (x, g_2(x, \eta))$ such that $g_2(x, 0) = 0$ and for some $p \in \{1, 2, \ldots, n\}$

$$\Phi \circ \Psi_2(x, \eta) = \Phi(x, 0) + \sum_{k=1}^{p-1} \varepsilon_k \eta_k^2 + \sum_{i,j \geqq p} \eta_i \eta_j H_{ij}(x, \eta). \tag{h}$$

Taking into account that $(x, 0)$ is a critical point of Φ, we obtain by differentiating

$$\left.\frac{\partial^2 \Phi \circ \Psi_2}{\partial \eta_i \partial \eta_j}\right|_{(x,0)} = \sum_{a,b} \left(\frac{\partial g_{2,a}}{\partial yi} \frac{\partial^2 \Phi}{\partial \eta^a \partial y^b} \frac{\partial y_{2,b}}{\partial \eta j} \right)(x, 0). \tag{i}$$

On the other hand the right-hand side of (h) implies (i) to be $\varepsilon_j \delta_{ij}$ if either $i < p$, or $j < p$, and $H_{ij}(x, 0)$ if $i, j \geqq p$. Hence $\{H_{ij}, 1 \leqq i, j \leqq n - p\}$ is nondegenerated, and at $(x, 0)$:

$$\left| \det \left\{ \frac{\partial^2 \Phi}{\partial y^i \partial y^j} \right\} \right| \left(\det \frac{\partial g_{2i}}{\partial \eta j} \right)^2 = |\det\{H_{ij}\}|, \tag{j}$$

$$\operatorname{sign} \left\{ \frac{\partial^2 \Phi}{\partial y^i \partial y^j} \right\} = \sum_{k=1}^{p-1} \varepsilon_k + \operatorname{sign}\{H_{ij}\}. \tag{k}$$

Applying an appropriate orthogonal transformation in the space of $\{\eta_i, p \leqq i \leqq n\}$-

variables, we get $H_{pp}(x_0, 0) \neq 0$. Consider the map $\Psi_3 : (x, \eta) \longmapsto (x, h(x, \eta))$ where $h_i(X, \eta) = \eta_i$ if $i \neq p$ and

$$h_p(x,\eta) = \sqrt{|H_{pp}(x,\eta)|}\left(\eta_p + \sum_{i>p} \eta_i \frac{H_{ip}(x,\eta)}{H_{pp}(x,\eta)}\right).$$

In some sufficiently small neighbourhood of $(x^0, 0)$ this map is well defined and is a local diffeomorphism. Hence there is a neighbourhood of $(x_0, 0)$ of the form $V_3 \times W_3$ suchthat the inverse map Ψ_2^{-1} is defined on $V_3 \times W_3$ and maps $V_3 \times W_3$ diffeomorphical onto some subset of $V_2 \times W_2$ containing $(x_0, 0)$. One immediately verifies that

$$\Phi \circ \Psi_2 \circ \Psi_3^{-1}(x, \xi) = \Phi(x, 0) + \sum_{k=1}^{p} \varepsilon_k \xi_k^2 + \sum_{i,j>p} \xi_i \xi_j H_{ij}(x, \xi)$$

where $\varepsilon_p = \operatorname{sign} H_{pp}(x_0, 0)$ and H_{ij} are other smooth functions. We obtained (h) with a greater value of p. Starting with (g) which corresponds to $p = 0$ and making the described transformations $p \longmapsto p + 1$, we achieve the complete representation (d). After the last step the matrix H_{ij} becomes the trivial matrix with zero signature and determinant 1. So (j) gives (e) and (k) gives (f). □

Passing to the immediate proof of Theorem AII.7, we introduce the coordinates (x, η) in a small neighbourhood $V' \times W'$ of a given point $(x_0, y(x_0)) \in C_\Phi$ by means of the map Ψ of Lemma AIII.8. It is sufficient to prove the Theorem for $x \in V'$, since the projection onto V of supp A may be covered by a finite number of such V'"s, and $I(A, \Phi, h, x) = 0$ if x belongs to the complement of the union of V'"s. Furthermore, introducing the appropriate partition of unity and taking into account (AIII.5), one may reduce the problem of evaluating (AIII.1) to that with the additional condition supp $A \subset V' \times W'$. So, we may use the coordinates (x, η) which Morse's Lemma suggests. Rewriting (AII.1) in these coordinates we have

$$\text{(AIII.9)} \qquad I(A, \Phi, \hbar, x) = e^{(i/\hbar)\Phi(x, y(x))} \int_{W'} e^{(i/\hbar)\sum_{k=1}^{n} \varepsilon_k \eta_k^2} A \circ \Psi(x, \eta) \left| \left\{ \det \frac{\partial g_i(x, \eta)}{\partial \eta k} \right\} \right| d\eta.$$

Let us introduce in the integral (AII.9) an additional stretching multiplier $\chi(\eta) = \prod_{n=1}^{n} \chi_1(\eta_k)$ where $\chi_1(\xi)$ is a smooth function of one real variable satisfying the conditions $(\chi 1)$ $\chi_1(\xi) = 1$ if $\xi \in]a, b[\ni 0$, $(\chi 2)$ supp χ_1 is a compact subset of W'. The change of the integral after this introduction is of order $\mathcal{O}(\hbar^\infty)$ due to Lemma (AII.5).

The next step is to apply the Taylor expansion:

$$\begin{aligned}
\text{(AII.10)} \qquad A \circ \Psi(x, \eta) | \det \left\{ \frac{\partial g_i(x, \eta)}{\partial \eta k} \right\} \Bigg| &= A(x, y(z)) \left| \det \left\{ \frac{\partial^2 \Phi_{(x,y)}}{\partial y^i \partial y^k} \right\} \right|^{-1/2} \Bigg|_{y = y(x)} \\
&\quad + \sum_{m=1}^{N} \frac{1}{k!} \sum_{|m|=k} D_\eta^m (A \circ \Psi) \left| \det \left\{ \frac{\partial g_i}{\partial \eta k} \right\} \right| \Bigg|_{\eta=0} \eta^m \\
&\quad + \sum_{|m|=N+1} \eta^m \varphi_m(x, \eta).
\end{aligned}$$

Here $\varphi_m = \varphi_{m_1, m_2, \ldots, m_n}$ are smooth functions. Substituting (AII.10) in the integral

(AIII.9) (with additional multiplier $\chi!$), we represent it as a sum of the corresponding terms. To evaluate these terms, we use the following two Lemmas.

Lemma AIII.11. *Let χ_1 be a smooth function of one real variable satisfying* $(\chi 1), (\chi 2)$. *Then, given $m \in \mathbb{Z}_+$ and $\varepsilon \in \{\pm 1\}$,*

$$\text{(AIII.11a)} \quad \chi_1(\xi)\, e^{i(\varepsilon/\hbar)\xi^2} \xi^m d\xi = \begin{cases} \mathcal{O}(\hbar^\infty) & \text{if } m \text{ is odd,} \\ \hbar^{(m+1)/2}\Gamma\left(\dfrac{m+1}{2}\right) e^{i\varepsilon(\pi/4)(m+1)} + \mathcal{O}(\hbar^\infty) & \text{if } m \text{ is even.} \end{cases}$$

Proof. Let us continue χ_1 on the complex plane $\mathbb{C}$ letting $\chi(\eta) = 1$ if $\operatorname{Im} \eta \neq 0$. We may rewrite (AII.11a) as a contour integral in the complex plane, the integration being over the path $C = C_1 \cup C_2 \cup C_3$ drawn for the case $\varepsilon = 1$ in Fig. 66. The picture for $\varepsilon = -1$ may be obtained from that of Fig. 66 by reflection with respect to the real axis. The integrals over C_i, $i = 1, 3$, are of order $\mathcal{O}(\hbar^\infty)$. The proof is the same as that of Lemma AII.5. As for the integral over C_2, the result is either zero, if m is odd because of the antisymmetry of the integrand, or the corresponding expression in the right-hand side of (AIII.11a) if m is even, as one may easily calculate. □

Lemma AIII.12. *Let $\varphi: \mathbb{R}^n \longrightarrow \mathbb{C}$ be a smooth function with compact support, $m = (m_1, m_2, \ldots, m_n) \in \mathbb{Z}^n_+$, and $\varepsilon_k \in \{\pm 1\}$, $1 \leq k \leq n$. Then*

$$\text{(AIII.12a)} \qquad \int_{\mathbb{R}^n} \eta^m \varphi(\eta) \exp\left\{(i/\hbar) \sum_{k=1}^n \varepsilon_k \eta_k^2\right\} d\eta = \mathcal{O}(\hbar^{(1/2)(n+|m|)}).$$

The constant in the estimation depends only on n, m, diam supp φ *and* $\|\varphi\|_r$, $r = 2[(n+1)/2] + |m|$. Here $\|\varphi\|_r$ denotes the C^r-norm of φ (see §8).

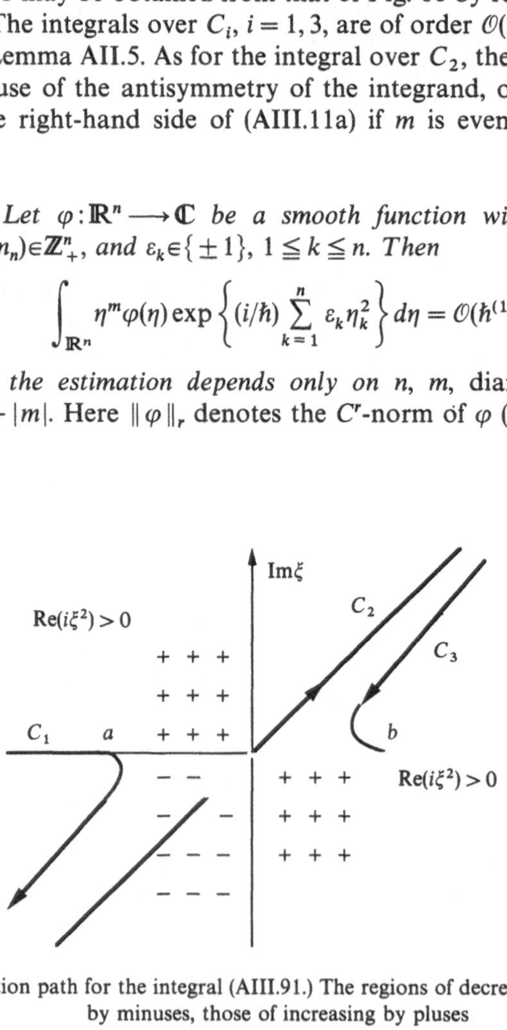

Fig. 66. The integration path for the integral (AIII.91.) The regions of decreasing exponent are shown by minuses, those of increasing by pluses

Proof. We use induction on $|m|$. Let us first prove (AII.12a) for the case $m=0$. Denote $I(\varphi)=\int\varphi(y)\exp\{(i/\hbar)\sum_{k=1}^{n}\varepsilon_k\eta_k^2\}\,d\eta$. Obviously

(a′) $I(\varphi)=\mathcal{O}(1)$, the constant in the estimate depending only on $\|\varphi\|_0$ and diam supp φ.
Apply to $\varphi(\eta)$ the Auxiliary Lemma AII.8′ and use AII.11. The expression for $I(\varphi)$ then reads

(b′) $I(\varphi)=c\hbar^{n/2}+\sum_{k=1}^{n}I(\eta_k\varphi_k)$ and $\|\varphi_k\|_r$ may be estimated from above by $\|\varphi_{r+1}$, $r=0, 1, 2,\ldots$ Since

(c′) $$\eta_j\exp\left(\frac{i}{\hbar}\sum_{k=1}^{n}\varepsilon_k\eta_k^2\right)=-\frac{i\hbar}{2}\varepsilon_k\frac{\partial}{\partial\eta_j}\exp\left(\frac{i}{\hbar}\sum_{k=1}^{n}\varepsilon_k\eta_k^2\right)$$

integration by parts, being applied to each integral in the right-hand side of (b′), yields

(d′) $$I(\varphi)=c\hbar^{n/2}+\frac{i\hbar}{2}I\left(\sum_{k=1}^{n}\varepsilon_k\frac{\partial\varphi_k}{\partial\eta_k}\right)=\mathcal{O}(\hbar^{\min(n/2,1)})\quad\text{(we used (a′)).}$$

The constant in the estimate (d′) depends on $\|\varphi\|_2$ and diam supp φ. Suppose

(e′) $$I(\varphi)=\mathcal{O}(\hbar^{\min(n/2,k)})$$

is already proven, the constant in $\mathcal{O}$ depending on $\|\varphi\|_{2k}$. Apply the estimate (e′) to the integral in the right-hand side of (d′) to obtain (e′) for greater values of k. Hence (AIII.12a) for $m=0$ may be obtained by a finite number of steps up to the value $k=[(n+1)/2]\geqq n/2$. Suppose (AIII.12a) is proven for $|m|\leqq N$, and let $|m|=N+1$. There is $j\in\{1,2,\ldots,n\}$ such that $m_j\geqq 1$. Using (c′) and integrating by parts with respect to η_j, one obtains

(f′) $$I(\eta^m\varphi)=\frac{i\hbar}{2}I\left(\frac{\partial}{\partial\eta_1}(\eta^m\eta_j^{-1}\varphi)\right).$$

The integrand in the right-hand side satisfies the induction assumption so (f′) may be estimated by *constant* multiplied by $\hbar^{1+(1/2)(n+|m|-2)}=\hbar^{(n+|m|)/2}$, *constant* depending on $\|\varphi\|_r$, $r=[(n+1)/2]+|m|$, and on diam supp φ. □

We are close to the end of the proof of Theorem AII.7. Substituting (AIII.10) into the integral $I(\chi A,\Phi,\hbar,x)$, evaluating each term by use of the above Lemmata, and taking into account the formulae (d), (e), (f) of Morse's Lemma (AII.8), one obtains (AIII.7a). The explicit formulae for $R_{pk}(\Phi)$ may be obtained, in view of (AIII.10) and (AIII.11), by calculation of the derivatives of the composite function $(A\circ\Psi)\det\{\partial g_i/\partial\eta_k\}$.

References

Abraham, R., Marsden J. E. (1972): Foundations of mechanics. Benjamin, New York

Ahlfors, L. V., Sario, L. (1960): Riemann surfaces. University of Princeton, Princeton, NJ

Alenitsyn, A. G. (1982): Splitting of the spectrum generated by a potential barrier in the problem with a symmetric potential. Diff. Uravenia, *18*, 1971–1975 (Russian)

Alexandroff, P., Hopf, H. (1935): Topologie. I. Springer, Berlin

Amiran, E. Y. (1991): Smooth convex planar domains for which the billiard ball map is integrable are ellipses. Math. Department Western Washington University, Bellingam WA 98225. Preprint

Arnol'd, V. I. (1963): Small denominators and problem of stability of motion in classical and selestial mechanics. Russian Math. Survey *18*, 85–191

Arnol'd, V. I. (1964): Instability of dynamical systems with several degrees of freedom. Soviet Math. Doklady *5*, 581–585

Arnol'd, V. I. (1967): On a characteristic class entering the quantum conditions. Funktional. Anal. i Prilozen. *1*, 1–14 (Russian)

Arnol'd, V. I. (1972): Modes and quasimodes. Funktional. anal. i Prilozen. *6*, 12–20 (Russian). [English transl.: Functional Anal. Appl. *6* (1972)]

Arnol'd, V. I. (1978): Mathematical methods of classical mechanics. Springer, New York Berlin Heidelberg. [Russian original: Moskow, 1974]

Arnol'd, V. I. Kozlov, V. V., Neishtadt, A. I. (1987): Mathematical aspects of classical and selestial mechanics. Dynamical Systems III. (Encylopaedia of Mathematical Sciences vol. 3). Springer, New York Berlin Heidelberg.

Arnol'd V. I., Sevrjuk, M. B. (1986): Reversible systems. In: Sagdeev R. S. (ed.) Nonlinear phenomena in plasma physics and hydrodinamics. Mir, Moskov, pp. 31–64

Aubry, S. (1983): The twist map, the extended Frenkel–Kontorova model and the Devil's staircase. Physica *D7*, 240–258

Aubry, S., Le Daeron (1983): The discrete Frenkel–Kontorova model and its extensions. Physica *D8*, 381–422

Babitch, V. M., Bouldirev, V. S. (1972): Asymptotic methods in problems of short wave diffraction. Nauka, Moskow (Russian)

Benettin, G., Galgani, L., Giorgilli, A., Strelcyn, J.-M. (1983): A proof of Kolmogorov's theorem on invariant tori using canonical transformations defined by the Lie method. Il Nuovo Cimento *79*, 201–223

Benettin, G., Strelcyn, J.-M. (1978): Numerical experiments on the free motion of a point mass moving in a plane convex region: stochastic transition and entropy. Phys. Rev. *A17*, 773–785

Berdon, A. F. (1983): The geometry of discrete groups. Springer, New York Berlin Heidelberg.

Beresin, F. A., Shubin, M. A. (1983): Schrödinger equation. Moskow university press, Moskow (Russian)

Bernstein, D., Katok, A. (1987): Birkhoff periodic orbits for small perturbations of completely integrable Hamiltonian systems with convex Hamiltonians. Inventiones Math. *88*, 225–241

Berry, M. V. (1981): Semiclassical mechanics of regular and irregular motion. In: Chaotic behaviour of deterministic systems. Les Houches session *36*, North-Holland Publishing Company, Amsterdam (1983), pp. 171–271

Birman, M. S., Solomjak, M. Z. (1977): The asymptotic of the spectra of differential equations. In Mathematical Analysis *14* (Itogi Nauki i Techniki) VINITI, Moskow pp. 5–58 (Russian)

Birman, M. S., Solomjak, M. Z. (1987): Spectral theory of self-adjoint operators in Hilbert space. D. Reidel Publishing Company, Dordrecht Boston Lancaster Tokyo Holland

Bogolubov, N. N., Krilov, N. M. (1937): La theorie génerale de la mesure dans son application à l'etude des systèmes dynamiques de la mécanidue nonlinéare. Annals of Math. *38*, 65–113
Boldrigini, C., Keane, M., Marchetti, F. (1978): Billiards in polygons. The Annals of Probability *6*, 532–540
Bost, J.-B. (1986): Tores invariants des systèmes dynamiques hamiltoniens (Séminaire Bourbaki 1984–85, No. 639, Fèvrier 1985), S.M.F. Astérisque 133–134, pp. 113–157
Bourbaki, N. (1963): Éléments de mathématique. Livre III. Topologie gènèrale. Ch. 5–10. Hermann, Paris
Bourbaki, N. (1965): Éléments de mathématique. Livre III. Topologie gènèrale. Ch. 1–4. Hermann, Paris
Bourbaki, N. (1966): Éléments de mathématique. Livre V. Espaces vectoriels topologiques. Paris: Hermann
Bourbaki, N. (1970): Éléments de mathématique. Livre II. Algèbre. Paris: Hermann.
Bunimovich, L. A. (1974): On the ergodic properties of some billiards. Funct. Anal. Appl. *8*, 73–74
Bunimovich, L. A. (1979): On the ergodic properties of nonwhere dispersing billiards. Comm. Math. Physics *65*, 29–312
Bunimovich, L. A. (1988): Many-dimensional nonwhere dispersing billiards with chaotic behaviour. Physica *D33*, 58–64
Chenciner, A. (1985): La dynamique au voisinage d'un point fixe elliptique conservatif: de Poincaré et birkhoff à Aubry et Mather. In: Séminaire Bourbaki, 1983–84, no. 622, Février 1984. S.M.F. Astérisque 121–122, pp. 147–170
Chierchia, L., Gallavotti, G. (1982): Smooth prime integrals for quasiintegrable Hamiltonian systems. Il Nuovo Cimento *67B*, 277–295
Chirikov, B. V. (1979): A universal instability of many-dimensional oscillator systems. Physics Reports *52*, 263–379
Churchill, R. C., Kummer, M., Rod, D. L. (1983): On averaging, reduction, and symmetry in Hamiltonian systems, Journal of Diff. Eq. *49*, 359–414
Colin de Verdière, Y. (1973): Spectre du laplacien et longueurs des géodesiques périodiques I, II. Compositio mathematica *27*, 83–106
Colin de Verdière, Y. (1977): Quasi-modes sur les variétés Riemanniennes. Inventiones mathematicae *3*, 15–52
Cornfeld, I. P., Fomin, S. V., Sinai, Ya. G. (1980): Ergodic Theory. Nauka, Moskow (Russian). [English translation Springer, New York Berlin Heidelberg, 1981)
Delshams, A. (1984): Por que la diffusion de arnold aparece genericamente en los sistemas Hamiltonianos con mas de 2 grade de libertad. These, Universitàt de Barcelona
Delshams, A., Gutiérres, P. (1991): Effective stability for nearly integrable Hamiltonian systems. to appear In: Proceedings of EQUADIFF 91, Barcelona
Dold, A. (1972): Lectures on algebraic topology. Springer, New York Berlin Heidelberg
Douady, R. (1982): Applications du théorème des tores invariants. These 3-ème cycle. Université Paris VII
Douady, R. (1988a): Stabilité ou instabilité des pointes fixes elliptiques. Ann. Scient. Éc. Norm. Sup. 4-e serie, *21*, 1–46
Douady, R. (1988b): Regular dependence of invariant curves and Aubry–Mather sets of twist maps of an annulus. Ergodic Theory and Dynam. Systems *8*, 555–584
Douady, R., Le Calvez, P. (1983): Exemple de point fixe elliptique non topologiquement stable en dimension 4. C.R. Acad. Sci. Paris, *296*, sér. I, 895–898
Dubrovin, B. A., Krichever, I. M., Novikov, S. P. (1988): Integrable systems. In: Dynamical systems IV (Encylopaedia of Mathematical Sciences vol. 4). Springer, New York Berlin Heidelberg.
Duistermaat, J. J. (1980): On global action-angle coordinates. Comm. Pure. Appl. Math. *33*, 687–706
Duistermatt, J. J., Guillemin, V. W. (1975): The spectrum of positive elliptic operators and periodic bicharacteristics. Inventiones mathematicae *29*, 39–79
Dunford, N., Schwartz, J. (1958): Linear operators, Part I. Wiley, New York
Einstein, A. (1917): Zum Quantensatz von Sommerfeld und Epstein. Verhandl. Dtsch. Phys. Ges., *19*, 82–92
Escande, D. F. (1985): Stochasticity in classical Hamiltonian systems: universal aspects. Physics Reports *121*, 165–261

Fedoryuk, M. V. (1966): Asimptotics of the discrete spectrum of the operator $w'' + \lambda^2(x)w$. Mat. Sbornik *68*, 81–110 (Russian)

Friedrichs, K. O. (1934): Spektraltheorie halbbeschränkter Operatoren und Anwendung auf die Spectralzerlegung von Differentialoperatoren. Math. Ann. *109*, 685–713

Fuks, D. B., Rokhlin, V. A. (1984): Beginner's course in topology. Springer, New York Berlin Heidelberg

Giorgilli, A., Galgani, L. (1985): Rigorous estimates for the series expansions of Hamiltonian perturbation theory. Celestial Mechanics *37*, 95–112

Godbillon, C. (1969): Géometrie différentielle et mécanique analytique. Hermann, Paris

Golubitsky, M., Guillemin, V. (1973): Stable mappings and their singularities. Springer, New York Berlin Heidelberg

Goroff, D. L. (1985): Hyperbolic sets for twist maps. Ergodic theory and dynamical systems *5*, 337–339

Greene, J. M. (1979): A method for determining a stochastic transition. J. Math. Phys. *20*, 1183–1201

Greene, J. M., MacKay, R. S., Stark, J. (1986): Boundary circles for area preserving maps. Physica *D21*, 26–295

Gromov, M. L. (1985): Pseudo holomorphic curves in symplectic manifolds. Inventiones mathematicae *82*, 307–347

Gromov, M. L., Eliashberg, Ya. M. (1973): Construction of a smooth mapping with a prescribed Jacobian I. Funktional. Anal. i ego Prilozen. 7, 33–40

Gruber, P. M. (1990): Convex billiards. Geometriae Deciata *33*, 205–226

Guillemin, V., Sternberg, S. (1977): Geometric asymptotics. American Math. Society, Providence Rhode Island

Gutkin, E. (1986): Billiards in polygons. Physica *D19*, 311–333

Halpern, B. (1977): Strange billiard tables. Transactions of AMS, *232*, 297–305

Harrel, E. M. (1980): Double wells. Comm. Math. Phys. *75*, 239–261

Hartman, P. (1964): Ordinary differential equations. Wiley, New York London Sydney

Hayli, A., Dumont, T. (1986): Experiences numeriques sur des billiards C^1 formes de quatre arcs de cercles. Celestial Mechanics *38*, 23–66

Heding, J. (1962): An introduction to phase-integral methods. Wiley, New York

Hejhal, D. A. (1976): The Selberg trace formula for PSL(2,R) vol. 1. (Lecture Notes in Mathematics, vol. 548). Springer, New York Berlin Heidelberg

Hejhal, D. A. (1983): The Selberg trace formula for PSL(2,R) vol. 2. (Lecture Notes in Mathematics, vol. 1001). Springer, New York, Berlin Heidelberg

Helfer, B., Robert, D. (1981): Comportement semi-classique du spectre des Hamiltoniens quantiques elliptiques. Ann. de l'Institut Fourier Grenoble *33*, 169–233

Helfer, B., Robert, D. (1983): Calcul fonctionnel par la transform de Mellin et opérateurs admissibles. Journal of Functional Analysis *53*, 246–268

Herman, M. (1983): Sur les courbes invariantes par le difféomorphismes de l'anneau, Volume I: Astérisque 103–104

Hirsch, M. W. (1976): Differential topology. Springer, New York Berlin Heidelberg

Holmes, P., Marsden, J., Scheurle, J. (1988): Exponentially small splittings of separatrices with application to KAM theory and degenerate bifurcations. Contemporary Mathematics *81*, 213–244

Ivrii, V. Ya. (1980): On the second term of the spectral asymptotic for the Laplace–Beltrami operator on manifold with boundary. Funktional. Anal. i Prilozen. *16*, 25–34 (Russian)

Ivrii, V. Ya. (1984): Exact classical and semiclassical asymptotics of eigenvalues for the spectral problem on a manifold. Soviet Doklady *277*, 1307–1210 (Russian)

Ivrii, V. Ya. (1986a): Estimations pour le nombre de valeurs propres négatives de l'opérateur de Schrödinger avec potentiels singuliers. C. R. Acad. Sc. Paris, *302*, sér. I, 467–470

Ivrii, V. Ya. (1986b): Estimations pour le nombre de valeurs propres négatives de l'opérateur de Schrödinger avec potentiels singuliers et application au comportement asymptotique des grandes valeurs propres. C. R. Acad. Sc. Paris, *302*, sér. I, 491–494

Ivrii, V. Ya. (1986c): Estimations pour le nombre de valeurs propres négatives de l'opérateur de Schrödinger avec potentiels singuliers et application au comportement asymptotique des valeurs propres s'accumulant vers -0, aux asymptotiques à deux paramètres et à la densité des états. C. R. Acad. Sc. Paris, *302*, sér. I, 535–538

Jona-Lasinio, G., Martinelli, F., Scoppola, E. (1981): New approach to the semiclassical limit of quantum mechanics. Comm. Math. Phys. *80*, 223–254
Katok, A. (1982): Some remarks on Birkhoff and Mather twist map theorems. Ergodic Theory and Dynam. Systems *2*, 185–194
Katok, A. (1982a): More about Birkhoff periodic orbits and Mather sets for twist maps, preprint I.H.E.S.
Katok, (1989): Small perturbations of completely integrable Hamiltonian systems. Mathematics 253–37. California Institute of Technology, Pasadena, CA 91125. Preprint
Katok, A., Strelcyn, J.-M. (1986): Invariant manifolds, entropy and billiards; smooth maps with singularities. (Lecture Notes in Mathematics, vol. 1001). Springer, New York Berlin Heidelberg
Keller, J. B., Rubinov, S. (1960): Asymptotic solution of eigenvalue problems. Ann. of Physics *9*, 27–75
Kelley, J. L. (1955): General topology, D. van Nostrand, New York
Khanin, K. M., Sinai, Ya. G. (1986): Renormalisation group theory and KAM-theory. In: Sagdeev, R. S. (ed) Nonlinear phenomena in plasma physics and hydrodinamics. Mir, Moskow, pp. 93–118
Kirillov, A. A. (1976): Elements of the theory of representations. Springer, New York Berlin Heidelberg
Kolmogorov, A. N. (1954): On conservation of conditionally periodic motion for a small change in Hamilton's function. Soviet Doklady *98*, 527–530 (Russian). [English transl. In: Casati, G., Ford, J. (eds) Lecture Notes in Physics vol. 93, Berlin (1979), pp. 51–54]
Krygin, A. B. (1971): Extension of volume-preserving diffeomorphisms. Funktional. Anal. i ego Prilozen. *5*, 72–75
Landau, L. D., Lifschitz, E. M. (1963): Quantum mechanics. Part I: Nonrelativistic theory, 2-nd rev. ed., Fismatgiz, Moskow (Russian). [German transl., Akademic Verlag, Berlin, 1966; English transl. of 1st ed., Pergamon Press, Oxford, and Addison-Wesley Reading, Mass., 1958]
Lazutkin, V. F. (1973): The existence of caustics for a billiard problem in a convex domain. Math. USSR Isvestija *7*, 185–214
Lazutkin, V. F. (1973a): Asymptotics of the eigenvalues of the Laplacian and quasimodes. A series of quasimodes corresponding to a system of caustics close to the boundary of the domain. Math. USSR Isvestija *7*, 439–466
Lazutkin, V. F. (1974): On Moser's theorem on invariant curves. In: Voprsoy raspr. seism. voln. vyp. 14. Nauka, Leningrad, pp. 105–120 (Russian)
Lazutkin, V. F. (1984): Splitting of the separatrices of Chirikov's standard map. Deposited in VINITI, September 24, 1984, no. 6372–84. Moskow (Russian)
Lazutkin, V. F. (1988): On the signature of symmetric reversible matrices. Mat. Zametki *44*, 202–207 (Russian)
Lazutkin, V. F. (1990): On the width of the instability zone near the separatrices of the standard map. Soviet Doklady. *313*, 268–272 (Russian)
Lazutkin, V. F., Schachnannski, I. G., Tabanov, M. B. (1989): Splitting of separatrices for standard and semistandard mappings. Physica *D 40*, 235–248
Le Calvez, P. (1988): Les ensembles d'Aubry–Mather d'un difféomorphim conservatif de l'anneau déviant la vertical sont en général hyperboliques. CRAS sér I. Math. *306*, 51–54
Lion, G., Vergne, M. (1980): The Weil representation, Maslov index and Theta series. Birkhäuser, Boston
Liouville, J. (1855): Note sur l'integration des équations differéntielles de la dinamiqe, présentée au Bureau des Longitudes le 29 juin 1853. J. Math. Pures Appl. *20*, 137–138
Lochak, P. (1991): Canonical perturbation theory via simultaneous approximations. Ecole Normal Supérieur, Paris. Preprint
MacKay, R. S. (1983): A renormalisation approach to invariant circles in area-preserving maps. Physica *D 7*, 283–300
MacKay, R. S. (1987): Hyperbolic cantori have dimension zero. J. Phys. *A 20*, L559–L561
MacKay, R. S., Meiss, J. D., Percival, I. C. (1984): Transport in Hamiltonian systems. Physica *D 13*, 55–81
MacKay, R. S., Percival, I. C. (1985): Converse KAM: theory and practice. Comm. Math. Phys. *88*, 469–512
Malgrange, B. (1966): Ideals of differentiable functions. Oxford University Press, Bombay

Markus, L., Meyer, K. R. (1978): Generic Hamiltonian systems are neither integrable nor ergodic. Memoirs of the AMS *144*

Maslov, V. P. (1972): Théorie des perturbations et méthodes asymptotiques. Dunod, Paris. [Russian original: Moskow 1965]

Maslov, V. P., Fedorjuk, M. V. (1981): Semiclassical approximation in quantum mechanics. Reidel, Boston. [Russian original: Moskow 19]

Mather, J. N. (1982): Existence of quasi-periodic orbits for twist homeomorpisms of the annulus. Topology *21*, 457–467

Mather, J. N. (1988): Destruction of invariant circles. Ergodic theory and Dynamical systems *8*, 199–214

McDonald, S. W., Kaufman, A. N. (1979): Spectrum and Eigenfunctions for a Hamiltonian with stochastic trajectories. Phys. Rev. Lett., *42*, 1189–1191

Moser, J. (1956): The analytical invariants of an area-preserving mapping near hyperbolic point. Communications Math. Phys., *9*, 673–692

Moser, J. (1962): On invariant curves of area-preserving mappings of an annulus, Nachr. Akad. Wiss., Göttingen, Math. Phys. Kl., 1–20

Moser, J. (1965): On the volume element on a manifold. Transactions of the AMS *120*, 286–294

Moser, J. (1973): Stable and random motions in dynamical systems. (Annals of Mathematics Studies Number 77). Princeton University Press, Princeton, New Jersey

Neishtadt, A. I. (1984): The separation of motions in systems with rapidly rotating phase. P.M.M. USSR *48*, 193–139

Nekhoroshev, N. N. (1972): Acton-angle variables and their generalisation. Trudy Mosk. Mat. O-va *26*, 181–198 (Russian) [English transl.: Trans. Moskow Math. Soc. *26*, 180–198 (1972)]

Nekhoroshev, N. N. (1977): An exponential estimate of the time of stability of nearly-integrable Hamiltonian systems I. Uspekhi Math. Nauk *32* (Russian). [English transl.: Russ. Math. Surveys *32*, 5–66 (1977)]

Nekhoroshev, N. N. (1979): An exponential estimate of the time of stability of nearly-integrable Hamiltonian systems II. Trudy Sem. Petrovs. *5*, 5–50 (Russian)

Newhouse, S. E. (1983): Generic properties of conservative systems. In: Iooss, G., Helleman, R. H. G., Stora, R. (eds) Chaotic behaviour of deterministic systems. Les houches, session 36, 29 Juin-31 Juillet 1981. North-Holland Publishing company, Amsterdam, pp. 443–454

Pankratova, T. F. (1984): Quasimodes and splitting of eigenvalues. Soviet. Math. Doklady *29*, 597–601

Percival, I. C. (1980): Variational principles for invariant tori and cantori. In: Mouth, M., Herrara, J. C. (eds) Symp. on nonlinear dynamics and beam–beam interactions, Amer. Inst. of Physics Conf. Proc. no. *57*, pp. 310–320

Petkov, V. M., Stojanov, L. N. (1985): Generic properties and the spectrum of the Poincaré map for multiple reflecting rays. Institute of Mathematics at Bulgarian Acad. of Sci. Preprint

Poincaré, H. (1892): Les Méthodes Nouvelles de la Méchanique Céleste. Vols 1–3, Gauthier-Villars, Paris [Reprint: Dover, New York, 1957]

Pöschel, J. (1982): Integrability of Hamiltonian systems on Cantor sets. Comm. on Pure and Appl. Math. *33*, 653–695

Pöschel, J. (1991): On Nekhoroshev's estimate for quasi-convex Hamiltonians. Forschungsinstitute für math. ETH Zürich. Preprint

Robinson, R. C. (1970): Generic properties of conservative systems I, II. Amer. Journ. Math. *92*, 562–603, 897–906

Rüssmann, H. (1970): Kleine Nemmer I: Über invarianten Kurven differezierbarer Abbildungen eines Kreisruneneines Kreisringes. Nachr. Akad. Wiss., Göttingen, Math. Phys. Kl., 67–105

Rüssmann, H. (1983): On the existence of invariant curves of twist mappings of an annulus. (Lecture Notes in Mathematics, vol. 1007). Springer, New York Berlin Heidelberg, pp. 677–712

Sevryuk, M. B. (1986): Reversible systems. (Lecture Notes in Mathematics, vol. 1211). Springer, New York Berlin Heidelberg

Sikorav, J.-C. (1989): Rigidité symplectique dans le cotangent de T^n. Duke Math. Journal *59*, 759–763

Simon, B. (1984): Semiclassical analysis of low lying eigenvalues II-Tunneling. Ann. Math. *120*, 89–118

Sinai, Ya. G. (1970): Dynamical systems with elastic reflections. Ergodic properties of dispersing billiards. Usp. Mat. Nauk *25*, 141–192 (Russian). [English transl.: Russian Math. Surv. *25*, 137–198 (1970)]

Sinai, Ya. G., Khanin, K. M. (1987): Renorm-group method in the theory of dynamical systems. In: Meeting "Renormgroup 86". Joint Institute of Nuclear Investigations, Doubna, pp. 203–222 (Russian)

Slavjanov, S. Yu. (1989): Asymptotics of singular Sturm–Liouville problems with respect to a large parameter in the case of nearby transition points. Differentsialnye uravcnenia *5*, 313–325 (Russian)

Spanier, E. H. (1966): Algebraic topology. McGraw-Hill, New York

Suzuki, H. (1985): Canonical relations defined by intersecting hypersurfaces in a symplectic manifold. Math. Ann. *271*, 479–484

Svanidze, N. V. (1980): Small perturbations of an integrable dynamic system with integral invariant. Tr. Mat. Inst. Steclova *147*, 124–146 (Russian). [English transl. Proc. Steclov. Inst. of Math. No. *2*, 127–151 (1981)]

Thom, R., Levin, H. (1959): Singularities of differentable mappings I. Bonn. Preprint.

Turaev, V. G. (1985): The first Chern symplectic class and Maslov's indices. In: Zapiski nauchnyh seminarov LOMI vol. 143. Investigations in topology, pp. 110–129 (Russian)

Vasil'ev, D. (1987): Asymptotics of the spectrum of a boundary value problem. Trans. Moskow Math. Soc. *49*, 173–245 (Russian)

Vecheslavov, V. V., Chirikov, B. V. (1989): With what velocity goes Arnol'd diffusion? Nuclear physics Institute of Siberian branch of Soviet Acad. of Sci. Preprint 89–72 (Russian)

Veerman, J. J. P., Tangerman, F. M. (1990): On Aubry Mather sets. Preprint 1990/3 SUNY Stony Brook Institute for Math. Sciences

Weinstein, A. (1971): Symplectic manifolds and their Lagrangian submanifolds. Advances in Math. *6*, 329–346

Weinstein, A. (1977): Lectures on symplectic manifolds. (Regional conference series in mathematics No. 29). American Math. Soc., Providence Rhode Island

Wojtkowski, M. P. (1985): Invariant families of cones and Lyapunov exponents. Ergodic Theory Dyn. Systems *5*, 145–161

Wojtkowski, M. P. (1986): Principle of the design of billiards with nonvanishing Lyapunov exponents. Comm. Math. Phys. *105*, 391–414

Wolf, J. A. (1972): Spaces of constant curvature. University of California, Berkley California

Zaslavski, Sagdeev, R. Z., Usikov, D. A., Chernikov, A. A. (1988): Minimal chaos, stochastic web, and structures with a symmetry of a "quasicristall" type. Uspekhi Physichesk, nauk, *156*, 193–251 (Russian)

Zehnder, E. (1973): Homoclinic points near elliptic fixed points. Comm. Pure. Appl. Math. *26*, 131–182

Zehnder, E. (1975): Generalized implicit function theorem with applications to some small divisor problems I. Comm. Pure. Appl. Math. *28*, 91–140

Zehnder, E. (1976): Generalized implicit function theorem with applications to some small divisor problems II. Comm. Pure. Appl. Math. *29*, 49–113

Zemljakov, A. N., Katok, A. B. (1975): Topological transitivity of billiards in polygons. Mat. Zametki *1*, 291–300 (Russian)

Subject Index